AF771025

Handbook of Robotics

Edited by
Lauren Barrett

Willford Press
www.willfordpress.com

Published by Willford Press,
118-35 Queens Blvd., Suite 400,
Forest Hills, NY 11375, USA

Copyright © 2020 Willford Press

This book contains information obtained from authentic and highly regarded sources. Copyright for all individual chapters remain with the respective authors as indicated. All chapters are published with permission under the Creative Commons Attribution License or equivalent. A wide variety of references are listed. Permission and sources are indicated; for detailed attributions, please refer to the permissions page and list of contributors. Reasonable efforts have been made to publish reliable data and information, but the authors, editors and publisher cannot assume any responsibility for the validity of all materials or the consequences of their use.

Trademark Notice: Registered trademark of products or corporate names are used only for explanation and identification without intent to infringe.

ISBN: 978-1-68285-776-2

Cataloging-in-Publication Data

Handbook of robotics / edited by Lauren Barrett.
p. cm.
Includes bibliographical references and index.
ISBN 978-1-68285-776-2
1. Robotics. 2. Robots. 3. Automation. 4. Machine theory. I. Barrett, Lauren.
TJ211 .H36 2020
629.892--dc23

For information on all Willford Press publications
visit our website at www.willfordpress.com

Willford Press

Printed in China.

Handbook of Robotics

Contents

Preface

I am honored to present to you this unique book which encompasses the most up-to-date data in the field. I was extremely pleased to get this opportunity of editing the work of experts from across the globe. I have also written papers in this field and researched the various aspects revolving around the progress of the discipline. I have tried to unify my knowledge along with that of stalwarts from every corner of the world, to produce a text which not only benefits the readers but also facilitates the growth of the field.

Robotics is an interdisciplinary field of science and engineering which is concerned with the design, construction, use and operation of robots and computer systems. These are developed to substitute for humans and act in a way that is useful in any environment that can pose a risk to human safety, such as bomb detection and deactivation. Although robots may differ widely in terms of application and form, they are designed with three basic similarities. Robots have a mechanical form, frame or construction that aids in their function, have electrical power components, and are driven by some level of computer programming. The basic components of a robot comprise of a power source, actuators, sensors, mobile manipulators, and an architecture that supports locomotion. The mechanical structure of a robot has to be handled in a controlled manner for it to perform a task. This requires perception, processing and action. This book outlines the processes and applications of robotics in detail. It strives to provide a fair idea about this discipline and to help develop a better understanding of the latest advances within this field. Researchers and students in robotics will be assisted by this book.

Finally, I would like to thank all the contributing authors for their valuable time and contributions. This book would not have been possible without their efforts. I would also like to thank my friends and family for their constant support.

Editor

1

A Novel Path Control Algorithm for Networked Underwater Robot

Minghong She[1,2] **and Liyu Tian**[2]

[1]*School of Automation, Chongqing University, Chongqing, China*
[2]*Computer College, Chongqing College of Electronic Engineering, Chongqing, China*

Correspondence should be addressed to Minghong She; mh_she@126.com

Academic Editor: Keigo Watanabe

Under the network environment, the traditional control method of underwater Robot path has the disadvantages of low control accuracy, large error, and inefficiency. This paper proposes a novel path control method for underwater Robots based on the NURBS (nonuniform rational B-spline, NURBS) curve fitting method, which utilizes a sensor or camera to detect the static and dynamic obstacles, establishes the kinematics model of underwater Robots, gets the target function of rob shortest path, and analyzes underwater Robot constraints. According to the basic fluid mechanics, the resistance of the underwater Robot is determined. The filter function is used to smooth the process, and the NURBS curve fitting method is applied to control the path of the underwater Robot. Experimental results show that the improved method that proved to be practical is superior to the traditional one in the aspect of control time and accuracy.

1. Introduction

With the depletion of land resources, money, time, and manpower are invested to develop underwater resources [1]. Therefore, underwater Robots are a concern for people of many countries [2]. In order to complete an underwater task for one time, a variety of underwater Robots complex operation and safe navigation are to be carried out in accordance with the corresponding navigation path [3, 4].

To achieve the best navigation plan and keep the underwater Robot running on the scheduled course [5], it is necessary to ensure the maneuverability of the underwater Robot [6, 7]. When the underwater Robot has good handling performance, it cannot only maintain stable driving course, depth, and speed but also quickly change the heading, depth, and speed and correctly perform all kinds of operations [8]. The path control of underwater Robot affects the performance of underwater Robot [9]. The method of networking can effectively control the running path of underwater Robots and shorten the control time [10, 11]. Therefore, the underwater Robot path control under the networked environment has become the focus of research in this field and has attracted the attention of many scholars.

Mahmoodabadi, Mohammad Javad et al. [12] mainly aims at controlling the path of underwater Robot when the underwater Robot is avoiding the obstacle and the direction of movement is not corrected. Based on the establishment of integrated navigation model and the detection based dynamic obstacle avoidance strategy, the discrete time multithread concurrent simulation framework and model are used to design the effective control of underwater Robot path. However, the method wastes much time. Zanoli, Silvia M., and Giuseppe Conte [13] aim at resolving the problem of low control speed or zero speed state in traditional control method, and the frame control moment gyroscope is introduced as the path control executive mechanism to realize the path control of the underwater Robot. This method can effectively shorten the control time of the path, but with a problem of low control accuracy [14].

To solve the above problems, a path control method of underwater Robot based on NURBS curve fitting method is proposed. The experimental results show that the control accuracy and efficiency are improved by adopting the presented control method.

2. Construction of Kinematics Model of Underwater Robot

When the underwater Robot is effectively controlled under the condition of network, it is necessary to have a good understanding of the underwater Robot. The use of sensors or cameras to detect static and dynamic obstacles and the establishment of underwater Robot kinematics model can improve the path control method to provide the basis.

When the kinematic model of the underwater Robot is established, the head adjustment of the target underwater Robot can be obtained by adjusting the amount of integration processing [15]. The kinematic model is given by

$$H_{g1} = \sum K_{r2} C_r \tag{1}$$

where C_r is output value and K_{r2} is weighted coefficient.

For the existence of the interference of underwater Robots, the use of arithmetic mean filtering method for filtering to eliminate the Gaussian noise has a positive effect [16]. When using the most single arithmetic mean filter, the data received at each time is replaced by the average of the received data in the neighborhood of the length of time. The expression is

$$y(n) = \frac{1}{m} \sum_{i=1}^{m} x(n-i) \tag{2}$$

where m is constant, i is the number of filters, and $x(n-i)$ is the wavelength of $n-i$ filter.

In the filter, the high frequency of interference removal effect will not achieve the best results. In this case, it is necessary to suppress a higher frequency of interference by adding a virtual damping force [17, 18]. This makes the response of the underwater Robot relatively slow, increasing its stability. The expression is

$$u_i = \frac{\Delta t e_i + T u_{i-1}}{\Delta t + T} \tag{3}$$

where Δt is the sampling period, T is the inertia constant, e_i is the time deviation of i, and u_{i-1} is the response time of time i.

In the case of satisfying the absolute stability of the underwater Robot, the positive moment is used as the stability storage device [18] and then the balance coefficient can be changed as follows:

$$C_v = 1 - \frac{M_w \left(m' + Z'_q\right)}{M'_q Z'_w} \tag{4}$$

During the operation of the underwater Robot, the obstacle in the underwater environment will form a certain potential field and the repulsion force of F_{rep} is generated for the underwater Robot. The closer the distance is, the larger F_{rep} will; that is to say, the closer the underwater Robot and the obstacle, the greater the potential energy of the underwater Robot and vice versa [19]. This potential field is similar to the potential field, inversely proportional to the distance, so the repulsive potential function can be expressed as

$$\begin{matrix} \dfrac{K_r}{\rho(X, X_{obs})} & \rho \le \rho_m \\ 0 & \rho < \rho_m \end{matrix} \tag{5}$$

where K_r is the repulsive potential field constant, X is the position vector of the underwater Robot, X_{obs} is the obstacle position vector in the water, ρ is the shortest distance between the underwater Robot and the obstacle, and ρ_m is the range of the repulsive potential field. When the distance between the underwater Robot and the obstacle is greater than ρ_m, the repulsive potential field does not interfere with the operation of the underwater Robot.

To avoid the collision between the underwater Robot and the obstacle, it is required to set a minimum safe distance ρ_0, when $\rho \to \rho_0$ and $F_{rep} \to \infty$, and to ensure that F_{rep} is continuous, the expression of F_{rep} is

$$F_{rep} = \begin{cases} K_r \left(\dfrac{1}{\left(\rho(X, X_{obs}) - \rho_0\right)^2} - \dfrac{1}{\left(\rho_m - \rho_0\right)^2} \right) & \rho \le \rho_m \\ 0 & \rho > \rho_m \end{cases} \tag{6}$$

Based on the above, sensors or cameras are used to monitor static and dynamic underwater obstacles and to obtain the position, speed, and acceleration information of obstacles in real time. These messages can fuse into moving obstacles. A number of navigation paths represented by time t can be used in the obstacle motion constraints of underwater Robots. Suppose that obs_k is a rigid body with a center of $\alpha_k(t) = (\alpha_{1k}(t), \alpha_{2k}(t))$, $k = 1, 2, \ldots, q$, $t \in [0, t_k]$ and $\alpha_k(.) = 1, 2, \ldots, q$ is a continuous real vector valued function, also the path of motion obstacles. Suppose that the underwater Robot and obstacle k can be represented by T_{rob} and T_k, respectively, $k = 1, 2, \ldots, q$. When the constraint of the underwater Robot is $\dot{x}_1 \sin\theta - \dot{x}_2 \cos\theta = 0$ without lateral collision and the general coordinate system is $q(t) = [x_1(t) x_2(t) \theta(t)]^T$, the kinematic model of the underwater Robot is expressed as follows:

$$q = \begin{bmatrix} \cos\theta t & 0 \\ \sin\theta t & 0 \\ 0 & 0 \end{bmatrix} \begin{bmatrix} v(t) \\ w(t) \end{bmatrix} \tag{7}$$

where $v(t)$ and $\omega(t)$ are, respectively, tangential velocity and angular velocity. In the navigation process, the speed of underwater Robot should be regulated within the maximum allowable speed to prevent the underwater Robot from colliding; meanwhile the underwater Robot must follow the rules of acceleration.

3. Improvement of Path Control Method for Underwater Robot

3.1. Constraint Analysis. According to the kinematic model of the underwater Robot, the objective function of minimizing the path length is obtained:

$$f(h) = \min \sum \frac{h\left(2\sum_{f=1}^{F} E(fh) - E(0) - E(tf)\right)}{2} \quad (8)$$

On this basis, various constraints are introduced into the objective function as the penalty term and the expression of the self-adaptation function is obtained:

$$\min F_{pcna} = \sum \frac{h\left(2\sum_{f=1}^{F} E(fh) - E(0) - E(tf)\right)}{2} + c_n \sum_{f=0}^{F} \max\left(0, X^l - E(t')\right) \quad (9)$$

where $t' = fh, c_n$ is a larger value that is penalized for breaking the constraint, $n = 1, ..., 9$; X^l is the boundary area of its *rob* activity. Their constraint types are as follows:

(i) Boundary Constraint. Assuming that X is the boundary of the selected area, the underwater Robot should be subject to the following motion boundary constraints during the moving process:

$$p(0) = (x_0, \theta_0)$$
$$p(t_f) = (x_f, \theta_f) \quad (10)$$
$$X_0, X_f \in X$$

(ii) Movement Restriction. In the process of underwater Robot navigation, there is a certain motion constraint. When the ship is sailing along the planning path, its acceleration and speed are prevented from colliding in the prescribed area. In literature [20], its motion constraint is expressed as follows:

$$\upsilon(t) \geq \upsilon_{\min}$$
$$\upsilon(t) \leq \upsilon_{\max}$$
$$\omega(t) \geq \omega_{\min} \quad (11)$$
$$\omega(t) \leq \omega_{\max}$$
$$t \in [0, t_f]$$

(iii) Obstacle Avoidance Constraint. To ensure that underwater Robots maintain a minimum safe distance from other static or moving underwater obstacles during navigation, d_{orb_k} and d_{ok} are setting as the given safety distance and actual distance of *rob* and α_k, respectively; therefore, d_{orb_k} is expressed as

$$d_{orb\_k} = \|x(t) - \alpha_k(t)\|_2 - (r + r_k) \quad (12)$$

where $1 \leq k \leq q$. Thus, in X, *rob* should satisfy the following barrier constraints:

$$d_{obs\_k} - d_{ok}(t) \leq 0, \quad \forall k = 1, \ldots, q. \; t \in \left[0, \max\left(t_f, t_k\right)\right] \quad (13)$$

3.2. Implementation of the Proposed Method. Based on the analysis of constraints for underwater Robot and the basic fluid mechanics, the resistance of underwater Robot is determined, the filter functions are smoothed, and underwater Robot path is controlled by using NURBS curve fitting method

In view of the basic fluid dynamics, the total resistance R_t of the underwater Robot can be divided into frictional resistance R_f and residual resistance R_r. When the underwater Robot is in water for a long period of time, the wave resistance is negligible and the resistance is as follows:

$$R_m = 0.5\rho V_s^2 S\left(C_f + \Delta C_f + C_{pv}\right) \quad (14)$$

where ρ is the density of water, V_s is the speed of underwater Robot, S is wet surface area, C_f is friction resistance coefficient, ΔC_f is friction resistance subsidy coefficient, and C_{pv} is viscosity resistance coefficient.

Supposing that the running path of an underwater Robot can be represented by a weighted directed graph $G(V, \{E\})$, in which the set V is a set of fixed points, n is the fixed-point number of the fixed-point set, E is the arc set of the m arc, $\langle v_i, v_j \rangle$ is the arc from v_i to v_j in E, and $d\langle v_i, v_j \rangle$ is the nonnegative weight of the arc $\langle v_i, v_j \rangle$. Assuming that v_i and v_j are the paths from vertex v_i to v_j in V, the expression is as follows:

$$P = \left\{v_0 = v_i, \ldots, v_n = v_j\right\} \quad (15)$$

For a path sequence consisting of n path points, the length of the $[p_0, p_1, ..., p_{n-1}, p_n]$ path is

$$L_n = \sum_{k=1}^{n} d_{p_{k-1}, p_k} \quad (16)$$

where d_{p_{k-1}, p_k} is the straight-line distance of the p_{k-1} and p_k path segments. The evaluation function is $f_L = L_n/d_{\min}$, including the fact that $d_{\min}$ is the linear length of p_0 and p_n. When $f_L = 1$, the path length is the shortest; that is, the underwater Robot can sail straight from p_0 to p_n. Considering the operating state of underwater Robot as a discrete process, the position of some obstacles will change, causing it to be controlled in the wrong direction. So the filtering function is used to smooth the filtering function:

$$h_k = \frac{h_{k-l} + 2h_{k-l+1} + \cdots + 2h_{k+l-1}}{2l + 1} \quad (17)$$

where h_k is the density of obstacles after smoothing and l is the filter length.

It is assumed that the entire underwater Robot path using the *DPSS* algorithm is $D = \{d_1, d_2, ..., d_n\}$, including the fact

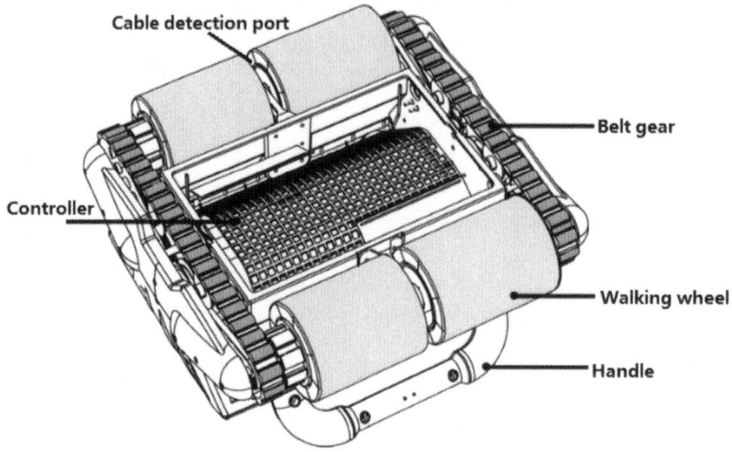

FIGURE 1: The structure of the cable detection underwater Robot.

that d_n is the n obstacle node on the path, and the Euclidean distance between all two adjacent obstacles is defined as

$$L = \sum_{n=1}^{l} \sqrt{(x_{n+1} - x_n)^2 + (y_{n+1} - y_n)^2} \quad (18)$$

where (x_n, y_n) is the coordinates of the grid node d_n.

The minimum turning radius $R_{\min}$ is the limit condition which cannot be ignored in underwater Robot path planning. Path control adopts the NURBS curve fitting method to determine the minimum turning radius of the underwater Robot, so as to realize the rapid and safe navigation of the underwater Robot in water. Expression is as follows:

$$f_R = \begin{cases} \sum_{k=1}^{n} \frac{1}{(r_k - R_{\min})}, & r_k \geq R_{\min} \\ \infty, & r_k < R_{\min} \end{cases} \quad (19)$$

In (19), r_k is the turning radius of the path k point. When $r_k < R_{\min}$, the path control caused by the turning is infinite, not worth the candle.

In order to satisfy the minimum turning radius of an underwater Robot, the maximum curvature $k_{\max}$ of the optimal path generated must satisfy the following constraints:

$$k_{\max} < \frac{1}{R_{\min}} \quad (20)$$

For the underwater Robots with different running paths, it is possible to satisfy the minimum turning radius requirement of the underwater Robot by selecting the appropriate parameter R to set up the upper model. The path control method obtained from the previous analysis is in accord with the control characteristics and motion characteristics of the underwater Robot.

4. Simulation Results

In order to verify the effectiveness and feasibility of the improved control method, it is necessary to make a comparative analysis of the improved methods. The experiment is simulated by MATLAB, and the result is compared by NURBS curve fitting method. The simulated cable detection underwater Robot is shown in Figure 1. The hardware test platform is under the condition of single-phase inverter unit and controlled rectifier of model PM201CL1A061. Platform identification algorithm is DSP27334 main control chip. The production company is TI.

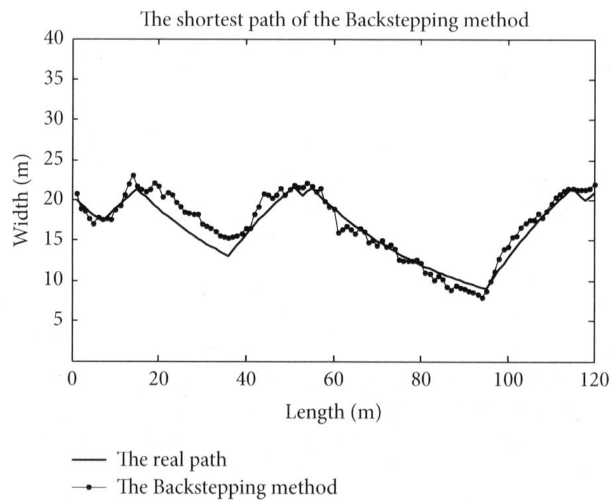

FIGURE 2: The shortest path of the Backstepping method.

4.1. Navigation Path Comparison. In order to verify the effectiveness of the improved method, the experimental range is set at a certain number of underwater faults in 100∗40. Backstepping method, immune fuzzy PID method, and improved method are adopted to control the shortest path, respectively. The results are shown in Figures 2, 3, 4, and 5.

In Figure 2, the black dotted line represents the control point of the underwater Robot Backstepping method in the case of underwater obstacles.

In Figure 3, the blue cross line represents the control point of the underwater Robot immune fuzzy PID method in the case of underwater obstacles.

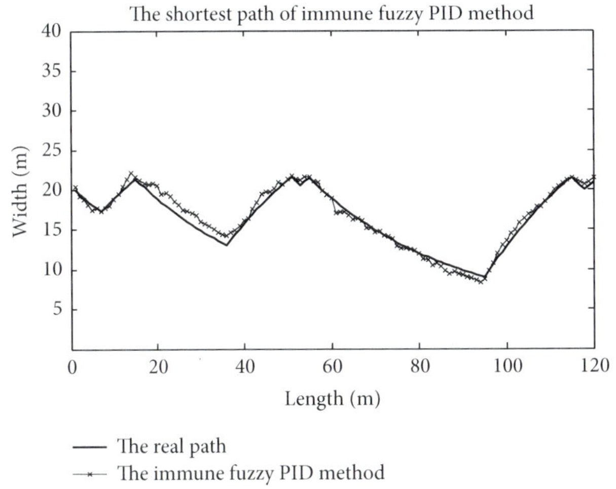

FIGURE 3: The shortest path of immune fuzzy PID method.

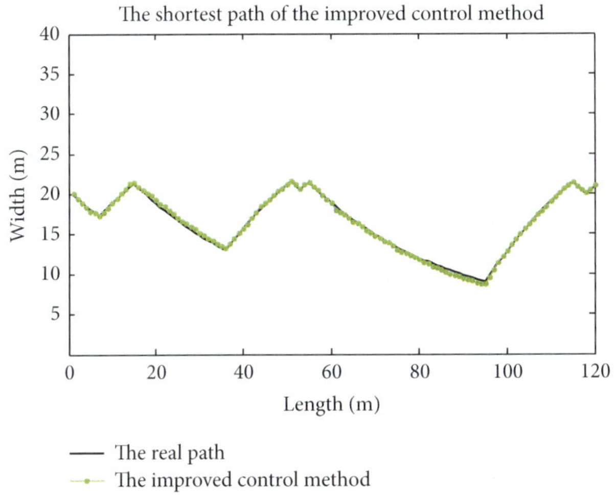

FIGURE 4: The shortest path of the improved control method.

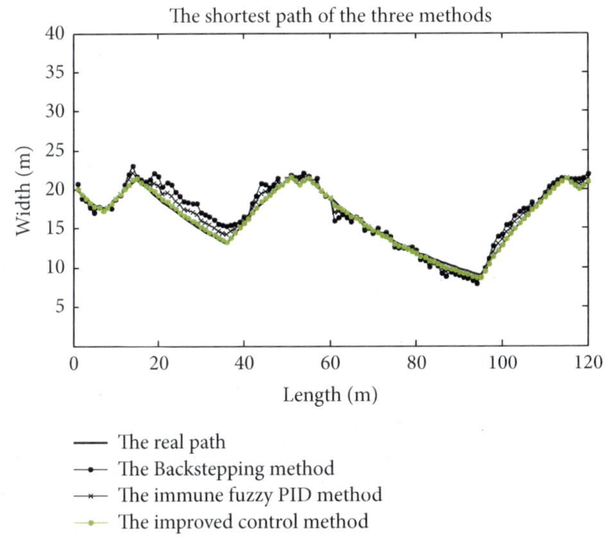

FIGURE 5: The comparison of the shortest paths of the three-control methods.

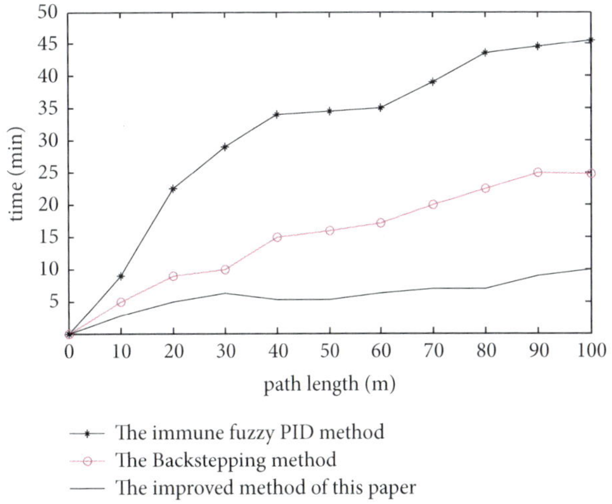

FIGURE 6: Time comparison analysis.

The light green dotted line in Figure 4 shows the control point of the underwater Robot of the improved control method in the case of underwater obstacles.

In Figure 5 shows the result of the comparison of the shortest paths of the three-control methods.

As can be seen from Figures 2–5, the results of three methods are different under the condition of underwater Robot that has a certain range of motion. The Backstepping method was controlled 62 times, of which only 29 were on the best path for the underwater Robot. When the obstacles in the water increase, the trajectory of the Robot deviates from the optimal path and the control is not effective enough. The immune fuzzy PID method increases the number of control accordingly, and the control increases to 85 times with the increase of the obstacles in water. The path of navigation deviates from the optimum path, only 4 times returning to the best path, but rapidly deviating. This shows that the stability is poor; the improved method can make the number of control 100 times; the whole path of the underwater Robot deviates little from the optimal path; and it coincides with the optimal path many times. It shows good stability. From this we can see that the presented method has certain advantages.

4.2. Complexity Analysis. In the optimal path control method of underwater Robot, the degree of underwater Robot path control is described by the running time of underwater Robot under the same path of the same length. This quantitative description of underwater Robot navigation path control method is highly operational and can be used to determine the trajectory control of underwater Robot directly. As a result, the time of the voyage can be calculated directly by the length of the section and the speed of sailing. The expression for travel time is

$$Rtime = \frac{Rdist}{Rspace} \tag{21}$$

Rdist is the length of the current section, and *Rspace* is the average speed of the underwater Robot. The time required for the Robot to run in three ways is shown in Figure 6.

As you can see from Figure 6, when the path length is constant, the three methods vary in time. By using the Backstepping method, the running time is about 21.89 min and the time spent in 0-20M is rapidly increasing while the time spent in 20-30 does not increase. After 30m, it had been on the rise and had no tendency to decline. The immune fuzzy PID method takes about 32.5 min time and takes a long time. As the path of the ship increases, the time of sailing increases accordingly. Only when the region is stable at 40-60 m does the rest of the road become on the rise. The time taken by the improved method is about 4.98 min, and the running time increases with the increase of path length, but the rising trend is not significant. This method has good stability and has some advantages.

5. Concluding Remarks

The traditional control methods always have the problems of poor effect, nonfixed path selection, and long time consuming. Hence, a path control method for underwater Robots based on NURBS curve fitting method is proposed. Experimental comparison results are illustrated:

(i) Although the number of control is greater than the Backstepping method and the immune fuzzy PID method when navigating in the same path length area, the overall control is better than the Backstepping and the immune fuzzy PID method.

(ii) Compared with the Backstepping and the immune fuzzy PID method, the navigation time of the improved method is shortened by 16.91 min and 27.52 min, respectively, with good stability and advantages.

Conflicts of Interest

The authors declare that they have no conflicts of interest.

Acknowledgments

This work is supported by the Science and Technology Research Program of Chongqing Municipal Education Commission (Grant no. KJ1729403); the National Natural Science Foundation of China (no. 61403055); Chongqing Education Science and Planning Issues, project name: the Current Situation Analysis and Countermeasure Research of Modern Apprenticeship Practice in Vocational Colleges (no. 2017-GX-422); the Fundamental Research Funds for the Central Universities (Grant no. CDJXS12 17 11 01); Intelligent Robot Technology Research Center of Chongqing College of Electronic Engineering (no. XJPT201705).

References

[1] F. A. Azis, M. S. M. Aras, M. Z. A. Rashid, M. N. Othman, and S. S. Abdullah, "Problem identification for Underwater Remotely Operated Vehicle (ROV): A case study," in *Proceedings of the 2nd International Symposium on Robotics and Intelligent Sensors 2012, IRIS 2012*, pp. 554–560, mys, September 2012.

[2] W. Zheng, Q. Zou, and W. Ni, "Design and application of underwater unmanned vehicle simulation system for navigation and obstacle avoidance," *Xitong Fangzhen Xuebao / Journal of System Simulation*, vol. 28, no. 1, pp. 91–98, 2016.

[3] L. Li, X. Yibo, L. Li, and et al., "Path control of submarine swing arm tracked mobile platform based on Backstepping method," *Mining and Metallurgy Engineering*, vol. 35, no. 6, pp. 1–5, 2015.

[4] Z. Chaohua, T. Guoyuan, H. Daomin, and et al., "Design of DSP control system for attitude control moment gyros group of autonomous underwater Robot," *China Ship Research*, vol. 11, no. 5, pp. 107–112, 2016.

[5] O. Cerman and P. Hušek, "Adaptive fuzzy sliding mode control for electro-hydraulic servo mechanism," *Expert Systems with Applications*, vol. 39, no. 11, pp. 10269–10277, 2012.

[6] M. J. Mahmoodabadi, M. Taherkhorsandi, and A. Bagheri, "Optimal robust sliding mode tracking control of a biped robot based on ingenious multi-objective PSO," *Neurocomputing*, vol. 124, pp. 194–209, 2014.

[7] A. Bagheri, T. Karimi, and N. Amanifard, "Tracking performance control of a cable communicated underwater vehicle using adaptive neural network controllers," *Applied Soft Computing*, vol. 10, no. 3, pp. 908–918, 2010.

[8] S.-Y. Li, C.-H. Yang, C.-T. Lin, L.-W. Ko, and T.-T. Chiu, "Adaptive synchronization of chaotic systems with unknown parameters via new backstepping strategy," *Nonlinear Dynamics*, vol. 70, no. 3, pp. 2129–2143, 2012.

[9] J. Javadi Moghaddam, M. H. Farahani, and N. Amanifard, "A neural network-based sliding-mode control for rotating stall and surge in axial compressors," *Applied Soft Computing*, vol. 11, no. 1, pp. 1036–1043, 2011.

[10] M. J. Mahmoodabadi, S. Momennejad, and A. Bagheri, "Online optimal decoupled sliding mode control based on moving least squares and particle swarm optimization," *Information Sciences*, vol. 268, pp. 342–356, 2014.

[11] M. J. Mahmoodabadi, S. Arabani Mostaghim, A. Bagheri, and N. Nariman-zadeh, "Pareto optimal design of the decoupled sliding mode controller for an inverted pendulum system and its stability simulation via Java programming," *Mathematical and Computer Modelling*, vol. 57, no. 5-6, pp. 1070–1082, 2013.

[12] J. J. Valera García, V. Gómez Garay, E. Irigoyen Gordo, F. Artaza Fano, and M. Larrea Sukia, "Intelligent Multi-Objective Nonlinear Model Predictive Control (iMO-NMPC): Towards the 'on-line' optimization of highly complex control problems," *Expert Systems with Applications*, vol. 39, no. 7, pp. 6527–6540, 2012.

[13] S. M. Zanoli and G. Conte, "Remotely operated vehicle depth control," *Control Engineering Practice*, vol. 11, no. 4, pp. 453–459, 2003.

[14] J. Javadi-Moghaddam and A. Bagheri, "An adaptive neuro-fuzzy sliding mode based genetic algorithm control system for under water remotely operated vehicle," *Expert Systems with Applications*, vol. 37, no. 1, pp. 647–660, 2010.

[15] B. W. Hobson, A. D. Sherman, and P. R. McGill, "Imaging and sampling beneath free-drifting icebergs with a remotely operated vehicle," *Deep-Sea Research Part II: Topical Studies in Oceanography*, vol. 58, no. 11-12, pp. 1311–1317, 2011.

[16] Z. Chaohua, T. Guoyuan, T. Daomin et al., "Design of DSP control system for attitude control moment gyros group of autonomous underwater Robot," *China Ship Research*, vol. 11, no. 5, pp. 107–112, 2016.

[17] Z. Yang, F. Liu, and Y. Wang, "Path following of underactuated surface vessels based on neural sliding mode," *Ship Building of China*, vol. 56, no. 2, pp. 45–55, 2015.

[18] W. Yanqiong and C. Shiping, "Dual path routing admission control protocol for smart road wireless networks," *Computer Engineering*, vol. 41, no. 11, pp. 24–29, 2015.

[19] S. Wei, H. Yulong, and L. Baoshan, "Non symmetric unmanned amphibious platform path tracking control," *Acta Armamentarii*, vol. 37, no. 7, pp. 1161–1169, 2016.

[20] Z. Chu, D. Zhu, and S. X. Yang, "Observer-based adaptive neural network trajectory tracking control for remotely operated vehicle," *IEEE Transactions on Neural Networks and Learning Systems*, vol. 28, no. 7, pp. 1633–1645, 2017.

2

Filtered Medial Surface based Approach for 3D Collision-Free Path Planning Problem

Karima Benzaid,[1] Romain Marie,[2] Noura Mansouri,[1] and Ouiddad Labbani-Igbida[2]

[1] *LARC Laboratory, University Frères Mentouri Constantine 1, Constantine, Algeria*
[2] *X-Lim Institute, UMR CNRS 7252, University of Limoges, Limoges, France*

Correspondence should be addressed to Karima Benzaid; kari_benzaid@yahoo.fr

Academic Editor: Farrokh Janabi-Sharifi

This paper introduces an original 3D path planning approach for Unmanned Aerial Vehicle (UAV) applications. More specifically, the core idea is to generate a smooth and collision-free path with respect to the vehicle dimension. Given a 3D grid representation of the environment, the Generalized Voronoi Graph (GVG) is first approximated using a filtered medial surface (FMS) algorithm on the corresponding navigable space. Based on an efficient pruning criterion, the produced FMS excludes GVG portions corresponding to narrow passages unfitting safe UAV navigation constraints, and thus it defines a set of guaranteed safe trajectories within the environment. Given a set of starting and destination coordinates, an adapted A-star algorithm is then applied to compute the shortest path on the FMS. Finally, an optimization process ensures the smoothness of the final path by fitting a set of 3D Bézier curves to the initial path. For a comparative study, the A-star algorithm is applied directly on the input environment representation and relevant comparative criteria are defined to assert the proposed approach using simulation results.

1. Introduction

This paper deals with the issue of 3D path planning for UAVs in a known environment with stationary obstacles. The problem is to define an efficient method for finding the safest path from a starting point to a target, taking into account the dimension of the UAV.

To deal with this problem, many methods are proposed in the literature. Sampling based algorithms [1], graph based search algorithms [2, 3], mathematic model based algorithms [4, 5], and bioinspired algorithms [6–8] are the fundamental families of path planning algorithms [9]. Most of the path planning solutions were originally developed for 2D problems and then extended to the third dimension, thus increasing the complexity of the approach and its computational cost.

Among the most known sampling based algorithms, we can cite the Rapidly Exploring Random Tree (RRT) [10] and the Probabilistic Road-Maps (PRM) [11] algorithms. These methods are initially designed for 2D path planning problems based on the sampling of the configuration space. The drawback of this kind of algorithms is that they return nonoptimal solutions [12]. To improve the quality of the solution, many variants of these sampling based algorithms have been proposed [12–16]. For aerial vehicles path planning problem, some adapted versions [17–20] are applied. Few methods based on Voronoi diagram have been used to deal with the problem of 3D path planning [21].

The A-star is the most popular graph based search algorithm. Developed by Hart et al. [22], it is based on Dijkstra's algorithm [23] and has been widely implemented in robotics and video games. The conventional A-star is used to find the shortest (possible) path connecting edges of 8 neighbors in 2D grids and 26 neighbors in 3D grids. However, this algorithm cannot find the path in any neighbor connection angle. This implies a suboptimal solution represented as a path with continuous heading changes. With the aim of overcoming this limit and/or improving results in terms of time computation, memory requirement, and path length, some variants of the A-star algorithm have been developed [24–26]. For 3D applications, we can cite [2, 3, 27, 28]. Even though the solution was improved, the planned path is not unruffled enough and needs more refinements to make it smooth for aerial vehicles tracking.

In this paper, we propose a novel approach based on the skeletonization of the navigable space along with an adapted A-star path planner for aerial vehicles. The environment is represented as a discrete 3D occupancy grid supposedly given (its construction is out of scope here). Based on a pruning parameter, the skeletonization algorithm is able to discard noisy skeleton branches and thus exclude narrow passages with respect to the UAV dimensions. This constraint is of major importance but is rarely considered in the literature.

The key idea is to extract the *filtered medial surface* of the grid representation according to the dimension of the aerial vehicle. In the next step, the use of the A-star algorithm aims to find the shortest path on the FMS. Since the returned path is not smooth due to the continuous heading changes of the A-star path, a smoothing method is finally applied.

This may appear similar to the idea adopted in [29]. However, in this cited paper, the authors propose a 2D solution for a 3D path planning problem. The method is based on a 2D skeleton represented as a thin line. The 3rd dimension is not considered at any time.

The main contributions can be summarized in three essential points:

(i) We introduce a 3D skeletonization algorithm for direct applications in aerial robotics context based on a 2D shape representation algorithm.

(ii) In order to eliminate oscillations and tight turns in the generated path, we propose a new safety corridor based smoothing approach using Bézier curves.

(iii) We perform a comparative study of the proposed approach by applying the A-star algorithm directly on an inflated occupancy grid and defining a set of quantitative criteria.

The paper is organized as follows. In Section 2, we present the two steps constituting the proposed solution. The first is the extraction of the FMS on the navigable space, and the second is the computation of the shortest path on the FMS. In Section 3, we introduce a new Bézier curve based method to smooth the initial planned path. In Section 4, our solution is compared to the A-star solution applied on the 3D grid representation. We finally conclude the paper and introduce the future work.

2. Path Planning on the Environments GVG

This section is dedicated to the problem of accurate and real-time path planning on the Generalized Voronoi Graph (GVG), i.e., the set of points in the environment at equal distance from their two closest obstacles. Given a 3D occupancy grid $\mathcal{F} \cup \mathcal{O}$ delimiting the navigable space $\mathcal{F}$ from the obstacles $\mathcal{O}$, it is approximated here by the medial surface of $\mathcal{F}$, filtered with respect to the robot dimension.

2.1. Skeletonization and Filtering of the Navigable Space. Let $\mathcal{X}_{\mathbb{R}}$ be a subset of $\mathbb{R}^n$ and $\mathcal{X}^c_{\mathbb{R}}$ be its complement. Initially defined by Blum [30], the real medial axis $S_{\mathbb{R}}$ of $\mathcal{X}_{\mathbb{R}}$ (also referred to as skeleton) is the set of points $\mathbf{x} \in \mathcal{X}_{\mathbb{R}}$ at equal distance from their two closest neighbors in $\mathcal{X}^c_{\mathbb{R}}$. As pointed out in [31], this definition is conceptually equivalent to the GVG of an environment and is thus used here as an accurate approximation. In the specific case of 3D shapes, it generates a set of surfaces and is thus also referred to as medial surface.

For each shape point $\mathbf{x} \in \mathcal{X}_{\mathbb{R}}$, let $\mathcal{P}_{\mathbf{x}} = \operatorname{argmin}_{\mathbf{y} \in \mathcal{X}^c_{\mathbb{R}}} d(\mathbf{x}, \mathbf{y})$ denote the set of its closest points on $\mathcal{X}^c_{\mathbb{R}}$, referred to as projection of $\mathbf{x}$. The corresponding Euclidean distance is denoted as $\mathcal{D}_{\mathbf{x}} = d(\mathbf{x}, \mathcal{P}_{\mathbf{x}})$. Accordingly, the skeleton $S_{\mathbb{R}}(\mathcal{X}_{\mathbb{R}})$ can then be defined by

$$S_{\mathbb{R}}(\mathcal{X}_{\mathbb{R}}) = \{\mathbf{x} \in \mathcal{X}_{\mathbb{R}}, |\mathcal{P}_{\mathbf{x}}| \geq 2\}. \quad (1)$$

To efficiently approximate and filter the environment GVG, two important concerns have to be considered. First, the navigable space is expressed in this work in a discrete domain (the occupancy grid $\mathcal{F} \cup \mathcal{O}$). Applying (1) directly on $\mathcal{F}$ would fail, since, in most cases, $|\mathcal{P}_{\mathbf{x}}| = 1$ even when a given cell $\mathbf{x}$ actually belongs to the environment GVG.

Therefore, the definition of the discrete medial axis has to be slightly modified. Although several solutions were proposed over the years, this work is based upon the Integer Medial Axis (IMA) [32]. Let $\mathcal{P}^*_{\mathbf{x}}$ denote the reduced projection of $\mathbf{x}$, corresponding to the first element in $\mathcal{P}_{\mathbf{x}}$ with respect to a lexical ordering. Given a digital shape $\mathcal{X}_{\mathbb{Z}} \subset \mathbb{Z}^n$ and its complement $\mathcal{X}^c_{\mathbb{Z}}$, the discrete skeleton $S_{\mathbb{Z}}$ of $\mathcal{X}_{\mathbb{Z}}$ can then be expressed as

$$\begin{aligned} S_{\mathbb{Z}}(\mathcal{X}_{\mathbb{Z}}) = \Big\{ & \mathbf{x} \in \mathcal{X}_{\mathbb{Z}}, \exists \mathbf{y} \in N_{\mathbf{x}} \mid d(\mathcal{P}^*_{\mathbf{x}}, \mathcal{P}^*_{\mathbf{y}}) \\ & > 1, d(\mathbf{m}, \mathcal{P}^*_{\mathbf{x}}) \geq d(\mathbf{m}, \mathcal{P}^*_{\mathbf{y}}) \Big\}, \end{aligned} \quad (2)$$

where $N_{\mathbf{x}} = \{\mathbf{y} \in \mathcal{X}_{\mathbb{Z}}, d(\mathbf{x}, \mathbf{y}) = 1\}$ is the set of direct neighbors of $\mathbf{x}$ and $\mathbf{m}$ is the midpoint of line segment $[\mathbf{x}, \mathbf{y}]$. The first condition measures the (Euclidean) distance between the projections of $\mathbf{x}$ and $\mathbf{y}$ to remove skeleton points related to aliasing, while the second condition is used to produce a skeleton as thin as possible, since it only selects the best approximation to the real skeleton. Note that it is theoretically proven [32] that this definition efficiently approximates the medial axis of the corresponding shape in $\mathbb{R}^n$ while removing skeleton points induced by discretization.

The second major concern is the question of boundary noise: Skeletons are known to be very sensitive to small shape deformations and tend to produce spurious branches. The IMA efficiently approximates Blum's real medial axis but is not designed to filter noisy branches. This issue is addressed here using the delta medial axis (DMA) [33], a real-time pruning strategy able to identify and discard spurious skeleton branches in real time. The algorithm was initially designed for 2D shapes but can straightforwardly be extended to higher dimensions.

The core concepts of the algorithm are illustrated in Figure 1.

Given $\mathbf{x} \in \mathcal{X}_{\mathbb{Z}}$, $\mathbf{y} \in N_{\mathbf{x}}$, let $r_{\mathbf{x},\mathbf{y}}$ denote the radius of the largest disk disjoint from $\mathcal{X}^c$ and centered along $[\mathcal{P}_{\mathbf{x}}, \mathcal{P}_{\mathbf{y}}]$:

$$r_{\mathbf{x},\mathbf{y}} = \sup \left\{ \mathcal{D}_{\mathbf{z}} \mid \mathbf{z} \in [\mathcal{P}_{\mathbf{x}}, \mathcal{P}_{\mathbf{y}}] \right\}. \quad (3)$$

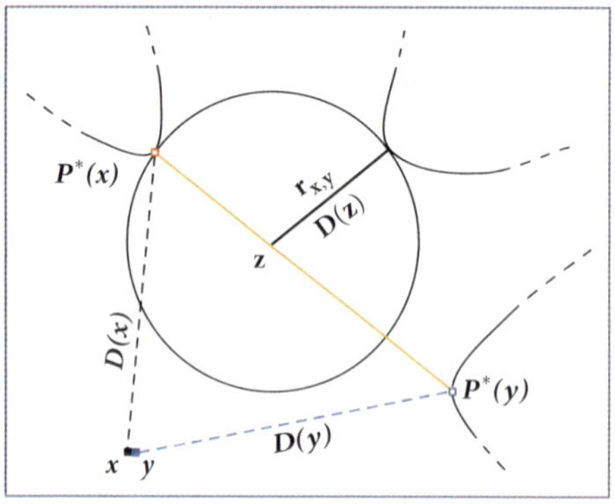

FIGURE 1: Illustrative representation of the concepts involved in DMA computation.

The pruning criterion $\delta^*(\mathbf{x})$ is then defined as the maximal value of $r_{\mathbf{x},\mathbf{y}}$ among its direct neighbors $\mathbf{y} \in N_{\mathbf{x}}$:

$$\delta^*(\mathbf{x}) = \sup\left\{r_{\mathbf{x},\mathbf{y}} \mid \mathbf{y} \in N_{\mathbf{x}},\ d\left(\mathbf{m}, \mathscr{P}_{\mathbf{x}}\right) \geq d\left(\mathbf{m}, \mathscr{P}_{\mathbf{y}}\right)\right\}. \quad (4)$$

Note that the second condition is similar to the IMA. It contributes to generating a thin skeleton, and thus a better approximation to the GVG. The DMA is then defined as follows.

Definition 1 (delta medial axis). Let $\delta \in \mathbb{R}_+^*$ and $\mathscr{X}_{\mathbb{Z}} \subset \mathbb{Z}^n$. The delta medial axis $MA_\delta(\mathscr{X}_{\mathbb{Z}})$ of $\mathscr{X}_{\mathbb{Z}}$ is the set of points $\mathbf{x} \in \mathscr{X}_{\mathbb{Z}}$ with a maximal delta value $\delta^*(\mathbf{x})$ of at least δ:

$$MA_\delta(\mathscr{X}_{\mathbb{Z}}) = \{\mathbf{x} \in \mathscr{X}_{\mathbb{Z}} \mid \delta_M(\mathbf{x}) \geq \delta\}. \quad (5)$$

Intuitively, the DMA discards skeleton branches generated by shape deformations smaller than parameter δ. Based on the robot dimension, it can thus be tuned to retain only skeleton branches (thus GVG navigable paths) large enough to ensure safe navigation.

Applied on the input environment representation $\mathscr{F} \cup \mathscr{O}$, the GVG is thus finally approximated and filtered for a given UAV size δ, by computing the corresponding FMS $MA_\delta(\mathscr{F})$. Examples for different δ values are depicted in Figure 2.

In order to illustrate how the FMS extracted by skeletonization depends on the UAV size, we have considered the same environment representation for two different UAV sizes. As it can be seen in Figure 2, the FMS narrows when δ increases. This means that there are more candidate paths to reach the destination from an initial position when the UAV size is small.

2.2. Shortest Path Computation. The FMS defines a subspace from the navigable space that guarantees safe paths according to the dimension of the UAV. The A-star algorithm is then used to compute a 3D refined path on the FMS. It allows finding the path with the smallest cost from a set of candidate paths leading to a destination. The A-star is based on a cost function $F(n)$:

$$F(n) = G(n) + H(n), \quad (6)$$

where n is the current node, $G(n)$ denotes the moving cost from the *starting point (node)* P_S to the current node n, and $H(n)$ denotes the moving cost from the current node n to the *target point (node)* P_T.

Given the coordinates of P_S and P_T, the aim here is to compute the shortest path on the FMS. The cost function is then the Euclidean distance in the 3D space. A tolerable distance D_t has to be defined to consider two points in the FMS connected. In a 3D occupancy grid, the A-star algorithm uses searching restricted to 26-connectivity (giving the closest cells in 3D environment) or in every angle for some variants of the A-star. Here, the search forwards on the FMS based on D_t. For a current node, neighbors are determined according to their distance to the current node. They will be examined if their distance to the current node is less than or equal to D_t.

The generated path is a set of ordered waypoints defined on the FMS from P_S to P_T. We consider the two previous cases of UAV size δ (20 cm and 40 cm). Figure 3 shows the shortest path found for each case, with the vertical position z. One can see that the planned path is suboptimized with oscillations, which requires a postprocessing step to refine it.

3. Smoothing the Planned Path

The generated path that the UAV is supposed to track must be elegant and unruffled without oscillations, but the path computed on the FMS is not necessarily smooth. To solve this problem, we propose to add an additional processing step based on the use of 3D Bézier curves.

3.1. Bézier Curves. Bézier curves are parametric curves frequently used in vehicle design, fonts representation, games, and 3D animations. Consider $P_0, P_1, \ldots, P_n$ $(n \geq 1)$, $n+1$ points. The Bézier curve of n degrees associated with these points is the curve C defined by $M(t)$ with $t \in [0,1]$:

$$M(t) = B_0(t)P_0 + B_1(t)P_1 + \cdots + B_n(t)P_n \quad (7)$$

P_i $(i = 1 : n)$ are called the control points of the Bézier curve and define the control polygon. B_i $(0 \leq i \leq n)$ coefficients are the *Bernstein polynomials* of degree n given as follows:

$$B_{i,n}(t) = C_n^i t^i (1-t)^{n-i}, \quad i = 0, \ldots, n$$
$$\left(\text{with } C_n^i = \frac{n!}{i!(n-i)!}\right). \quad (8)$$

In a 3D environment, P_i is the point (x_i, y_i, z_i). This means that the point $M(t)$ has as coordinates

$$x(t) = \sum_{i=0}^{n} B_i(t) x_i,$$
$$y(t) = \sum_{i=0}^{n} B_i(t) y_i, \quad (9)$$
$$z(t) = \sum_{i=0}^{n} B_i(t) z_i.$$

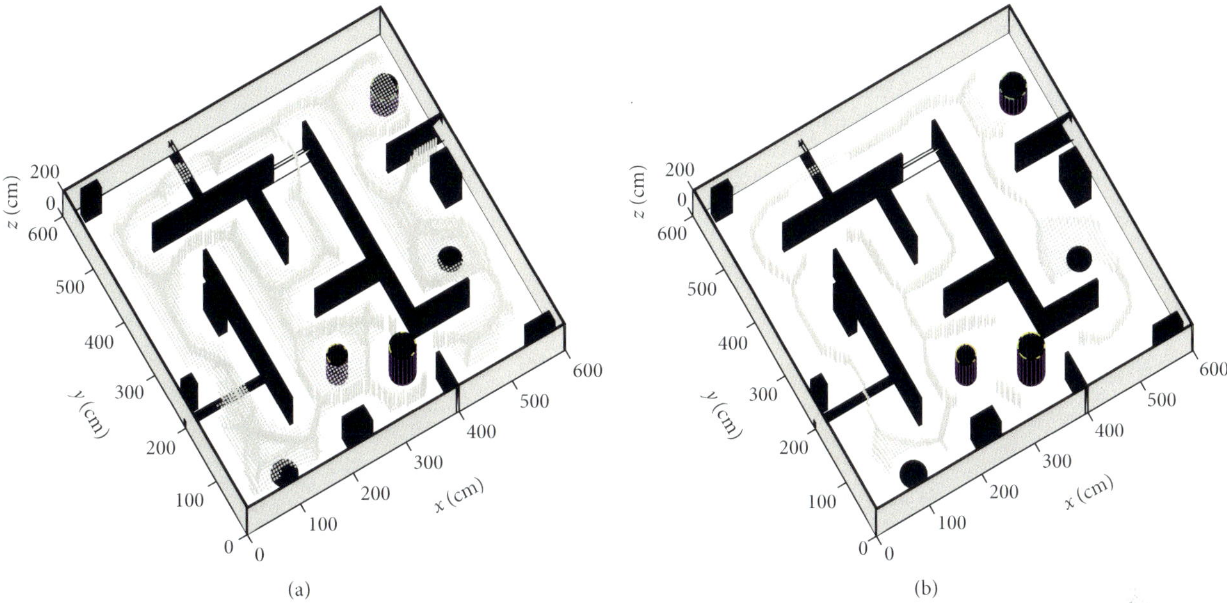

FIGURE 2: Two FMS of the environment based on different UAV sizes ((a) $\delta = 20$ cm, (b) $\delta = 40$ cm).

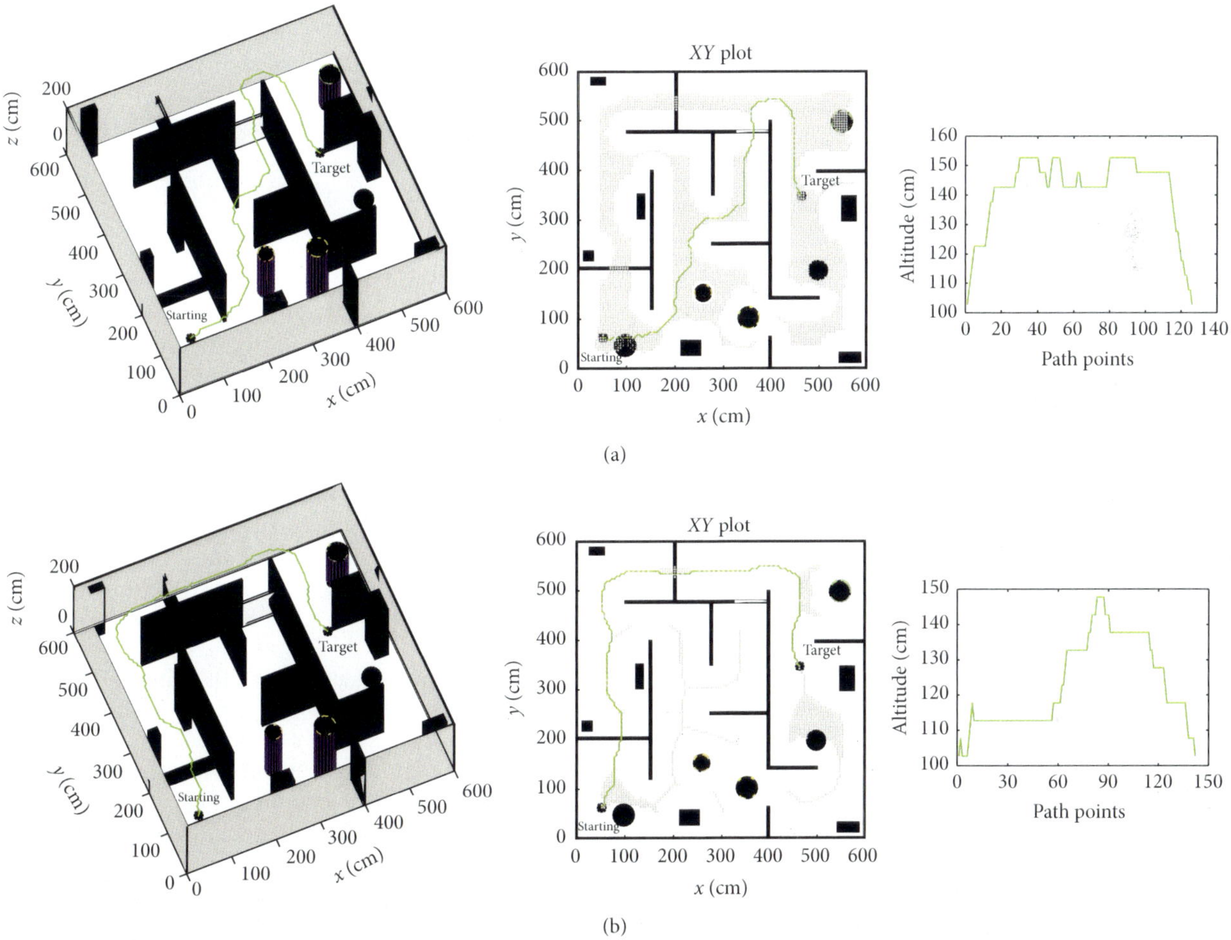

FIGURE 3: Shortest path computed on the FMS based on different UAV sizes (20 cm (a), 40 cm (b)) with the altitude (right column) plotted for each case.

Because they are elegant and easy to use and can describe any shape, Bézier curves are appropriate to solve the smoothing problem of a set of ordered data. When it is about a path between two specific points P_S and P_T, one more reason to use Bézier curves is their *endpoint interpolation property*. This means that the first and last control points (denoted, resp., by P_S and P_T) are the endpoints of the curve. To form a complex smooth path, cubic Bézier curves can be smoothly connected thanks to the *endpoint tangent property*. This property means that the cubic Bézier curve C is tangent to $[P_0, P_1]$ ($[P_2, P_3]$) in P_0 (P_3) and $\dot{M}(0) = 3(P_1 - P_0)$ ($\dot{M}(1) = 3(P_3 - P_2)$). We will give more details on this point thereafter.

3.2. Bézier Curves Smoothing Strategy. The issue of smoothing of the obtained path is formulated as an optimization problem with constraints. The objective is to compute an approximated path to the real one which allows the UAV to make the least maneuvers. Note that there is a compromise between the feasibility of the path and its optimality. By smoothing, the path optimality in terms of safety and path length may degrade. The idea we propose here is to define a safety corridor around the original path as follows:

$$\left(Q_j - p\left(t_j\right)\right) \le \left(D_j - \delta\right), \tag{10}$$

where Q_j are the original data, $p(t_j)$ are values of approximated data computed by interpolation, and D_j is the Euclidean distance to the nearest obstacle in each original path point.

The smoothed path has to be in the safety corridor. For that, we have to determine (by solving the optimization problem) the control points defining the shape of the smoothed path formed by connected patches of Bézier curves. The endpoint interpolation property of Bézier curves ensures that the path starts in P_S and ends in P_T. The objective function to be minimized is the mean of the squared distance between real data and smoothed data. Suppose that there are m waypoints Q_j ($j = 1 : m$) in the original data. For smoothed data $p(t_j)$, we note that $0 \le t_j \le 1$ with uniform division, so $Q_1 = p(0) = P_S$ and $Q_m = p(1) = P_T$. The objective function is given by

$$\min_{p(t_j)} \left(\frac{1}{m-2} \sqrt{\sum_{j=2}^{m-1} \left(Q_j - p\left(t_j\right)\right)^2} \right). \tag{11}$$

The original data can be divided into several segments by giving the initial set of break points. Each segment will be approximated by a cubic Bézier curve (patch) by identifying the middle control points. If there are l segments (i.e., $l - 1$ break points), by solving the problem of optimization (see Figure 4), $3l - 1$ control points must be determined between P_S and P_T to define l Bézier curves. It is important to ensure that the control points joining Bézier curves are not fixed and must be determined as well.

The C^1-continuity between patches of Bézier curve can be translated to constraints of the optimization. Let P_0^k, P_1^k, P_2^k, and P_3^k denote the control points of the kth Bézier curve. kth

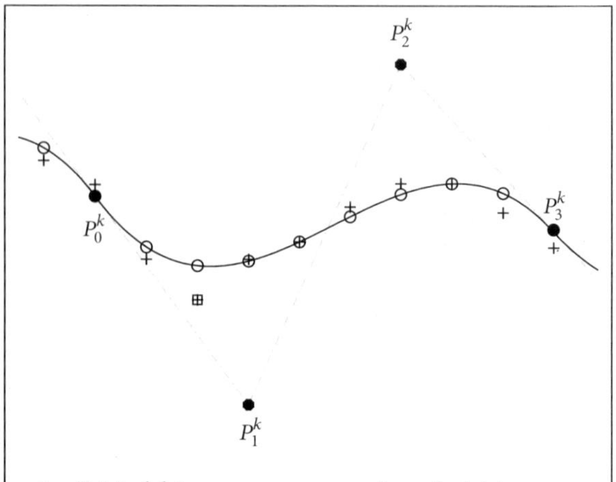

FIGURE 4: Smoothing strategy illustration for the kth segment.

and $(k + 1)$th cubic Bézier curves are C^1 continuous at the joining point if

$$P_3^k = P_0^{k+1}, \\ P_3^k - P_2^k = P_1^{k+1} - P_0^{k+1}. \tag{12}$$

We note that it is possible to add the constraints of UAV heading in the starting and target points, basing on the endpoint tangent property. But this is not considered in this work.

In the optimization process, we consider the case where no solution is found while respecting the constraint that the smoothed path must not be outside the safety corridor. In this case, we split the segment which leaves the safety corridor in two segments and we add their joining point to the break points list. This gives one more segment (consequently, one more Bézier curve to be defined) by replacing one segment with two new segments. We try again to find a solution of the optimization problem defining an approximated path in the safety corridor. The split step is repeated until this is done. Figure 5 shows the refined path after using the Bézier curves based approach. We can see that the path planned by the proposed method is optimal in terms of safety and smoothness.

Table 1 illustrates the results of the gap between original and smoothed paths. Intuitively, this gap can reach its maximum value where the safety corridor width is the maximum. We note that the smoothing method satisfies the condition on the maximum permitted smoothing gap. The smoothing gap mean does not exceed.

4. Results and Discussions

To provide a comparative study of the proposed approach, we apply the A-star directly on the 3D occupancy grid to find the shortest possible path. For that, we supply an inflated occupancy grid (OG) that takes into account the different

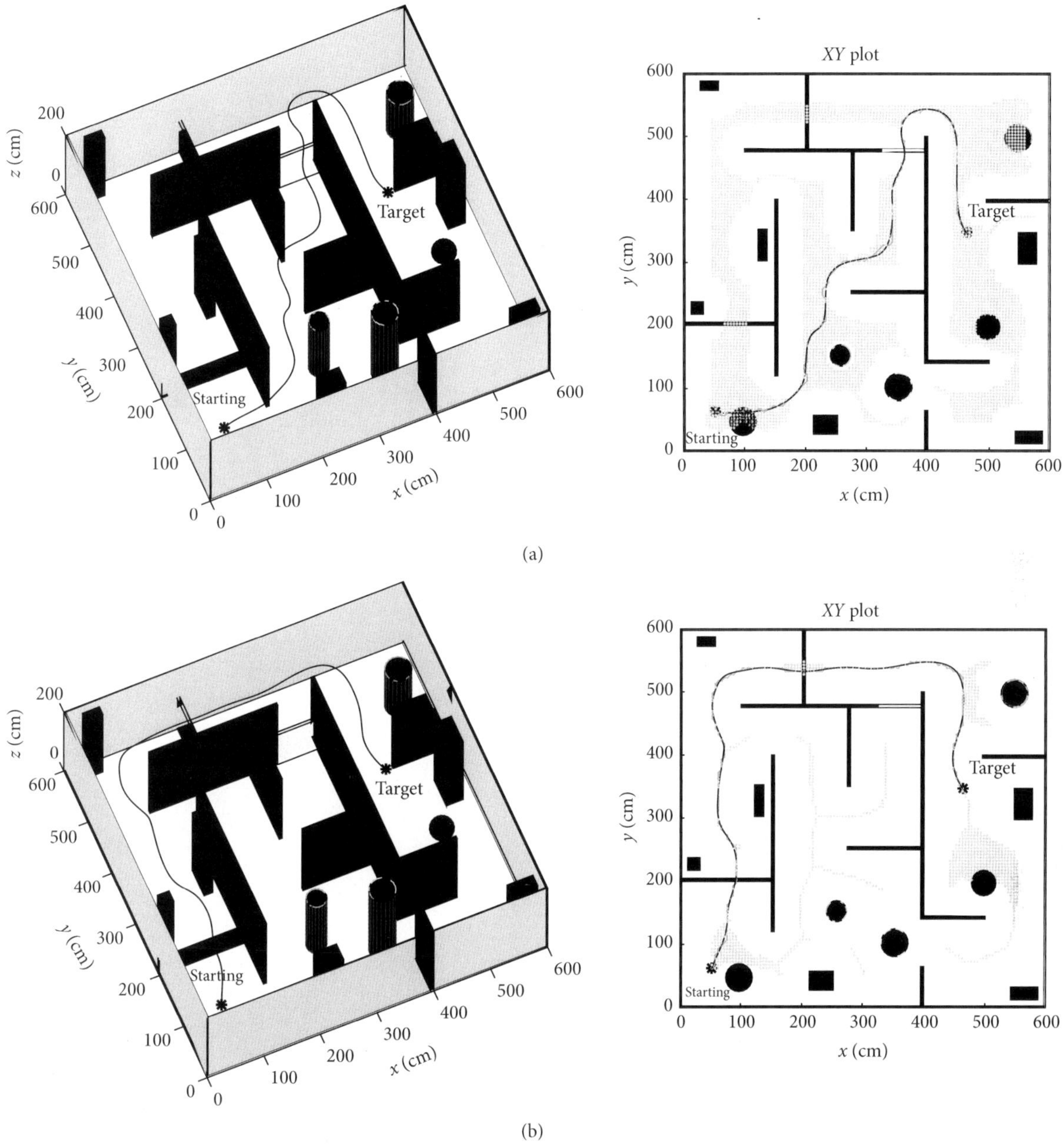

FIGURE 5: Final path for different UAV sizes (20 cm (a), 40 cm (b)).

TABLE 1: Smoothing results.

	case 1 $\delta = 20$ cm	*case 2* $\delta = 40$ cm
Maximum allowed smoothing gap (cm)	42.59	32.58
Maximum smoothing gap (cm)	8.13	8.98
Smoothing gap mean (cm)	2.70	2.55

UAV sizes considered before ($\delta = 20$ cm and $\delta = 40$ cm). The results are shown in Figure 6 compared with the solutions depicted in Figure 5.

To efficiently analyze the results, we define a set of quantitative criteria for the comparative study. The comparison results are shown in Table 2.

The input data (see Table 2) of the search algorithm depends on the type of the approach applied and the UAV dimension. When directly using the A-star on the occupancy grid (without skeletonization), the input data is the navigable space in the inflated occupancy grid. Using the proposed approach, it is the FMS of the navigable space in the original grid. The observations drawn from Table 2 are as follows:

(i) The proposed approach offers a major decrease in the volume of the input data in both cases (90.15% and 87.21% of decrease, resp.) for an increase of,

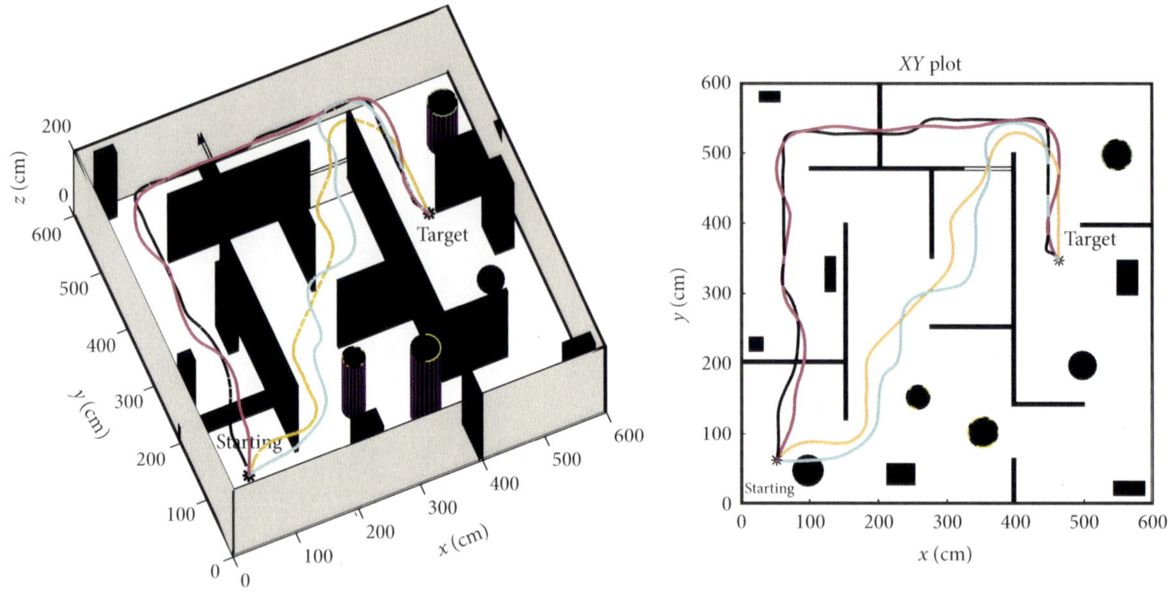

FIGURE 6: Comparison between the proposed approach based on different UAV sizes (20 cm (cyan line), 40 cm (magenta line)) and the A-star applied on the occupancy grid taking into account different UAV sizes (20 cm (orange line), 40 cm (black line)).

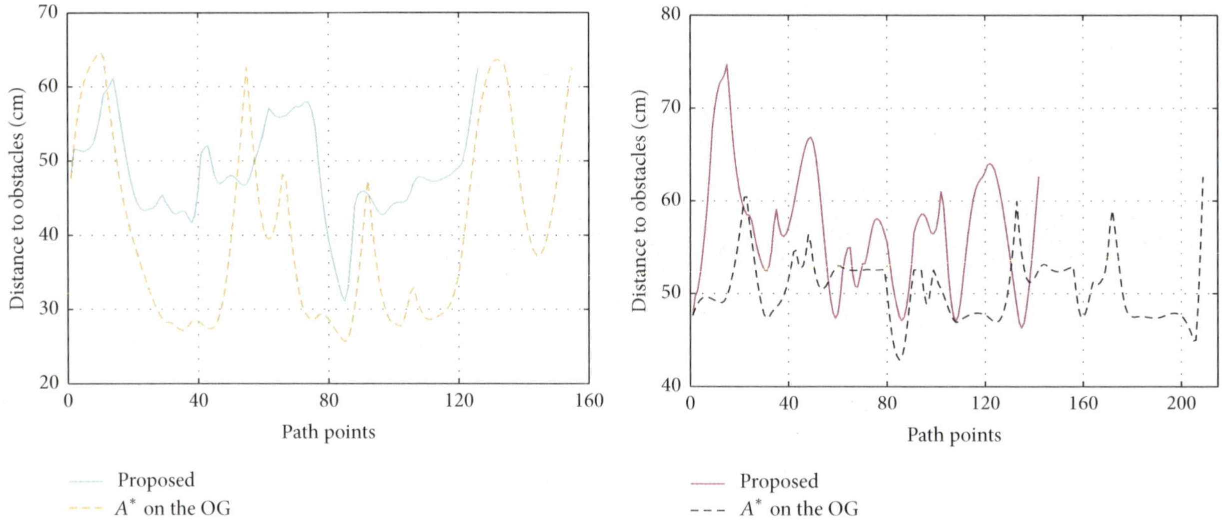

(a) Case 1 (b) Case 2

FIGURE 7: Evolution of the distance to obstacles ((a) δ = 20 cm, (b) δ = 40 cm).

TABLE 2: Quantitative comparison between the A-star applied on the OG and the proposed approach based on two different UAV sizes (δ = 20 cm (case 1), δ = 40 cm (case 2)).

	A-star on the OG		Proposed approach	
	case 1	case 2	case 1	case 2
Input data number	236644	61841	23324	7913
Path length (m)	9.03	11.08	9.95	11.41
Number of path points	155	209	126	142
Smallest dist. to obstacles (cm)	25.63	42.86	31.13	46.28
Mean of dist. to obstacles (cm)	40.18	50.51	48.42	56.88

respectively, 10.18% and 2.97% in the path length. The reduction in the research space is thanks to the skeletonization step. This is of great benefit when addressing real-time applications.

(ii) For all cases, the smallest distance to obstacles is greater than the UAV size, which confirms that all planned paths are collision-free. The proposed approach gives more distance to obstacles supplying better path in terms of safety than the A-star algorithm applied directly on the OG. The evolution of the distance to obstacles along the path that the UAV is supposed to track is shown in Figure 7.

We perceive that the planned path provided by the proposed approach is not the optimum with regard to its length. However, it is more optimum with regard to the safety criterion than the path generated using the A-star directly on the occupancy grid.

The vehicle dynamics have been detailed in [34]. In this previous work, we have presented a generalized dynamical model of miniature multirotor vehicles. These UAVs are known for their high maneuverability. A control approach has been proposed and applied to track (user-designed) trajectories for an example of quadrotor, hexarotor, and octorotor UAVs.

As the smoothing step is performed to compute an approximate path requiring the least maneuvers of UAV without affecting its safety, we define two new quantitative parameters to better evaluate the proposed solution:

(i) $\Delta Z_{\max}$: maximal change value of the altitude.

(ii) $\Delta \psi_{\max}$: maximal change value of the heading angle.

For the two UAV sizes considered, the following numbers are obtained:

(i) *In case 1* ($\delta = 20\,cm$): $\Delta Z_{\max} = 08.78$ cm, $\Delta \psi_{\max} = 28.43$ deg.

(ii) *In case 2* ($\delta = 40\,cm$): $\Delta Z_{\max} = 05.41$ cm, $\Delta \psi_{\max} = 23.84$ deg.

These parameter values confirm that there are no aggressive changes in altitude and heading angle between two waypoints in the obtained path.

We note that, for all simulated cases, we consider the particular case where starting and destination points are on the FMS. This is not necessarily the case. If one of the initial and target points or both of them are not on the FMS, the idea is to reach (from a starting point) the closest point on the FMS and/or give up the FMS at the closest point to the destination. These additional portions at the beginning and/or the end of the path can be generated using 3D Bézier curves. An optimization process can be applied to guarantee the smoothness of the path at the connecting points.

5. Conclusion and Future Works

In this paper, we present an efficient solution for 3D path planning problem in aerial robotics. It is based on the extraction of the FMS by a 3D skeletonization algorithm. Including a pruning parameter δ, the solution is given based on the UAV dimension discarding small branches. In the second step, we used the A-star algorithm to find the shortest path on the FMS. For smoothing, we proposed a method based on the use of Bézier curves.

In order to evaluate the performance of our approach, we compared the results to the A-star algorithm applied directly on the 3D occupancy grid. The quantitative criteria of comparison reveal that the proposed approach meets interesting performance criteria. Reducing interestingly the dimensionality of the searching space (by more than 87.21%) reduces memory requirement and is beneficial in terms of computational cost. It is also proved that this approach attaches greater importance to the path safety compared to the path length, so it provides the shortest safest path.

Currently, the proposed approach uses the conventional A-star for the planning of the shortest path. An improved version of the A-star is obviously required especially when the ratio *UAV size/environment dimensionality* is smaller. In future works, we intend to consider curvature constraints of the planned path in the cost function. Furthermore, we propose dealing with the dynamic constraints of the UAV in order to find feasible trajectories within its maneuverability ability.

Conflicts of Interest

The authors declare that there are no conflicts of interest regarding the publication of this paper.

References

[1] S. M. LaValle, "Sampling-based motion planning," in *Planning Algorithms*, chapter 5, pp. 185–248, Cambridge University Press, 2006.

[2] L. De Filippis, G. Guglieri, and F. Quagliotti, "Path planning strategies for UAVS in 3D environments," *Journal of Intelligent & Robotic Systems*, vol. 65, no. 1–4, pp. 247–264, 2012.

[3] J. Carsten, D. Ferguson, and A. Stentz, "3D field D*: improved path planning and replanning in three dimensions," in *Proceedings of the IEEE/RSJ International Conference on Intelligent Robots and Systems*, pp. 3381–3386, October 2006.

[4] B. Miller, K. Stepanyan, A. Miller, and M. Andreev, "3D path planning in a threat environment," in *Proceedings of the 50th IEEE Conference on Decision and Control and European Control Conference*, vol. 6, pp. 6864–6869, IEEE, Orlando, Fla, USA, December 2011.

[5] A. Chamseddine, Y. Zhang, C. A. Rabbath, C. Join, and D. Theilliol, "Flatness-based trajectory planning/replanning for a quadrotor unmanned aerial vehicle," *IEEE Transactions on Aerospace and Electronic Systems*, vol. 48, no. 4, pp. 2832–2848, 2012.

[6] H. Duan and P. Li, *Bio-Inspired Computation in Unmanned Aerial Vehicles*, Springer, Berlin, Germany, 1st edition, 2014.

[7] H. Ergezer and K. Leblebicioglu, "Path planning for UAVs for maximum information collection," *IEEE Transactions on Aerospace and Electronic Systems*, vol. 49, no. 1, pp. 502–520, 2013.

[8] V. Roberge, M. Tarbouchi, and G. Labonte, "Comparison of parallel genetic algorithm and particle swarm optimization for real-time UAV path planning," *IEEE Transactions on Industrial Informatics*, vol. 9, no. 1, pp. 132–141, 2013.

[9] L. Yang, J. Qi, J. Xiao, and X. Yong, "A literature review of UAV 3D path planning," in *Proceedings of the 11th World Congress on Intelligent Control and Automation (WCICA '14)*, pp. 2376–2381, IEEE, Shenyang, China, July 2014.

[10] S. M. LaValle, "Rapidly-exploring random trees: a new tool for path planning," Tech. Rep., Computer Science Department, Iowa State University, 1998.

[11] L. E. Kavraki, P. Švestka, J.-C. Latombe, and M. H. Overmars, "Probabilistic roadmaps for path planning in high-dimensional configuration spaces," *IEEE Transactions on Robotics and Automation*, vol. 12, no. 4, pp. 566–580, 1996.

[12] S. Karaman and E. Frazzoli, "Sampling-based algorithms for optimal motion planning," *The International Journal of Robotics Research*, vol. 30, no. 7, pp. 846–894, 2011.

[13] R. Bohlin and L. E. Kavraki, "Path planning using lazy PRM," in *Proceedings of the International Conference on Robotics and Automation*, pp. 521–528, April 2000.

[14] E. Frazzoli, M. A. Dahleh, and E. Feron, "Real-time motion planning for agile autonomous vehicles," *Journal of Guidance, Control, and Dynamics*, vol. 25, no. 1, pp. 116–129, 2002.

[15] K. E. Bekris, B. Y. Chen, A. M. Ladd, E. Plaku, and L. E. Kavraki, "Multiple query probabilistic roadmap planning using single query planning primitives," in *Proceedings of the IEEE/RSJ International Conference on Intelligent Robots and Systems*, pp. 656–661, October 2003.

[16] D. Aarno, D. Kragic, and H. I. Christensen, "Artificial potential biased probabilistic roadmap method," in *Proceedings of the IEEE International Conference on Robotics and Automation*, pp. 461–466, April 2004.

[17] P. O. Pettersson and P. Doherty, "Probabilistic roadmap based path planning for an autonomous unmanned helicopter," *Journal of Intelligent & Fuzzy Systems: Applications in Engineering and Technology*, vol. 17, no. 4, pp. 395–405, 2006.

[18] M. Kothari and I. Postlethwaite, "A probabilistically robust path planning algorithm for UAVs using rapidly-exploring random trees," *Journal of Intelligent & Robotic Systems*, vol. 71, no. 2, pp. 231–253, 2013.

[19] F. Yan, Y.-S. Liu, and J.-Z. Xiao, "Path planning in complex 3D environments using a probabilistic roadmap method," *International Journal of Automation and Computing*, vol. 10, no. 6, pp. 525–533, 2013.

[20] X. Wu, C. Guo, Y. Li, W. Chen, and Y. Wang, "The research of RRT route planning algorithm for UAV that based on kinematic equation," *International Journal of Control and Automation*, vol. 8, no. 1, pp. 287–296, 2015.

[21] Y. V. Pehlivanoglu, "A new vibrational genetic algorithm enhanced with a Voronoi diagram for path planning of autonomous UAV," *Aerospace Science and Technology*, vol. 16, no. 1, pp. 47–55, 2012.

[22] P. E. Hart, N. J. Nilsson, and B. Raphael, "A formal basis for the heuristic determination of minimum cost paths," *IEEE Transactions on Systems, Man, and Cybernetics*, vol. 4, no. 2, pp. 100–107, 1968.

[23] E. W. Dijkstra, "A note on two problems in connexion with graphs," *Numerische Mathematik*, vol. 1, no. 1, pp. 269–271, 1959.

[24] S. Koenig and M. Likhachev, "D* Lite," in *Proceedings of the 18th National Conference on Artificial Intelligence*, pp. 476–483, August 2002.

[25] A. Nash, S. Koenig, and M. Likhachev, "Incremental Phi*: incremental any-angle path planning on grids," in *Proceedings of the 21st International Joint Conference on Artificial Intelligence*, pp. 1824–1830, IEEE, July 2009.

[26] A. K. Guruji, H. Agarwal, and D. K. Parsediya, "Time-efficient A* algorithm for robot path planning," *Procedia Technology*, vol. 23, pp. 144–149, 2016.

[27] A. Nash, S. Koenig, and C. Tovey, "Any-angle path planning and path length analysis in 3D," in *Proceedings of the 24th AAAI Conference on Artificial Intelligence*, pp. 147–154, July 2010.

[28] Z. Qi, Z. Shao, Y. S. Ping, L. M. Hiot, and Y. K. Leong, "An improved heuristic algorithm for UAV path planning in 3D environment," in *Proceedings of the 2nd International Conference on Intelligent Human-Machine Systems and Cybernetics*, pp. 258–261, Nanjing, China, August 2010.

[29] X. Liu, C. Zhou, M. Ding, and C. Cai, "Skeleton-based fast path planning for UAV," in *Remote Sensing and GIS Data Processing and Other Applications*, October 2009.

[30] H. Blum, "A transformation for extracting new descriptors of shape," in *Models for the Perception of Speech and Visual Form*, pp. 362–380, MIT Press, 1967.

[31] R. Ogniewicz and M. Ilg, "Voronoi skeletons: theory and applications," in *Proceedings of the IEEE Computer Society Conference on Computer Vision and Pattern Recognition*, pp. 63–69, Champaign, Ill, USA, 1992.

[32] W. H. Hesselink and J. B. T. M. Roerdink, "Euclidean skeletons of digital image and volume data in linear time by the integer medial axis transform," *IEEE Transactions on Pattern Analysis and Machine Intelligence*, vol. 30, no. 12, pp. 2204–2217, 2008.

[33] R. Marie, O. Labbani-Igbida, and E. M. Mouaddib, "The delta medial axis: a fast and robust algorithm for filtered skeleton extraction," *Pattern Recognition*, vol. 56, pp. 26–39, 2016.

[34] K. Benzaid, N. Mansouri, and O. Labbani-Igbida, "A generalized dynamical model and control approach applied to multirotor aerial systems," in *Proceedings of the 8th International Conference on Modelling, Identification and Control*, pp. 225–230, November 2016.

3

Path Planning for a Space-Based Manipulator System based on Quantum Genetic Algorithm

Zhengcang Chen[1,2] and Weijia Zhou[2]

[1]State Key Laboratory of Robotics, Shenyang Institute of Automation, The Chinese Academy of Sciences (CAS), Shenyang 110016, China
[2]University of Chinese Academy of Science, Beijing, China

Correspondence should be addressed to Zhengcang Chen; chenzhengcang@sia.cn

Academic Editor: Shahram Payandeh

In this study, by considering a space-based, n-joint manipulator system as research object, a kinematic and a dynamic model are constructed and the system's nonholonomic property is discussed. In light of the nonholonomic property unique to space-based systems, a path planning method is introduced to ensure that when an end-effector moves to the desired position, a floating base achieves the expected pose. The trajectories of the joints are first parameterized using sinusoidal polynomial functions, and cost functions are defined by the pose deviation of the base and the positional error of the end-effector. At this stage, the path planning problem is converted into a target optimization problem, where the target is a function of the joints. We then adopt a quantum genetic algorithm (QGA) to solve this objective optimization problem to attain the optimized trajectories of the joints and then execute nonholonomic path planning. To test the proposed method, we carried out a simulation on a six-degree-of-freedom (DOF) space-based manipulator system (SBMS). The results showed that, compared to traditional genetic optimization algorithms, the QGA converges more rapidly and has a more accurate output.

1. Introduction

During space missions, astronauts face a harsh and life-threatening environment in space, especially in the course of extravehicular activities. A growing number of researchers think it inevitable that space manipulators will be used, autonomously or semiautonomously, in future missions to perform tasks either completely or incompletely [1–3]. With the rapid development of space technologies, a number of tasks can be implemented by space manipulator systems. At present, some manipulators are installed on a base whose position or attitude is controlled by reaction wheels or thrusters, taking a satellite-manipulator system, for instance. This kind of system has the following disadvantages: (1) the position and pose control system (PACS) consumes fuel, which shortens the lifespan of the entire space mission. (2) Gas emitted by the thruster might negatively impact the surrounding objects. (3) When the PACS is used, a transient response appears that may impact the dynamic performance of the system. (4) The pose control system can easily attain a saturated state.

In order to solve this problem, many researchers use space-based systems to replace the ones whose base is installed with PACS. However, there exists a highly dynamic coupling phenomenon between a floating base and a manipulator in a space-based system, especially when both have a similar mass. This affects the control precision of the end-effector of the manipulator because the space-based manipulator system (SBMS) is nonholonomic. This study focuses on using the nonholonomic property to plan the path of joints such that the end-effector and the floating base simultaneously arrive at the desired state.

Unlike fixed systems, the position and pose of the end-effector of a manipulator in space-based systems are dependent not only on the given joint angles, but also on the joint paths. In other words, for the same joint configuration, different joint paths correspond to different poses. The authors of [4, 5] studied the course of the development and the status of nonholonomic systems as well as their principles of dynamics and control methods. Rathee and Pathak [6] present path planning of dual arm free flying space robot using smooth functions of time. Results of their works demonstrate that if the base mass and inertia are increased to a certain limit keeping in mind the singularities, then base disturbance can be decreased and the end effectors follow the required circular path. Zeng et al. [7] present a subsection algorithm for obstacle avoidance path planning of free-floating space manipulator for orbit observation. The approach could guarantee that the end-effector reaches the desired point and realize the obstacle avoidance. Kaigom et al. [8] propose a Constrained Particle Swarm Optimization algorithm to steer the end-effector to a target pose with minimized base disturbance. This approach is based on Reaction Null-Space concept. And its simulation results demonstrate the effectiveness of the planning method. Liu et al. [9] analyze the path planning of redundant, free-floating space manipulators with revolute joints, and 7 degrees of freedom. They provide a tractable planning method to avoid the pseudo-inverse of the Jacobian matrix. Nenchev [10] summarizes results based on the application of Reaction Null-Space approach. Via the null-space and pseudo-inverse mapping of the coupling inertia, the joint space is decomposed into two complementary orthogonal subspaces which is useful for motion analysis and control of various unfixed-base systems. Jia et al. [11] present a novel path planning method to avoid obstacle based on A* algorithm for space manipulators. In this article, a concept of "collision-free workspace" of a space manipulator is provided. Shi et al. [12] propose a nonholonomic motion planning methodology for free-floating space robot (FFSR) using swarm optimization algorithm. Also, an optimal control technique is applied for FFSR trajectory planning. Zhang [13] proposed a path planning method for a free-floating space robot in Cartesian space. Shi [14] proposed a quantum particle swarm optimization method to plan path for space robots, and Jia et al. [15] developed a straight-line trajectory optimization method for a redundant robot with nine degrees of freedom (DOF).

In this study, based on an analysis of nonholonomic properties, we first determine the coupling relationship between a base and a manipulator. We consider the pose of the base and the position of the end-effector as the optimized targets. We then parameterize the joint trajectories using sinusoidal polynomial functions, set ranges of value for the joint angles, rate, and acceleration, and convert the path planning problem into a target optimization problem. Finally, we make use of a QGA to resolve and obtain the optimal parameters as well as an expression of the joint trajectory. To test the effectiveness of the proposed method, we conducted a forward simulation to solve for the optimal target.

2. Kinematic Model of a Space-Based Manipulator System

According to the modeling theory of unrooted multibody systems, with respect to coordinate transformation, a new modeling approach is proposed.

2.1. Description of SBMS. An SBMS consists of an n-joints manipulator and a floating base. Its simplified schematic diagram is shown in Figure 1:

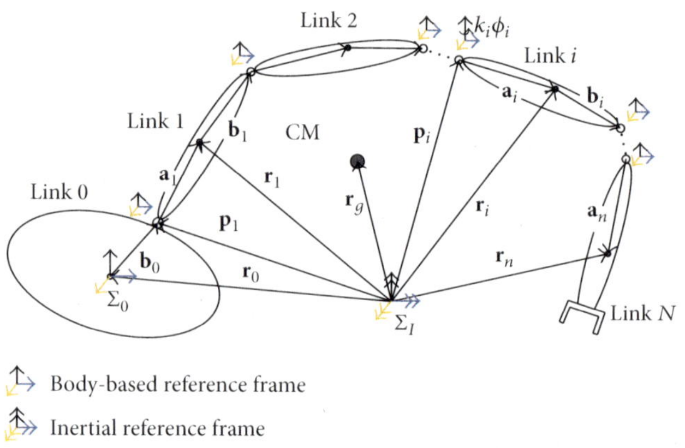

FIGURE 1: Structure diagram of a space-based manipulator system.

(1) Each component of the system is rigid and numbered from 0 to N starting from the base.

(2) For simplicity, each joint between link i and link $i+1$ has one DOF.

(3) The entire system is in microgravity without an external force exerting on it.

(4) The system has a tree topology.

(5) Σ_i and Σ_I are, respectively, the body-fixed reference frame and the inertial reference frame. In general, the origin of the inertial reference frame and the center of mass of the SBMS overlap to some extent in the initial state, and the direction of Σ_0 coincides with that of Σ_I.

(6) $\mathbf{a}_i$ represents the position vector relating joint i to the CM of link i, and $\mathbf{b}_i$ represents that relating the CM of link i to the next joint.

The selection of coordinate systems is a precondition of modeling. A mixed coordinate system consisting of related and reference coordinates is chosen in this study. In particular, $\mathbf{x}_0 \in R^3$ and $\theta \in R^3$ represent the position and pose of the base, and $\boldsymbol{\phi} \in R^n$ represents the joint angles here.

2.2. Kinematic Modeling. We define dependent coordinates $\mathbf{Y}_{ci} = [\mathbf{x}_i^T, \boldsymbol{\theta}_i^T]^T$, $i = 0, \ldots, n$ (called the positional screw), mixed independent coordinates $\mathbf{y}_p = [\mathbf{x}_0^T, \boldsymbol{\theta}_0^T, \boldsymbol{\phi}^T]^T$, and the dependent velocity screw as $\dot{\mathbf{Y}}_\mathbf{i} = [\mathbf{v}_\mathbf{i}^\mathbf{T}, \boldsymbol{\omega}_\mathbf{i}^\mathbf{T}]^\mathbf{T}$.

The constraint equation of an n-joint open-loop serial rigid body system is written as

$$\Phi(\mathbf{Y}_c) = 0 \in \mathbf{R}^m, \tag{1}$$

where $\mathbf{Y}_c = [\mathbf{Y}_{c0}^T, \ldots, \mathbf{Y}_{cn}^T]^T \in \mathfrak{R}^{6n+6}$ and m represents the number of constraint equations. The velocity screw is not just a differential of positional screw. It is

$$\dot{\mathbf{Y}}_c = \mathbf{U}(\mathbf{Y}_c)\,\dot{\mathbf{Y}}. \tag{2}$$

When a different reference frame is selected, the expression of $\mathbf{U}$ also differs: see [16].

We differentiate (1) and substitute (2) into it:

$$\mathbf{\Phi}_Y \dot{\mathbf{Y}} = 0, \tag{3}$$

where $\mathbf{\Phi}_Y = \partial\mathbf{\Phi}/\partial\mathbf{Y} \in \mathfrak{R}^{m\times(6n+6)}$ represents the Jacobian of the constraint equations.

Since the DOF of a system is equal to the difference between the number of dependent coordinates and that of the constraint equations—that is, $f = (6n+6) - m$—the kernel of $\mathbf{\Phi}_Y$ is spanned by f linearly independent column vectors. The basic solutions of this vector space are defined as $\{\mathbf{S}_1, \ldots, \mathbf{S}_f\}$. Hence, the dependent velocity screw can be expressed as

$$\dot{\mathbf{Y}} = \mathbf{S}_1 t_1 + \mathbf{S}_2 t_2 + \ldots + \mathbf{S}_f t_f, \tag{4}$$

where $S = [\mathbf{S}_1, \ldots, \mathbf{S}_f] \in R^{(6n+6)\times(n+6)}$ is called the basic solution matrix. Let $\dot{\mathbf{y}} = \mathbf{t} = [t_1, \ldots, t_{n+6}]^T$, which is an independent velocity vector. The velocity screw of the nth link can be written as

$$\dot{\mathbf{Y}}_n = \mathbf{B}_{n,0}\dot{\mathbf{Y}}_0 + \mathbf{J}_n\dot{\boldsymbol{\phi}}. \tag{5}$$

Since the system is in free-floating state, momentum is conserved and is assumed to be zero:

$$\dot{\mathbf{Y}}_0 = \begin{bmatrix} \mathbf{J}_{Tb\phi} \\ \mathbf{J}_{Rb\phi} \end{bmatrix} = \begin{bmatrix} \dfrac{-\mathbf{J}_{T\omega}}{w} + \tilde{\mathbf{r}}_{g0}^{T}\mathbf{I}_s^{-1}\mathbf{I}_m \\ -\mathbf{I}_s^{-1}\mathbf{I}_m \end{bmatrix} \dot{\boldsymbol{\phi}}. \tag{6}$$

Connecting (5) and (6), the velocity-level differential equation is written as

$$\dot{\mathbf{Y}}_n = \mathbf{J}^* (\boldsymbol{\phi})\, \dot{\boldsymbol{\phi}}, \tag{7}$$

where $\mathbf{J}^*$ represents the generalized Jacobian matrix, the concrete representation of which is

$$\mathbf{J}^* = \left\{ \begin{bmatrix} \mathbf{E}_3 & \tilde{\mathbf{r}}_{n,0}^{T} \\ 0 & \mathbf{E}_3 \end{bmatrix} \begin{bmatrix} \dfrac{-\mathbf{J}_{\mathbf{T}\boldsymbol{\omega}}}{\mathbf{w}} - \tilde{\mathbf{r}}_{g0}^{\mathbf{T}}\mathbf{I}_{\mathbf{s}}^{-1}\mathbf{I}_{\mathbf{m}} \\ -\mathbf{I}_{\mathbf{s}}^{-1}\mathbf{I}_{\mathbf{m}} \end{bmatrix} + \mathbf{J}_n \right\},$$

$$\mathbf{J}_i = \begin{bmatrix} \mathbf{J}_{Ti} \\ \mathbf{J}_{Ri} \end{bmatrix} = [\mathbf{B}_{i,1}\boldsymbol{\eta}_1 \ \cdots \ \mathbf{B}_{i,i-1}\boldsymbol{\eta}_{i-1} \ \ \boldsymbol{\eta}_i \ \ 0 \ \cdots \ 0] \in \mathfrak{R}^{6\times n},$$

$$\mathbf{B}_{i,j} = \begin{bmatrix} \mathbf{E}_3 & \tilde{\mathbf{r}}_{i,j}^{T} \\ 0 & \mathbf{E}_3 \end{bmatrix},$$

$$\boldsymbol{\eta}_i = \begin{bmatrix} \mathbf{k}_i \times \mathbf{a}_i \\ \mathbf{k}_i \end{bmatrix},$$

$$\mathbf{J}_{\mathbf{T}\boldsymbol{\omega}} = \sum_{\mathbf{i}=1}^{\mathbf{n}} \mathbf{m}_{\mathbf{i}}\mathbf{J}_{\mathbf{Ti}},$$

$$\mathbf{r}_{i,j} = \mathbf{r}_i - \mathbf{r}_j,$$

$$\mathbf{I}_{\mathbf{m}} = \sum_{\mathbf{i}=1}^{\mathbf{n}} \left(\mathbf{I}_{\mathbf{i}}\mathbf{J}_{\mathbf{Ri}} - \mathbf{m}_{\mathbf{i}}\tilde{\mathbf{r}}_{\mathbf{i}}^{\mathbf{T}}\mathbf{J}_{\mathbf{Ti}}\right) - \tilde{\mathbf{r}}_{\mathbf{g}}\mathbf{J}_{\mathbf{T}\boldsymbol{\omega}},$$

$$\mathbf{I}_{\mathbf{s}} = \mathbf{I}_0 + \sum_{\mathbf{i}=1}^{\mathbf{n}} \left(\mathbf{I}_{\mathbf{i}} + \mathbf{m}_{\mathbf{i}}\tilde{\mathbf{r}}_{\mathbf{i}}\tilde{\mathbf{r}}_{\mathbf{i}0}^{\mathbf{T}}\right) - w\tilde{\mathbf{r}}_g\tilde{\mathbf{r}}_{0g}^{T}. \tag{8}$$

During nonholonomic path planning, the integral and iterative processes based on (6) and (7) can implement path following a floating base.

2.3. Analysis of Nonholonomic Properties. In order to better understand the coupling phenomenon between a floating base and a manipulator installed on it, an SBMS is here demonstrated as being nonholonomic [17].

Definition 1. A linear subspace spanned by one or more n-dimensional unsteady vectors is called a distribution.

Definition 2. If the result of a Lee bracket operation between two random vectors belonging to a distribution also belongs to it, the distribution is deemed involutive.

Frobenius' Theorem. The necessary and sufficient condition for the integrality of the Pfaffian constraint equations is that their solution space is an involutive distribution.

The state variables are defined as the angle velocity of the base and the joint angle velocity of the manipulator; that is, $\dot{\mathbf{x}}_z = [\boldsymbol{\omega}_0^T, \dot{\boldsymbol{\phi}}^T]^T$. The input value is defined as $\mathbf{u} = \dot{\boldsymbol{\phi}}$. From (2) and (6),

$$\dot{\mathbf{x}}_z = \mathbf{K}\mathbf{u}, \tag{9}$$

where the coefficient matrix $\mathbf{K}$ can be written as $\mathbf{K} = \begin{bmatrix} \mathbf{U}^{-1}(\boldsymbol{\theta}_0)\mathbf{J}_{Rb\phi} \\ \mathbf{E}_{\mathbf{n}} \end{bmatrix}$ and $\boldsymbol{\theta}_0$ represents the pose angle.

A generalized velocity distribution Δ consists of n columns of $\mathbf{K}$; that is, $\Delta = \text{span}(\mathbf{K}_1, \ldots, \mathbf{K}_n)$, where $\mathbf{K}_i = (k_{1i}, k_{2i}, k_{3i}, \mathbf{e}_i{}^T)^T$ and $\mathbf{e}_i{}^T$ represents the standard unit vector. The result of the Lee bracket between $\mathbf{K}_i$ and $\mathbf{K}_j$ is

$$\left[\mathbf{K}_i, \mathbf{K}_j\right] = \left[\mathbf{k}_{1ij}, \mathbf{k}_{2ij}, \mathbf{k}_{3ij}, \mathbf{0}^T\right]^T \notin \Delta. \tag{10}$$

Hence, Δ is not involutive. According to the Frobenius theorem, SBMS has been shown to be a nonholonomic system.

3. Transformation of Path Planning Problem to Objective Optimization Problem

3.1. Solving the Objective Function. As we know, the whole system is composed of a floating base, a macroarm, and a dexterous robotic hand. During objective capturing process, the macroarm and dexterous hand are fixed together. The main topic of this text is to make the dexterous hand (or a special point of the macroarm) arrive a desired position by planning the joint path of the macroarm. As for how and with what attitude to capture an objective by the dexterous hand, it is not included in the topics. Also, attitude of the floating base has an important influence on communication quality and solar conversion efficiency. So, the focus of this study is to plan joint trajectory in order to enable the posture of the base and the position of the end-effector arrive at their desirable states simultaneously. The relevant variables include the quaternion of the base and the position vector of the end link, which are written as $\mathbf{X} = [\mathbf{Q}_b^T, \mathbf{P}_e^T]^T \in \mathbf{R}^6$. $\mathbf{X}_0$, $\mathbf{X}_d$, and $\mathbf{X}_f$ are, respectively, the initial state, the desired state, and the final state.

Steps of path planning are as follows:

(1) Transformation of state variables ($\mathbf{X} \rightarrow \mathbf{X}' = [\mathbf{Q}_b^T, \boldsymbol{\phi}^T]^T$). End-effector positional vector with respect to the base-linked reference frame is only dependent on joint angles vector $\boldsymbol{\phi}$; that is, ${}^0\mathbf{P}_e^T = f_1(\boldsymbol{\phi})$, while the homogenous transformation matrix from inertial reference frame to base-linked one is related to attitude of the base and the base's positional vector $\mathbf{r}_b$; that is, ${}^IA_0 = f_2(\mathbf{Q}_b^T, \mathbf{r}_b)$. Assume that $\mathbf{r}_b = \mathbf{r}_{b0}$ (a constant) at $t = 0$. According to momentum conservation law, vector $\mathbf{r}_b$ is related to history of joint angles; that is, $\mathbf{r}_b = \mathbf{r}_b(\boldsymbol{\phi})$. So state variable is transformed by $\mathbf{P}_e^T = {}^IA_0 \cdot {}^0\mathbf{P}_e^T = f_3(\mathbf{Q}_b^T, \boldsymbol{\phi})$.

(2) Parameterized joint trajectory using sinusoidal polynomial functions.

(3) Design of objective function based on the deviation between the final state and the desire state.

The destination of path planning is as follows: the planned joint trajectory $\boldsymbol{\phi}(t)$ can render true the condition whereby the final state variables can infinitely approach the desired variables; that is, $\|\mathbf{X}'_d - \mathbf{X}'_f\| \rightarrow 0$.

3.2. Parameterization of Joint Angles. Using the sinusoidal function to parameterize the joint angles can better constrain their scope. To ensure that the velocity and acceleration curves are smooth, we define

$$\phi_i(t) = a_i \sin\left(b_{i7}t^7 + \cdots + b_{i1}t + b_{i0}\right) + c_i. \quad (11)$$

There are six constraint conditions: $\phi_i(0) = \phi_0$, $\dot{\phi}_i(0) = 0$, $\ddot{\phi}_i(0) = 0$, $\phi_i(t_f) = \phi_{id}$, $\dot{\phi}_i(t_f) = 0$, and $\ddot{\phi}_i(t_f) = 0$. The range of ϕ_i is $\phi_i \in [\phi_i \mid \phi_{i\min} < \phi_i < \phi_{i\max}]$.

So, $a_i = (\phi_{i\max} - \phi_{i\min})/2$ and $c_i = (\phi_{i\max} + \phi_{i\min})/2$.

Then, joint angle velocity and acceleration can be written as

$$\begin{aligned}\dot{\phi}_i(t) &= a_i \cos\left(b_{i7}t^7 + \cdots + b_{i1}t + b_{i0}\right) \cdot \left(7b_{i7}t^6 + \cdots + b_{i1}\right), \\ \ddot{\phi}_i(t) &= -a_i \sin\left(b_{i7}t^7 + \cdots + b_{i1}t + b_{i0}\right) \cdot \left(7b_{i7}t^6 + \cdots + b_{i1}\right)^2 \\ &\quad + a_i \cos\left(b_{i7}t^7 + \cdots + b_{i1}t + b_{i0}\right) \cdot \left(42b_{i7}t^5 + \cdots + 2b_{i2}\right).\end{aligned} \quad (12)$$

Equation (11) contains eight unknown parameters: $b_{i0} \sim b_{i7}$. With six constraint conditions, six of the eight unknowns can be eliminated. Let $\mathbf{b}_i = [b_{i7}, b_{i6}]$ and $\mathbf{b} = [\mathbf{b}_1, \ldots, \mathbf{b}_n]$. The deviation of the state variables can be written as

$$\Delta\mathbf{X}' = \mathbf{X}'_d - \mathbf{X}'_f = \left(\left[\Delta\mathbf{Q}_b^T, \Delta\boldsymbol{\phi}^T\right]^T\right). \quad (13)$$

Then, the objective function is defined as

$$f(\mathbf{b}) = \frac{\operatorname{norm}\left(\Delta\mathbf{Q}_b^T\right)}{K_Q} + \frac{\operatorname{norm}\left(\Delta\mathbf{Q}_b^T\right)}{K_\phi}. \quad (14)$$

At this time, the path planning problem in SBMS has been transformed into an optimization problem of nonlinear function $f(\mathbf{b})$. We choose the QGA to solve it.

4. Using QGA to Solve Optimization Problem

The quantum genetic algorithm [18] is a method that combines quantum computation and a genetic algorithm, which is a probabilistic evolutionary algorithm.

4.1. Theory of QGA. The genetic algorithm is a common method to solve complicated optimization problems. In 1998, Dr. Bagley proposed the concept of the "genetic algorithm" in his doctoral dissertation and developed a self-adaptation genetic algorithm [19]. The basic idea underlying this algorithm is to learn the jungle law in biological evolution and selection, as well as the crossover or mutation of chromosomes. It is robust and has a wide range of applicability. However, its iterations is much, its convergent rate is low, and it is easily caught in a local optimum.

On the contrary, quantum computation involves solving an NP problem in classical computation using superposition, entanglement, and the interference of quantum-state information units. In 1994, Shor first proposed the quantum algorithm [20].

The QGA is a genetic algorithm based on the expression of a quantum-state vector. The probability amplitude expression of the quantum bit is applied to a chromosome's encoding process, because of which the chromosome can be expressed through the superposition of quantum states whose update process is conducted by means of a quantum rotate gate. Target optimization is thus completed.

The quantum bit is a two-state system in a superposed state, such as $|\varphi\rangle = \alpha|0\rangle\beta|1\rangle$, where (α, β) is a set of amplitude constants that satisfies $|\alpha|^2 + |\beta|^2 = 1$. States $|0\rangle$ and $|1\rangle$, respectively, represent the spin-up and spin-down states. That is, a quantum bit contains information concerning States $|0\rangle$ and $|1\rangle$.

The QGA is updated through the quantum rotate gate, which is expressed as $\mathbf{U}(\boldsymbol{\theta}_i) = \begin{bmatrix} \cos(\boldsymbol{\theta}_i) & -\sin(\boldsymbol{\theta}_i) \\ \sin(\boldsymbol{\theta}_i) & \cos(\boldsymbol{\theta}_i) \end{bmatrix}$. The principle of update is $\begin{bmatrix} \alpha_i \\ \beta_i \end{bmatrix}_{\text{update}} = \mathbf{U}(\boldsymbol{\theta}_i) \begin{bmatrix} \alpha_i \\ \beta_i \end{bmatrix}$, where $[\alpha_i, \beta_i]^T$ and $[\alpha_i, \beta_i]^T_{\text{update}}$ represent probability amplitudes before and after the update, respectively, and $\boldsymbol{\theta}_i$ represents the spin angle.

Thus, the key differences between the quantum genetic algorithm and traditional genetic algorithm are that the former has more abundant population diversity and better parallel calculating abilities with larger search range and convergence rate. This is why its adaptation and efficiency is much higher.

4.2. Brief Introduction to QGA Performance

(1) Compared to the genetic algorithm, QGA has several new features.

(2) It uses the global searching method, which enlarges the search scope and increases search efficiency.

(3) It effectively resists premature phenomena and has better convergence and precision.

(4) With increase in the number of iterations, the global search capability decreases and local search ability increases, which can cause the system to converge to the global optimum.

(5) The quantum system is a complex nonlinear system that can adapt to practical issues.

4.3. Path Planning Based on QGA. The path planning process of an SBMS based on the QGA is as follows:

(1) Set known parameters, such as mass, inertia, length, and the bound range of the joint angles. Input the initial state variable, the desired state variable, and the planning time.

(2) Parameterize the joint trajectory, and determine the type and number of parameters.

(3) Solve the final state variable. Compute the difference between the final state and the desired state.

(4) Define parameters mentioned in (2) as chromosomes, and initialize them randomly to obtain a population composed of n quantum bits.

(5) Measure the population and encode it.

(6) Design the objective function based on state difference.

(7) Solve the fitness of each individual and record the results.

(8) Judge whether the iterative process has end. The basis is the fitness constraint condition and the maximum number of iterations.

(9) Update individuals by quantum rotate gate, and perform the circuit operation along (5)~(8) until it jumps from the loop.

The corresponding flowchart is shown in Figure 2.

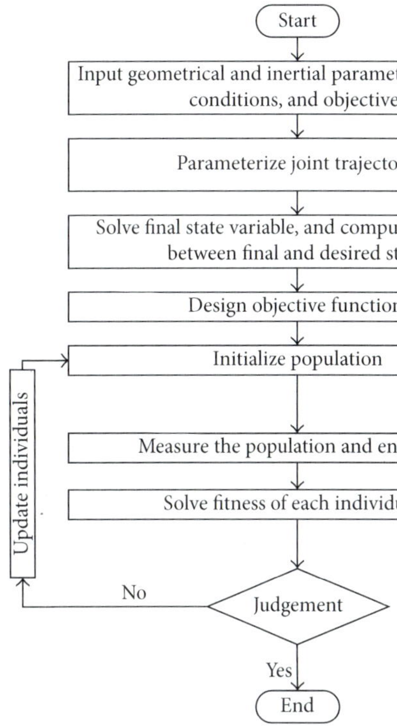

FIGURE 2: Flowchart of QGA.

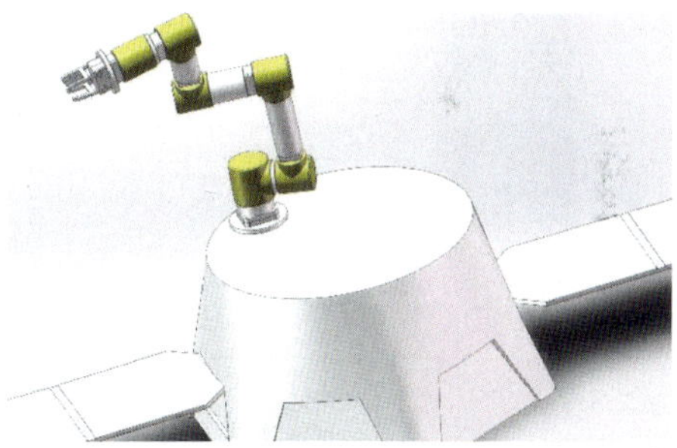

FIGURE 3: A space-based, 6-DOF manipulator system.

5. Simulation Test

In this section, we consider a 6-DOF SBMS as an example, the 3D model of which is shown in Figure 3, for a simulation test to test the proposed path planning method. We also compared it with the traditional genetic algorithm.

The geometrical and inertial parameters of the system are shown in Figure 3 and Table 1. The D-H model is shown in Figure 4: it depicts the deployable state, also called the zero state. The D-H parameters are shown in Table 2. The initial and desired poses of the base are represented by the quaternion as follows:

$$\mathbf{Q}_{b0}^T = [1, 0, 0, 0],$$
$$\mathbf{Q}_{bd}^T = [0.9727, -0.1877, 0.0782, 0.112]. \quad (15)$$

Table 1: Geometrical and inertial parameters of a system.

Number	Mass/kg	Length/m		Inertia tensor/kg*m^2					
				Ixx	Iyy	Izz	Ixy	Iyz	Ixz
0	100.2	— $\mathbf{b}_0$	— 1.3	10.12	10.05	9.88	0	0	0
1	10.1	$\mathbf{a}_1$ $\mathbf{b}_1$	0.554 0.554	1.012	0.099	0.102	0	0	0
2	10.3	$\mathbf{a}_2$ $\mathbf{b}_2$	0.554 0.554	1.012	0.099	0.102	0	0	0
3	9.5	$\mathbf{a}_3$ $\mathbf{b}_3$	0.554 0.554	1.012	0.099	0.102	0	0	0
4	9.8	$\mathbf{a}_4$ $\mathbf{b}_4$	0.554 0.554	1.012	0.099	0.102	0	0	0
5	9.9	$\mathbf{a}_5$ $\mathbf{b}_5$	0.489 0.489	1.01	0.110	0.103	0	0	0
6	7.5	$\mathbf{a}_6$ $\mathbf{b}_6$	0.412 0.412	0.808	0.1	0.099	0	0	0

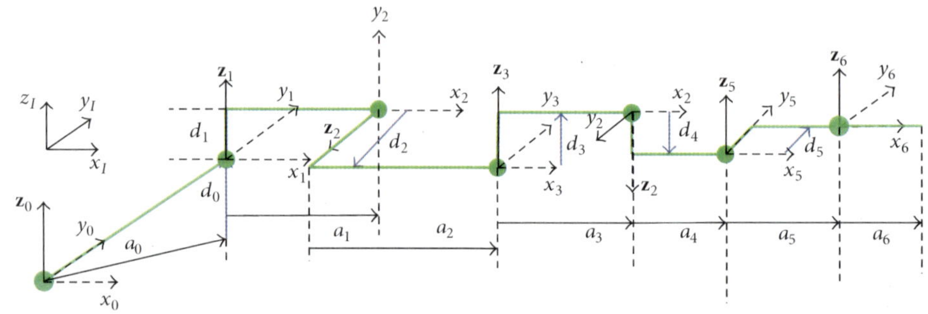

Figure 4: D-H frames of space-based systems.

Table 2: D-H parameters of a space-based manipulator system.

Link number	θ_i/rad	α_i/rad	a_i/m	d_i/m
0	0	0	1.3	0
1	$\pi/2$	0	1.02	0.433
2	$-\pi/2$	0	1.02	0.433
3	π	0	1.02	0.433
4	$-\pi$	0	1.02	0.433
5	0	$124/180*\pi$	0.81	0.548
6	0	0	0.808	0

The initial and desired positions of the end-effector are

$$\mathbf{P}_{e0}^{T} = [5.5, 1.05, 0],$$
$$\mathbf{P}_{ed}^{T} = [4.936, 2.7622, -0.1242]. \quad (16)$$

According to the two preconditions above, we transformed them into another form that incorporated joint configuration. Based on the requests (positional error of end-effector $k_{pe} \le 0.05$ m and attitude error of the base $k_{rb} \le 5^\circ$) of positional precision of the floating base and the attitude precision of the end-effector, we then defined the positional coefficient $k_1 = 0.4$, $k_{pe} = 0.02$, and attitude coefficient $k_2 = 0.0175$. According to formula (12) and the boundary values of it, the value ranges for each parameter were set as

$$b_{i7} \in \left[-3\times 10^{-3}, 3\times 10^{-3}\right],$$
$$b_{i6} \in \left[-3\times 10^{-3}, 3\times 10^{-3}\right]. \quad (17)$$

The optimization of the objective functions was executed by the QGA and the genetic algorithm. The results, including the objective values and optimal parameters, are given below.

Results of the Genetic Algorithm. The optimized parameters were $\mathbf{b}_{GA} = [0.448, 1.973, 2.309, -1.274, 2.327, 1.887, -0.215, 0.855, 0.786, 0.518, -1.888, 0.714] \times 10^{-4}$. The corresponding objective value was $J_{GA} = 0.385$.

Figure 5 shows that the simulation process converged after 152 iterations.

Results of QGA. The optimized parameters were $\mathbf{b}_{QGA} = [-1.9319, -1.7996, -2.0789, 1.6905, 0.46052, -2.1949, 1.3569, -0.42342, 0.98057, -0.28985, 1.877, -0.4173] \times 10^{-4}$. The corresponding objective value was $J_{QGA} = -2.2\times 10^{-4}$.

Figure 6 shows that the simulation converged after 80 iterations. Compared to the GA, the QGA converged much

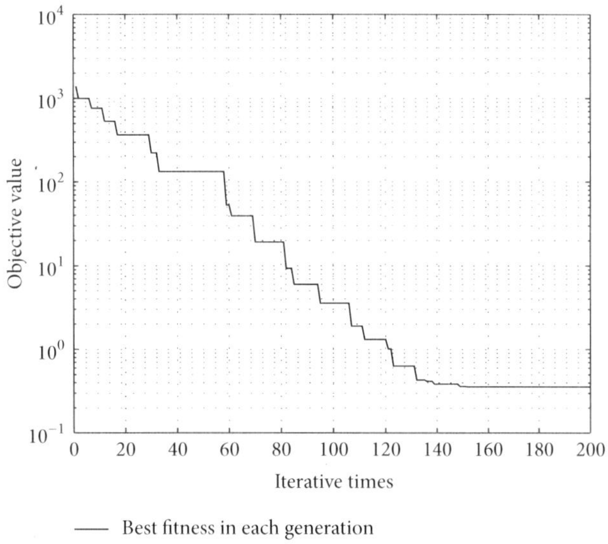

FIGURE 5: Fitness curve of global optimal points for GA.

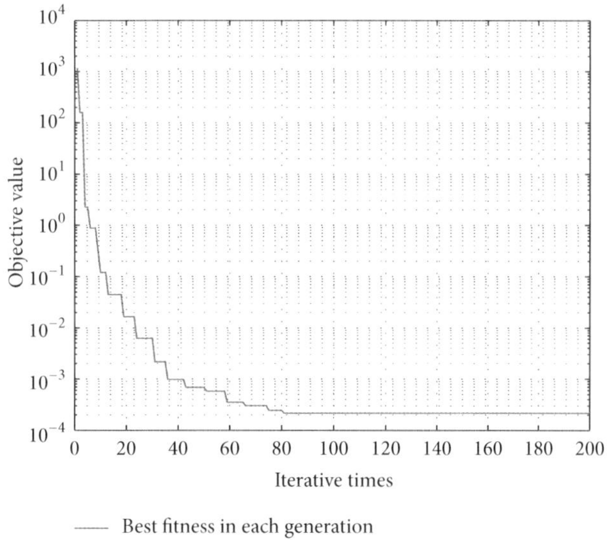

FIGURE 6: Fitness curve of global optimal points of QGA.

more rapidly with higher precision. Moreover, step width of ladder-shaped curve shown in Figure 5 is larger than that shown in Figure 6. This implies that the genetic algorithm can easily fall into a local optimum. Hence, the QGA is superior to the genetic algorithm.

We then solved the optimized joint trajectory according to the parameters obtained from the QGA, as shown in Figure 7. Finally, we used the joint trajectories in Figure 7 as inputs to the motion of the control joint and observed whether the base and the end-effector arrived at the desired state simultaneously. Figure 8 shows the RPY angle curves of the floating base. Positional curves of the floating base and the end-effector are, respectively, described in Figures 9 and 10. Figure 11 shows the distance between the actual and desired positions of the end-effector. The deviation as 2.78×10^{-2}, which was acceptable. From Figure 7, we see that the curves are smooth, and this is satisfactory for the control of the joints.

6. Conclusion

In this study, we built a kinematical model for a SBMS and analyzed its nonholonomic properties. Moreover, we also parameterized the joint trajectory and the design objective function and hence transformed the path planning problem into a nonlinear objective optimization problem. We used the genetic algorithm and the QGA to solve above problem. The results showed that the proposed method can obtain a superior solution compared to prevalent systems and converges much more rapidly.

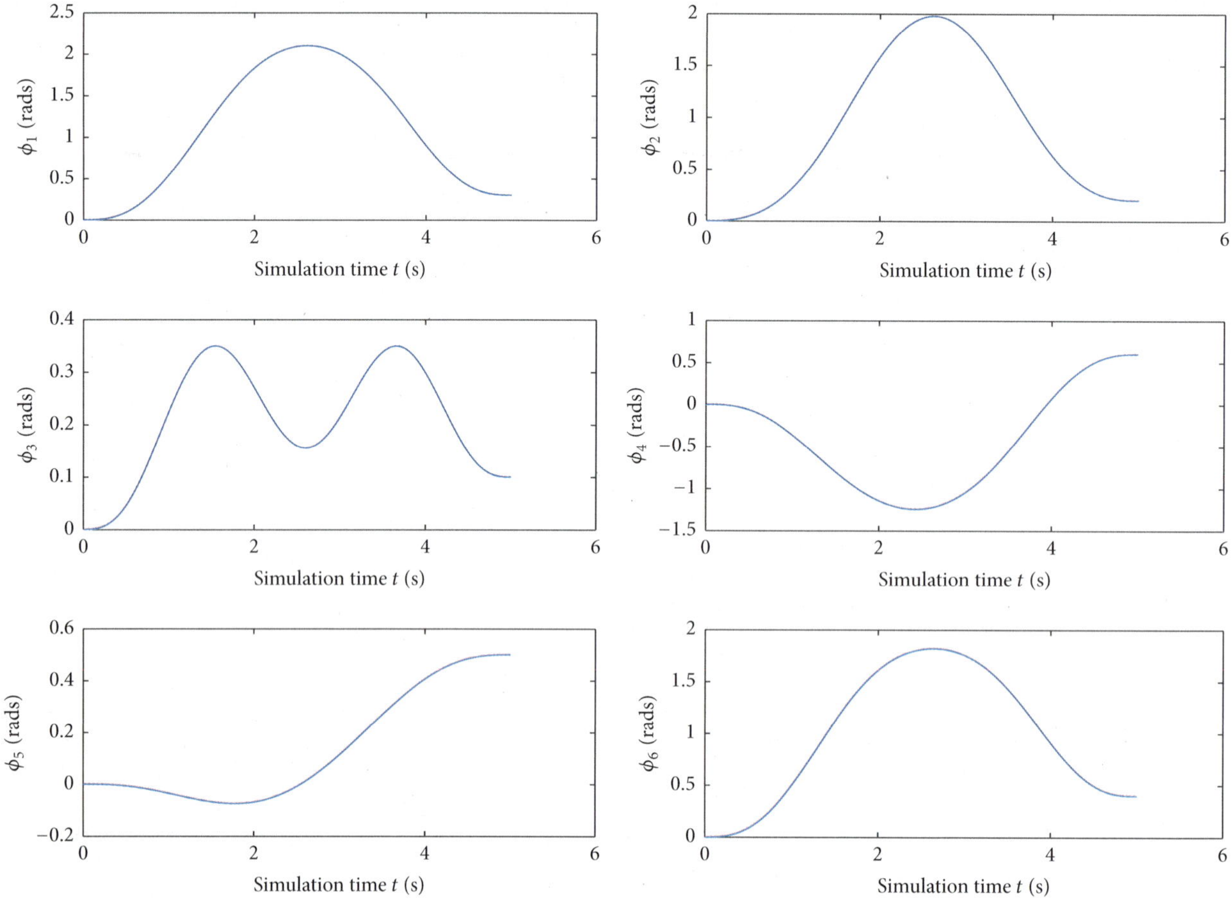

FIGURE 7: Optimal joint curves.

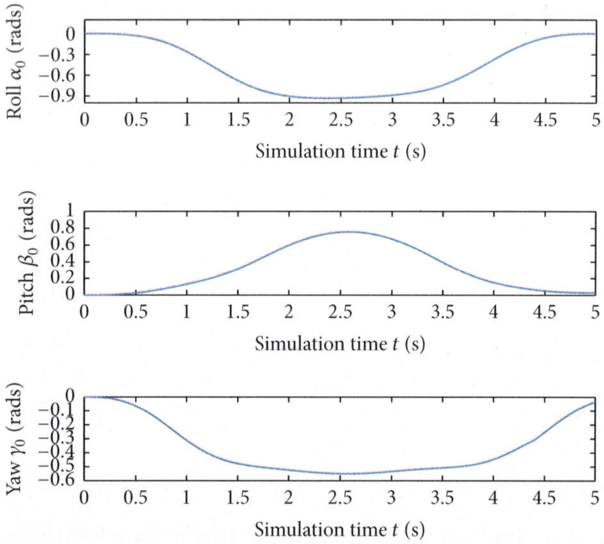

FIGURE 8: Pose error curves of the floating base.

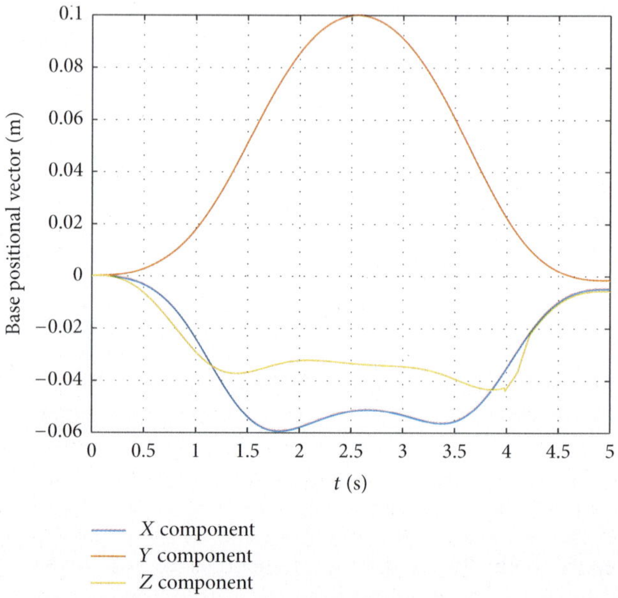

FIGURE 9: Positional curves of the floating base.

We also obtained joint trajectories based on the optimized parameters and drew pose curves of the base and the position curve of the end-effector. The results indicated that the proposed method can arrive at the desired state.

Our method can efficiently reduce the energy consumption and time in a coordinated control process between a floating base and a manipulator.

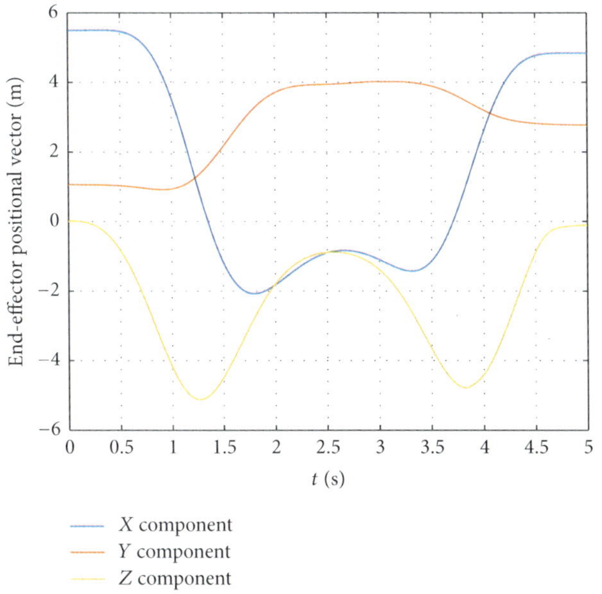

FIGURE 10: Positional curves of the end-effector in inertial reference frame.

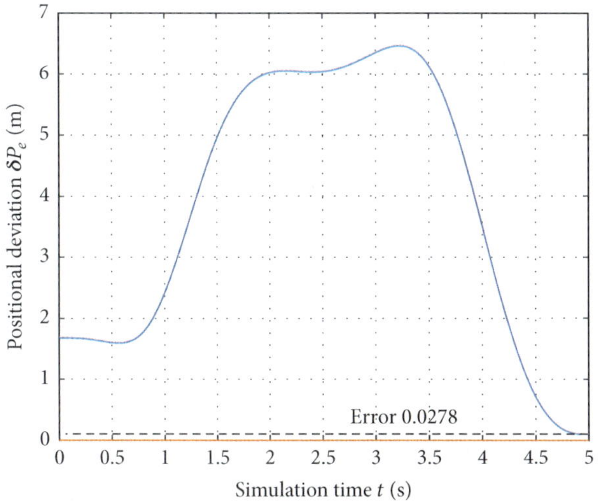

FIGURE 11: Position error curve of end feature point.

Conflicts of Interest

The authors declare that there is no conflict of interests regarding the publication of this paper.

Acknowledgments

Support for this work was provided by the National Natural Science Foundation of China (Grant no. 51505470) and the Dr. Start-Up Fund in Liaoning Province (20141152).

References

[1] Y. Wang, L. Sun, W. Yan, and J. Liu, "An analytic and optimal inverse kinematic solution for a 7-DOF space manipulator," *Jiqiren/Robot*, vol. 36, no. 5, pp. 592–599, 2014.

[2] A. Flores-Abad, O. Ma, K. Pham, and S. Ulrich, "A review of space robotics technologies for on-orbit servicing," *Progress in Aerospace Sciences*, vol. 68, pp. 1–26, 2014.

[3] Y. Liu, Q. Jia, G. Chen, and H. Sun, "Load maximization trajectory optimization for free-floating space robot using multi-objective particle swarm optimization algorithm," *Robot*, vol. 36, no. 4, pp. 402–410, 2014.

[4] A. M. Bloch, J. E. Marsden, and D. V. Zenkov, "Nonholonomic dynamics," *Notices of the American Mathematical Society*, vol. 52, no. 3, pp. 324–329, 2005.

[5] H. Li and R. Li, "Zero-disturbance control of free-floating space manipulators using integral-type sliding mode control," *Mathematical Problems in Engineering*, vol. 2015, Article ID 308305, 7 pages, 2015.

[6] R. Rathee and P. M. Pathak, "Dual arm free flying space robot trajectory planning using polynomial," *Journal of Robotics*, vol. 2015, Article ID 324831, 11 pages, 2015.

[7] C. Zeng, Q. Song, and Z. Li, "Obstacle avoidance path planning of free-floating space manipulator for on-orbit observation," in *Proceedings of the International Conference on Artificial Intelligence and Robotics and the International Conference on Automation, Control and Robotics Engineering*, p. 3, Kitakyushu, Japan, July 2016.

[8] E. G. Kaigom, T. J. Jung, and J. Roßmann, "Optimal motion planning of a space robot with base disturbance minimization," in *Proceedings of the 11th Symposium on Advanced Space Technologies in Robotics and Automation*, pp. 1–6, April 2011.

[9] X. Liu, H. Baoyin, and X. Ma, "Optimal path planning of redundant free-floating revolute-jointed space manipulators with seven links," *Multibody System Dynamics*, vol. 29, no. 1, pp. 41–56, 2013.

[10] D. N. Nenchev, "Reaction null space of a multibody system with applications in robotics," *Mechanical Sciences*, vol. 4, no. 1, pp. 97–112, 2013.

[11] J. Hanxu and Z. Shuangqi, "Path planning for space manipulator to avoid obstacle based on A~ * algorithm," *Journal of Mechanical Engineering*, vol. 13, p. 17, 2010.

[12] Z. Shi, Y.-Z. Wang, and Q.-L. Hu, "A polynomial interpolation based particle swarm optimization algorithm for trajectory planning of free-floating space robot," *Journal of Astronautics*, vol. 32, no. 7, pp. 1516–1521, 2011.

[13] F. Zhang, Y. Fu, and S. Wang, "A path planning method for free-method space robot in Cartisian space," *Robot/Jiqiren*, vol. 31, p. 6, 2009.

[14] Y. Shi, B. Liang, X. Wang, and W. Xu, "Cartesian non-holonomic path planning of space robot based on quantum-behaved particle swarm optimization algorithm," *Journal of Mechanical Engineering*, vol. 47, no. 23, pp. 65–73, 2011.

[15] Q.-X. Jia, M. Chu, H.-X. Sun, and L. Hong, "Research on the optimal algorithm for trajectory planning of a 9-DOF hyper-redundant robot," *Beijing Youdian Daxue Xuebao/Journal of Beijing University of Posts and Telecommunications*, vol. 31, no. 2, pp. 20–25, 2008.

[16] P. C. Hughes, *Spacecraft Attitude Dynamics*, Courier, 2012.

[17] Z. Li and J. F. Canny, *Nonholonomic Motion Planning*, vol. 192, Springer, 2012.

[18] C.-Y. Liang, H. Bai, M.-J. Cai, and W.-X. Lu, "Advances in quantum genetic algorithm," *Jisuanji Yingyong Yanjiu*, vol. 29, pp. 2401–2405, 2012.

[19] J. Bagley, "The behavior of adaptative systems wich emply genetic and correlation algorithms. Doctoral diseertation, University of Michigan," *Dissertation Abstracts International*, vol. 29, no. 12, 1967.

[20] P. W. Shor, "Algorithms for quantum computation: discrete logarithms and factoring," in *Proceedings of the 35th Annual Symposium on Foundations of Computer Science*, pp. 124–134, November 1994.

4

Motion Planning of Redundant Manipulators for Painting Uniform Thick Coating in Irregular Duct

Yan Chen,[1] Wenzhuo Chen,[1] Ken Chen,[2] and Bo Li[1]

[1]*Department of Petroleum Supply Engineering, Logistical Engineering University, Chongqing 401331, China*
[2]*Department of Mechanical Engineering, Tsinghua University, Beijing 100084, China*

Correspondence should be addressed to Yan Chen; yansohucom@sohu.com

Academic Editor: Yangmin Li

The paper presents a motion planning method of redundant manipulator for painting uniform thick coating on the interior of irregular duct of some aircrafts. Discontinuous peripheral painting method is employed by analyzing the restrictions during painting the duct. For improving the longitudinal uniformity of thick coating, the interlacing painting method plans two sets of sweeping paths and an interlacing distance between the starting paths of the two times of painting. The interlacing distance and overlapping distance are optimized by establishing and analyzing the model of longitudinal uniformity. To enhance the transverse uniformity, sweeping speeds for curved surfaces are calculated by the ratio of transfer efficiency after the basic sweeping speed for the plane is determined. The intertwining method, minimizing the sum of the weighed distances between the duct centerline and key points of the manipulator links, is employed for the joint trajectory planning without collision. The simulation and experiment results show that the redundant manipulators can finish painting the internal surface of the irregular S-shaped duct without collision. The maximum relative deviation is 16.3% and the thicknesses of all measurement points satisfy the acceptance criteria of the factory.

1. Introduction

Narrow inlet ducts of some airplanes have S-shaped centerline, whose cross sections transform from circular to irregular quadrangle. Its interior can be seen as a closed surface compounded by several surfaces with different curvatures. It is one of important manufacture processes to paint some inner surfaces of the irregular duct [1, 2]. Finishing this process by a painting robot can enhance the quality of the coating, improve process speed, reduce pollution, and take workers out of hazardous jobs [3]. Presently, redundant robots are used to paint the interior of the irregular duct by extending a manipulator into it [4, 5]. The uniformity of thick coating on the duct interior and the collision-free motion of the manipulator are two key aims needed to be achieved by motion planning.

Motion planning including task planning and joint trajectory planning is the base for designing, operating and controlling redundant manipulators for painting the duct interior. Task planning determines the path (including position and orientation) and speed of spray gun during painting tasks. Joint trajectory planning, which determines the time history of position and velocity for each joint, is the key for realizing collision-free motion during painting operation.

There are numerous researchers dealing with task planning problems of painting robot. Their bases for painting task planning are paint deposition models such as Gaussian model [6], parabolic model [7], β distribution model [8], and elliptical dual-β model [9]. It is hypothesized that coating thickness distribution would satisfy a certain function for establishing these models. Parabolic model and β distribution model can well reflect real situation with simple form and thus are more widely used by the researchers. Paint deposition models are fitted by measuring the paint deposition thickness on a plane. Only when the conditions of painting operation accord with or are approximate to the premises of the above models, the task planning will achieve ideal results.

Based on paint deposition model, researchers proposed several methods for path planning of thin coating. Suh et al. [10] developed an automatic trajectory planning system for painting robots. They chose the sweeping direction with least curvature. This method would cause worse coating uniformity because of the larger curvature in direction orthogonal to the sweeping direction. Asakawa and Takeuchi [11] developed a motion planning method using the parametric surface to paint a bumper. But how to find the spray overlap distance and the painting velocity was not addressed. The coating thickness was between 13 and 28 μm, and maximum relative deviation of the thickness was about 58%. Sheng et al. [12] presented a bounding box method for trajectory generation, which was modified a little by Chen et al. [13]. Its basic concept is that the free surface of the workpiece is divided into many smaller triangles and these generate into some spraying patches with the maximum angular deviation of each area no more than its critical angle, satisfying the requirements of coating thickness deviation. The bounding box method stresses the automation of path planning and the paint coverage on whole surface with little consideration of the optimization of coating uniformity and the influence of shape of complex surfaces. Atkar et al. [14] solved path planning problem on curved surfaces by decomposing it into three subproblems: selecting a seed curve, determining a speed profile along each pass, and selecting the spacing between successive passes. They pointed out that constant sweeping speed caused nonuniform paint deposition near surface boundaries, and in order to solve this problem, they used the paint deposition flux to obtain optimal speed along each pass. But its complex form of the objective function and the presence of the speed and acceleration constraints limit the scope of determining solutions using calculus of variations on a wide class of surfaces, and the experiments showed that the effects of their method on improving coating uniformity seemed not ideal. These researchers only studied the task planning methods for painting once without considering the thick coating which requires multiple times of painting. Therefore, it is desirable to develop an ideal task planning method for uniform thick coating in narrow irregular duct.

Researchers proposed many methods of joint trajectory planning for redundant robots. Potential field is one of the common methods to plan path for redundant manipulators but its biggest problem is that the manipulators can get stuck at local minima. In order to get rid of local minima, search methods are proposed, but when the number of obstacles increases, these methods may cause a manipulator to vibrate between the obstacles [15]. Yoo et al. [16] used a method of computing distance in polyhedron to study motion planning of redundant manipulators. This method is prone to computational complexity, because there would be many points where collision potentially occurs, and these points are subject to change depending on the configuration of the manipulators. Duguleana et al. [17] proposed motion planning of redundant manipulators using neural networks based on reinforcement learning. The principle of neural network method determines that its "solution" can be suboptimal, and even be wrong, but the collision-free problem in the narrow S-shaped duct needs the "solution" to be at least correct. The optimization of the gradient projection method [18] is better than that of the weighted minimum norm solution [19] so that it has been more widely used.

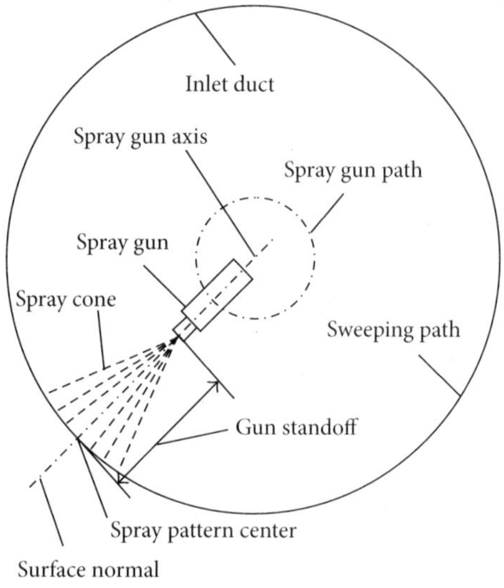

FIGURE 1: Basic principle of painting task planning.

When using the gradient projection method to find inverse kinematic solutions of redundant manipulator, one of the key problems to be solved is to find appropriate optimal functions according to task requirements. In this paper, the intertwining method is proposed to establish the optimal function.

The paper aims to find an appropriate motion planning scheme for redundant painting manipulators to achieve uniform thick coating on the interior of irregular S-shaped duct without collision during operation. It is organized as follows. In Section 2, painting task planning scheme is proposed by analyzing two kinds of uniformities. In Section 3, the intertwining method is proposed to generate the trajectories of robot joint. In Section 4, simulations and experiments are carried out to verify the proposed methods. Section 5 concludes the whole study.

2. Painting Task Planning

2.1. Basic Principle of Painting Task Planning. The basic principle of painting task planning is that the axis of spray gun must be perpendicular to the plane of the workpieces or in line with the normal line of the curved surface, with a constant gun standoff. In addition, the paint density must be kept uniform. These principles are important to obtain good surface finish and uniform coating thickness.

Figure 1 shows a sketch of a spray gun painting along a circular path. Painting task planning needs to firstly determine the spray gun orientation, standoff, and sweeping path, and speed and then transfer them to the path and speed of spray gun during painting tasks through coordinate transformation. Sweeping path and speed are defined, respectively, as path and speed of spray pattern center, not those of the spray gun. Gun standoff is determined by the performance of spray gun and paint property, which can be tested through spray experiments. Minimal sweeping speed is restricted by paint property and must insure that the paint would not sag on the surface. Maximum speed is confined by paint property and robot performance and must insure that the vibration of the spray gun is acceptable. The basic sweeping speed for the plane can be determined by calculation formula of coating average thickness and painting experiments. Rotation angle of spray gun when painting the interior of a duct must be bigger than that when painting normal workpieces (such as a flat surface) because the painting cross sections are circle, irregular quadrangle, and so on.

2.2. Path Planning of Painting Task. In this part, by analyzing the restrictions during painting the interior of the irregular S-shaped duct, the discontinuous peripheral painting method is proposed. The coating uniformity is divided into transverse uniformity (along the peripheral sweeping path) and longitudinal uniformity (in the direction perpendicular to the peripheral sweeping path). The interlacing method is developed for better longitudinal uniformity, and the ratio of transfer efficiency of two surfaces is used to plan sweeping speed for better transverse uniformity.

2.2.1. Discontinuous Peripheral Painting Method. When painting a curved surface, the sweeping direction of spray cone must be chosen according to the optimal objective. In order to realize uniform coverage, the curvature of the sweeping direction should be the largest, which lessens the influence of curved surface on the adopted paint deposition model, for the spray pattern is usually elliptically shaped. In addition, the paint and the compressed air pipes connected to the spray gun restrict the rotation of the spray gun about the duct centerline to approximately 420 degrees (clockwise or anticlockwise), so that it cannot paint along a thread trajectory which needs rotating arbitrarily. Lastly, the irregular duct cannot rotate during painting process; otherwise, collision would occur. Therefore, the discontinuous peripheral painting method is proposed to paint the interior of the irregular duct in a set of discrete cross sections.

The sweeping paths generated by discontinuous peripheral method, such as A1, A2, and A3 shown in Figure 2, are a series of discrete intersecting lines of the inner surface of the duct and the cross sections that are perpendicular to the duct centerline. The spray pattern center should successively move along these intersecting lines whose shape varies from circle to irregular quadrangle. The transitional paths such as B1 and B2 shown in Figure 2 bridge the discrete sweeping paths.

The spray pattern center must move clockwise (or anticlockwise) along a peripheral path such as A1 (Figure 2) during a single painting. Then the spray gun switches off and moves to another peripheral path A2 along a transition line B1. After that, the spray pattern center moves anticlockwise (or clockwise) along the path A2. Other cross sections can be painted in accordance with the above method. The spray gun should switch on and off once for uniform painting when it paints along each intersecting line. In addition, the spray gun enters into the duct and exits from it along the duct centerline.

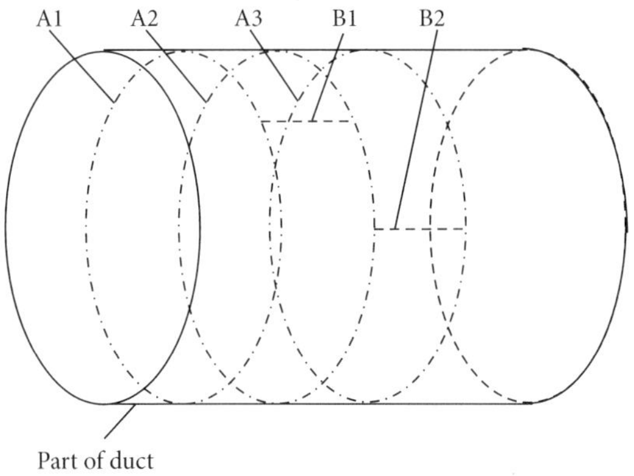

FIGURE 2: Sketch of sweeping path planned by discontinuous peripheral method.

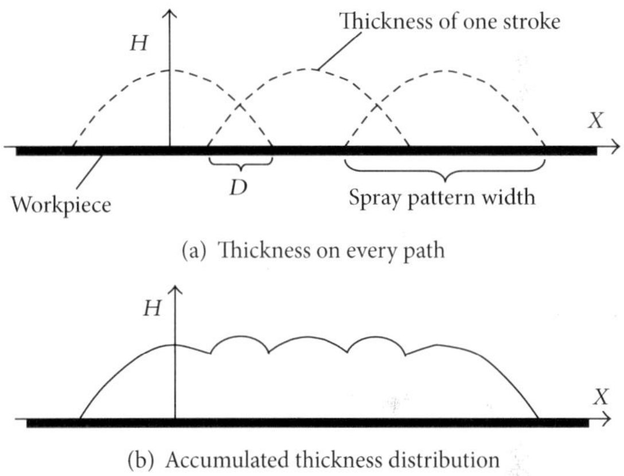

FIGURE 3: Coating thickness distribution of painting once.

2.2.2. Sweeping Path for Interlacing Painting. The thickness distribution of the coating cross section on the plane resulting from painting once will be like what is shown in Figure 3, where H represents the coating thickness, X axis is perpendicular to the direction of sweeping path, and D refers to the overlapping distance. There are periodic crests and troughs in the cross section of the coating. When painting surface with fixed curvature once, if the sweeping direction is along the largest curvature direction, there will also be similar crests and troughs. When painting the interior of the duct, multiple times of painting must be done to achieve the coating with enough thickness. But when planning a set of sweeping paths and painting the plane for multiple times, even with an optimized magnitude of D, there exist similar crests and troughs of the coating thickness (Figure 3(b)), and the coating uniformity thus is undesirable. Aiming at this problem, the paper proposes the interlacing painting method to improve the uniformity of the coating by interlacing crests and troughs in two sequential paintings. This method plans two sets of sweeping paths and makes that the position of crests and troughs resulting from one set of sweeping paths appears at the position of troughs and crests resulting from the other. The workpiece is painted along the two sets of paths in turn.

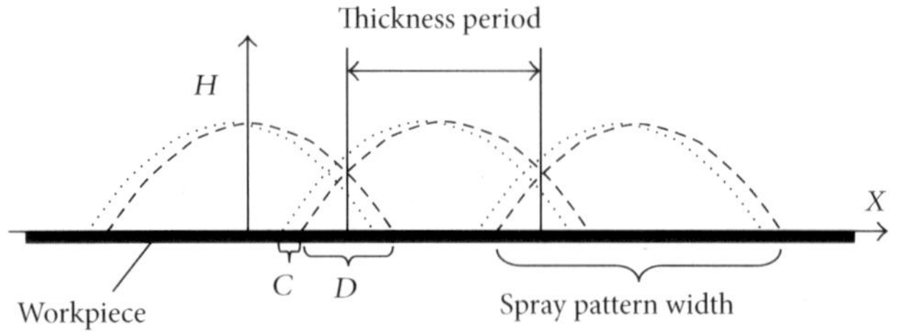

FIGURE 4: Model of interlacing painting.

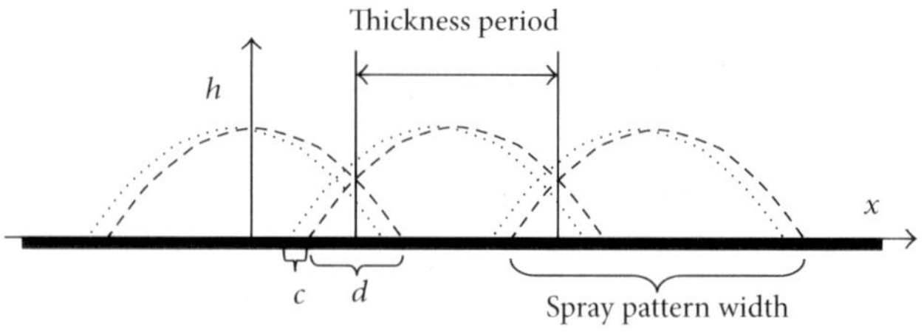

FIGURE 5: Dimensionless interlacing model.

The model of the interlacing painting is shown as Figure 4, where the heavy line is the cross section of the plane workpiece. The sweeping direction is in line with the cross section normal of the plane workpiece. The broken line and dot line, respectively, represent the coating thickness of every stroke of two times of painting with different sets of sweeping paths. The interlacing painting method adopts two sequential times of painting with the same gun standoff (which makes the same spray pattern width) and the same overlapping distance D, but there is an interlacing distance C between the starting paths of the two times of painting.

2.2.3. Dimensionless Interlacing Painting Model. The paint thickness distribution resulting from a single painting stroke can be expressed by parabolic distribution model shown as

$$H(x) = H_{max}\left(1 - \frac{4X^2}{W^2}\right), \tag{1}$$

where $H(x)$ is the paint thickness distribution of a cross section of the coating, X is the coordinate axis perpendicular to the sweeping paths which are a set of lines, H_{max} is the maximum thickness, and W is the width of spray pattern.

In this paper, parameters such as H_{max} and W are determined through experiment and remained constant during painting operation. The parabolic deposition model is nondimensionalized to simplify the calculation process.

Dimensionless parabolic deposition model $h(x)$ can be described as

$$h(x) = 1 - 4x^2, \tag{2}$$

where $h(x)$ is dimensionless coating thickness of painting once, $h(x) = H(x)/H_{max}$, and $h(x) \in [0,1]$; $x = X/W$ and $x \in [-0.5, 0.5]$.

Dimensionless interlacing painting model can be expressed as Figure 5. d represents the dimensionless overlapping distance, $d = D/W$, and $d \in [0, 0.5]$. c is the dimensionless interlacing distance, $c = C/W$, and $c \in [0, 1-d]$.

2.2.4. Optimized Parameters of Interlacing Painting. Indexes for evaluating the uniformity of coating thickness include relative thickness deviation and thickness variance. Relative thickness deviation is simpler and more explicit, so it is often used as an evaluation method for factories. While thickness variance is more precise and can well reflect the coating uniformity and thus usually employed as coating uniformity evaluation function in theoretical researches. Therefore, based on thickness variance, longitudinal uniformities of interlacing painting with different parameters (c and d) are evaluated. Then the influences of these parameters on the variance of coating thickness are analyzed, and optimized parameters thus are obtained.

Dimensionless longitudinal thickness variance can be described as

$$v = \frac{1}{l}\int_l \left(f - \overline{f}\right)^2 dl, \tag{3}$$

where f is the dimensionless coating thickness at a point, l represents the dimensionless width of the workpiece, and $\overline{f}$ is the dimensionless mean coating thickness.

The average film thickness $\overline{h}$ of interlacing painting in one coating thickness period is

$$\overline{h} = \frac{4}{1-d}\left\{\int_0^{0.5-d} h(x)\,dx + \int_{0.5-d}^{0.5-0.5d} \left[h(x) + h(x-1+d)\right] dx\right\}. \tag{4}$$

When $c \leq 0.5d$ and $c + d \leq 0.5$, shown in Figure 5,

$$\begin{aligned} v = \frac{1}{1-d}\Bigg\{ &\int_{0.5d-0.5}^{d-0.5-c} \left[h(x) + h(x+c) + h(x+1-d) + h(x+1-d+c) - \overline{h}\right]^2 dx + \int_{d-0.5-c}^{d-0.5} \left[h(x) + h(x+c) + h(x+1-d) - \overline{h}\right]^2 dx \\ &+ \int_{d-0.5}^{0.5-d-c} \left[h(x) + h(x+c) - \overline{h}\right]^2 dx + \int_{0.5-d-c}^{0.5-d} \left[h(x) + h(x+c) + h(x-1+d+c) - \overline{h}\right]^2 dx \\ &+ \int_{0.5-d}^{0.5-0.5d} \left[h(x) + h(x+c) + h(x-1+d+c) + h(x-1+d) - \overline{h}\right]^2 dx \Bigg\}. \end{aligned} \tag{5}$$

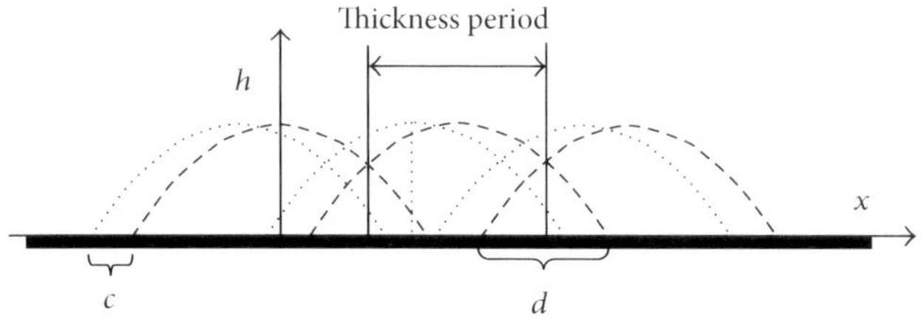

FIGURE 6: Coating thickness distribution when $c \leq 0.5d$, $c + d > 0.5$ and $c + 2d \leq 1$.

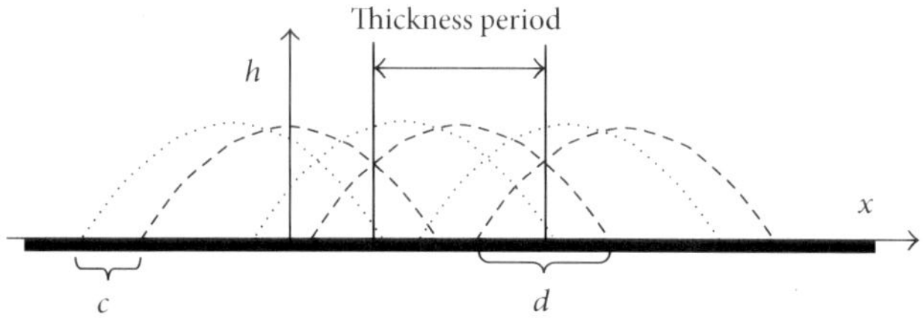

FIGURE 7: Coating thickness distribution when $c \leq 0.5d$, $c + d > 0.5$ and $c + 2d > 1$.

When $c \leq 0.5d$, $c + d > 0.5$ and $c + 2d \leq 1$, shown in Figure 6,

$$\begin{aligned} v = \frac{1}{1-d} \Bigg\{ & \int_{0.5d-0.5}^{d-0.5-c} \left[h(x) + h(x+c) + h(x+1-d) \right. \\ & \left. + h(x+1-d+c) - \overline{h} \right]^2 dx + \int_{d-0.5-c}^{d-0.5} \left[h(x) \right. \\ & \left. + h(x+c) + h(x+1-d) - \overline{h} \right]^2 dx \\ & + \int_{d-0.5}^{0.5-d-c} \left[h(x) + h(x+c) - \overline{h} \right]^2 dx \\ & + \int_{0.5-d-c}^{0.5-d} \left[h(x) + h(x+c) + h(x-1+d+c) \right. \\ & \left. - \overline{h} \right]^2 dx + \int_{0.5-d}^{0.5-0.5d} \left[h(x) + h(x+c) \right. \\ & \left. + h(x-1+d+c) + h(x-1+d) - \overline{h} \right]^2 dx \Bigg\}. \end{aligned} \tag{6}$$

When $c \leq 0.5d$, $c + d > 0.5$ and $c + 2d > 1$, shown in Figure 7,

$$\begin{aligned} v = \frac{1}{1-d} \Bigg\{ & \int_{0.5d-0.5}^{d-0.5-c} \left[h(x) + h(x+c) + h(x+1-d) \right. \\ & \left. + h(x+1-d+c) - \overline{h} \right]^2 dx + \int_{d-0.5-c}^{0.5-d-c} \left[h(x) \right. \\ & \left. + h(x+c) + h(x+1-d) - \overline{h} \right]^2 dx \\ & + \int_{0.5-d-c}^{d-0.5} \left[h(x) + h(x+c) + h(x+1-d) + h(x \right. \\ & \left. -1+d+c) - \overline{h} \right]^2 dx + \int_{d-0.5}^{0.5-d} \left[h(x) + h(x \right. \\ & \left. + c) + h(x-1+d+c) - \overline{h} \right]^2 dx \\ & + \int_{0.5-d}^{0.5-0.5d} \left[h(x) + h(x+c) + h(x-1+d+c) \right. \\ & \left. + h(x-1+d) - \overline{h} \right]^2 dx \Bigg\}. \end{aligned} \tag{7}$$

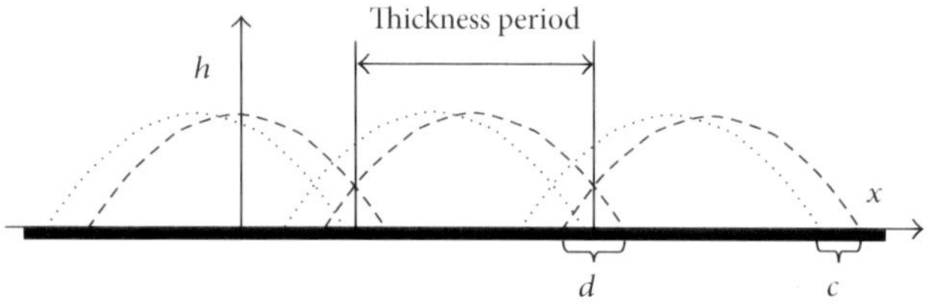

FIGURE 8: Coating thickness distribution when $0.5d < c \leq d$ and $c + d \leq 0.5$.

When $0.5d < c \leq d$ and $c + d \leq 0.5$, shown in Figure 8,

$$\begin{aligned} v = \frac{1}{1-d} \Bigg\{ & \int_{0.5d-0.5}^{d-0.5} \left[h(x) + h(x+c) + h(x+1-d) \right. \\ & \left. - \overline{h} \right]^2 dx + \int_{d-0.5}^{0.5-d-c} \left[h(x) + h(x+c) \right. \\ & \left. - \overline{h} \right]^2 dx + \int_{0.5-d-c}^{0.5-d} \left[h(x) + h(x+c) + h(x \right. \\ & \left. -1+d+c) - \overline{h} \right]^2 dx + \int_{0.5-d}^{0.5-c} \left[h(x) + h(x \right. \\ & \left. + c) + h(x-1+d+c) + h(x-1+d) - \overline{h} \right]^2 dx \\ & + \int_{0.5-c}^{0.5-0.5d} \left[h(x) + h(x-1+d+c) \right. \\ & \left. + h(x-1+d) - \overline{h} \right]^2 dx \Bigg\}. \end{aligned} \tag{8}$$

When $0.5d < c \leq d$, $c + d > 0.5$ and $c + 2d \leq 1$, shown in Figure 9,

$$\begin{aligned} v = \frac{1}{1-d} \Bigg\{ & \int_{0.5d-0.5}^{d-0.5} \left[h(x) + h(x+c) + h(x+1-d) \right. \\ & \left. - \overline{h} \right]^2 dx + \int_{d-0.5}^{0.5-d-c} \left[h(x) + h(x+c) \right. \\ & \left. - \overline{h} \right]^2 dx + \int_{0.5-d-c}^{0.5-d} \left[h(x) + h(x+c) + h(x \right. \\ & \left. -1+d+c) - \overline{h} \right]^2 dx + \int_{0.5-d}^{0.5-c} \left[h(x) + h(x \right. \\ & \left. + c) + h(x-1+d+c) + h(x-1+d) - \overline{h} \right]^2 dx \end{aligned}$$

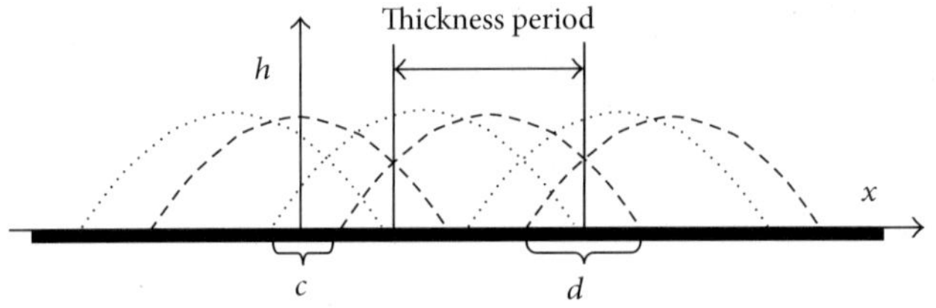

FIGURE 9: Coating thickness distribution when $0.5d < c \le d, c+d > 0.5$ and $c + 2d \le 1$.

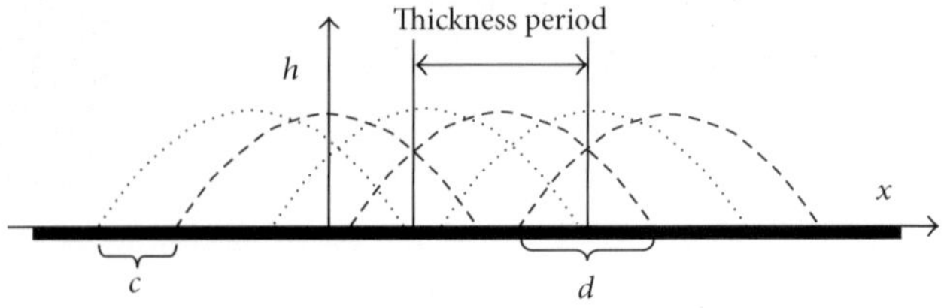

FIGURE 10: Painting when $0.5d < c \le d$, $c + d > 0.5$, $c + 2d > 1$ and $c + 1.5d \le 1$.

$$+ \int_{0.5-c}^{0.5-0.5d} \left[h(x) + h(x-1+d+c) + h(x-1+d) - \overline{h}\right]^2 dx \Bigg\}. \tag{9}$$

When $0.5d < c \le d, c+d > 0.5, c+2d > 1$ and $c+1.5d \le 1$, shown in Figure 10,

$$v = \frac{1}{1-d} \Bigg\{ \int_{0.5d-0.5}^{0.5-d-c} \left[h(x) + h(x+c) + h(x+1-d) - \overline{h}\right]^2 dx + \int_{0.5-d-c}^{d-0.5} \left[h(x) + h(x+c) + h(x-1+d+c) + h(x+1-d) - \overline{h}\right]^2 dx + \int_{d-0.5}^{0.5-d} \left[h(x) + h(x+c) + h(x-1+d+c) - \overline{h}\right]^2 dx + \int_{0.5-d}^{0.5-c} \left[h(x) + h(x+c) + h(x-1+d+c) + h(x-1+d) - \overline{h}\right]^2 dx + \int_{0.5-c}^{0.5-0.5d} \left[h(x) + h(x-1+d+c) + h(x-1+d) - \overline{h}\right]^2 dx \Bigg\}. \tag{10}$$

When $0.5d < c \le d, c+d > 0.5, c+2d > 1$ and $c+1.5d > 1$, shown in Figure 11,

$$\begin{aligned} v = \frac{1}{1-d} \Bigg\{ & \int_{0.5d-0.5}^{d-0.5} \left[h(x) + h(x+c) + h(x+1-d) + h(x-1+d+c) - \overline{h}\right]^2 dx \\ & + \int_{d-0.5}^{0.5-d} \left[h(x) + h(x+c) + h(x-1+d+c) - \overline{h}\right]^2 dx \\ & + \int_{0.5-d}^{0.5-c} \left[h(x) + h(x+c) + h(x-1+d+c) + h(x-1+d) - \overline{h}\right]^2 dx \\ & + \int_{0.5-c}^{1.5-2d-c} \left[h(x) + h(x-1+d+c) + h(x-1+d) - \overline{h}\right]^2 dx \\ & + \int_{1.5-2d-c}^{0.5-0.5d} \left[h(x) + h(x-1+d+c) + h(x-1+d) + h(x-2+c+2d) - \overline{h}\right]^2 dx \Bigg\}. \end{aligned} \tag{11}$$

When $c > d$ and $c + d \le 0.5$, shown in Figure 12,

$$v = \frac{1}{1-d} \Bigg\{ \int_{0.5d-0.5}^{d-0.5} \left[h(x) + h(x+c) + h(x+1-d) - \overline{h}\right]^2 dx + \int_{d-0.5}^{0.5-d-c} \left[h(x) + h(x+c) - \overline{h}\right]^2 dx + \int_{0.5-d-c}^{0.5-c} \left[h(x) + h(x+c) + h(x-1+d+c) - \overline{h}\right]^2 dx + \int_{0.5-c}^{0.5-d} \left[h(x) + h(x-1+d+c) - \overline{h}\right]^2 dx + \int_{0.5-d}^{0.5-0.5d} \left[h(x) + h(x-1+d+c) + h(x-1+d) - \overline{h}\right]^2 dx \Bigg\}. \tag{12}$$

When $c > d, c+d > 0.5$ and $c+2d \le 1$, shown in Figure 13,

$$v = \frac{1}{1-d} \Bigg\{ \int_{0.5d-0.5}^{d-0.5} \left[h(x) + h(x+c) + h(x+1-d) - \overline{h}\right]^2 dx + \int_{d-0.5}^{0.5-d-c} \left[h(x) + h(x+c)\right.$$

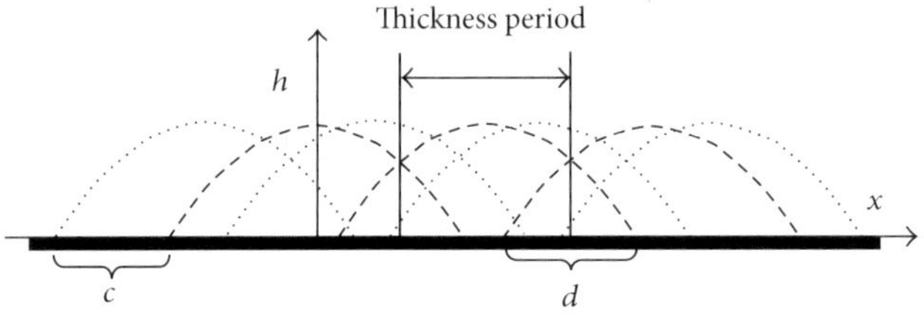

FIGURE 11: Painting when $0.5d < c \le d$, $c + d > 0.5$, $c + 2d > 1$ and $c + 1.5d > 1$.

$$-\overline{h}\Big]^2 dx + \int_{0.5-d-c}^{0.5-c}\Big[h(x) + h(x+c) + h(x - 1 + d + c) - \overline{h}\Big]^2 dx + \int_{0.5-c}^{0.5-d}\Big[h(x) + h(x - 1 + d + c) - \overline{h}\Big]^2 dx + \int_{0.5-d}^{0.5-0.5d}\Big[h(x) + h(x-1+d+c) + h(x-1+d) - \overline{h}\Big]^2 dx\Bigg\}. \tag{13}$$

When $c > d$, $c + d > 0.5$ and $c + 1.5d \le 1$, shown in Figure 14,

$$\begin{aligned} v = \frac{1}{1-d}\Bigg\{\Bigg[&\int_{0.5d-0.5}^{0.5-c-d}\Big[h(x) + h(x+c) + h(x+1-d) - \overline{h}\Big]^2 dx + \int_{0.5-d-c}^{d-0.5}\Big[h(x) + h(x+1-d) \\ &+ h(x-1+c+d) + h(x+c) - \overline{h}\Big]^2 dx + \int_{d-0.5}^{0.5-c}\Big[h(x) + h(x-1+c+d) + h(x+c) - \overline{h}\Big]^2 dx \\ &+ \int_{0.5-c}^{0.5-d}\Big[h(x) + h(x-1+d+c) - \overline{h}\Big]^2 dx + \int_{0.5-d}^{0.5-0.5d}\Big[h(x) \\ &+ h(x-1+d+c) + h(x-1+d) - \overline{h}\Big]^2 dx\Bigg\}. \end{aligned} \tag{14}$$

When $c > d$, $c + d > 0.5$ and $c + 1.5d > 1$, shown in Figure 15,

$$\begin{aligned} v = \frac{1}{1-d}\Bigg\{\Bigg[&\int_{0.5d-0.5}^{d-0.5}\Big[h(x) + h(x+c) + h(x+1-d) + h(x-1+d+c) - \overline{h}\Big]^2 dx \\ &+ \int_{d-0.5}^{0.5-c}\Big[h(x) + h(x+c) + h(x-1+d+c) - \overline{h}\Big]^2 dx + \int_{0.5-c}^{0.5-d}\Big[h(x) + h(x-1+d+c) - \overline{h}\Big]^2 dx \\ &+ \int_{0.5-d}^{1.5-2d-c}\Big[h(x) + h(x-1+d+c) + h(x-1+d) - \overline{h}\Big]^2 dx \\ &+ \int_{1.5-2d-c}^{0.5-0.5d}\Big[h(x) + h(x-1+d+c) + h(x-1+d) + h(x-2+c+2d) - \overline{h}\Big]^2 dx\Bigg\}. \end{aligned} \tag{15}$$

v is such a complicated function of c and d that it is difficult to perceive their relation. Their relation is shown in Figure 16 by numerical calculation through (4) to (15). The value of v is smaller in 4 areas: when $c \in [0.495(1-d), 0.505(1-d)]$ and $d \in [0.1800, 0.1925]$, v is around 0.0009; when $c \in [0.220(1-d), 0.270(1-d)]$ and $d \in [0.2950, 0.3075]$, $v \in [0.0007, 0.0009]$; when $c \in [0.730(1-d), 0.780(1-d)]$ and $d \in [0.2950, 0.3075]$, $v \in [0.0007, 0.0009]$; when $c \in [0.475(1-d), 0.525(1-d)]$ and $d \in [0.4100, 0.4375]$, $v \in [0.0005, 0.0009]$. Better thickness uniformity can be realized theoretically by choosing path parameters within these areas.

From the analysis of Figure 16, c and d are key influencing factors of longitudinal thickness uniformity. There are three advantages if d is set around 0.3 during painting process. It can lessen the influence of starting path on coating uniformity, the influence of path deviation on coating uniformity, and the bad effect on coating uniformity of odd number of painting times. Therefore, the area of $c \in [0.220(1-d), 0.270(1-d)]$ and $d \in [0.2950, 0.3075]$ or the area of $c \in [0.730(1-d), 0.780(1-d)]$ and $d \in [0.2950, 0.3075]$ should be chosen. The sweeping paths of the two areas are same in essence.

2.3. Planning Painting Speed. In this part, the transverse uniformity is considered to plan sweeping speed. And this uniformity is achieved through keeping the average coating thickness $\overline{H}$ unchanged along a sweeping path.

Suppose that one-stroke painting is done on a surface with constant curvature with the painting flux Q at the sweeping speed V. The stroke length is L. The spray pattern width is W. The transfer efficiency is η, which is the percentage of total coating solids employed by a coating applicator which adheres to the object being coated. During the period of t, the average coating thickness on the surface $\overline{H}$ can be expressed as

$$\overline{H} = \frac{Q\zeta(\eta t)}{WL}. \tag{16}$$

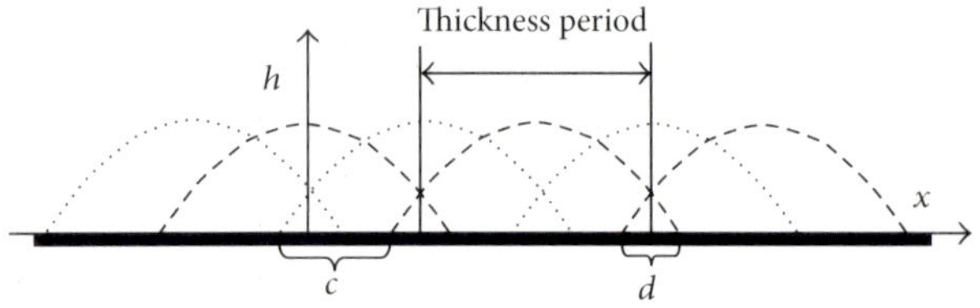

FIGURE 12: Coating thickness distribution when $c > d$ and $c + d \leq 0.5$.

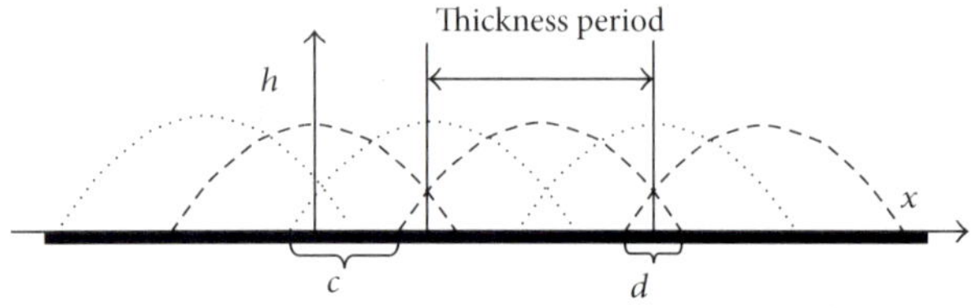

FIGURE 13: Coating thickness distribution when $c > d$, $c + d > 0.5$ and $c + 2d \leq 1$.

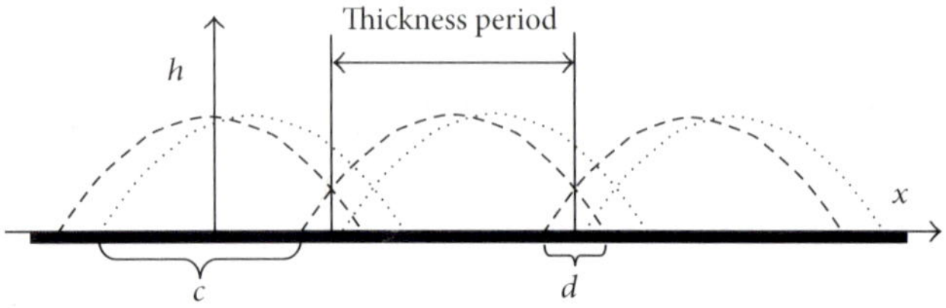

FIGURE 14: Coating thickness distribution when $c > d$, $c + d > 0.5$ and $c + 1.5d \leq 1$.

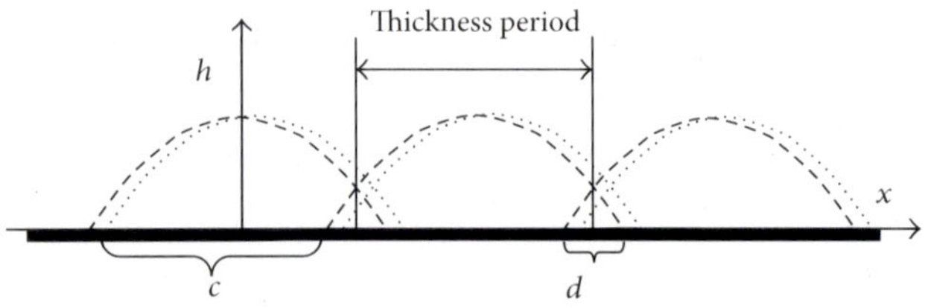

FIGURE 15: Coating thickness distribution when $c > d$, $c + d > 0.5$ and $c + 1.5d > 1$.

By substituting $V = L/t$ in to (16), we obtain

$$\overline{H} = \frac{Q\zeta\eta}{WV}. \quad (17)$$

It can be seen that, according to (17), the average coating thickness on the workpiece is proportional to the transfer efficiency η and inversely proportional to sweeping speed V. Because the painting process adopts constant painting flux and unchanged gun standoff, the volume fraction of solid constituent ζ and the spray pattern width W can be seen as invariable. Therefore, the average coating thickness $\overline{H}$ is only related to the values of η and V. In order to find a way to maintain constant $\overline{H}$ on a surface with changing curvature, the factors affecting the value of η should be found.

Transfer efficiency η is mainly influenced by the painting angle, gun standoff, sweeping speed, and shape (especially

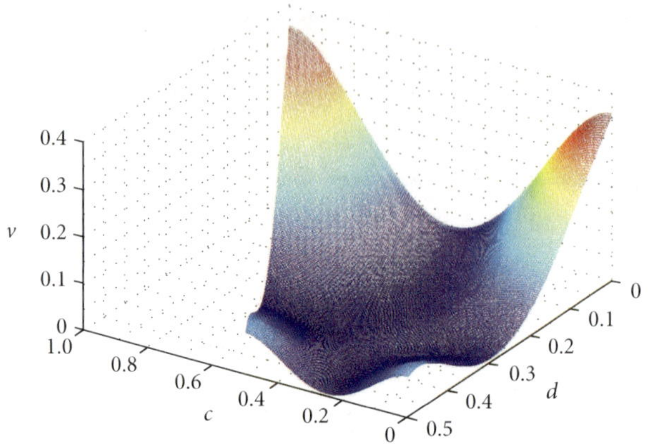

FIGURE 16: Relationship of v with d and c.

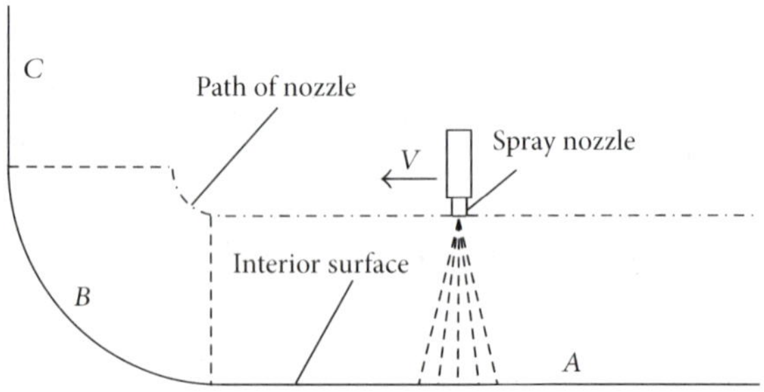

FIGURE 17: Typical sketch of peripheral painting path.

curvature in our research) of workpiece. The painting angle and gun standoff are constant in this research, so they can be neglected. The influence of minor changes of the sweeping speed on η can be neglected, because the sweeping speed is usually no more than 20 cm/s when painting the duct and is much lower than the speed of compressed air and paint particulate (close to the sound speed). Therefore, it is enough only to consider the influence of the curvature on the magnitude of η. The inner surface with bigger curvature makes greater transfer efficiency.

Figure 17 shows a typical sketch of a spray gun painting along a peripheral path. If the coating thicknesses on sections A and B are equal, the following equations can be obtained.

$$\frac{Q\zeta\eta_A}{WV_A} = \frac{Q\zeta\eta_B}{WV_B}, \quad (18)$$

$$\frac{\eta_A}{V_A} = \frac{\eta_B}{V_B}. \quad (19)$$

Assume that the curvature of section B is greater than that of section A, then η_B is greater than η_A. Define the ratio of transfer efficiency of two surfaces:

$$\beta = \frac{\eta_B}{\eta_A}. \quad (20)$$

Combining (19) and (20), we obtain

$$V_B = \beta V_A. \quad (21)$$

The ratio of transfer efficiency can be obtained by experiments or simulation of computational fluid dynamics. After the basic sweeping speed for the plane is determined, the sweeping speeds of curved surfaces can be calculated.

3. Joint Trajectory Planning

Joint trajectory refers to the position, velocity, and acceleration for each DOF of manipulators at run time. Its planning is the bases for the operation and control of robot.

Coordinate systems from $\{0\}$ to $\{i\}$ are attached, respectively, on redundant painting manipulators with i joints by using D-H approach. Then, the description of the position and orientation of spray gun in the coordinate system $\{0\}$, ${}^{0}_{p}\mathbf{T}$, can be obtained as

$$ {}^{0}_{p}\mathbf{T} = {}^{0}_{1}\mathbf{T}\,{}^{1}_{2}\mathbf{T}\,{}^{2}_{3}\mathbf{T}\,{}^{3}_{4}\mathbf{T}\,{}^{4}_{5}\mathbf{T}\,{}^{5}_{6}\mathbf{T}\cdots{}^{i-1}_{i}\mathbf{T}\,{}^{i}_{p}\mathbf{T}, \quad (22) $$

where ${}^{i-1}_{i}\mathbf{T}$ is the position and orientation of the coordinate system $\{i\}$ in the coordinate system $\{i-1\}$, and ${}^{i}_{p}\mathbf{T}$ is the position and orientation of spray gun in the coordinate system $\{i\}$.

The differential relation between Cartesian velocities of the redundant manipulators and the joint velocities is

$$ \dot{\mathbf{X}} = \mathbf{J}\dot{\mathbf{q}}, \quad (23) $$

where $\dot{\mathbf{X}}$ is the vector of translational and rotational velocities of the spray gun with reference to base coordinates, $\dot{\mathbf{q}}$ is the vector of n joint rates, and $\mathbf{J}$ is the Jacobian matrix (m by n, $m < n$). m denotes the space DOF.

The gradient projection method is used to solve (23), and the result can be expressed as

$$ \dot{\mathbf{q}} = \mathbf{J}^{+}\dot{\mathbf{X}} + (\mathbf{I} - \mathbf{J}^{+}\mathbf{J})\,\mathbf{E}, \quad (24) $$

where $\mathbf{J}^{+} = \mathbf{J}^{T}(\mathbf{J}\mathbf{J}^{T})^{-1}$ is the Moore-Penrose pseudoinverse of $\mathbf{J}$, $\mathbf{I}$ is the unit matrix ($n \times n$), and $\mathbf{E} = -\lambda\nabla F(\mathbf{q})$. λ is a real scalar constant coefficient, and $\nabla F(\mathbf{q})$ is the gradient vector of $F(\mathbf{q})$ which is the objective function to be optimized.

As the number of the solution of (24) is infinite, some restrictions are needed to obtain a unique solution which is optimistic. The intertwining method is proposed to obtain a unique solution of (24): the distance between the centerline and each part of the manipulators will be controlled so that the manipulators can intertwine the duct centerline closely without collision in the duct. The distance between the inner surface of the duct and each part of the manipulators should be limited to ensure collision-free operation. It is hard to calculate the distance due to the complexity of the duct. But it is much simpler to calculate the distance between the duct centerline and each part of the manipulators. And controlling the distance between each part of the manipulators and the centerline is equivalent to controlling the distance between the inner surface and the parts. If the manipulators intertwine the duct centerline closely, they would work without collision in the duct. Therefore, it can build the collision-free restriction to calculate the distance sum of the duct centerline and all joints of the manipulators. The midpoint and ends of each link are defined to be key points. If these points were all near to the duct centerline, there would be no collision. The weighed distance sum of the duct centerline and the link key points, G, can be calculated as

$$ G = \sum_{i=1}^{\alpha}\left[k_x(x_i - x_{Oi})^2 + k_y(y_i - y_{Oi})^2 + k_z(z_i - z_{Oi})^2\right], \quad (25) $$

where x_i, y_i, and z_i are the coordinate of the link key point i in the base, x_{Oi}, y_{Oi}, and z_{Oi} are the coordinate of the center point of the duct cross section, k_x, k_y, and k_z are the weighed coefficient in the direction of x, y, and z, and α is the key point number.

Then, an objective function F can be obtained by minimizing the weighed distance sum of the duct centerline and the link key points:

$$ F = \min(G). \quad (26) $$

After (25) and (26) are substituted into (24), we obtain

$$ \dot{\mathbf{q}} = \mathbf{J}^{+}\dot{\mathbf{X}} - \lambda(\mathbf{I} - \mathbf{J}^{+}\mathbf{J})\nabla\sum_{i=1}^{\alpha}\left[k_x(x_i - x_{Oi})^2 + k_y(y_i - y_{Oi})^2 + k_z(z_i - z_{Oi})^2\right]. \quad (27) $$

By solving (27), optimized joint trajectories can be obtained.

Before task planning, preprocessing should be set up. The trajectory of the spray nozzle and the duct center line should be calculated. Suppose that the path points are $\mathbf{x}_i^d$ ($i = 0, 1, 2, \ldots, n$) and that the forward kinematics is $\mathbf{X}_0 = f(\mathbf{q}_0)$. Iteration step size Δt and iteration error should be set, and initial index i should be set as 0. Then, the process of joint trajectory planning is as follows.

Step 1. Calculate the velocity of the nozzle in task space by using difference quotient of two consecutive path points, which is $\dot{\mathbf{x}} = (\mathbf{x}_{i+1}^d - \mathbf{x}_i^d)/\Delta t$.

Step 2. Set the weighed coefficient λ and real scalar constant coefficient k.

Step 3. In the current configuration $\mathbf{q}_i$, calculate the Jacobian matrix, its pseudo inverse $\mathbf{J}^{\dagger} = \mathbf{J}^{T}(\mathbf{J}\mathbf{J}^{T})^{-1}$, and the optimized vector E.

Step 4. Compute the joint velocity $\dot{\mathbf{q}}$ by using (25) and the current configuration $\mathbf{q}_i'$, where $\mathbf{q}_i' = \mathbf{q}_i + \Delta t \cdot \dot{\mathbf{q}}$.

Step 5. Calculate position and orientation $\mathbf{x}$ of the spray nozzle with robot configuration as $\mathbf{q}_i'$, where $\mathbf{x} = f(\mathbf{q}_i')$; calculate the error vector $\boldsymbol{\xi}$, where $\boldsymbol{\xi} = \mathbf{x}_{i+1}^d - \mathbf{x}$. If $\|\boldsymbol{\xi}\| < \varepsilon$, the iteration will break and $\mathbf{q}_i = \mathbf{q}_i'$; if $\|\boldsymbol{\xi}\| > \varepsilon$, let $\mathbf{q}_i = \mathbf{q}_i'$ and $\dot{\mathbf{x}} = \boldsymbol{\xi}/\Delta t$, and then return to Step 3.

Step 6. Check if there is collision during motion. If collision exists, return to Step 2; If not, let $i = i + 1$ and return back to Step 1, until $i = n$.

FIGURE 18: Painting robot.

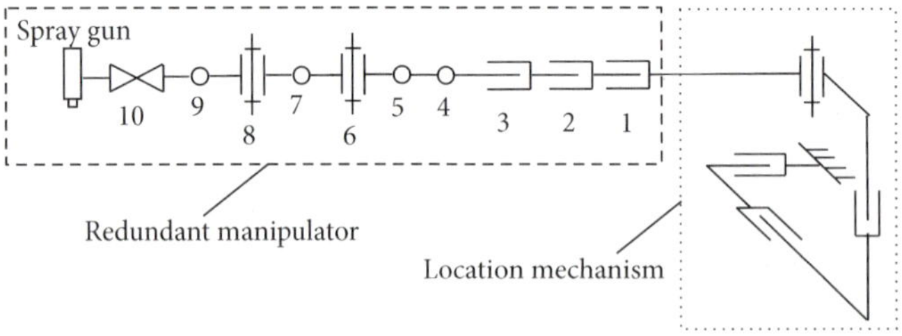

FIGURE 19: Robot mechanism sketch.

Step 7. After obtaining the path points in joint space, calculate the path $\{\mathbf{q}(t), \dot{\mathbf{q}}(t), \ddot{\mathbf{q}}(t)\}$ by using third-order polynomial.

4. Simulation and Experiment

The simulation and experiment adopted a painting robot with a 10-DOF manipulator for painting uniform thick coating in an irregular S-shaped duct. The robot is shown as Figure 18 and its mechanism sketch is shown as Figure 19. The 10-DOF manipulator consists of three prismatic joints (joints 1, 2, and 3) and seven revolute joints (joints 4–10). The axes of joints 1, 2, and 3 are in the same direction. The axes of joints 6, 8, and 10 are perpendicular to the axes of joints 4, 5, 7, and 9. Joint 10 is designed as the last joint of the painting manipulator linking the spray gun.

The motion planning method proposed above was implemented so that the painting path and joint trajectory were generated, and the motion of the robot was firstly simulated in the software. The 3D models of the painting manipulator and the duct were also built according to their real size (Figure 20) by Pro/E. The generated trajectory data were transferred into the Pro/E to check if there would be any motion interference during painting operation. The result of simulation indicated that no collision would occur. Then, during the operation of the painting manipulator in the duct, the minimal distance between the duct and the manipulators was 18 mm, which satisfied the safety requirement.

Three typical parts including the bottom of the front, the top of the front, and the back of the irregular S-shaped duct were chosen to be painted and no collision occurred. The painting experiment was conducted as shown in Figure 21. The pressure for both spray pattern and the atomization was 0.28 MPa. The parameters of dynamic surge suppressor and the paint agitator were set as 0.25 MPa and 0.30 MPa, respectively. The paint flux of the spray gun was 0.17 L/min and the viscosity of the paint was 14.5 s. The operational parameters were set as 23.1 cm of the painting width, 7.1 cm of interlacing distance, 6.0 cm of overlapping distance, 16 cm of the gun standoff, and 12.5 cm/s of the sweeping speed in plane area.

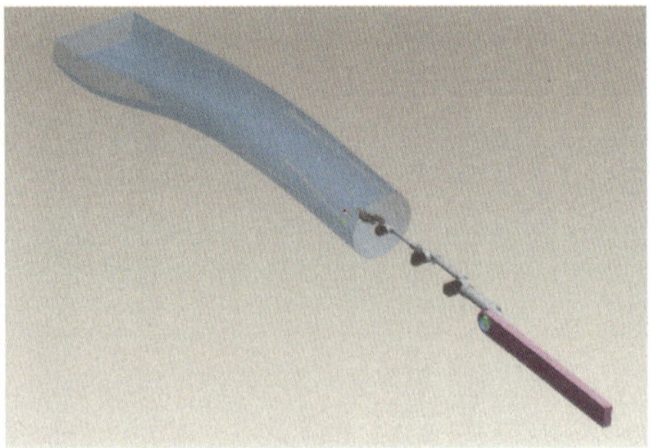

FIGURE 20: Simulation model for painting motion in Pro/E.

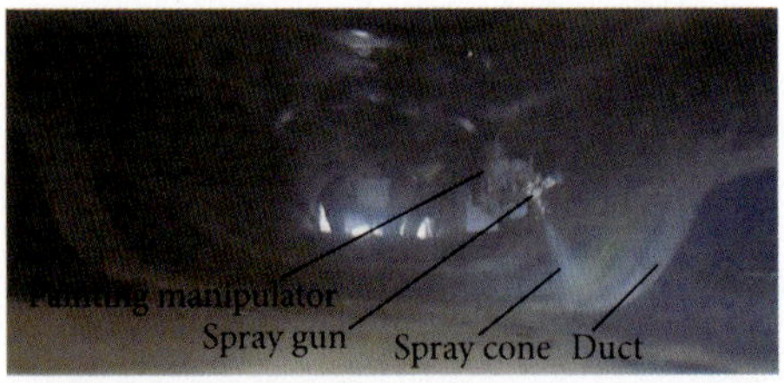

FIGURE 21: Painting experiment in irregular duct.

TABLE 1: Dimensionless thickness on different parts.

Position	Dimensionless minimum thickness	Dimensionless maximum thickness	Average thickness	Maximum positive relative deviation (%)	Maximum negative relative deviation (%)	Intolerance (%)
Bottom of the front	0.84	1.12	1.00	13.1	15.2	100
Top of the front	0.80	1.08	0.96	13.0	16.3	100
Back	0.92	1.16	0.53	8.9	13.6	100

After the coating on the interior had been dry out, thickness gauge was used to measure different parts of the duct. The dimensionless thickness of the coating on different parts of the duct is shown in Table 1. Dimensionless thickness which is defined as the ratio of the measured thickness and designed thickness is used to analyze the experiment data. The dimensionless value of designed thickness is 1. The maximum relative deviation is 16.3% at the front top of the duct. The thicknesses of all measurement points are in tolerance and satisfy the acceptance criteria proposed by the factory.

Although agreeable coating uniformity was obtained, relative deviation of the thickness distribution in the experiment was higher than that in theory, because of the following several reasons. The process of paint deposition on workpieces is influenced by some basic factors which involve spray factors including paint pressure, spray air pressure, and spray pattern pressure, paint factors including paint flow rate, viscosity, and uniformity, and workpiece factors including shape and temperature of the workpiece. But spray factors and paint factors cannot be precisely controlled during the painting operation. In addition, the adopted paint deposition model is obtained through fitting the data of experimental measurements of coating thickness distribution on planar work piece; therefore, some deviations occur when it is applied to painting curved surfaces. Lastly, the precision error of robot can cause some deviations of the actual paths and speeds off from the planned ones. Therefore, the optimal interlacing distance and overlapping distance were affected to some extent.

When painting the back of the duct, the extension of the manipulators is shorter than that painting the front, so the stiffness is higher and the dynamic accuracy is better, which makes the coating uniformity of the back part better. Therefore, the maximum thickness deviation of the front (both the top and bottom part) of the duct is higher than that of the back.

5. Conclusion

It is one of the important manufacture processes to paint some inner surfaces of the irregular duct of some aircrafts. Redundant robots are used to paint the duct by extending a manipulator into it. The paper presents a motion planning method of redundant manipulator for painting uniform thick coating on the interior of the irregular duct.

After analyzing the restrictions during painting the irregular duct, discontinuous peripheral painting method is presented. In order to improve the uniformity of the thick coating on the duct interior, based on the analysis of the longitudinal uniformity and transverse uniformity, the paper proposes the interlacing painting method for sweeping path planning, optimizes its parameters including overlapping distance and interlacing distance, and then presents that sweeping speeds of curved surfaces can be calculated by the ratio of transfer efficiency after the basic sweeping speed for the plane is determined.

The motion of the manipulators is controlled by restricting the distance between the links and the centerline. Joint trajectory is planned through the intertwining method, namely, minimizing the sum of the distances between the centerline of the duct and the key points on the links.

The experiment shows that the redundant manipulators can finish painting the internal surface of the S-shaped irregular duct without collision. And the paint path generated by using the interlacing method can lead to uniform coating.

Although the interlacing method is used for painting the interior of the narrow duct, it can also be practical for painting other products. And the trajectory planning method proposed by the paper can also be useful for the trajectory planning for redundant manipulators in other narrow areas.

Conflict of Interests

The authors declare that there is no conflict of interests regarding the publication of this paper.

Acknowledgment

This project is supported by National Natural Science Foundation of China (Grant no. 51475469).

References

[1] K. Liu, R. Glenn, T. J. Lawley, and F. L. Lewis, "Steward-platform-based inlet duct painting system," in *Proceedings of the IEEE International Conference on Robotics and Automation*, pp. 106–113, Atlanta, Ga, USA, May 1993.

[2] Q. Lin, C. Sha, B. Zhang et al., "The development of aircraft inlet automatic spraying equipment," *Manufacturing Automation*, vol. 35, no. 1, pp. 92–93, 2013 (Chinese).

[3] N. A. Seegmiller, J. A. Bailiff, and R. K. Franks, "Precision robotic coating application and thickness control optimization for f-35 final finishes," *SAE International Journal of Aerospace*, vol. 2, no. 1, pp. 284–290, 2010.

[4] Y. Chen, J. Shao, C. Zhang, C. Fu, and K. Chen, "Redundant-robot-based painting system for variable cross-section S-shape

pipe," in *Proceedings of ASME/IFToMM International Conference on Reconfigurable Mechanisms and Robots*, pp. 743–749, London, UK, June 2009.

[5] M. D. Prouty, *Robotic Coating Thickness Measurement System*, NAVAIR, Washington, DC, USA, 2008.

[6] H. Hyotyniemi, "Minor moves-global results: robot trajectory planning," in *Proceedings of the 2nd International IEEE Conference on Tools for Artificial Intelligence*, pp. 16–22, IEEE, Herndon, Va, USA, November 1990.

[7] W. Persoons and H. Van Brussel, "CAD-based robotic coating of highly curved surfaces," in *Proceedings of the 24th International Symposium on Industrial Robots*, pp. 611–618, Tokyo, Japan, November 1993.

[8] T. Balkan and M. A. S. Arikan, "Modeling of paint flow rate flux for circular paint sprays by using experimental paint thickness distribution," *Mechanics Research Communications*, vol. 26, no. 5, pp. 609–617, 1999.

[9] Y. Zhang, Y. Huang, F. Gao, and W. Wang, "New model for air spray gun of robotic spray-painting," *Chinese Journal of Mechanical Engineering*, vol. 42, no. 11, pp. 226–233, 2006.

[10] S.-H. Suh, I.-K. Woo, and S.-K. Noh, "Development of an automatic trajectory planning system (ATPS) for spray painting robots," in *Proceedings of the IEEE International Conference on Robotics and Automation*, pp. 1948–1955, Sacramento, Calif, USA, April 1991.

[11] N. Asakawa and Y. Takeuchi, "Teachless spray-paintingof sculptured surface by an industrial robot," in *Proceedings of the IEEE International Conference on Robotics and Automation*, vol. 3, pp. 1875–1879, Albuquerque, NM, USA, April 1997.

[12] W. Sheng, N. Xi, M. Song, Y. Chen, and P. Macneille, "Automated CAD-guided robot path planning for spray painting of compound surfaces," in *Proceedings of the IEEE International Conference on Intelligent Robots and Systems*, pp. 1918–1923, IEEE, Piscataway, NJ, USA, November 2000.

[13] H. Chen, W. Sheng, N. Xi, M. Song, and Y. Chen, "CAD-based automated robot trajectory planning for spray painting of free-form surfaces," *Industrial Robot*, vol. 29, no. 5, pp. 426–433, 2002.

[14] P. N. Atkar, A. Greenfield, D. C. Conner, H. Choset, and A. A. Rizzi, "Uniform coverage of automotive surface patches," *The International Journal of Robotics Research*, vol. 24, no. 11, pp. 883–898, 2005.

[15] E. S. Conkur, "Path planning using potential fields for highly redundant manipulators," *Robotics and Autonomous Systems*, vol. 52, no. 2-3, pp. 209–228, 2005.

[16] Y. J. Yoo, K. J. Oh, Y. J. Choi, and S. C. Won, "Obstacle avoidance for redundant manipulator without information of the joint angles," in *Proceedings of the 9th Asian Control Conference (ASCC '13)*, pp. 1–6, Istanbul, Turkey, June 2013.

[17] M. Duguleana, F. G. Barbuceanu, A. Teirelbar, and G. Mogan, "Obstacle avoidance of redundant manipulators using neural networks based reinforcement learning," *Robotics and Computer-Integrated Manufacturing*, vol. 28, no. 2, pp. 132–146, 2012.

[18] A. Liegeois, "Automatic supervisory control of the configuration and behavior of multibody mechanisms," *IEEE Transactions on Systems, Man and Cybernetics*, vol. 7, no. 12, pp. 868–871, 1977.

[19] T. F. Chan and R. V. Dubey, "A weighted least-norm solution based scheme for avoiding joint limits for redundant joint manipulators," in *Proceedings of the IEEE International Conference on Robotics and Automation*, pp. 395–402, Atlanta, Ga, USA, May 1993.

5

Information-Fusion Methods based Simultaneous Localization and Mapping for Robot Adapting to Search and Rescue Postdisaster Environments

Hongling Wang,[1] Chengjin Zhang,[1] Yong Song,[2] and Bao Pang[1]

[1] *School of Control Science and Engineering, Shandong University, Jinan 250101, China*
[2] *School of Mechanical, Electrical and Information Engineering, Shandong University at Weihai, Weihai 264209, China*

Correspondence should be addressed to Chengjin Zhang; cjzhang@sdu.edu.cn

Academic Editor: Shahram Payandeh

The first application of utilizing unique information-fusion SLAM (IF-SLAM) methods is developed for mobile robots performing simultaneous localization and mapping (SLAM) adapting to search and rescue (SAR) environments in this paper. Several fusion approaches, parallel measurements filtering, exploration trajectories fusing, and combination sensors' measurements and mobile robots' trajectories, are proposed. The novel integration particle filter (IPF) and optimal improved EKF (IEKF) algorithms are derived for information-fusion systems to perform SLAM task in SAR scenarios. The information-fusion architecture consists of multirobots and multisensors (MAM); multiple robots mount on-board laser range finder (LRF) sensors, localization sonars, gyro odometry, Kinect-sensor, RGB-D camera, and other proprioceptive sensors. This information-fusion SLAM (IF-SLAM) is compared with conventional methods, which indicates that fusion trajectory is more consistent with estimated trajectories and real observation trajectories. The simulations and experiments of SLAM process are conducted in both cluttered indoor environment and outdoor collapsed unstructured scenario, and experimental results validate the effectiveness of the proposed information-fusion methods in improving SLAM performances adapting to SAR scenarios.

1. Introduction

Mobile robots perform simultaneous localization and mapping (SLAM) task in search and rescue (SAR) postdisaster scenarios, while building an exploration global map and localizing themselves with this map. Modeling decision on information-fusion was earlier put up by Torra and Narukawa [1], and parallel tracking and mapping (PTAM) strategy [2] and keyframes subblocks matching [3] were used for robotics mapping and localization. Murphy et al. [4] proposed that SLAM ability is necessary for mobile robots conducting SAR task. Motivating by the researches, we developed that information-fusion SLAM (IF-SLAM) of multirobots and multisensors (MAM) autonomously explores an SAR area without recourse to global positioning system (GPS). Multiple sensors are mounted on board multirobots. The information from several sources or from different times of the same sensor is integrated into one single datum to implement SLAM assignment for each robot. An image matching solution incorporates extraction of overlap regions and key frames extraction of overlap regions to rebuild through subblocks matching, and the imagine fusion of matching features transforms 3D object of interest in the real world into 2D image in global map. The multisensors integration block provides continuous online information about the mobile robots and its local environment.

The algorithms of integration particle filter (IPF) and optimal improved extended Kalman filter (IEKF) combine traditional methods of consistent Rao-Blackwellized (R-B) particle filters embedded in MAM IF-SLAM architecture. This architecture consists of multirobots and multisensors including laser range finder (LRF) sensors, localization sonars, gyro odometry, Kinect-sensor, RGB-D camera, and other proprioceptive sensors mounted on board multiple robots.

Mobile robots have a great potential to assist in the underground operations, searching ahead of rescue teams and reporting conditions that may be hazardous to the teams. Robots can become the invaluable tool. The SLAM challenge appears when a mobile robot attempts to build a map of cluttered SAR environment while simultaneously localizing itself relative to this map. Robotic SLAM in SAR environment confronts the difficulties on random-exploration trajectories. These difficulties include the relationships between the map accuracy relative to its steady-state motion behavior and a simultaneous estimate of both robot and landmark locations. Underground mine rescue confronts the same issues in the use of belays and tethers for operating down voids or on slopes.

Generally, mobile robots do not have the luxury of GPS localization while exploring underground [5]. Multiple robots on-board multisensors SAR SLAM combining measurements and trajectories has improved SLAM performances in object identification, area coverage, and loop-closure, which is necessary for robotic search and rescue tasks.

Multiple robots cooperatively acquired the surrounding's information in wide research work [1, 6–9], and an integration strategy of different kinds of measurements was used in multimode sense of 3D perception. Pfingsthorn et al. [7] updated the local submaps through their graph-based structures, and made it be efficiently communicated among mobile robots' sensors. Using a group of mobile robots equipped with sensors for source concentration measurement, the mobile robots team cooperatively estimates the gradient of the exploration field, moves to the objects of interest, and keeps a predefined formation in movement [8]. Atanasov et al. [9] developed an acquisition method of multisensors active information capturing the common characteristics of these scenarios and controlling mobile sensing systems to improve the accuracy and efficiency of collecting information autonomously.

We developed robotics SLAM of fusion data provided by multiple robots on-board multisensors, vision fusion of images, and integration of different kinds of measurements. The on-board proprioceptive sensors perceive the surroundings to supply various information and measurements, and then the mobile robot fuses the achieved data and the exploration trajectories for SLAM processing. Usually, the landmarks are difficult to assignment and to determine their poses in SAR disaster site, collapsed and cluttered. In the proposed methods, the mobile robots explore coverage SAR collapsed area and take relative observations of landmarks, and the sensors' data fusion approaches made a few achieved landmarks localize maximally consistent with mobile robot's pose.

According to Torra and Narukawa and Milisavljevic's theory [1, 10], information-fusion is the processes which embedded the particular function mathematical algorithms in an integration architecture, which needs a special techniques for combining the information. Sensors and data fusion theory in [10] illustrated the strategy of fusion of motion targets trajectories and sensors' measurements. Literature [2, 3] proposed a monocular single camera SLAM and keyframes localization approaches; however, these methods are not suitable for SAR SLAM in urgent postdisaster areas.

Inspired by the overview research means, we designed novelty multirobots and multisensors information-fusion SLAM (IF-SLAM) architecture, in which the keyframes and environmental data obtained from MAM architecture are processed by fusion strategies, and one single datum is output for performing SLAM. Multiple robots explore SAR postdisaster area and perform SLAM task, maintaining a graph with sensor observations stored in vertices and pose differences including uncertainty information stored in edges. Integration particle filter (IPF) and optimal improved extended Karman filter (IEKF) algorithms embedded MAM platform for improving motion trajectories accuracy, which is greatly adapting to SAR postdisaster environment. The proposed SAR SLAM methods of information-fusion employ hybrid map representations to fuse the strengths of topological maps and occupancy grids; these representations facilitated multirobots mapping.

The remainder parts are arranged as follows: in Section 2, we describe the information integration architecture. In Section 3, information-fusion algorithms are elaborated and derived, and the virtual robots conduct SLAM simulations with information-fusion method in Section 4. In Section 5, the experiments are performed employing Pioneer LX robot (Adept MobileRobots Pioneer 3, Mobile Robots Inc., http://www.mobilerobots.com) and crawler mobile robot in both indoor cluttered and outdoor collapsed scenarios, respectively. In Section 6, the results are discussed, and in Section 7, we summarized our conclusions and future work.

2. Information Integration Architecture

2.1. Multiple Robots. Mobile robots installed with multisensors and initial preset metric map have been developed. According to Saeedi et al. [11], multiple robots conduct mapping by fusion of two occupancy grid submaps; Deshpande et al. [6] proposed a robotic platform equipped with camera and various sensors for multimode 3D perception. In the designed architecture of multiple robots, SLAM is achieved through the process of mapping out an unknown environment and generating a consistent map. In general, single-robot SLAM is achieved through extended Kalman filter (EKF) data fusion; this approach is then extended to multirobots SLAM by using MobileRobot Operating System (MOS). The multirobots apply to conduct exploration missions and provide a perception of its surroundings for performing SLAM in the SAR environment.

2.2. Multisensors On-Board Robots. A multisensors robotic platform intended for deployment in environmental hazardous situations. Thrun et al. [12] equipped several LRFs to register consecutive scans; 2D/3D information of scans was acquired from pointed forward LRFs. Multisensors system was developed in for acquiring accurate environmental maps of SAR environments. The system is based on an autonomous robot (i.e., a Pioneer LX) installed with LRFs and sensors; other mobile robots (i.e., Amigo robots) are equipped with various sensors. As displayed in Figure 1, the designed multiple sensors (LRFs, gyros, sonars, and cameras) are on-board multirobots for integration information.

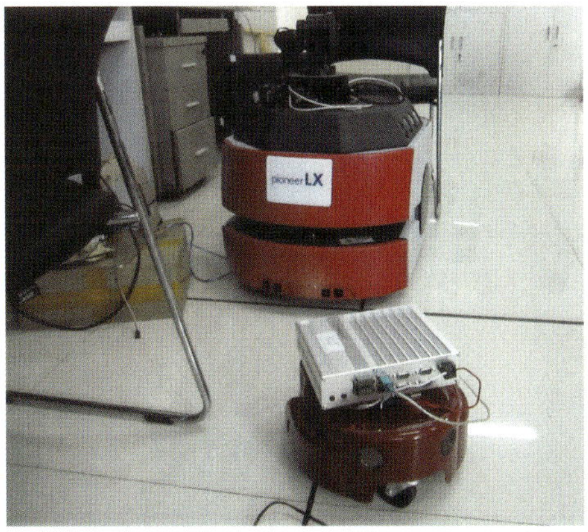

FIGURE 1: Multirobots and multisensors integration architecture.

2.3. Integration Strategies. Fusion algorithms and SLAM programs are embedded in multirobots and multisensors architecture. Zhou and Roumeliotis [13] merged local submaps that are generated by different robots independently; a robots team cooperatively built a joint environmental map; Henry et al. [14] used single RGB-D camera to build dense 3D map, as the researcher; only a two-dimensional map is lack of any depth information of surrounding objects.

Locally consistent submaps are incapable of global consistency. An optimal algorithm for fusion of local submaps which are created is based on multiple robots and sensors. The fusion of the visual and shape information achieved from cameras and laser scans is used to construct dense 3D objects mapping. The rich visual features along with random sample consensus (RANSAC) verification are given fixed data, which is fused into the iterative closest point (ICP) optimization. To build consistent maps of SAR postdisaster environment, the ICP algorithm is used for estimating global correspondences and aligning robot exploration trajectories. With fusion algorithms by integration of local occupancy grid submaps, the global consistent map is formed. The integration operators are developed to combine N measurements within an area A and output a processed value δ. Represent integration operators as Γ functions form:

$$\delta = \Gamma(a_1, \ldots, a_N), \tag{1}$$

where $a_1, \ldots, a_N$ indicate N measurement inputs, which correspond to mathematical strategies used for information-fusion.

3. Information-Fusion Algorithms Adapting to SAR SLAM

3.1. Modeling Predefined. A map of environment can be parametrized as a set of spatially located landmarks, occupancy grids, surface terrain, and sensors' measurements. The IF-SLAM builds a joint state which constitutes of the multiple robots poses and each landmark location, employing a huge state vector, which is arranged on the order of landmarks number maintained in the map, to compute the landmarks' scaling. In SLAM process, the true locations of a mobile robot are never known or measured directly. The observations are employed to estimate between true locations of robot and landmark locations. The vectors' sets are defined as follows.

(1) The previous locations of mobile robot:

$$\begin{aligned} X_{0::k} &= \{x_0, x_1, \ldots, x_k\} \\ &= \{X_{0::k-1}, x_k\}. \end{aligned} \tag{2}$$

(2) The previous step of control inputs:

$$\begin{aligned} U_{0::k} &= \{u_0, u_1, \ldots, u_k\} \\ &= \{U_{0::k-1}, u_k\}. \end{aligned} \tag{3}$$

(3) The all landmarks' set:

$$m = \{m_0, m_1, \ldots, m_n\}. \tag{4}$$

(4) All landmark observations' set:

$$\begin{aligned} Z_{0::k} &= \{z_0, z_1, \ldots, z_k\} \\ &= \{Z_{0::k-1}, z_k\}, \end{aligned} \tag{5}$$

where x_k is the location and orientation state vector. u_k is the control vector, which drives the mobile robot state from x_{k-1} of time $k-1$ to x_k of time k. m_i is the vector of the ith landmark location, which is time invariant. z_{ik} is the ith landmark observation, which is corresponding with the mobile robot location. z_k is the multiple landmarks observations.

In integration particle filters, a degree of convergence between the mobile robots locations based EKFs and the probabilistic localization and mapping is expressed as

$$\begin{aligned} &P(X_{0:k}, m \mid Z_{0:k}, U_{0:k}, x_0) \\ &\quad = P(m \mid X_{0:k}, Z_{0:k}) P(X_{0:k} \mid Z_{0:k}, U_{0:k}, x_0), \end{aligned} \tag{6}$$

where the robot locations' probability distribution is $P(x_k \mid Z_{0:k}, U_{0:k}, m)$ and $P(x_k, m \mid Z_{0::k}, U_{0::k}, x_0)$ is the probable distribution.

The robot searches an object in the current scanning by employing a greedy iterative algorithm with heuristic constraints, and this object is matched with the nearest unmatched object of the last scanning. Assuming the current list objects' set Q, then the calculation time from the current location of mobile robot to each available landmark within SAR exploration area $A(t)$ is represented in (7), and the measurement distances $B(d)$ between mobile robot and the objects of interest are expressed as (8), respectively,

$$A(t) = \{m \in Q \mid T(m) > t\}, \tag{7}$$

$$B(d) = \{m \in Q \mid \text{norm}(m, r)_{\text{inf}} > d\}, \tag{8}$$

where $T(m)$ is the time difference between the current time and the time that the object was last measured. $\text{norm}(m, r)_{\text{inf}}$ is the minimum distance from the robot's current location to the boundary of an object m.

Equations (2)–(8) establish the predefined models of mobile robots' SLAM adapting to SAR environment.

3.2. IF-SLAM of Integration Particle Filter Algorithms. The on-board proprioceptive sensors incorporate laser measurement system (LMS), omnidirectional stereo camera, inertial measurement unit (IMU), and Kinect, which enable a mobile robot to build a map of the environment and to localize itself at the same time within this map. Olofsson et al. [15] discussed the data information and map fusion of two workspaces; the state space is estimated based on EKF and Rao-Blackwellized.

In robots IF-SLAM approach, a sequence of motion trajectories is constructed submaps by fusion of their on-board sensors data. Sensors' information algorithms consider a minimum of robot modeling; the first is based on improved extended Kalman filter (IEKF), and another is based on the Rao-Blackwellized (R-B) particle filter; an integration particle (IPF) algorithm was combination of IEKF and R-B particle filters, the mobile robot motion in the probabilistic form:

$$P(x_k \mid x_{k-1}, u_k) \Leftrightarrow x_k = f(x_{k-1}, u_k) + w_k, \quad (9)$$

where x_k is the robot's pose state vector from (2), u_k is the control vector, and w_k represents Gauss white noise; the observation model and motion model and the prior and posterior distributions are modeled as the follows:

$$P(x_1, x_2) = P(x_2 \mid x_1) P(x_2), \quad (10)$$

$$x_{k+1}^l = F^l x_k^l + B^l n_k^l, \quad (11)$$

$$q_{k+1} = f(\varpi_k - n_k^w) q^k, \quad (12)$$

$$y_k^q = q_k + e_k^q, \quad (13)$$

$$\begin{aligned} y_k &= \begin{bmatrix} y_k^p & y_k^v & y_k^a \end{bmatrix} \\ &= h(q_k) + C(q_k) x_k^l + \begin{bmatrix} e_k^p & e_k^v & e_k^a \end{bmatrix}^T \end{aligned} \quad (14)$$

and the margined joint distribution is as follows:

$$P(x_2) = \frac{1}{N} \sum_{i=1}^{N} (x_2 \mid x_1^{(i)}), \quad (15)$$

where $\begin{bmatrix} e_k^p & e_k^v & e_k^a \end{bmatrix}^T \in (0, R^l)$ and $R^l = \operatorname{diag}[R^p \ R^v \ R^a]$, $x_1^{(i)} \sim P(x_1)$, and $P(x_1)$ is sampled probability from state vector set x_k; x_k^l is a linear part; q_k is a nonlinear part; N is a set of weighted particles; h is evaluated particles weights.

3.3. Sensors' Measurements Data Fusion. The measurements data fusion algorithm matches relative frame-frame poses of mobile robot independently. The authors in [16–18] fuse stereo camera and inertial measurement units (IMUs) for robots estimation in GPS-denied large-scale unstructured environments, which enables the robot exploration using combination algorithms of stereo model-based outlier rejection and spatial-temporal filtering for 3D mapping and localization. In [19], the robot captures 3D scans of points of interest and combines these key points into a topological map.

Multisensors system fuses a stereo-camera sensor, inertial measurement units (IMU), and LRF odometry with an EKF algorithm to ensure robust, low-latency performance of SLAM. An initial metric map of SAR environment is preinstalled, and then this map is updated and reconstructed in real-time by fusing the multi-information achieved from their on-board sensors. A mobile robot explores in an unstructured postdisaster area while incrementally building a global consistent environmental map and locating itself in this map simultaneously. The points of interest are extracted from a group of LRFs. By fusing these interest points into a topological map, we obtain a valuable tool for exploration and localization in SAR cluttered environments. In Figure 2, three LRF sensors measurements data fusion compare each single-sensor trajectory with their fusion trajectory of measurements data.

The sensors' measurements fusion makes the nodes or particles maximally consistent with the poses of the mobile robot, and the large errors are minimized or even eliminated. A graph-based SLAM builds a simplified estimation by extracting the sensors' measurements. In a graph-based map, the nodes correspond to the poses of the robot at various points, and the edges represent the constraints between different connected poses. Such a graph-based map can be calculated by searching the spatial configuration of the nodes which are usually consistent with the data measurements modeled by the edges. A mobile robot performs perception autonomous exploration to reach goal positions in collapsed disaster scenarios by using fusion planar laser scans, stereo vision, and proprioceptive sensing system. As shown in Table 1, a measurements set corresponding to Figure 5 followed experiment is achieved from the mobile robot running SLAM in cluttered environment, which fuses multisensors of LRF, four pairs of sonars, and gyro odometry.

Table 1: A part of measurements data set of fusion multisensors achieved from robot SLAM process.

xdirection coordinates (m)	y direction coordinates (m)	Orientation angles (rad)
1.08	6.066	−0.9
0.475	5.095	179.9
−2.923	4.529	−55.3
−1.746	1.633	−80.89
0.415	0.517	−58.51
0.015	−0.185	128.6
−0.992	2.223	99.81
−0.840	3.179	42.19
0.454	3.132	−5.76
1.144	3.142	−5.71
1.616	3.204	4.07
3.245	3.21	3.90

3.4. Trajectories Fusion. According to [20], R-B particle filters fused robot's trajectory and local submap into a distribution over potential trajectories tracked by a particle filter and a distribution of potential map for each particle.

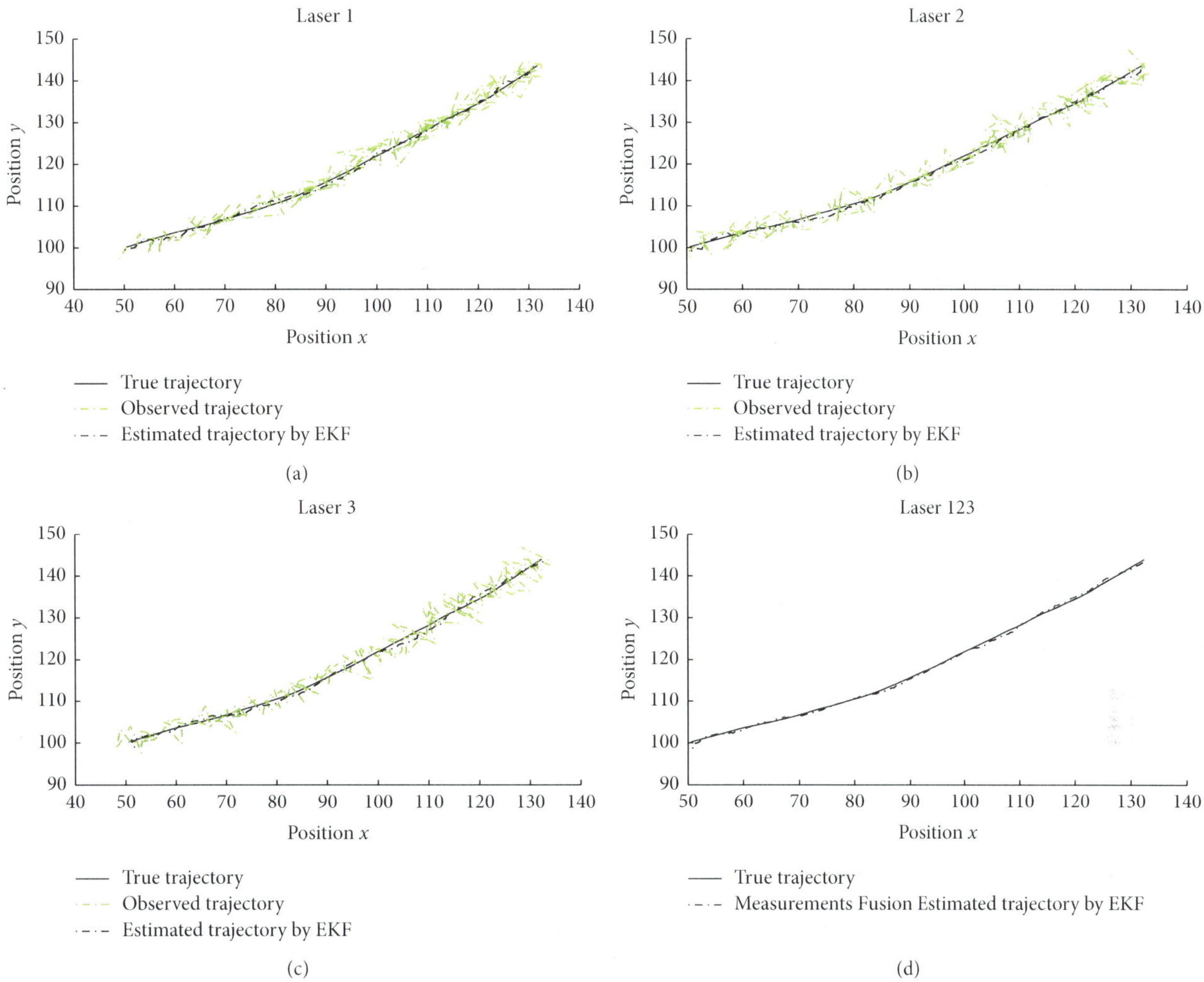

FIGURE 2: Sensors measurements data fusion: (a) is the laser sensor 1 trajectory, (b) is laser sensor 2 trajectory, (c) is laser sensor 3 trajectory, and (d) is all the three laser sensors' measurements fused into an integrated trajectory; the red lines represent true exploring trajectory of mobile robot, the green dash dotted lines represent the observed exploring trajectory, and the blue dash dotted lines indicate fusion trajectory of measurements data using information algorithms.

Researchers in [21, 22] detected an object by incorporating the LRFs and vision system measurements information for Kalman filter and built a submap by combined constraint data association (DDCA) approach constructing and searching a correspondence graph. Literature [23] considered fusion structures of the centralized and decentralized; however, the former increased computation and the latter partially reduced the system accuracy.

While initially motivated by problems in data association and loop-closure, these methods have resulted in qualitatively different methods of describing the SLAM problem, focusing on trajectory estimation rather than landmark estimation. We combined the two methods with centralized trajectory and decentralized trajectories of each robot. A mobile robot explores the disaster unstructured area while estimating the robot trajectory and mapping the environment, offering higher accuracy and improving the computation speed.

Trajectories fusion is a method of integrating all trajectories achieved from different sensors into an estimated trajectory, which has improved the accuracy and efficiency of reaching the objects of interest. In Figure 3, (a) displays a fusion trajectory of three different LRF sensors' information trajectories, in which the red solid line represents true trajectory and the blue dash dotted line indicates fusion trajectory by the proposed algorithms; (b) is sensor 1 trajectory error analysis relative to the mean value, (c) is sensor 2 trajectory error analysis, and (d) is sensor 3 trajectory error analysis, in which each red line represents the mean value, respectively. As shown in Table 2, the trajectory fusion can well match the true trajectory, and each single-sensor mean error is analyzed.

3.5. Error Analysis of Two Fusion Methods. In the proposed optimal IEKF and IPF algorithms, each particle performs to update on the observed landmarks as a simple mapping operation with known mobile robot pose. Optimal IEKF as the core of the pose filter results in a modular system to which new sensor measurements can easily be added. A set of criteria are used to assess the quality of 2D topological map

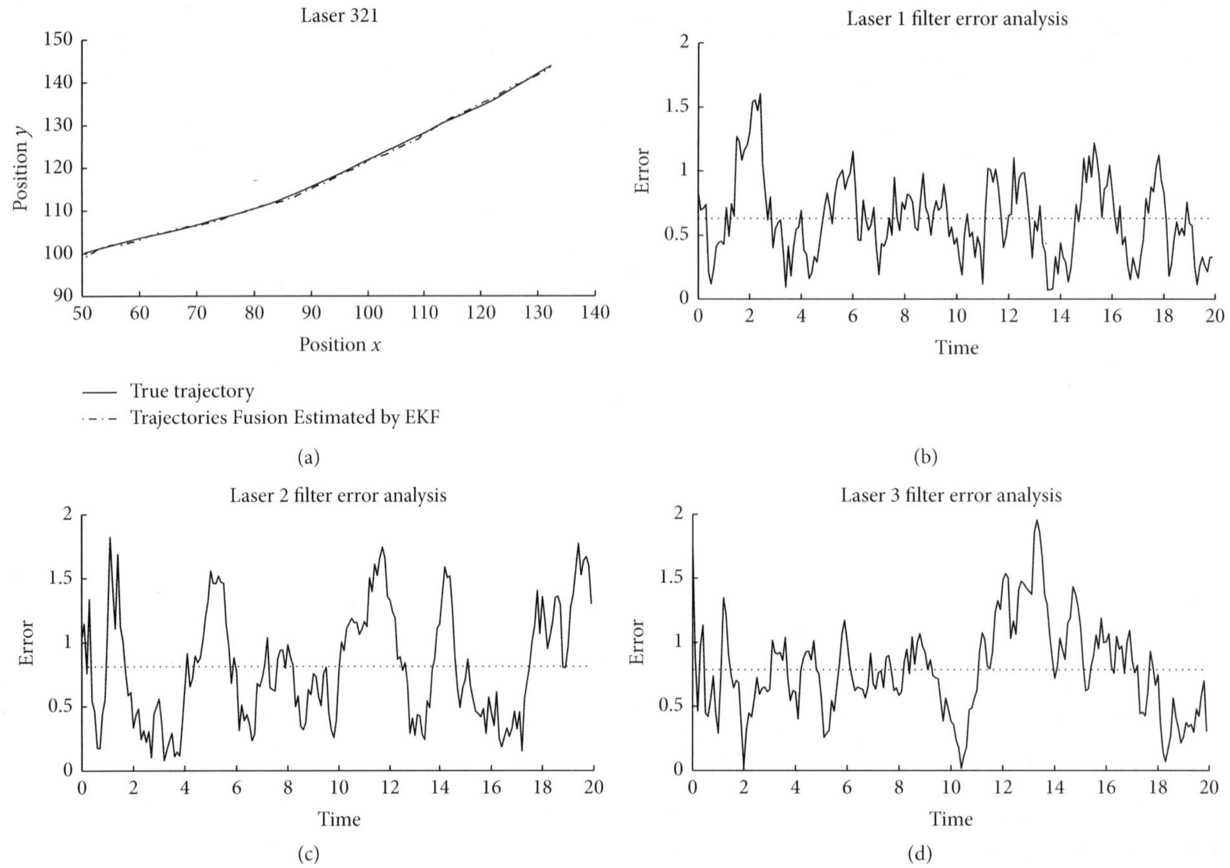

FIGURE 3: Mobile robot's trajectory fusion and error analysis: (a) is fusion trajectory (red solid represents true exploration trajectory of mobile robot, and the blue dash dotted line shows the fusion trajectory by our proposed algorithm); (b) is sensor 1 error analysis, (c) is sensor 2 error analysis, and (d) is sensor 3 error analysis (each red line indicates the mean value).

segmentation. Based on spectral clustering of an incremental generalized Voronoi decomposition of discretized metric maps, Liu et al. [24] extracted sparse spatial information from the maps and built an environmental model which aims at simplifying exploration task for mobile robots; in [25] the ICP-based and the stereo-based T and R system on the respective data conducted offline processing.

To analyze the localization accuracy, we recorded the sensors data (laser point clouds and stereo images) of all runs during the experiment, comparing the proposed IPF algorithm with a state-of-the-art stereo-vision-based technique, and show that the novel approaches have greatly increased robustness to trajectory deviations and are less dependent on environmental conditions. In errors analysis, the error state vector X is defined as follows:

$$X = [\delta_p \;\; \delta_v \;\; \delta_\theta \;\; g_a \;\; g_\theta], \qquad (16)$$

where δ_p is the errors in location; δ_v is the errors in velocity; δ_θ is the errors in attitude; g_a is sensors' biases in acceleration; g_θ is the pose biases in attitude.

The mobile robot's movement model is expressed as

$$X_k = A_k X_{k-1} + w_{k-1} \qquad (17)$$

and the error measurements model is expressed as

$$E_k = H_k X_{k-1} + v_k, \qquad (18)$$

where $w_k \sim N_1(0, \sigma_1^2)$, $v_k \sim N_2(0, \sigma_2^2)$ represent white Gaussian noise, respectively; E_k is error measurements; X_k is the predicted error states vector; A_k is error measurements system matrix; H_k is the observed system matrix. Algorithm 1 has listed the IPF algorithms procedure.

Figure 4 shows two fusion methods performance by analyzing the errors relative to mean values: (a) is the fusion of trajectories of sensor 1, sensor 2, and sensor 3; (b) is the fusion of measurements data achieved from these different sensors, in which each red line represents error mean value, and blue curves represent the biases relative to error mean value.

As displayed in Table 2, the error mean value of measurements data fusion is smaller than that of trajectories fusion; that is, the measurements fusion demonstrates a better accurate performance in getting to the objects of interest. It also can be seen that multisensors fusion of two methods are much smaller than that of a single sensors, which demonstrates that information-fusion is highly valuable in improving robots' SLAM performances for SAR environments.

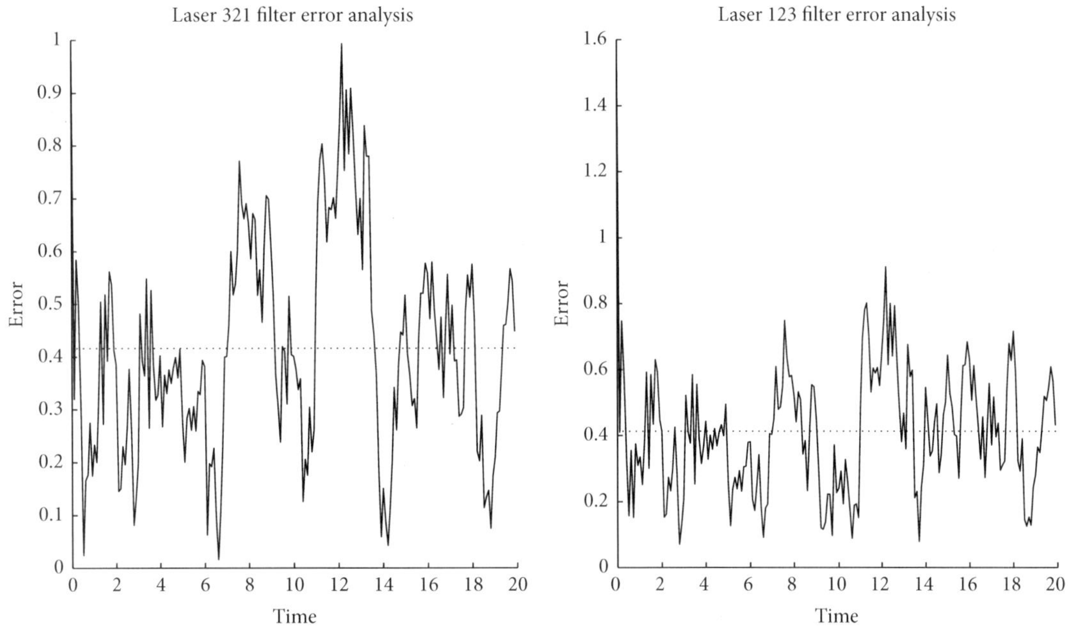

(a) (b)

FIGURE 4: The comparison of two fusion methods error analysis: (a) is trajectories fusion error analysis, (b) is measurements data fusion error analysis, in which each red line represents error mean value, and blue curves indicate error biases relative to error mean values.

TABLE 2: The comparison errors between mobile robot's trajectories and two approaches of fusion trajectory.

Different sensors	Single-sensor error mean (cm)	Measurements fusion error mean (cm)	Trajectories fusion error mean (cm)
Sensor 1	0.6301		
Sensor 2	0.8131	0.4125	0.4162
Sensor 3	0.7878		

Algorithm: Integration particle filter (IPF) procedures.
(1) Q_w = noise covariance matrix, eqs. (12) (16).
(2) R_v = observation matrix eqs. (1) (9) (11) (15) (17).
$X\_t = F * X\_t + W$;
X_Est1;
X_Est2;
X_Est3;
X_Est123;
X_Est321;
(3) Error_variance = Error_sum [ ], eqs. (13) (14) (18).

ALGORITHM 1: Information fusion algorithms.

4. SLAM Simulations with Fusion Information

The sensors' IF-SLAM can make a robot truly autonomous exploration in an uncertain collapsed area, in which both the exploration trajectory and the positions of all landmarks are estimated in real-time; a robot can build a map of an environment and simultaneously use this map to calculate its location.

Some researchers [26] have written software to simulate and implement SLAM task in Matlab, C++, and java available online. For implementation of a single robot's trajectory in particle filter SLAM, sampling the robot's poses, the observed landmarks are updated and the poses are accurate; hence, the local submap for a single particle is managed by the accuracy of trajectory. All observed trajectories provide a probabilistic location model of a robot. A robust six-degree-of-freedom relative exploration combines the iterative closest point (ICP) registration algorithm and a noise adaptive Kalman filter. This approach fuses measurements from a laser scanner and an inertial measurement unit (IMU) in a closed-loop configuration. Aghili and Su [27] integrated ICPs and Kalman filter using LRFs and IMUs to provide real-time accurate pose estimation even when poor vision data fail converging ICPs. The authors [28] presented multisensors data retrieval (MSDR) to share the measurements and information, which grants asynchronous assess to fusion system from different robots.

According to the researches, we designed simulations as shown in Figures 5(a)–5(f), in which a virtual mobile robot installed multisensors is running SLAM program with an information-fusion algorithm in a simulated SAR

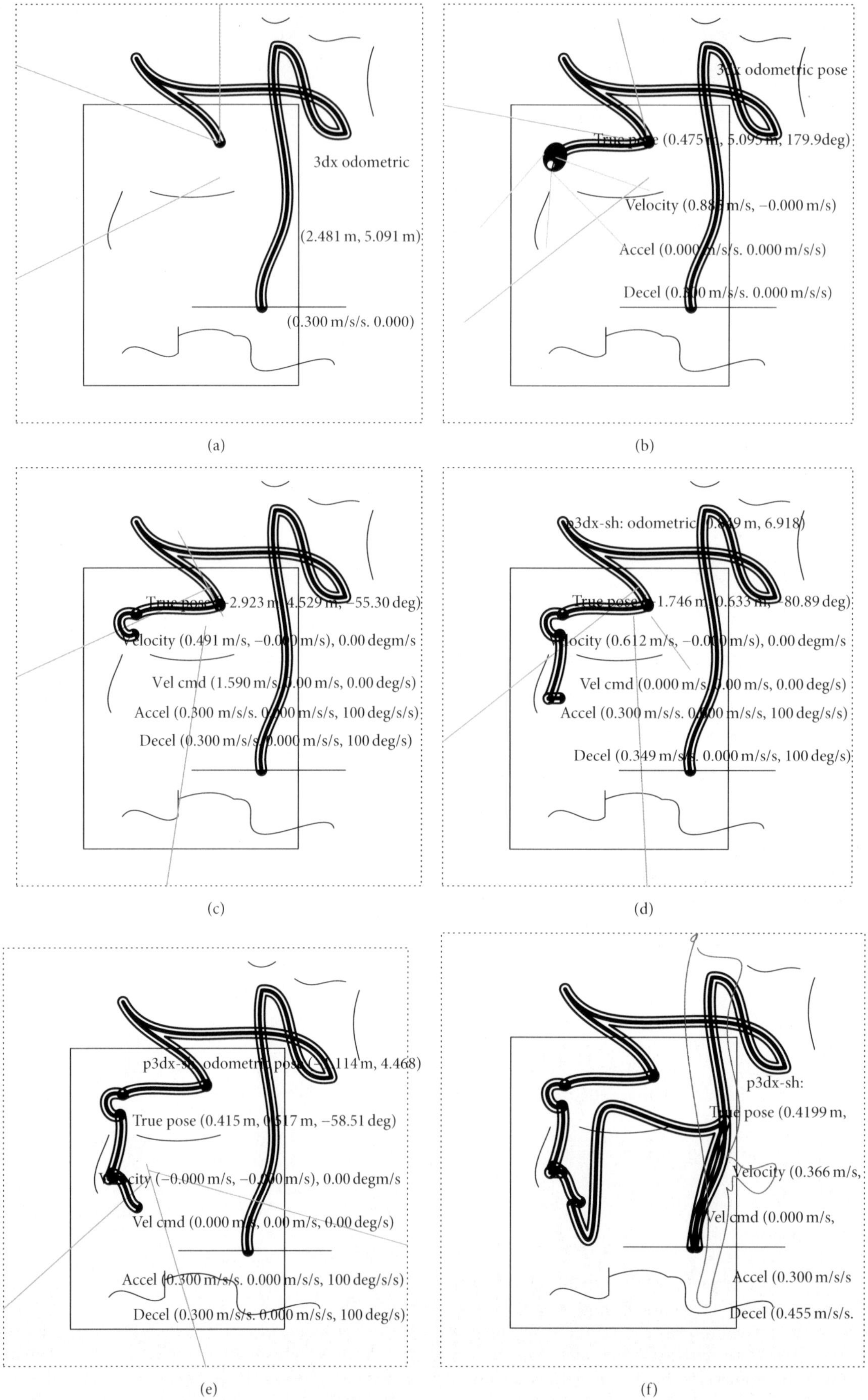

Figure 5: A virtue mobile robot performs information-fusion SLAM with LRF sensors, sonars, and gyro odometry; the red trajectories in (a)–(f) represent a loop-closure exploration.

environment; in Figure 5(b), the red trajectories represent exploration appointed loop-closure area, the red image at the ahead of trajectory is a virtual robot, the white dot on the virtual robot represents gyro odometry, the blue band represents LRF sensor information, and the grey rays indicate sonars' signal. The measurements of mobile robot' pose are marked around the robot in real-time. Table 1 displays the measurements data achieved from multisensors information-fusion.

Through comparison of the single-sensor trajectory analysis in Figures 2 and 3 and the exploration trajectories of robot in Figure 5, we can see that information-fusion methods improve the performance of a robot SLAM adapting to SAR environment.

5. The Experiments for Multi-Information-Fusion SLAM

Milisavljevic's work of sensor and data fusion [10] illustrated the multiple sensors platform for exploration environment with many targets, its associated mathematical model, and data fusion performance evaluation for various sensors. An IMU-camera sensor system developed by Panahandeh and Jansson [29] fuses inertial data from the IMU and key features of interest captured by camera.

According to the research, the experimental system consists of a ground-facing monocular camera mounted on an inertial measurement unit (IMU); the LRF data are processed to determine the distance and the relative speed of the objects (or obstacles) located in the SAR space by the laser scanning. These SLAM methods are adaptive to outdoor unstructured environments in addressing identification of SAR objects and detection of loop-closure.

A typical experiment platform used in 2D mapping is a Pioneer LX mobile robot equipped with a SICK-LMS range finder. With the embedded arnlServer software, Pioneer LX is self-guided and self-charging. The robot includes a complete robot control system and embedded personal computer (PC), differential drive system with encode system feedback, as well as LRFs, ultrasonic (sonar) sensors, and a bumper panel. As shown in Figure 6(a), a Pioneer LX mobile robot is conducting information-fusion SLAM task with installed LRF sensors, sonars, and gyro odometry. As a result, an improved updated map is formed in Figure 6(c) compared with the initial metric map in Figure 6(b).

As shown in Figures 7(a) and 7(b), a crawler mobile robot explores the SAR collapsed area and performs SLAM task, the LRF sensors and Kinect are equipped on the robot, and thus, the robot achieves multisensors information including image frames and LRF measurements data.

6. Results and Discussion

Information-fusion systems of mobile robots are designed and programmed to cover a map with their installed multisensors. In Figures 8(a), 8(b), and 8(c), the remote desk displays the robot SLAM process from PC control center.

Torra and Narukawa [1] illustrated modeling decision of information-fusion, and the quality evaluation of performance is set in Milisavljevic's work [10] by a series of criteria of data and fusion. Literature [30] discussed local sensors fusion technique with event-based SLAM tasks, [31] presented the strategy and flowchart of a hybrid image registration algorithms, according to Reid et al. [32] and Pappas [33]; with robot operation system, the exploration tasks and mapping are subdivided among multiple robots. The use of multirobots allows SLAM tasks to complete more efficiently. As Torra and Milisavljevic's theory [1, 10] and Blum and Liu's work [31] illustrate, information-fusion is an effective means of research, and the main contributions are that we first develop the novelty multirobots and multisensors (MAM) search and rescue SLAM mathematical algorithms based on this resultful research means. Comparing to general single-sensor SLAM, we improved robotic SAR SLAM performances of loop-closure, objects identification, and exploration area coverage.

According to the observed, the robot explores an appointed area and locates itself in this environment based on a preinstalled metric map; the robot processes the achieved multisensors information using the proposed algorithms. As a result, the robot forms an improved updated map comparing to the preinstalled initial metric map.

Information data of each sensor are collected to fuse for building the entire map of covered area. A motion estimation approach of information is introduced for a vision-aided inertial exploration system. A local multisensor fusion technique with a scene-based global position applies to improve the localization of a mobile robot with limited SAR scenario on site. The multirobot integration SLAM enables teams of robots to build joint maps without initial knowledge of their relative poses.

Some conventional example approaches of information-fusion are roughly divided into two groups, which can be described nonmultiscale-decomposition-based (NMDB) fusion and multiscale decomposition-based fusion methods. Distributed multisensors trajectories fusion is used to identify the information coming from different local particles which mapped the same object of interest. The multisensors information-fusion system can be easily adapted and deployed into a variety of SAR scenarios. On the basis of these useful achievements, we design the information-fusion systems for mobile robots' SLAM adapting to SAR environment.

In Figures 6, 7, and 8, the experimental results show the efficiency and accuracy of information-fusion methods in loop-closure, exploration trajectories of robotic SLAM. The trajectory mean errors are achieved by the sensors error analysis, measurement fusion, and trajectory fusion, which are displayed in Table 1 and Figures 2 and 3. The measurements data fusion estimation and trajectory fusion estimation are simulated, respectively. The error analysis shown in Algorithm 1 and Figure 4 demonstrated that the two discussed methods of information estimation fusion are much superior to single trajectory, and simultaneously, the measurements data fusion is a little better than the trajectory fusion in reaching the targets of interest. In measurements

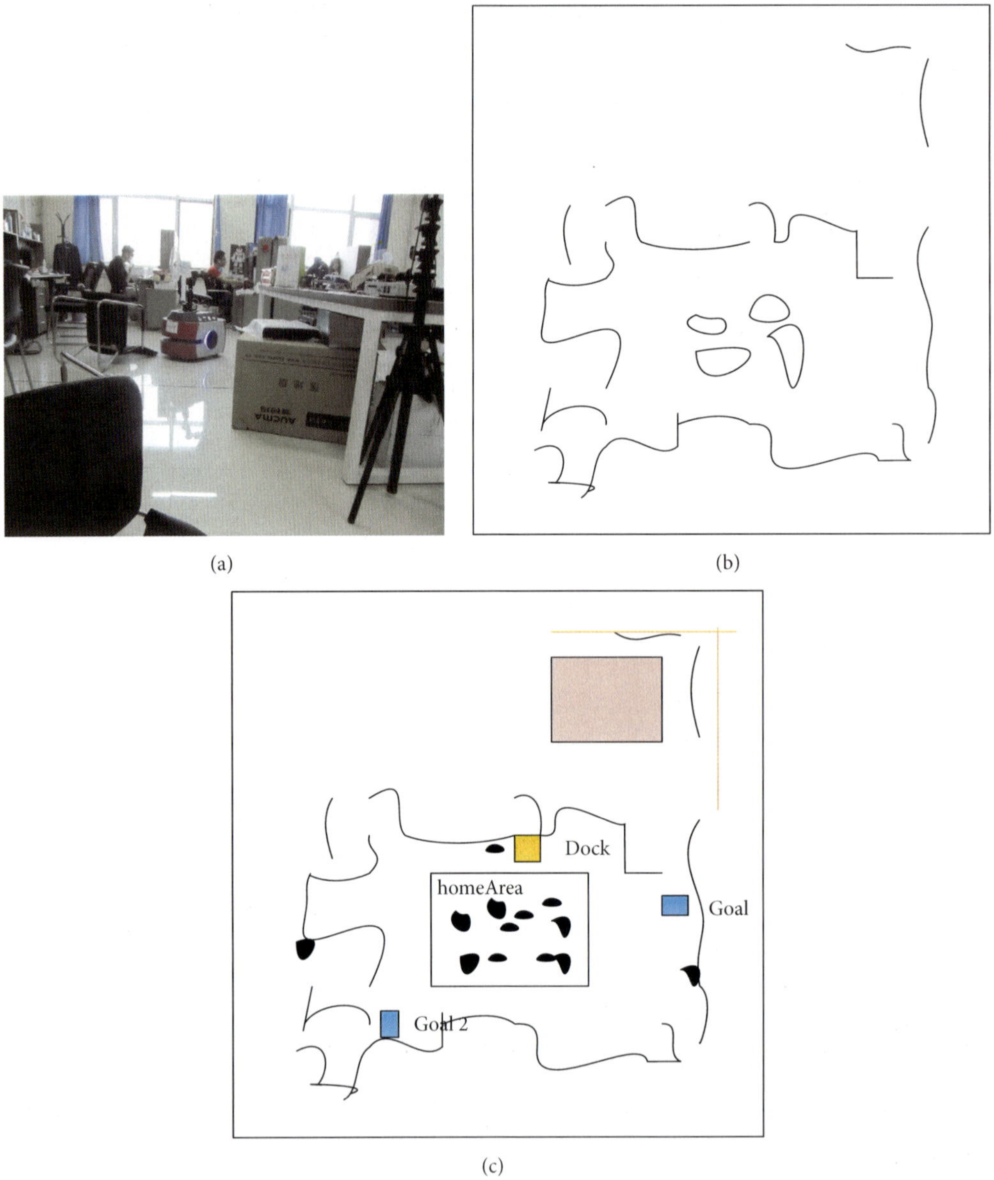

FIGURE 6: A Pioneer LX robot runs information-fusion SLAM in an experimental cluttered indoor environment. (a) The mobile robot installed LRF sensors, sonars, and gyro odometry is exploring a cluttered environment. (b) The initial metric map. (c) The generated updated map of this environment. the blue and yellow blocks marked are the objects of interest.

data fusion, essentially, there is no information loss. However, the computation of trajectory fusion method is even simpler and faster in exploring and reaching the targets of interest. Comparing Figure 9(a) with Figure 8(c) and comparing Figure 9(b) with Figure 6(c), the results demonstrate the deviation of robots motion trajectories and formed imprecise map with conventional SLAM; using the proposed IF-SLAM, Figures 6 and 8 formed accurate maps marked objects, loop-closure, and area coverage.

A challenge of information-fusion systems confronts that if any one of multisensors is bad to supply large error, the estimation exploring trajectory probably results in huge inaccuracy and inefficiency, even without getting to the appointed targets. Therefore, it requires highly and strictly each of multisensors should work well in SLAM while mobile robot is exploring an SAR postdisaster environment.

7. Conclusions and Future Work

Information-fusion systems are designed for mobile robots SLAM in unstructured collapsed postdisaster scenarios. The mobile robot achieves various pieces of information from multisensors' systems and then fuses the information into an exploration integrated trajectory by using the optimal algorithms. The information-fusion incorporates two methods of the measurement fusion and trajectory fusion, which improves the SLAM performance of mobile robots through

(a) (b)

FIGURE 7: Information-fusion of Kinect and LRF sensors applies to SLAM in a collapsed scene. (a) A crawler mobile robot equipped with the Kinect and laser sensors. (b) The RGB-D image observed from the Kinect while the robot is conducting information-fusion SLAM task.

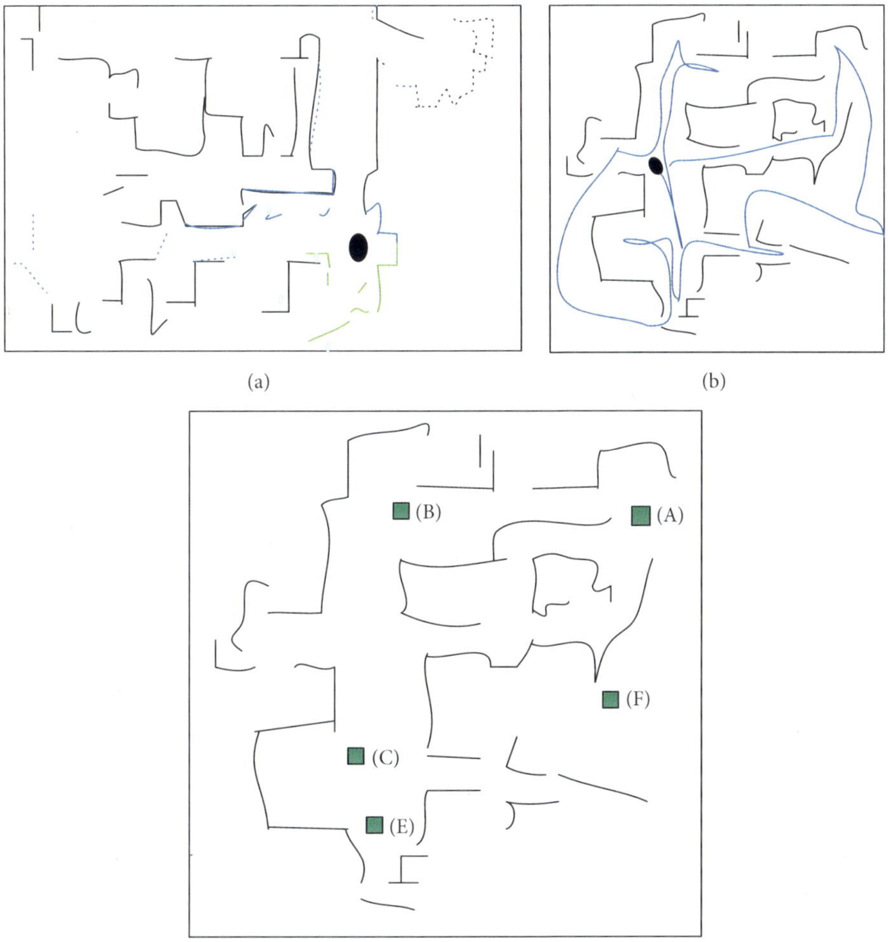

(a) (b) (c)

FIGURE 8: A robot SLAM process observed from remote PC control center. (a) The robot is exploring and locating itself with a preinstalled metric map. (b) Map processing with achieved multisensors' information. (c) The updated complete map of exploration site with marked objects (i.e., the blue blocks (A), (B), (C), (E), and (F)) of interest.

estimating the exploration trajectory so as to reaching the targets accurately and efficiently.

In this study, information-fusion systems are designed, which incorporated mobile robots, multisensors (e.g., LRF sensor, localization sonars, gyro odometry, and Kinect camera), and the algorithm that integrates R-B and EKF into an IPF procedure. We validated the proposed information-fusion methods by using a Pioneer LX and a crawler mobile

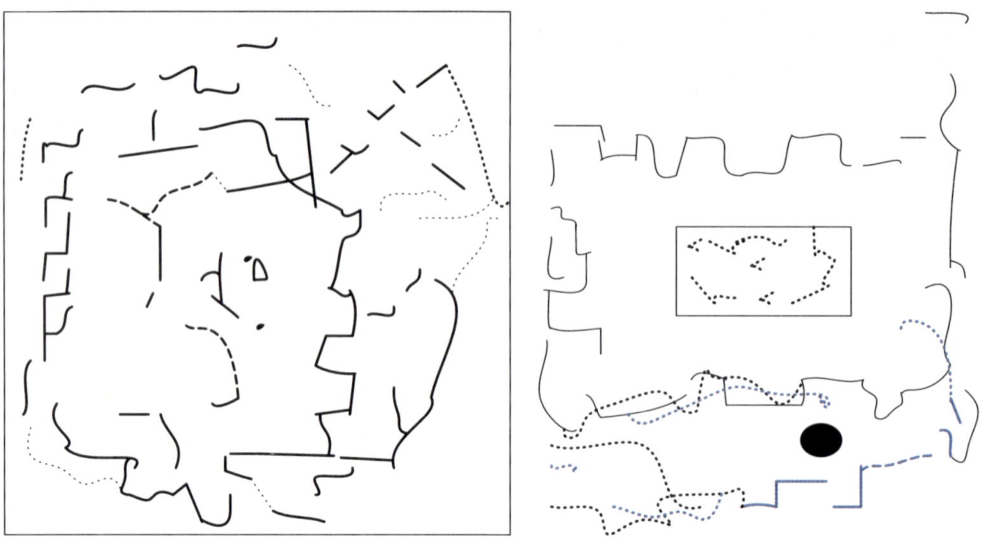

(a) (b)

Figure 9: The deviation of mobile robot trajectories and generated imprecise map. (a) Comparing with Figure 8(c). (b) Comparing with Figure 6(c).

robot installed multisensors. The results demonstrate that the designed information-fusion systems are well adapted to SAR scenarios and improve mobile robots SLAM performance of object identification, loop-closure, and exploration area coverage. In the future work, we will research the multi-sensors and the optimal fusion algorithms, which makes information-fusion method to integrate more and novel sensors and to obtain high precision estimation exploring trajectory adapted to the SAR environment.

Conflicts of Interest

The authors declare that there are no conflicts of interest regarding the publication of this paper.

Acknowledgments

This research is supported by the NSFC (National Nature Science Foundation of China) under Grants nos. 61573213, 61473174, and 61473179.

References

[1] V. Torra and Y. Narukawa, "Modeling Decisions: Information Fusion and Aggregation Operators," *Cognitive Technologies*, vol. 13, 2007.

[2] S. Jia, K. Wang, and X. Li, "Mobile robot simultaneous localization and mapping based on a monocular camera," *Journal of Robotics*, vol. 2016, Article ID 7630340, 11 pages, 2016.

[3] T. Cao, H. Cai, D. Fang, H. Huang, and C. Liu, "Keyframes Global Map Establishing Method for Robot Localization through Content-Based Image Matching," *Journal of Robotics*, vol. 2017, Article ID 1646095, 2017.

[4] R. R. Murphy, J. Kravitz, S. L. Stover, and R. Shoureshi, "Mobile robots in mine rescue and recovery," *IEEE Robotics and Automation Magazine*, vol. 16, no. 2, pp. 91–103, 2009.

[5] S. Thrun, S. Thayer, W. Whittaker et al., "Autonomous exploration and mapping of abandoned mines," *IEEE Robotics and Automation Magazine*, vol. 11, no. 4, pp. 79–91, 2005.

[6] P. Deshpande, V. R. Reddy, A. Saha, K. Vaiapury, K. Dewangan, and R. Dasgupta, "A next generation mobile robot with multi-mode sense of 3D perception," in *Proceedings of the 17th International Conference on Advanced Robotics, ICAR 2015*, pp. 382–387, Istanbul, Turkey, July 2015.

[7] M. Pfingsthorn, M. Slamet, and A. Visser, *A Scalable Hybrid Multi-robot SLAM Method for Highly Detailed Maps*, Springer, Berlin, Germany, 2008.

[8] S. Li, R. Kong, and Y. Guo, "Cooperative distributed source seeking by multiple robots: Algorithms and experiments," *IEEE/ASME Transactions on Mechatronics*, vol. 19, no. 6, pp. 1810–1820, 2014.

[9] N. Atanasov, J. Le Ny, K. Daniilidis, and G. J. Pappas, "Decentralized active information acquisition: Theory and application to multi-robot SLAM," in *Proceedings of the 2015 IEEE International Conference on Robotics and Automation, ICRA 2015*, pp. 4775–4782, Seattle, Wash, USA, May 2015.

[10] N. Milisavljevic, *Sensor and Data Fusion*, I-Tech Education and Publishing, Vienna, Austria, 2009.

[11] S. Saeedi, L. Paull, M. Trentini, and H. Li, "Occupancy grid map merging for multiple robot simultaneous localization and mapping," *International Journal of Robotics and Automation*, vol. 30, no. 2, pp. 149–157, 2015.

[12] S. Thrun, D. Hähnel, D. Ferguson et al., "A system for volumetric robotic mapping of abandoned mines," in *Proceedings of the 2003 IEEE International Conference on Robotics and Automation*, pp. 4270–4275, Taipei, Taiwan, September 2003.

[13] X. S. Zhou and S. I. Roumeliotis, "Multi-robot SLAM with unknown initial correspondence: The robot rendezvous case," in *Proceedings of the 2006 IEEE/RSJ International Conference on Intelligent Robots and Systems, IROS 2006*, pp. 1785–1792, Beijing, China, October 2006.

[14] P. Henry, M. Krainin, E. Herbst, X. Ren, and D. Fox, "RGB-D mapping: using Kinect-style depth cameras for dense 3D

modeling of indoor environments," *International Journal of Robotics Research*, vol. 31, no. 5, pp. 647–663, 2012.

[15] B. Olofsson, J. Antonsson, H. G. Kortier, B. Bernhardsson, A. Robertsson, and R. Johansson, "Sensor Fusion for Robotic Workspace State Estimation," *IEEE/ASME Transactions on Mechatronics*, vol. 21, no. 5, pp. 2236–2248, 2016.

[16] J. Ma, S. Susca, M. Bajracharya, L. Matthies, M. Malchano, and D. Wooden, "Robust multi-sensor, day/night 6-DOF pose estimation for a dynamic legged vehicle in GPS-denied environments," in *Proceedings of the IEEE International Conference on Robotics and Automation*, pp. 619–626, Saint Paul, Minn, USA, 2012.

[17] D. Wooden, M. Malchano, K. Blankespoor, A. Howardy, A. A. Rizzi, and M. Raibert, "Autonomous navigation for BigDog," in *Proceedings of the 2010 IEEE International Conference on Robotics and Automation, ICRA 2010*, pp. 4736–4741, Anchorage, Alaska, USA, May 2010.

[18] M. Bajracharya, J. Ma, M. Malchano, A. Perkins, A. A. Rizzi, and L. Matthies, "High fidelity day/night stereo mapping with vegetation and negative obstacle detection for vision-in-the-loop walking," in *Proceedings of the 2013 26th IEEE/RSJ International Conference on Intelligent Robots and Systems: New Horizon, IROS 2013*, pp. 3663–3670, Tokyo, Japan, November 2013.

[19] D. Silver, D. Ferguson, A. Morris, and S. Thayer, "Feature extraction for topological mine maps," in *Proceedings of the 2004 IEEE/RSJ International Conference on Intelligent Robots and Systems (IROS) (IEEE Cat. No.04CH37566)*, pp. 773–779, Sendai, Japan.

[20] C. Brand, M. J. Schuster, H. Hirschmüller, and M. Suppa, "Stereo-vision based obstacle mapping for indoor/outdoor SLAM," in *Proceedings of the 2014 IEEE/RSJ International Conference on Intelligent Robots and Systems, IROS 2014*, pp. 1846–1853, Chicago, Ill, USA, September 2014.

[21] S. Y. Chen, "Kalman filter for robot vision: a survey," *IEEE Transactions on Industrial Electronics*, vol. 59, no. 11, pp. 4409–4420, 2012.

[22] T. Bailey and H. Durrant-Whyte, "Simultaneous localization and mapping (SLAM): part II," *IEEE Robotics and Automation Magazine*, vol. 13, no. 3, pp. 108–117, 2006.

[23] A. Assa and F. Janabi-Sharifi, "Virtual visual servoing for multicamera pose estimation," *IEEE/ASME Transactions on Mechatronics*, vol. 20, no. 2, pp. 789–798, 2015.

[24] M. Liu, F. Colas, L. Oth, and R. Siegwart, "Incremental topological segmentation for semi-structured environments using discretized GVG," *Autonomous Robots*, vol. 38, no. 2, pp. 143–160, 2014.

[25] P. Krüsi, B. Bücheler, F. Pomerleau, U. Schwesinger, R. Siegwart, and P. Furgale, "Lighting-invariant adaptive route following using iterative closest point matching," *Journal of Field Robotics*, vol. 32, no. 4, pp. 534–564, 2015.

[26] H. Durrant-Whyte and T. Bailey, "Simultaneous localization and mapping: Part I," *IEEE Robotics and Automation Magazine*, vol. 13, no. 2, pp. 99–110, 2006.

[27] F. Aghili and C.-Y. Su, "Robust relative navigation by integration of ICP and adaptive Kalman filter using laser scanner and IMU," *IEEE/ASME Transactions on Mechatronics*, vol. 21, no. 4, pp. 2015–2026, 2016.

[28] L. Wang, M. Liu, and Q. H. Meng, "Real-Time Multisensor Data Retrieval for Cloud Robotic Systems," *IEEE Transactions on Automation Science & Engineering*, vol. 12, no. 2, pp. 507–518, 2015.

[29] G. Panahandeh and M. Jansson, "Vision-aided inertial navigation based on ground plane feature detection," *IEEE/ASME Transactions on Mechatronics*, vol. 19, no. 4, pp. 1206–1215, 2014.

[30] L. Marín, M. Vallés, Á. Soriano, Á. Valera, and P. Albertos, "Event-based localization in ackermann steering limited resource mobile robots," *IEEE/ASME Transactions on Mechatronics*, vol. 19, no. 4, pp. 1171–1182, 2014.

[31] R. S. Blum and Z. Liu, *Multi-Sensor Image Fusion and Its Applications*, Taylor & Francis Group, LLC, New York, NY, USA, 2006.

[32] R. Reid, A. Cann, C. Meiklejohn, L. Poli, A. Boeing, and T. Braunl, "Cooperative multi-robot navigation, exploration, mapping and object detection with ROS," in *Proceedings of the 2013 IEEE Intelligent Vehicles Symposium, IEEE IV 2013*, pp. 1083–1088, Gold Coast, Australia, June 2013.

[33] B. Pappas, *Multi-Robot Frontier Based Map Coverage Using the ROS Environment [Ph.D. thesis]*, Graduate Faculty of Auburn University, Auburn, Ala, USA, 2014.

6

A Novel Edge Detection Algorithm for Mobile Robot Path Planning

Rami Al-Jarrah,[1] **Mohammad Al-Jarrah**,[2] **and Hubert Roth**[3]

[1] *Mechanical Engineering Department, Hashemite University, P.O. Box 330127, Zarqa 13133, Jordan*
[2] *Department of Electrical and Computer Engineering, Khalifa University of Science and Technology, Abu Dhabi 127788, UAE*
[3] *Automatic Control Engineering Institute (RST), Siegen University, Siegen, Germany*

Correspondence should be addressed to Rami Al-Jarrah; ramia@hu.edu.jo

Academic Editor: Zhaojie Ju

A novel detection algorithm for vision systems has been proposed based on combined fuzzy image processing and bacterial algorithm. This combination aims to increase the detection efficiency and reduce the computational time. In addition, the proposed algorithm has been tested through real-time robot navigation system, where it has been applied to detect the robot and obstacles in unstructured environment and generate 2D maps. These maps contain the starting and destination points in addition to current positions of the robot and obstacles. Moreover, the genetic algorithm (GA) has been modified and applied to produce time-based trajectory for the optimal path. It is based on proposing and enhancing the searching ability of the robot to move towards the optimal path solution. Many scenarios have been adopted in indoor environment to verify the capability of the new algorithm in terms of detection efficiency and computational time.

1. Introduction

Path planning is an essential part of mobile robotic systems to find the optimal path [1–4]. The boundary values of velocity, acceleration, and deceleration of a mobile robot (MR) are defined as closed mathematical forms over the spatial path. These values can be calculated offline for any MR model [1]. In [2], an approach has been proposed to generate a time optimal velocity profile for any path in static or dynamic environments. The proposed algorithms are not time-consuming if the path is planned as smooth curves. A real-time robot path planning approach based on the utilization of inequality and optimization technique has been presented in [3]. The problem of finding a path has been transformed into finding a collision free path without calculating the configuration space obstacles. In addition, many heuristic algorithms have been developed recently to be used for path planning in real world. The fundamental similarities between these algorithms and the motivation behind each one have been introduced in [4]. However, these algorithms can be grouped in many categories such as static algorithms, anytime algorithms, and replanning algorithms [4].

However, computer vision (CV) has become an important tool in modern robot systems since it provides a useful tool for detection and decision-making during robotic navigation missions. In this context, many studies have applied CV for path planning tasks [5–9]. The object detection has been proposed for MRs to perform their missions in the environment [5]. The proposed algorithm in [5] is based on collecting data images from the surrounding environment during the training phase. Whilst the algorithm is measuring the errors in data image, it updates the classifier. As a result, it provides the means of classifying and more reliable object detection. The work in [6] addressed the problem of how making a robot learns natural terrain selectively and exploits its knowledge to estimate the terrain for optimal path planning. The author proposed a scheme that combines vision learning and interaction to give the robot an ability to understand the intricate environment. In addition, particle swarm optimization (PSO) algorithm was applied to search for the optimal path. In [7], a combination between CV and robot path planning has been presented aiming to program free robot applications. Image processing and path planning tools was presented using the Wise-Shop floor framework. The authors in [9] developed a CV system for autonomous MR navigation. A vision sensor was placed on system on-board to detect and localize another robot in the simulated environment. This proposed approach used A* algorithm to determine the path in order to find the best solution.

However, the classification of CV systems depends on the sophisticated degree of the system itself. Therefore, CV systems would be classified into several categories such as visual tracking and visual navigation [6–9]. Since visual tracking is highly dependent on the edge, many algorithms have been introduced in the literature to develop powerful edge detectors, that is, Canny and Sobel [10–13]. The modified Canny algorithm that uses different sigma values for different sections of the image has been proposed in [14]. Each image is divided into multi-sub-images to improve the accuracy level of the final output. The investigated pixel is calculated and processed by a Gaussian filter with different sigma values. The values of sigma were chosen carefully so that the output image has only the true edge. Moreover, it has been proved that the proposed method gives better results than the original algorithm. However, the histograms of the results were not distributed uniformly and the modified algorithm is still inappropriate for noisy images [15]. The authors of [15] proposed an edge detection process based on smoothing and morphological operations. This approach was executed based on Canny edge detection principle which is represented as a combination of erosion and dilation. In addition, they compared the results with other existing techniques and exhibited that their proposed technique can produce better results. However, the approach is based on probing an image with a 2 $\times$ 2 structuring element in which each of the pixels contains a value of zero or one. So, one of the limitations on this approach is that the frame rate cannot reach the required level for real-time applications. Although the proposed algorithms in [14, 15] can be successfully implemented to solve and optimize the edge detection problem, they have inaccurate results and some edges may be lost [14, 15].

Therefore, to avoid such limitations on these algorithms, many researchers have implemented intelligent algorithms such as the fuzzy edge detection. The method of fuzzy edge representation was proposed in [16]. This representation contains the ideal edges which are not affected by any degradation. This method is more general than the classical edge representation method because it defines the location of edges using the fuzzy set theory. Moreover, this model aims is to provide more information about the location of edges and enabling more flexibility in high level processing. However, further mathematical analysis and experiments with various membership functions are needed. In addition, fuzzy image processing is capable of dealing with the uncertainty in the given data by presenting a good mathematical framework to manipulate the drawbacks in edge detection [17, 18]. Fuzzy logic based edge detection has been considered for both smooth and noisy images in [19], where the proposed method utilizes a 3 $\times$ 3 mask guided by a set of fuzzy rules. The results were compared with other edge detection algorithms like Sobel, Canny, Laplace of Gaussian, Prewitt, and Roberts. In addition, the tools of fuzzy based edge detection have been described in [20]. In [21], an edge detection method has been proposed based on Sobel algorithm and the generalized type 2 fuzzy logic systems. In addition, to limit the complexity of handling the generalized type 2 fuzzy logic, α-planes theory has been applied [21]. Simulation results have been obtained with the Sobel operator and the generalized type 2 fuzzy logic system.

However, the GA is a very important technique in combinatorial optimization issues [22, 23]. The objective of such algorithm in mobile robot (MR) field is to search for the optimal path and trajectory. In [22], the authors developed the genetic operators for GA to be able to solve path planning problems. They claimed that their proposed algorithm has fast convergence towards the global optimum. Their algorithm depends on three ideas: new mutation operators, matrix coding, and visible space. Even though that algorithm is effective for static environment, it is not effective for dynamic one. Moreover, the modified A* algorithm has been presented in [23] in order to reduce the processing time. However, these researches showed that the GA can be improved in many ways in order to solve the path planning issues. In addition, some researchers are still investigating this problem to improve the algorithm and explore more solutions. However, all researches mentioned above were accomplished using MATLAB simulations and none of them was implemented in real-time applications. Also, most of them solved the optimal path in terms of single objective. In [24], the initial working path for the robot is obtained by using Dijkstra algorithm. Then ant colony algorithm is adopted to optimize this initial path and the global optimal path is obtained. The experiments are carried out but it was assumed to have a gas hazard area only if the gas concentration is greater than 2%. The global path planning algorithm, bidirectional simultaneous visibility graph construction, and path optimization by A* are presented in [25]. The algorithm is not constructing a visibility graph before the path optimization to improve the time efficiency of path planning. However, the algorithm was validated only using different simulation environments. In addition, it was assumed that the environments are known and static. Also, they assumed that obstacles positions should be known in advance.

In this paper, an intelligent edge detection algorithm is proposed for robot navigation systems. This work is extended to our previous work which takes the benefits of mixing robots system heterogeneity for navigation and path planning purposes [26]. The proposed algorithm is based on a combination between fuzzy edge detection and the bacterial algorithm. The bacterial algorithm is applied to optimize the membership functions and fuzzy rules to reduce the computational time. Moreover, the optimal path is generated by the modified GA with enhanced A*. This combination aims to convert the optimal path to trajectory path taking into account the most important objectives in path planning, that is, the minimum travelling time, safety, security, and energy consumption. Although the conversion process addressed in this paper has been ignored by most researchers, it is an essential part of any visual navigation system. In addition, the histogram equalization is presented to stretch the image histogram and allow the proposed approach to extract more details from the image. Unlike all researches mentioned above, real-time robot navigation system is adopted here.

This paper is organized as follows. After this introduction, the problem formulation is described in Section 2. The intelligent computer vision is described in Section 3. In Section 4, real-time results are shown and compared to a method used in [15]. Finally, conclusions and future work are addressed in Section 5.

2. Problem Formulation

The conceptual terms related to robot navigation systems are defined as follows.

Definition 1. The 2D map is the map that resulted from edge detection and includes information about the target object and/or any obstacle in the environment.

Definition 2. The path is the geometric trace that the vehicle follows to reach its destination avoiding obstacles.

Definition 3. Path planning is the problem of finding the geometric path from an initial point to a given terminating point (end point).

Definition 4. The trajectory is represented as a sequence of states visited by the vehicle based on time, velocity, and acceleration.

Definition 5. The trajectory planning for a robot concerns in the transition of one feasible state to another. It is referred to motion planning that is based on time, velocity, and acceleration. In addition, it should satisfy the vehicle's kinematic limits and avoid the obstacles.

In this study, edge detection based object recognition is considered to identify specific objects such as robot and obstacles in unstructured environment. In addition, it provides important data to reconstruct the environmental map. To detect the edge in an image more accurately and efficiently, we need to design a robust approach that should work fine and deal with the noise in the image. These edges are so helpful for real-time applications such as robot navigation. Consequently, the histogram equalization is presented to stretch the image histogram to extract more details from the image. In addition, the intelligent image processing algorithm is proposed based on a combination between fuzzy edge detection and the bacterial algorithm that is applied to optimize the membership functions and fuzzy rules and to reduce the computational time of the proposed approach. The results are used to obtain the 2D map that is used in the motion planning, which is considered as an optimization problem. This means that the next step in the research is to find the multiobjective function to minimize the navigation time and the navigation distance. Moreover, a beneficial optimization should consider the safety during the navigation and successfully achieve the robot missions. Furthermore, the optimal path is generated by the modified GA with enhanced A^*. This combination aims to convert the optimal path to trajectory path taking into account the previously mentioned objectives in path planning.

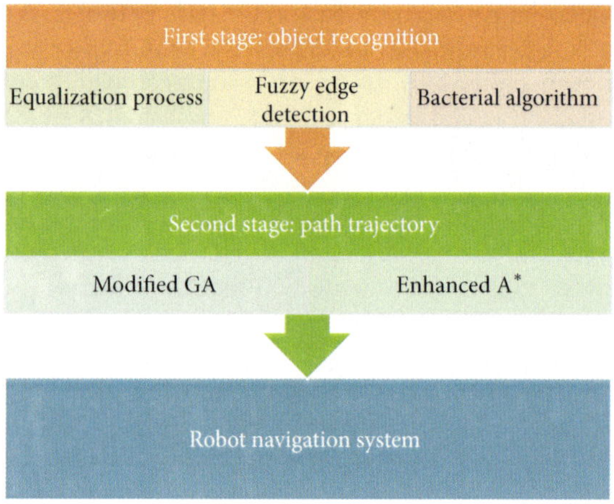

FIGURE 1: The proposed computer vision algorithm.

It should be noted that, during each planning cycle, the path planner module generates several trajectories from the vehicle's current location towards the final destination point. This process is executed depending on the speed of the vehicle's on-board sensors. In addition, each trajectory is evaluated with respect to some cost function to determine the optimal trajectory. It is assumed that the robot is placed on a plane environment and the contacts between the wheels and the ground have pure rolling and nonslipping conditions during the mission. The robot moves according to 2D map that might be static or dynamic. Also, the obstacles could be placed at any grid point in this map. The MR's mission is to search offline and online for the optimal path for the robot from a start point to a destination point.

3. Intelligent Computer Vision

The proposed CV algorithm has two main stages as it is described in the block diagram in Figure 1. First, fuzzy based intelligent image processing algorithm is combined with the bacterial algorithm. This stage is responsible for detecting the edges in the captured images, where the bacterial algorithm is used in order to optimize the membership functions and fuzzy rule. The second stage is the modified GA and the enhanced A^*, where this stage is applied in order to find the optimal path and optimal trajectory.

3.1. Intelligent Image Processing Algorithm (IIPA). As stated earlier, the main core for the proposed IIPA algorithm is the fuzzy edge detection that starts detecting the edge points of the input image by processing each pixel with its 8 neighbors. In addition, the bacterial algorithm is applied to control the membership functions.

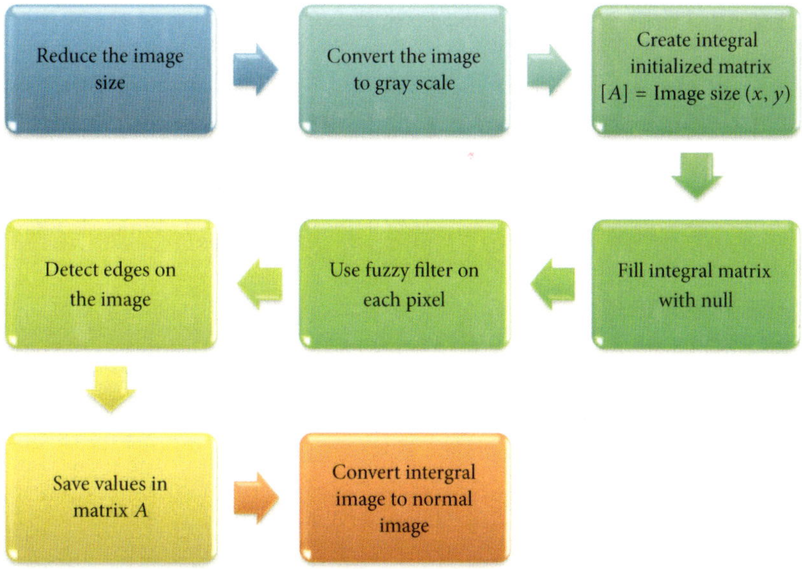

Figure 2: IIPA description.

The definition of fuzzy image processing is simply explained as an approach which is capable of processing integral images with their segments values as fuzzy sets [27]. Even though this process has three main steps like the normal fuzzy, it is based on image analysis rather than on input values. For simplicity, before applying the fuzzy logic on any image, it is necessary to transform that image to perception domain. Fuzzy image processing describes the image characteristics as linguistic variables and maps them to membership functions. Hence, the fuzzification process transfers the images from gray level plane to membership plane. Then, the fuzzy inference system (FIS) maps the input to fuzzy output using fuzzy knowledge rules based on fuzzy operators. The last step is the defuzzification in which the crisp output is calculated based on the center of gravity method. Hence, the fuzzy operator generates the useful data and maps the fuzzy values to visualize them. The IIPA proposed in this paper is described in Figure 2. It starts by reducing the input image size and converting it to the gray scale. Then, an initialized integral matrix is created in order to save the values of each pixel in this matrix. Thereafter, the whole structure is tuned into enhancing filter function that segments the input images. The filter slides over each pixel, and it is guided by fuzzy rules to ensure that the current pixel is an edge or not. After that, the detected edges are collected and the integral image is converted into normal image.

The image is encoded with $n = 8$ bits and converted to gray scale and then transformed to digital integral image. Each pixel consists of a value representing the intensity level. The histogram is the distribution of the discrete intensity levels with range $[0, L-1]$, where $L = 2^n$ is the length of image. This distribution is a discrete function (P_r) associated with intensity (r_k) and the number of pixels having this intensity (n_k). In order to normalize the histogram, we need to transform the discrete distribution of intensities into a discrete distribution of probabilities. The normalization gives the properties of the probability density function. Therefore, it is required to divide each value of the histogram according to the number of pixels. It is clear that the digital image is a discrete set of values that could be represented by a matrix, so it is equivalent to divide each n_k by the dimensions of the array.

$$n_{kn} = \frac{n_k}{L \times W} = p_r(r_k), \quad (1)$$

where L and W are the dimensions of the array.

In order to adjust the contrast of the image and modify the intensity distribution, the equalization of histograms is implemented. The main objective of this process is to give a linear trend to the cumulative distribution function (CDF) associated with the image. In other words, this technique aims to have an approximate uniform histogram and to assign a linear CDF to the resulting image. The CDF can be viewed as a cumulative sum of all the probabilities lying in its domain, and it is defined by

$$\begin{aligned} [0, L-1] &\longrightarrow N \\ x &\longrightarrow \mathrm{Card}(x) \\ [0, L-1] &\longrightarrow [0,1] \\ x &\longrightarrow \mathrm{CDF}(x) = \frac{\mathrm{Card}(x)}{\sum_{i=0}^{L-1} \mathrm{Card}(x_i)}, \end{aligned} \quad (2)$$

where Card is the cardinality of a set or the number of pixels.

In the tested image shown in Figure 3, the floor and some parts of the furniture are clearly appearing white. Also, there are a huge number of pixels having intermediate gray level. The furniture and the robot are black, and they are occupying a big part of the image. So, it makes sense to have three different regions on the histogram: a black region with high peak, a gray region, and a white region, as it

FIGURE 3: The tested image.

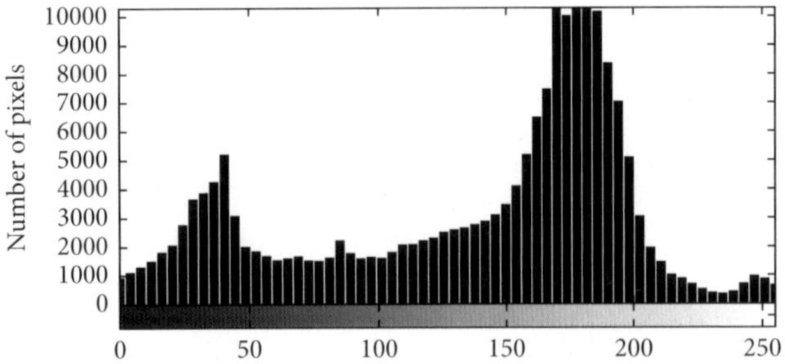

FIGURE 4: The histogram associated with the tested image.

is shown in Figure 4. If restriction on the range of the histogram is applied, it will display and focus only on the regions corresponding to the right range of the histogram. So, we have considered only the range of values between 150 and 200 as shown in Figure 5. As shown in Figure 4, the histogram values concentrated in the right side have a lot of white intensities. Therefore, the equalization can stretch the histogram allowing us to really obtain more details from the image that we were not able to see before. Figure 6 shows the equalized histogram. The next step is implementing the fuzzy edge detection algorithm. The trapezoidal fuzzy membership functions have been used in this research work because they are capable of covering a wide range of the fuzzy space. The trapezoidal fuzzy membership function has the following form [28]:

$$\mu_{A_{ij}}\left(x_j\right) = \begin{cases} \dfrac{x_j - a_{ij}}{b_{ij} - a_{ij}}, & a_{ij} \le x_j \le b_i \\ 1, & b_{ij} \le x_j \le c_i \\ \dfrac{d_{ij} - x_j}{d_{ij} - c_{ij}}, & c_{ij} \le x_j \le d_i \\ 0, & \text{otherwise,} \end{cases} \qquad (3)$$

where i and j represent the rule index and the input variable, respectively, and $a_{ij} \le b_{ij} \le c_{ij} \le d_{ij}$ are the breakpoints of the membership functions.

The fuzzy knowledge base has been designed to be a [3 × 3] fuzzy filter, as shown in Figure 7. The investigated pixel P_{ij} (or F_5) contains binary value that is compared to the

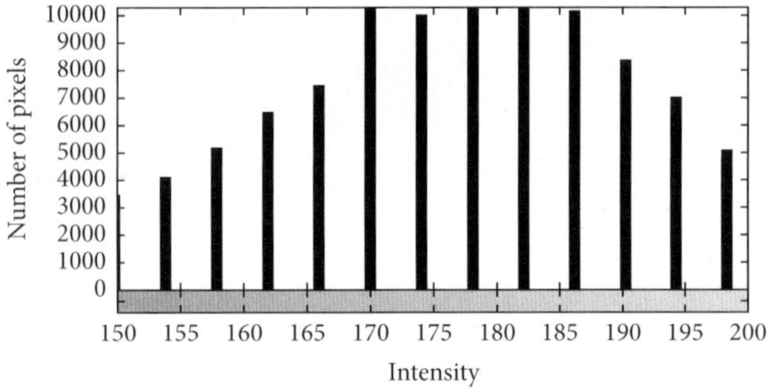

FIGURE 5: The histogram with min = 150 and max = 200.

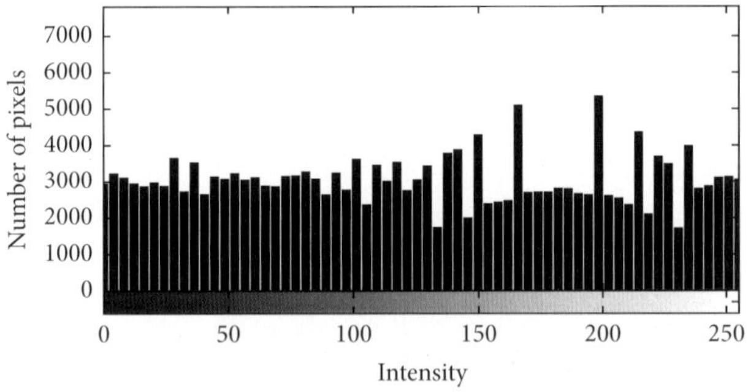

FIGURE 6: The equalization histogram.

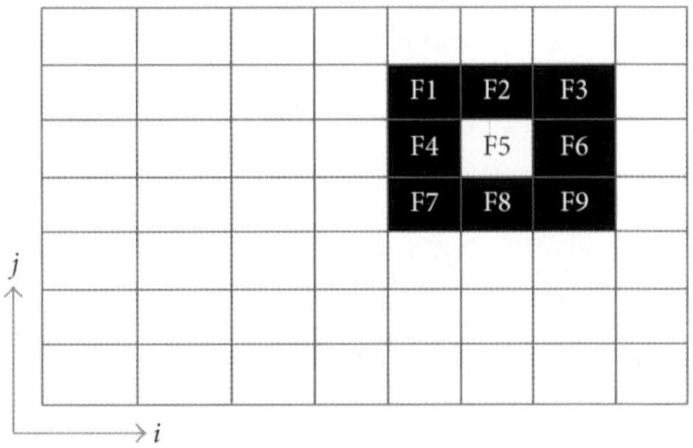

FIGURE 7: The fuzzy mask.

binary values of its eight neighbors. So, the fuzzy knowledge base has eight inputs and one output. Each input has two linguistic variables (B: black; W: white). The output is the edge with three linguistic variables (B: black; E: edge; W: white). The initial membership functions for inputs and output are given in Tables 1 and 2, respectively. Also, the initial rules are considered to be 35 rules.

After many experiments and researches, we had discovered that the computational time of IIPA is obviously longer than other techniques like Canny. As a result, we were looking for an optimization method that might be combined with fuzzy sets in order to reduce the computational time in addition to its capability of dealing with noisy images. The bacterial algorithm, which was proposed to discover the parameters of the fuzzy system and to design the fuzzy system, was the best choice for such combination [29]. This method discovers the combination of input variables of the

TABLE 1: Initial membership functions for inputs.

Color	Membership range
Black	$\{-10000\ \ 0\ \ 20\ \ 90\}$
White	$\{80\ \ 220\ \ 250\ \ 10000\}$

TABLE 2: Initial output membership functions.

Color	Membership range
Black	$\{-1000\ \ 0\ \ 10\ \ 20\}$
White	$\{235\ \ 235\ \ 255\ \ 1000\}$
Edge	$\{110\ \ 120\ \ 120\ \ 130\}$

TABLE 3: Final membership functions for inputs.

Color	Membership range
Black	$\{-10000\ \ 0\ \ 22.5\ \ 112\}$
White	$\{105\ \ 218\ \ 255\ \ 10000\}$

TABLE 4: Final output membership functions.

Color	Membership range
Black	$\{-1000\ \ 0\ \ 11\ \ 13\}$
White	$\{245\ \ 251\ \ 255\ \ 1000\}$
Edge	$\{110\ \ 115\ \ 115\ \ 118\}$

rules, the parameters of the membership functions of the variables, and a set of relevant rules. Consequently, we have optimized the IIPA by combining the bacterial algorithm with the fuzzy sets.

However, there are two operations that can be applied to optimize the process using the bacterial algorithm [30]. The first one is the annihilation which means deleting the fuzzy rule if the membership functions are too narrow. The second one is removing the rule and the membership function by a fusion operator when the medians difference between two membership functions is too small. The optimized membership functions for the inputs and output are given in Tables 3 and 4, respectively. The optimized rules are assumed to be 15 rules as presented in Table 5. It has been found that the best fitness function value for 1000 generations is 0.5206. The fitness function which describes the absolute errors is shown in Figure 8.

The second tested image and the results obtained from implemented fuzzy edge detection are illustrated in Figures 9 and 10, respectively.

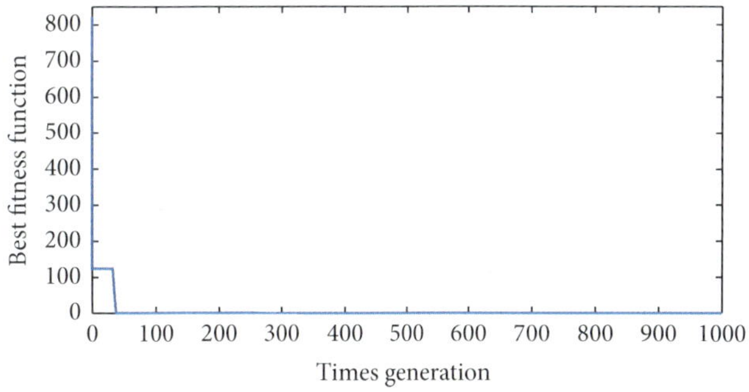

FIGURE 8: Fitness function value.

FIGURE 9: Second tested image.

FIGURE 10: The tested image result.

3.2. The Modified Genetic Algorithm. The soft computing GA can solve complex problems such as optimization of trajectory planning. It is based on proposing and enhancing the searching ability of the robot to move towards optimal solution. The formulation of such problem is concerned with two main steps. The first step is to find the multiobjectives optimal path from the start point to the end point. There are several purposes for introducing the multiobjectives optimal path such as minimizing the distance and time and ensuring the smoothness and safety of the path. Figure 11 describes these restricted tasks that should be done by GA and search A^* algorithm. The second step is dealing with complex environment by converting the optimal path to trajectory. Moreover, the path planner is required in trajectory to compute a collision free path that is later converted into a time-based trajectory [31]. However, the first step converts the images from previous sections into 2D maps, then describes this environment model, and constructs a closed workspace with different numbers of static obstacles. In order to eliminate obstacles' nodes from the map, a decreased operator has been used. The second step of GA is the initialization with population size 20, number of generations is 10, and mutation type is flip bits and crossover type with one point crossover. In the final step, the GA and the modified A^* are used to generate a set of the suboptimal feasible paths in the given maps. The flow chart for the proposed GA approach is shown in Figure 12. The fitness values and average distances between the individuals in 10 generations have been shown in Figures 13 and 14, respectively.

Table 5: The fuzzy knowledge rules.

Rule	R1	R2	R3	R4	R5	R6	R7	R8	Result
1	W	W	W	W	W	B	B	B	E
2	B	B	B	W	W	W	W	W	E
3	W	B	B	W	B	W	W	B	E
4	B	B	B	B	W	W	W	W	E
5	W	W	W	B	W	B	B	B	E
6	B	B	B	W	B	W	W	W	E
7	W	W	W	W	B	B	B	B	E
8	B	B	B	B	B	W	W	W	E
9	B	B	W	B	W	B	B	W	E
10	W	W	W	B	B	B	B	B	E
11	W	B	B	W	B	W	B	B	E
12	B	B	W	B	W	W	W	W	E
13	W	W	W	B	W	B	B	W	E
14	W	W	W	W	W	W	W	W	W
15	B	B	B	B	B	B	B	B	B

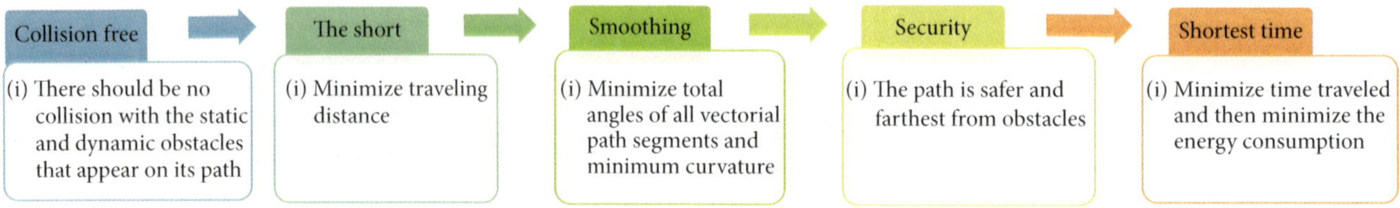

Figure 11: Multiobjectives for trajectory path planning.

4. Results and Comparison

In this section, we have illustrated the results of our proposed approach for different scenarios. Figure 15 shows the number of false edge pixels which are detected by other techniques: Sobel, Canny, and scan line approximation. These values are compared to our proposed fuzzy edge detection. When the proposed modified fuzzy edge detection was applied to two images with sizes, 512 × 512 and 256 × 256, false edge pixels were 78 and 128, respectively. As a result, it is obvious that the number of false edge detections is much less in our method.

Figure 16 shows a noisy version of the same image, and Figure 17 shows the histogram that describes the distribution of this image. However, the result of implementing the Canny edge detection on this noisy image was the unclear image shown in Figure 18. The equalization of this image is shown in Figure 19 whilst Figure 20 shows the result from our proposed algorithm. It is clear that Canny approach results in very bad edge detection when compared to our proposed approach.

Additionally, in order to compare our results with the existing methods, we have implemented the classical methods presented in [15]. Original image of the scene is shown in Figure 21. The detected edge of the image in Figure 21 using the detection algorithm of [15] is shown in Figure 22(a). On the other hand, the detected edge of the same image using our proposed approach is shown in Figure 22(b). It is clearly found that some of the edges were lost when the reference method (method of [15]) was implemented. In contract, these edges were represerved when our proposed approach was implemented.

Figure 23 shows a noisy version of the same image, whilst Figures 24(a) and 24(b) show the detected edges of the noisy image using the reference method and our approach, respectively. From these figures, it is clear that our approach is more immune to noise than the reference method. We have also measured the computational time for the considered two approaches (the reference approach and our approach) as illustrated in Figure 25. Both of them were tested 7 times for different images sizes. Regarding terms of computational time, it is shown that our method is faster than the reference one.

Moreover, depending on the shape and color features, the proposed algorithm will address the environment contents which contain some features such as the start and end points, target MR, and the environment obstacles as shown in Figures 26(a) and 26(b). The final result, which shows the optimal trajectory for the robot from its initial position to the target position, is shown in Figure 27. Another tested scenario was in our mechanical department corridor as it is illustrated in Figure 28. The final trajectory path for the mission inside the corridor is shown in Figure 29.

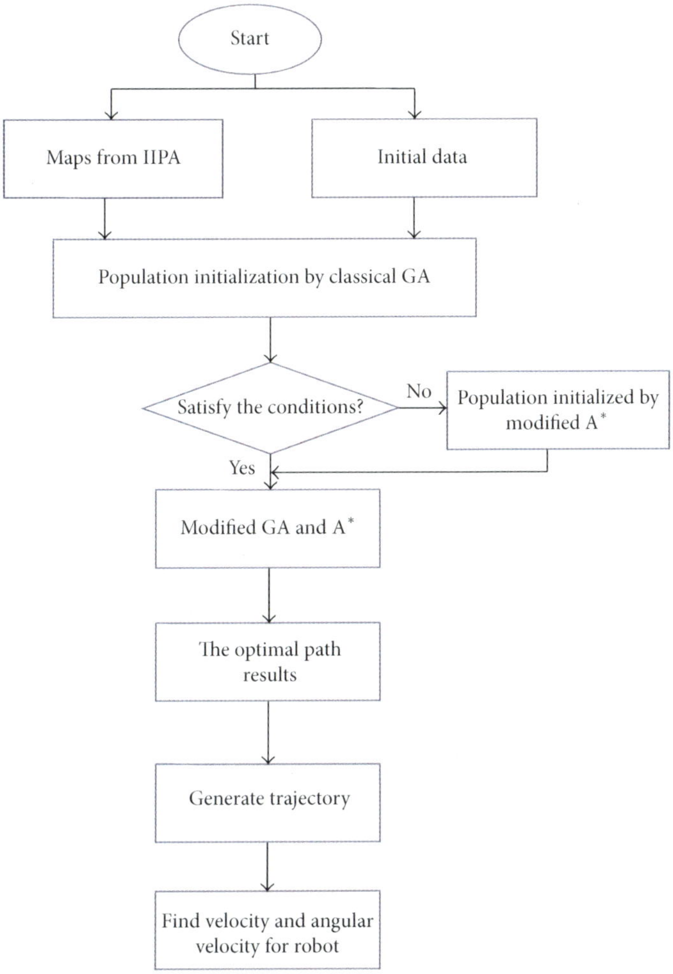

FIGURE 12: Flow chart of the proposed GA.

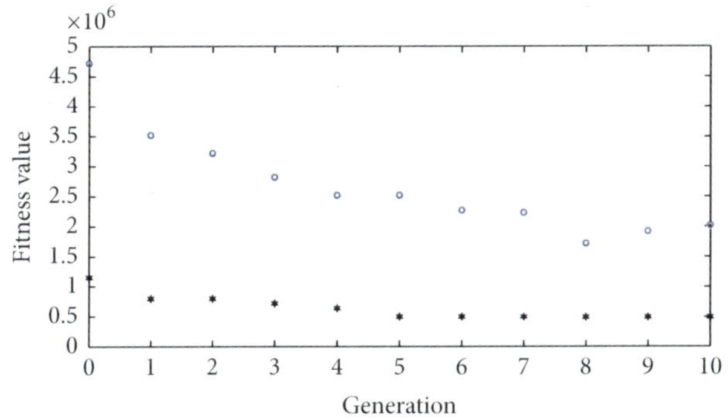

FIGURE 13: Fitness value of proposed approach.

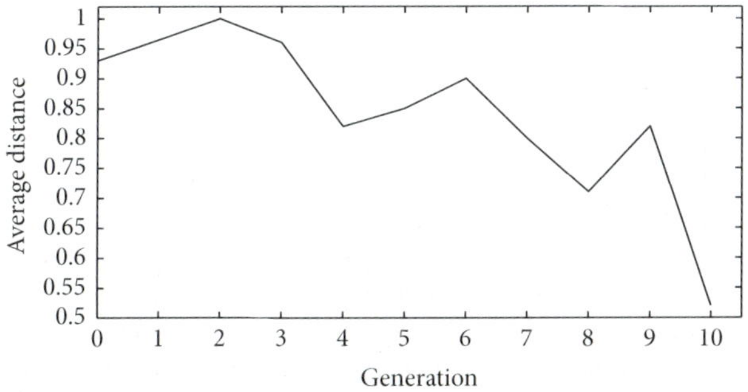

FIGURE 14: Average distance between individuals values.

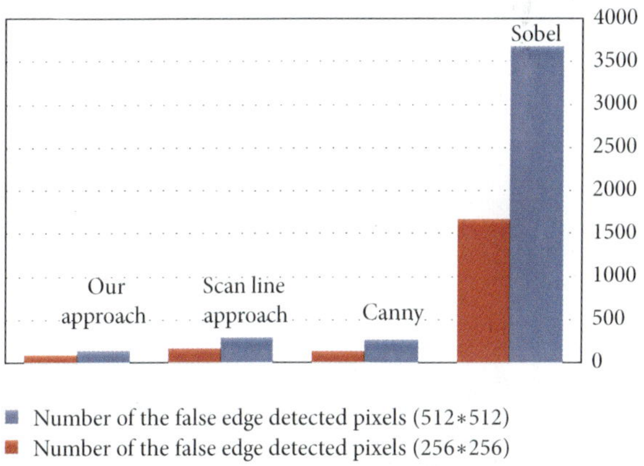

FIGURE 15: The number of false edge pixels.

5. Conclusion and Future Work

In this paper, a strategy that takes the advantages of using artificial intelligent techniques has been proposed. The equalization of the input images has been implemented to get more details from the captured images; then, the fuzzy edge detection was optimized by the bacterial algorithm to reduce

FIGURE 16: The tested image with noise.

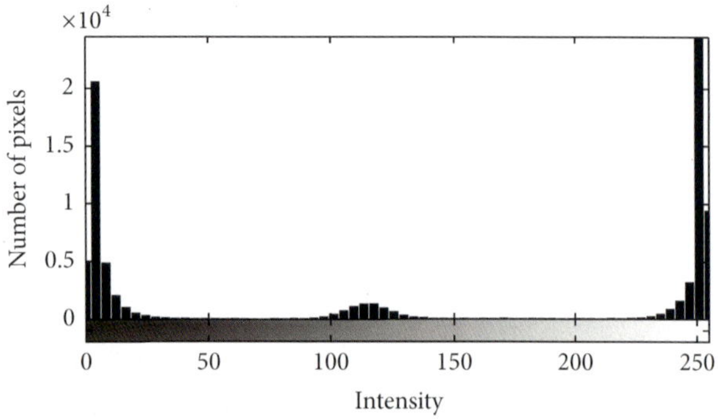

FIGURE 17: The histogram without equalization.

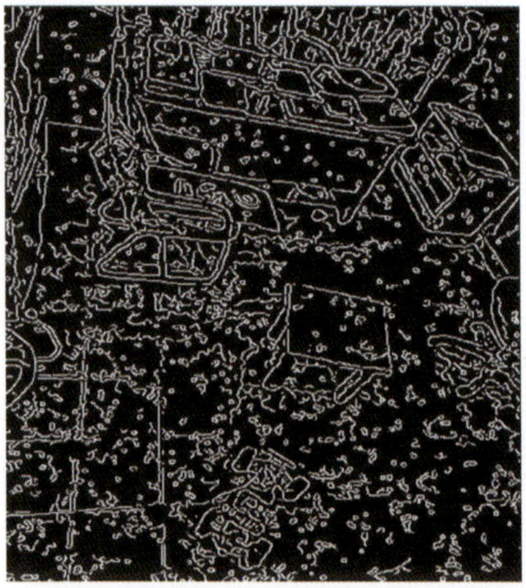

FIGURE 18: The result of Canny on noised image.

the computation time for the process. The algorithm has the ability to classify each pixel in the digital image as edge or not. Then, the results from this stage were used to construct an environmental and local 2D-map. This map is fed to the next stage that contains GA and search $\mathbf{A}^*$ algorithm to find the optimal trajectory between the robot initial position and the destination position. In addition, the

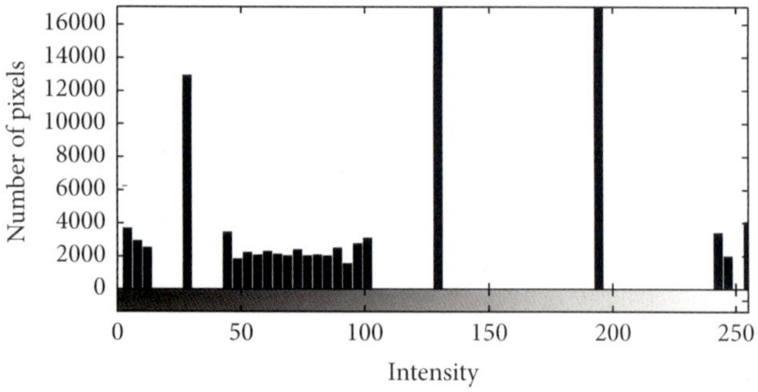

FIGURE 19: The histogram with equalization.

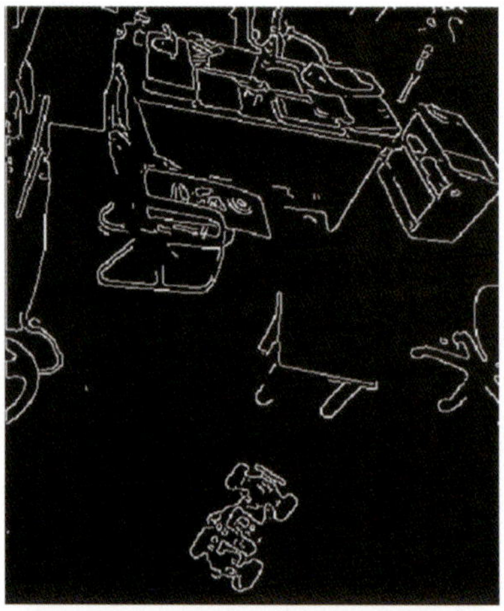

FIGURE 20: The result of proposed approach on noised image.

FIGURE 21: The tested image.

presented approach makes the robot capable of estimating its velocity based on the trajectory. It was noticeably shown that our approach is capable of detecting the edges more precisely along with reducing the noise in an image. Moreover, in terms of computing performance, the proposed method was better than the classical methods and the computational time was reduced. As part of future work, this method can be applied to optimize the speeded-up robust features technique (SURF)

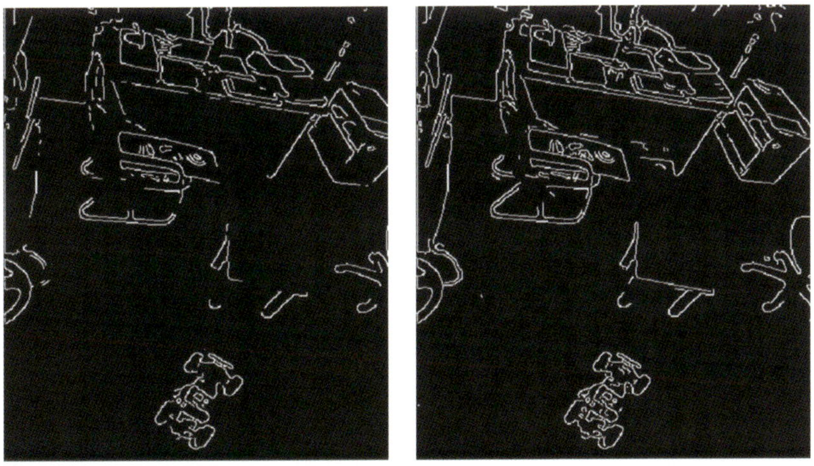

(a) (b)

FIGURE 22: Comparison results for third tested image.

FIGURE 23: Noisy tested image.

(a) (b)

FIGURE 24: Comparison results for noised tested image.

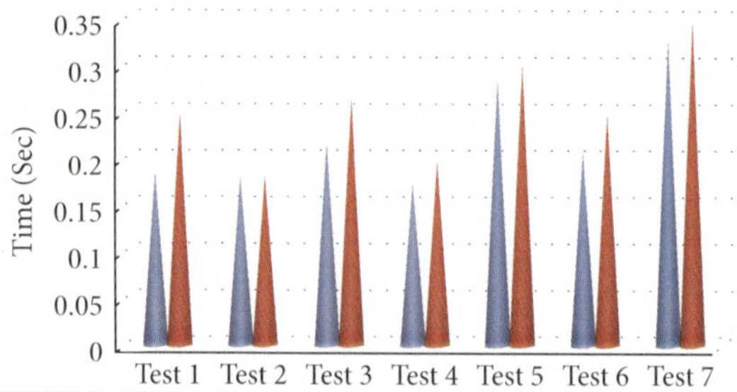

	Test 1	Test 2	Test 3	Test 4	Test 5	Test 6	Test 7
Our approach	0.187	0.175	0.221	0.173	0.286	0.209	0.329
Reference method	0.25	0.18	0.263	0.197	0.302	0.249	0.349

FIGURE 25: Comparison of computation time.

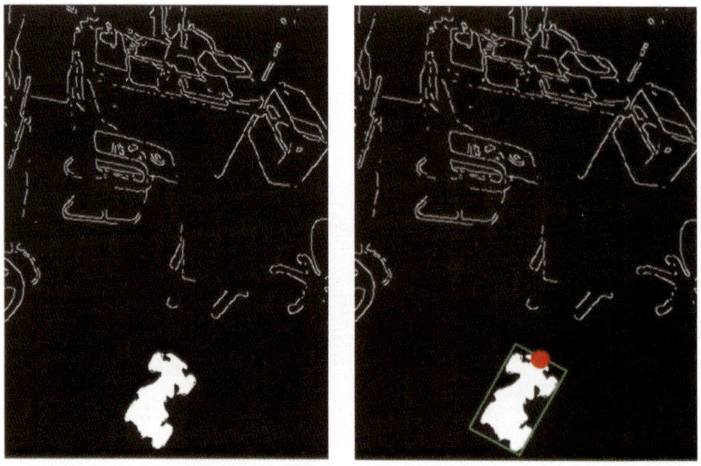

(a) (b)

FIGURE 26: Localize the target robot in the image.

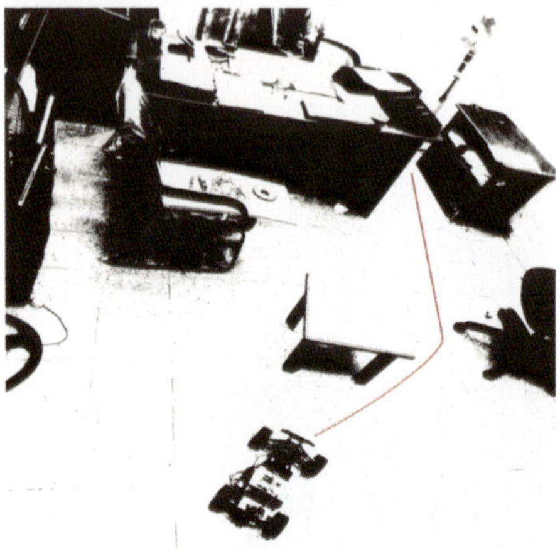

FIGURE 27: Trajectory path planning.

algorithm, that is, used for object detection and recognition. Moreover, the performance of the proposed algorithm can be studied in more complicated applications such as 3D-map reconstruction.

FIGURE 28: The second tested scenario.

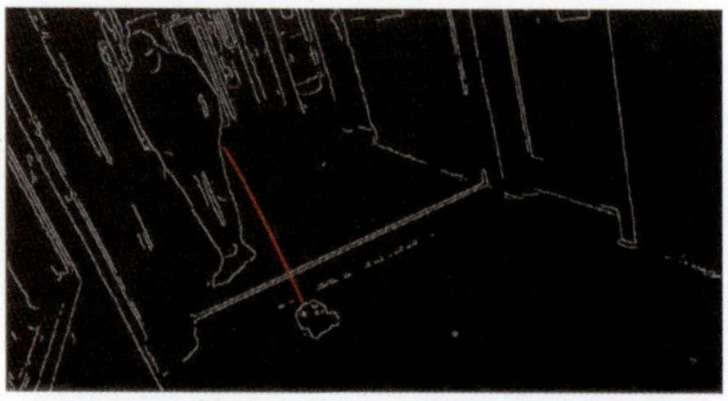

FIGURE 29: Trajectory path planning.

Conflicts of Interest

The authors declare that there are no conflicts of interest regarding the publication of this article.

References

[1] M. Prado, A. Simon, and F. Ezquerro, "Velocity, acceleration and deceleration bounds for a time-optimal planner of a wheeled mobile robot," *Robotica*, vol. 20, no. 2, pp. 181–193, 2002.

[2] M. Prado, A. Simon, E. Carabias, A. Perez, and F. Ezquerro, "Optimal Velocity Planning of Wheeled Mobile Robots on Specific Paths in Static and Dynamic Environments," *Journal of Robotic Systems*, vol. 20, no. 12, pp. 737–754, 2003.

[3] Y. Wang, L. Han, M. Li, Q. Wang, J. Zhou, and M. Cartmell, "A real-time path planning approach without the computation of Cspace obstacles," *Robotica*, vol. 22, no. 2, pp. 173–187, 2004.

[4] D. Ferguson, M. Likhachev, and A. Stentz, "A guide to heuristic-based path planning," in *Proceedings of the International Workshop on Planning under Uncertainty for Autonomous Systems, International Conference on Automated Planning and Scheduling (ICAPS)*, June 2005.

[5] Z. Jia, A. Saxena, and T. Chen, "Robotic Object Detection: Learning to Improve the Classifiers Using Sparse Graphs for Path Planning," in *Proceedings of the Twenty-Second International Joint Conference on Artificial Intelligence*, pp. 2072–2078, 2011.

[6] J. Bao, H. Tang, and A. Song, "Combining vision learning and interaction for mobile robot path planning," *International Journal of Advanced Robotic Systems*, vol. 9, no. 102, 2012.

[7] A. Mohammed, L. Wang, and R. X. Gao, "Integrated image processing and path planning for robotic sketching," in *Proceedings of the 8th CIRP International Conference on Intelligent Computation in Manufacturing Engineering, ICME 2012*, pp. 199–204, Italy, July 2012.

[8] R. Al-Jarrah and M. Al-Jarrah, "Visual fuzzy logic path planning controller for mobile robots," *International Journal of Enhanced Research in Science Technology & Engineering*, vol. 3, 2014.

[9] A. Geus, M. H. Stoppa, and S. Silva, "PathFinder: An autonomous mobile robot guided by computer vision," *International Conference for Artificial Intelligence*, pp. 55–60, 2015.

[10] R. Wang, L. Gao, S. Yang, and Y. Liu, "An edge detection method by combining fuzzy logic and neural networks," in *Proceeding of the IEEE 4th International Conference on Machine Learning and Cybernetics*, vol. 7, pp. 4539–4543, 2005.

[11] R. Gonzalez and R. E. Wood, *Digital Image Processing*, Prentice Hall, Upper Saddle River, NJ, USA, 3rd edition, 2007.

[12] P. S. Venugopala, H. Sarojadevi, A. Ankitha, and N. Niranjan, "An approach to improvise canny edge detection using morphological filters," *International Journal of Computer Applications*, vol. 116, no. 9, pp. 38–42, 2015.

[13] Lalrinawma and R. Kaur, "Edge detection of an image using an improved canny algorithm: a review," *International Journal of Advanced Research in Computer and Communication Engineering*, vol. 5, no. 8, pp. 587–590, 2016.

[14] B. Debbarma and D. Ghoshal, "A modified canny edge detection algorithm with variable sigma," *International Journal of Emerging Trends in Electrical and Electronics (IJETEE)*, vol. 1, no. 2, 2013.

[15] S. M. Hasan and K. Kon, "Depth edge detection by image-based smoothing and morphological operations," *Journal of Computational Design and Engineering*, vol. 3, no. 3, pp. 191–197, 2016.

[16] T. Y. Kim and J. H. Han, "A Fuzzy Approach to Edge Detection and Representation," in *Proceedings of the 6th IEEE International Conference on Fuzzy Systems*, pp. 69–74, 1997.

[17] C. Gonzalez, J. R. Castro, P. Melin, and O. Castillo, "An edge detection method based on generalized type-2 fuzzy logic," in *Proceedings of the 2013 IEEE Symposium on Advances in Type-2 Fuzzy Logic Systems, T2FUZZ 2013 - 2013 IEEE Symposium Series on Computational Intelligence, SSCI 2013*, pp. 39–44, Singapore, April 2013.

[18] P. Melin, C. I. Gonzalez, I. R. Castro, O. Mendez, and O. Castillo, "Edge detection method for image processing based on generalized type-2 fuzzy logic," *IEEE Transactions on Fuzzy Systems*, vol. 22, no. 6, pp. 1515–1525, 2014.

[19] I. Haq, S. Anwar, K. Shah, M. T. Khan, and S. A. Shah, "Fuzzy logic based edge detection in smooth and noisy clinical images," *PLoS ONE*, vol. 10, no. 9, Article ID e0138712, 2015.

[20] N. Peric, "Fuzzy logic and fuzzy set theory based edge detection algorithm," *Serbian Journal of Electrical Engineering*, vol. 12, no. 1, pp. 109–116, 2015.

[21] C. I. Gonzalez, P. Melin, J. R. Castro, O. Mendoza, and O. Castillo, "An improved sobel edge detection method based on generalized type-2 fuzzy logic," *Soft Computing*, vol. 20, no. 2, pp. 773–784, 2016.

[22] X. Zhang, Y. Zhao, N. Deng, and K. Guo, "Dynamic Path Planning Algorithm for a Mobile Robot Based on Visible Space and an Improved Genetic Algorithm," *International Journal of Advanced Robotic Systems*, vol. 13, no. 3, 2016.

[23] A. K. Guruji, H. Agarwal, and D. K. Parsediya, "Time-Efficient A* Algorithm for Robot Path Planning," in *the 3rd International Conference on Innovations in Automation and Mechatronics Engineering ICIAME 2016*, vol. 23, pp. 144–149, 2016.

[24] R. Mao and X. Ma, "Research on Path Planning Method of Coal Mine Robot to Avoid Obstacle in Gas Distribution Area," *Journal of Robotics*, vol. 2016, Article ID 4212076, 2016.

[25] T. Lv, C. Zhao, and J. Bao, "A Global Path Planning Algorithm Based on Bidirectional SVGA," *Journal of Robotics*, vol. 2017, Article ID 8796531, 11 pages, 2017.

[26] R. Al-Jarrah and H. Roth, "Towards a heterogeneous navigation team of aerial-ground robots based on fuzzy image processing," *International Journal of Materials Science and Engineering*, vol. 3, no. 1, pp. 60–65, 2015.

[27] I. Bloch, "Fuzzy sets for image processing and understanding," *Fuzzy Sets and Systems*, vol. 281, pp. 280–291, 2015.

[28] L. A. Zadeh, "Fuzzy sets," *Information and Control*, vol. 8, no. 3, pp. 338–353, 1965.

[29] N. E. Nawa and T. Furuhashi, "Bacterial Evolutionary Algorithm for Fuzzy System Design," in *IEEE International Conference on Systems, Man, and Cybernetics*, pp. 2424–2429, 1998.

[30] J. Botzheim, C. Cabrita, L. T. Kóczy, and A. E. Ruano, "Fuzzy rule extraction by bacterial memetic algorithms," *International Journal of Intelligent Systems*, vol. 24, no. 3, pp. 312–339, 2009.

[31] M. Brezak and I. Petrović, "Time-optimal trajectory planning along predefined path for mobile robots with velocity and acceleration constraints," in *Proceedings of the 2011 IEEE/ASME International Conference on Advanced Intelligent Mechatronics, AIM 2011*, pp. 942–947, Hungary, July 2011.

Dynamic Path Planning of Unknown Environment based on Deep Reinforcement Learning

Xiaoyun Lei, Zhian Zhang, and Peifang Dong

Key Laboratory of Intelligent Ammunition Technology, School of Mechanical Engineering, Nanjing University of Science and Technology, Nanjing 210094, China

Correspondence should be addressed to Zhian Zhang; zzayoyo@163.com

Academic Editor: Gordon R. Pennock

Dynamic path planning of unknown environment has always been a challenge for mobile robots. In this paper, we apply double Q-network (DDQN) deep reinforcement learning proposed by DeepMind in 2016 to dynamic path planning of unknown environment. The reward and punishment function and the training method are designed for the instability of the training stage and the sparsity of the environment state space. In different training stages, we dynamically adjust the starting position and target position. With the updating of neural network and the increase of greedy rule probability, the local space searched by agent is expanded. Pygame module in PYTHON is used to establish dynamic environments. Considering lidar signal and local target position as the inputs, convolutional neural networks (CNNs) are used to generalize the environmental state. Q-learning algorithm enhances the ability of the dynamic obstacle avoidance and local planning of the agents in environment. The results show that, after training in different dynamic environments and testing in a new environment, the agent is able to reach the local target position successfully in unknown dynamic environment.

1. Introduction

Since deep reinforcement learning [1] was first proposed in 2013 formally, tremendous progress has been made in the field of artificial intelligence [2]. Deep Q-network agent was demonstrated to be able to surpass the performance of all previous algorithms and achieve a level comparable to that of a professional human games tester in Atari 2600 games [3–7]. AlphaGo zero [8, 9] has defeated all previous AlphaGo versions by self-play without using any human chess spectrum. An agent can be trained to play FPS game receiving only pixels and game score as inputs [10]. The aforementioned examples fully demonstrate the great potential in autonomous decision-making field after the reinforcement learning of neural network solving the problem of the curse of dimensionality. In [11, 12], deep reinforcement learning has been applied to autonomous navigation based on the inputs of visual information, which has achieved remarkable success. In [11], Piotr Mirowski et al. highlighted the utility of un/self-supervised auxiliary objectives, namely, depth prediction and loop closure, in providing richer training signals that bootstrap learning and enhance data efficiency. The authors analyze the agent behavior in static mazes feature complex geometry, random start position and orientation, and dynamic goal locations. Their results show that their approach enable the agent navigate within large and visually rich environments that include frequently changing start and goal locations, but the maze layout itself is static. The authors did not test the algorithm in environments with moving obstacles. If there are moving obstacles in environments, this means the images collected by cameras may be unknown for each episode. In [12], Yuke Zhu et al. try to find the minimum length sequence of actions that move an agent from its current location to a target that is specified by an RGB image. To solve the problem of a lack of generalization, i.e., the network should be retrained for new targets, they specify the task objective (i.e., navigation destination) as inputs to the model and addresses problems by introducing shared Siamese layers to the network. But as it is mentioned in the paper, in the network architecture of deep Siamese actor-critic model, the ResNet-50 layers are pretrained on ImageNet and fixed during training. This means that they have to collect a large number of different target and scene images with a constrained background to pretrain the ImageNet before training the navigation model, indicating that the generalization ability is still conditioned to the information of maps in advance.

In this paper, we present a novel path planning algorithm and solve the generalization problem by means of local path planning with deep reinforcement learning DDQN based on lidar sensor information. In the aspect of the recent deep reinforcement learning models, the original training mode results in a large number of samples which are moving states in the free zone in the pool, and the lack of trial-and-error punishment samples and target reward samples ultimately leads to algorithm disconvergence. So, we constrain the starting position and target position by randomly setting target position in the area that is not occupied by the obstacles to expand the state space distribution of the pool of sample.

To evaluate our algorithm, we use TensorFlow to build the DDQN training frameworks for simulation and demonstrate the approach in real world. In simulations, the agent is trained in a lower-level and intermediate dynamic environment. The starting point and target point are randomly generated to ensure diversity and complexity of local environment, and the test environment is a high-level dynamic map. We show details of the agent's performance in an unseen dynamic map in the real world.

2. Deep Reinforcement Learning with DDQN Algorithm

The conventional Q-learning algorithm [1] cannot effectively plan a path in random dynamic environment because of the lack of generalization ability and a large Q table. To solve the problem of the curse of dimensionality in high dimensional state space, the optimal action value function Q in Q-learning can be parameterized by an approximate value function.

$$Q(s,a;\theta) \approx Q^*(s,a) \tag{1}$$

where θ is the Q-network parameter. We approximate the value of Q in a definite environment state by function equation, which is mainly linear approximation; thus, there is no need to build a large Q table to determine the corresponding Q values for different state-actions. Neural network is used to approximate the linear function which can obtain nonlinear approximation with generalization ability. (s, a) is regarded as the input of the neural network and the output is the value of Q, where θ is the weight of Q-network. Q table is replaced by Q-network. The training process is to constantly adjust the network weights to reduce the bias of the output of Q-network and the target value of Q. Assuming that the target value of Q is denoted by y, thus the loss function of Q-network is yielded:

$$L_i(\theta_i) = E_{s,a}\left[(y_i - Q(s,a;\theta_i))^2\right] \tag{2}$$

where iteration i is the current iteration times and (s, a) is a pair of state-action.

The update mode of the value of Q is similar to that of Q-learning; that is,

$$y_i = E_{s' \sim \varepsilon}\left[r + \gamma \max_{a'} Q(s', a'; \theta_{i-1}) \mid s, a\right] \tag{3}$$

where r is the current reward and γ is a discount factor.

Stochastic gradient descent algorithm is adopted to train the neural network. According to the back propagation of the derivative value of loss function, the network weights are constantly adjusted, resulting in the network output approaching the target value of Q. The network gradient can be derived according to (2) and (3).

$$\begin{aligned} \nabla_{\theta_i} L_i(\theta_i) &= E\left[\left(r + \gamma \max_{a'} Q(s', a'; \theta_{i-1}) - Q(s,a;\theta_i)\right)\right. \\ &\quad \left.\cdot \nabla_{\theta_i} Q(s,a;\theta_i)\right] \end{aligned} \tag{4}$$

Equation (4) indicates that the updates of neural network and Q-learning are simultaneous. The Q-network of the current iteration epoch is updated with the target value of Q updated from the previous iteration epoch.

Reinforcement learning is known to be unstable or even to diverge when a nonlinear function approximator such as a neural network is used to represent the action value (also known as Q) function [13]. This instability has several causes: the correlations present in the sequence of observations and the correlations between the action values (Q) and the target values. To address these instabilities, DeepMind presents a biologically inspired mechanism termed experience replay [14–16] that randomizes over the data. To perform experience replay, DeepMind stores the agent's experiences, environment state, action, and reward at each time-step t in a dataset. The pool of data samples is extended with running e-greedy policy to search the environment. During the learning, train the network with random data from the pool of stored samples, thereby removing correlations in the observation sequence and improving the stability of algorithm.

In 2015, DeepMind improved the original algorithm in the paper "Human-Level Control through Deep Reinforcement Learning" published in *Nature*. They added a target Q-network, which is delayed updating in comparison with the predicted Q-network, to update the target values of Q, thereby restraining a big bias of Q resulting from a fast dynamic updating of the target Q-network. The improved loss function is

$$L_i(\theta_i) = E\left[\left(r + \gamma \max_{a'} Q(s', a'; \theta_i^-) - Q(s,a;\theta_i)\right)^2\right] \tag{5}$$

where θ_i are the parameters of the Q-network at iteration i and θ_i^- are the network parameters used to compute the target at iteration i. The target network parameters θ_i^- are only updated with the Q-network parameters θ_i every C steps and are held fixed between individual updates, thus ensuring that the update of target network is delayed and the estimate of Q is more accurate.

The update mode of Q-learning algorithm results in a problem of overestimate action values. The algorithm estimates the value of a certain state too optimistically, consequently causing that the Q value of subprime action is greater than that of the optimal action, thereby changing the optimal action and affecting the accuracy of the algorithm. To address the overestimate problem, in 2016, DeepMind presented an improved algorithm in the paper "Deep Reinforcement Learning with Double Q-Learning", namely, Double Q-network algorithm [17], which selected an action by maximizing the value of the predicted Q-network and updating the predicted Q-network with the selected action value in Q-network instead of directly updating the predicted Q-network with the maximum value of target Q-network.

$$y_i = E\left[r + \gamma Q\left(s', \arg\max\left(Q\left(s', a'; \theta_i\right)\right); \theta_i^-\right) \mid s, a\right] \quad (6)$$

The loss function of the improved algorithm is

$$L_i(\theta_i) = E\left[\left(r + \gamma Q\left(s', \arg\max\left(Q\left(s', a'; \theta_i\right)\right); \theta_i^-\right) - Q(s, a; \theta_i)\right)^2\right] \quad (7)$$

The framework illustration of DDQN is shown in Figure 1.

3. Local Path Planning with DDQN Algorithm

In this paper, we achieve local path planning with deep reinforcement learning DDQN. Lidar is used to detect the environment information over 360 degrees. The sensor range is regarded as observation window. The accessible point nearest to the global path at the edge of the observation window is considered as the local target point of the local path planning. The network receives lidar dot matrix information and local target point coordinates as the inputs and outputs the direction of movement.

Considering the computation burden and actual navigation effect, we set angle resolution to be 1 degree and range limit to be 2 meters so each observation consists of 360 points indicating the distance to obstacles within a two-meter circle around the robot. The local target point is the intersection point of the observation window circular arc and the global path. If there were several intersection points, the optimal point would be chosen by heuristic evaluation.

As to the lidar dot matrix information, the angle and distance can be denoted by $[angle, distance]$. We set a rule that the angles of the lidar points increases clockwise relative to the front of the mobile robot. 360 points result in 720 inputs of data. The relative coordinate of local target point and the current center coordinate of the mobile robot body ($[\Delta x, \Delta y]$) are also regarded as the input. To achieve the network output weights of the relative target points in training and conveniently design the convolutional network, we make 40 copies of the relative target points as the inputs; namely, the total inputs are 800 datasets.

The outputs are fixed-length omnidirectional movements of eight directions, where the movement length is 10 cm, and the directions are forward, backward, left, right, left front, left rear, right front, and right rear, which are denoted by 1 to 8 in order as follows.

$$\boldsymbol{a} = [a_1, a_2, a_3, a_4, a_5, a_6, a_7, a_8] \quad (8)$$

The design of positive activation function mainly accounts for obstacle avoidance and approaching the target point. The shortest movement path which satisfied the two conditions is the most effective. The rewards and punishments function is designed as

$$r = \begin{cases} 1\,(p(x_1, y_1) = g(x, y)) \\ -0.01 \begin{pmatrix} p(x_1, y_1) \neq g(x, y) \\ p(x_1, y_1) \neq o(x, y) \end{pmatrix} \\ -1\,(p(x_1, y_1) = o(x, y)) \end{cases} \quad (9)$$

where p is the current position, g is a local target point, and o represents an obstacle with expansion processing. Equation (9) indicates that the reword would be -1 if the agent encountered an obstacle; if the agent arrived at the target point, it would get a reward +1; in the other states, each step of movement would cost -0.01. During the training, in order to continuously improve the cumulative reward, the agent gradually avoids the obstacle and reaches the target point in a short path by the method of the trial-and-error learning.

Each lidar detected point receives continuous measurement values changing from 0 cm to 200 cm, which means that the state space tends towards infinitude and it is impossible to represent all of the states with Q table. With DDQN, the generalization ability of neural network makes it possible that the network is able to approach all states after training; therefore, when the environment changed, the agent could plan a proper path and arrive at the target position according to the weights of the network.

To ensure the deep reinforcement learning training converging normally, the pool should be large enough to store state-action of each time-step and keep the training samples of neural network be independent identical distribution; besides, the environment punishment and reward should reach to a certain proportion. If the sample space was too sparse, namely, the main states were random movements in free space, it was difficult to achieve a stable training effect. As to the instability of DDQN in training and the reward sparsity of the state space, the starting point is randomly set in the circle with the target point as the center and the radius L. The initial value of L is small, thereby increasing the probability that the agent reaches the target point from the starting point in the random exploration and ensuring a positive incentive in the sample space. The value of L gradually increases with the update of neural network and the increase of greedy law probability. The local space searched by agent is expanded as follows.

$$L = \begin{cases} i_{\min} & (n \leq N_1) \\ i_{\min} + \sqrt{\dfrac{(n - N_1)}{m}} & (N_1 < n < N_2) \\ i_{\max} & (n \geq N_2) \end{cases} \quad (10)$$

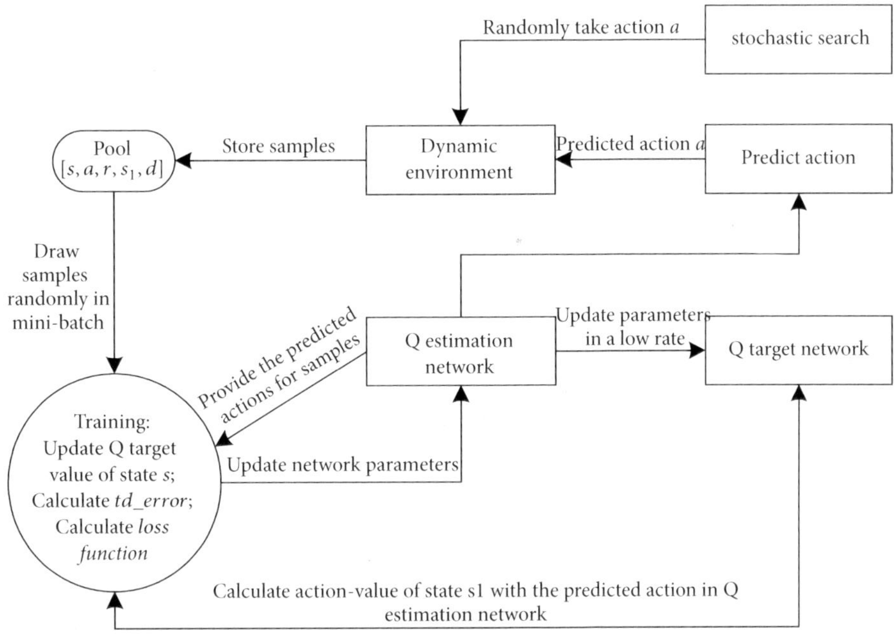

FIGURE 1: The framework illustration of DDQN.

where n is the current iteration time-step; i_{min} is the initial value of L; i_{max} is the maximum value of L; m is the search speed in space; N_1 and N_2 are thresholds of iteration times which need to be adjusted according to training parameters.

The termination of each episode is to achieve a fixed number of moving steps instead of directly terminating the current training episode when encountering obstacles or reaching the target point. The original training mode results in a large number of samples which are moving states in the free zone in the pool, and the lack of trial-and-error punishment samples and target reward samples ultimately leads to algorithm disconvergence.

4. The Neural Network in DDQN

Neural network is a primary method for reinforcement learning to enhance the ability of generalization. The optimization design of the neural network architecture is able to reduce overfitting and improve the prediction accuracy with high training speed and low computational cost.

Considering 800 inputs, if we directly adopted fully connected layers, the number of parameters needed to be trained would increase exponentially with the increase of the layers, resulting in a heavy computation burden and overfitting. Convolutional neural networks (CNNs) have made breakthrough progress in the field of image recognition in recent years. CNNs are featured with a unique network processing mode which is characteristic of local connection, pooling, sharing the weights, etc. The unique mode effectively reduces the computational cost, computational complexity, and the number of the trained parameters. With the method of CNNs, the image model in the translation, scaling, distortion, and other transformations is able to be invariant to a certain degree, thereby improving the robustness of system to failures. Based on these superior features, CNNs surpass fully connected neural network in information processing task where the data are relevant to each other [18].

The framework of CNNs is shown in Figure 2. Generally, the CNNs consist of input layer, convolutional layer, pooling layer, fully connected layer, and output layer. convolutional layer adopts local connection; namely, each node is connected with some nodes of the upper layer. Four adjacent pixels of input layer form a 2-by-2 local sensory field which is connected with a neuron in convolutional layer. Each local sensory field represents a local feature, and the size of the field can be adjusted artificially.

The next layer of convolutional layer is pooling layer, which can also reduce the matrix size. The spatial invariant feature is obtained by reducing the resolution of the feature surface [19]. The number of feature maps in the pooling layer is equal to the number of convolutional layers, and each neuron performs a pooling operation on the local sensory field. Pooling is suitable for separating the sparse features.

A fully connected layer is added after one or more convolutional layers and pooling layers. The fully connected layer integrates the partial information involved in category distinguishing in convolutional layers or pooling layers [20]. Besides, rectified linear unit (ReLU) is a recent used activation functions in CNNs, which can improve the performance of a CNN. The output value of the fully connected layer is directly transferred to the output layer. For deep reinforcement learning, the output layer is the action space.

The lidar data between adjacent points reflects the distribution of obstacles and the width of the free zone. The convolutional network can effectively extract the information

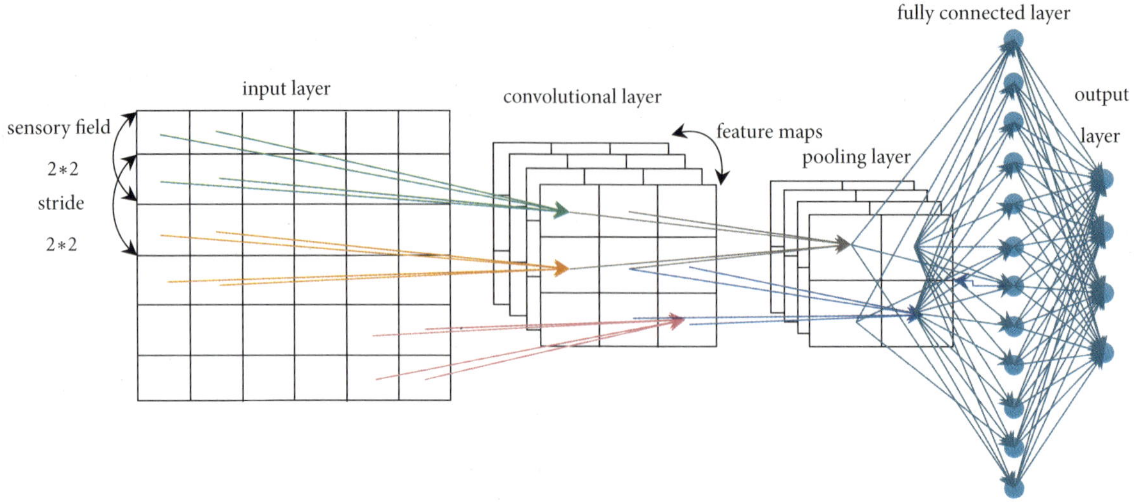

FIGURE 2: The architecture of CNNs.

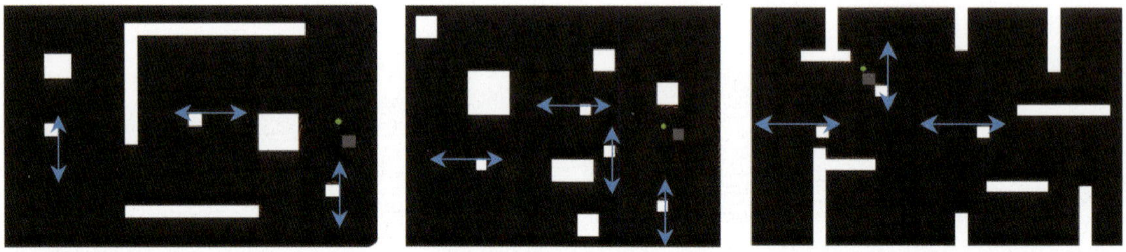

(a) Low level environment (b) Intermediate environment (c) High-level environment

FIGURE 3: Dynamic environments.

of these characteristics and reduce the network parameters greatly. Therefore, the CNNs are used to train the path planning of local dynamic environment.

In this paper, the architecture of Q-network consists of three convolutional layers and a fully connected layer. The input layer is a three-dimensional matrix with the size of 20×20×2 formed by a vector with 800 elements, where the data in the third dimension represents the angle and distance of a lidar point. The size of the input layer is 20×20×2. According to the size of the input layer, we design the first convolutional layer, of which the size of the receptive field is 2-by-2, the stride is 2-by-2, and the number of feature maps is set to be 16. Thus, the size of the output layer is 10×10×16. The kernel size of the second convolutional layer is 2-by-2; the stride is 2-by-2; and the number of the characteristic plane is 32. The size of the output of this convolutional layer is 5×5×32. The kernel size of the third convolutional layer is 5-by-5; the stride is 1-by-1; and the number of the characteristic plane is 128. The size of the output of the third convolutional layer is 1×1×128. Then a three-dimensional structure is transformed to a one-dimensional vector with 128 elements, which are connected to the fully connected layer, of which the size is 128-by-256. The size of the output layer is the number of the actions, namely, 8. We use ReLU activation function and Adam optimizer.

5. Local Path Planning Simulation with Pygame Module

We use the open source machine learning framework TensorFlow to build the DDQN training framework. Pygame module is to build dynamic environment. As shown in Figure 3, the white area denotes obstacles, with the size of 20-by-20 cm, are moving in a constant speed, and the corresponding blue arrow represents its moving range and directions. The green point denotes a local target position. The gray square object represents the mobile robot (agent). During the training, the local path planning is trained in the lower-level and intermediate environment. The starting point and target point are randomly generated to ensure diversity and complexity of local environment, and the test environment is a high-level dynamic map. Two CNNs of the same architecture are established, which are Q estimation network and Q target network. The network parameters are randomly initialized with normal distribution, of which the mean value is 0.

The training policy is a variable random greedy rule. At the beginning of the training, the pool of experience is updated by the method of random exploitation because of the lack of environmental information. The sample consists of five components, namely, $[s, a, r, s_1, d]$, where s denotes

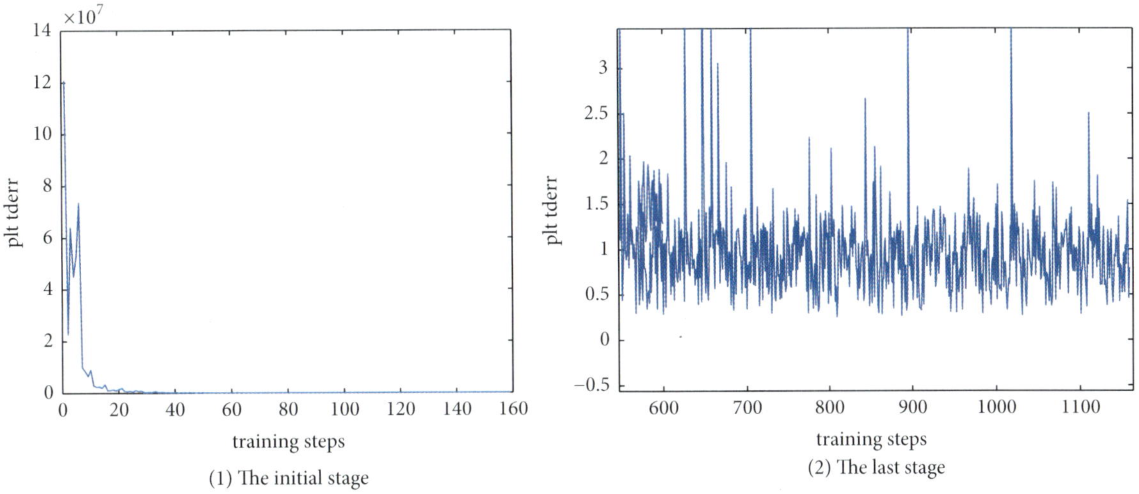

FIGURE 4: **Training curves of the loss function of Q target network.** Each point is the average loss function value achieved per ten epochs. The y-axis denotes the value of loss function and x-axis denotes iteration epoch.

the current state; a denotes the adopted action; s_1 is the state after taking action; R is the reward value obtained for the current state transition; d is a flag representing that whether the current iteration epoch ends. We set the size of the pool to 40000 samples. If the samples stored in the pool reached a certain number, the network was to be trained with the randomly selected samples in the pool. In the first 5000 time-steps of the random exploration, the network parameters are not updated, but the samples of the pool are increased. After the sample size reaches to 5000, the network is trained in every four time-step movements. We randomly draw 32 samples in sequence by means of minibatch to update the Q estimation network and update the Q target network with a low learning rate to make the parameters of Q target network approach the parameters of Q estimation network, thereby ensuring a stable learning of Q target network. If the sample size of the pool was up to the upper limit, the earliest sample would be excluded from the pool in the policy of first-in first-out after a new sample is added, thereby ensuring a continuous updating of the pool.

According to the actual experimental effect, if the position of agent and the target point were completely randomly set at first, there would be a large probability that the distance between the agent and target point was too large, resulting in that the agent was unable to reach the target point in fixed steps with random exploration. Consequently, to ensure the target point being detected by the agent, we set the initial distance between the agent and target point randomly in a range of L and gradually increase L with exploration and training. This process is also a process of constantly changing the environmental states which leads to a more extensive distribution of samples.

During the training process, the values of loss function of Q estimation network and Q target network decrease continuously. Figure 4 shows the training curves of the loss function of Q target network. Each point is the average loss function value achieved per ten epochs. Figure 4 (1) indicates that, in initial stage of the training, the magnitude of loss function value is 10^7. while, as we got from our data, after training for 1000 epochs, actually, the value dropped to about 10,000 (seen from the figure, the curve tends to be close to x-axis); after 7000 epochs, as we can see from Figure 4 (2) that it dropped to less than 1. With the whole process completed, the loss function value is less than 0.25, and the Q-network converges successfully. The cumulative reward of each iteration is gradually increased; that is, in the process of exploration, the agent gradually learns to take an action which can increase the reward. Figure 5 shows the average cumulative reward curves. Each point is the average cumulative reward achieved per hundred epochs. The figure indicates that the average reward gradually increases with the increase of training epochs. After training for 30000 epochs, the reward value tends to be stable and average cumulative reward is greater than 80. We set the fixed step times to be 100, which means the agent can avoid obstacles in a dynamic

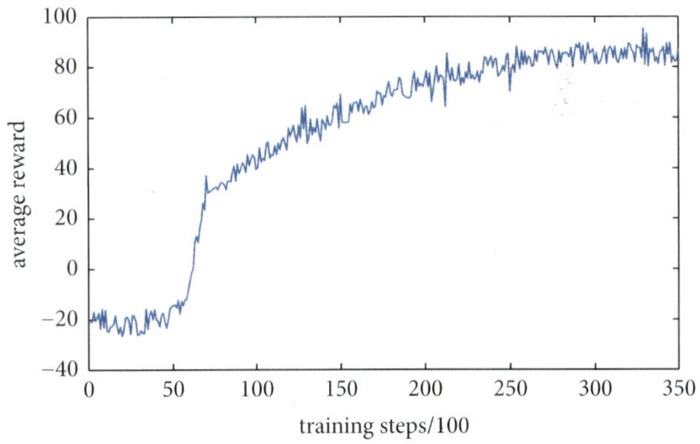

FIGURE 5: **The average cumulative reward curves.** Each point is the average cumulative reward achieved per hundred episodes. The y-axis denotes the average cumulative reward and x-axis denotes iteration epoch.

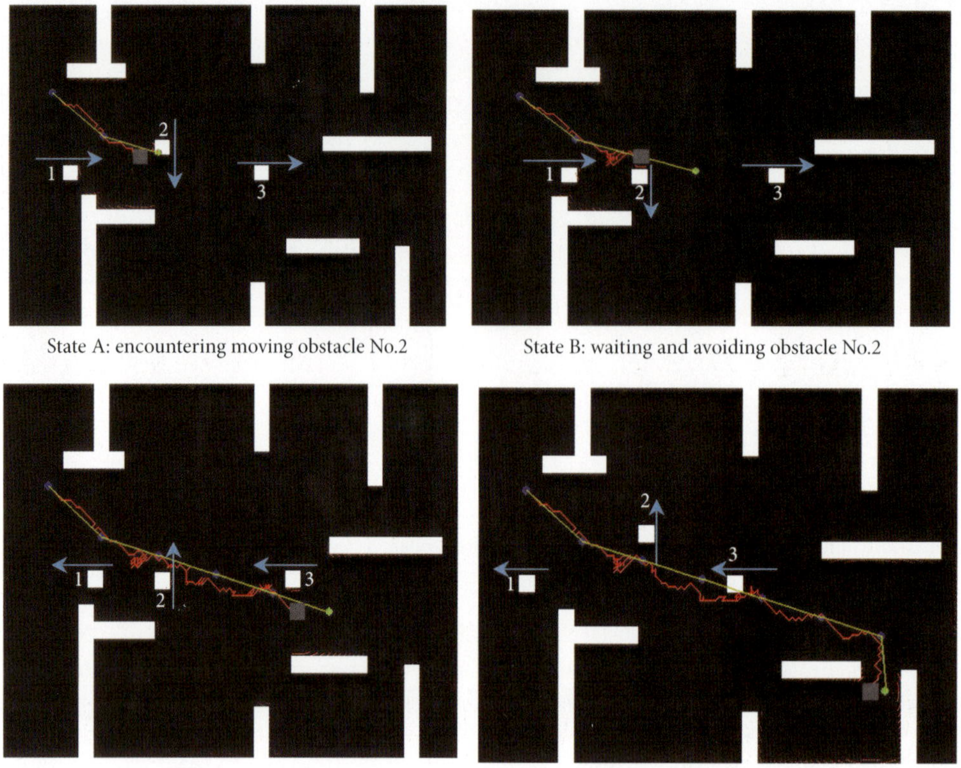

State A: encountering moving obstacle No.2

State B: waiting and avoiding obstacle No.2

State C: avoiding obstacle No. 3 towards down right

State D: arriving at target point

FIGURE 6: Local path planning in a test map.

environment and quickly reach the target point to obtain a continuous reward until the current training epoch ends.

After training for 40000 epochs, Q estimation network and Q target network converge. We store the network parameters and test in an experimental environment map. The test is designed as the following. Assuming a global path, we set several local target points in the lidar searching area regardless of the positions of the dynamic and static obstacles to test the agent's local path planning ability. Figure 6 is a new environmental map which has never tested in training. This map is used to evaluate the generalization ability and the path planning ability of DDQN in a new unknown dynamic environment. There are three free moving obstacles with constant speed and the moving directions are denoted by the blue arrows. In the figure, state A demonstrates that the local path is blocked by dynamic obstacle No. 2; then, the agent waits for the obstacle to move downwards. When the obstacle is out of the path, the agent moves towards upper right. As shown in state B, the agent successfully reaches the third local target. In state C, obstacle No. 3 moves towards the agent. The agent perceives obstacle No. 3 before collision and gets to the sixth local target point by moving toward the bottom right. In state D, the agent reaches the end point and completes the path planning without any accident collision with the obstacles. The whole process demonstrates that the agent after being trained by DDQN is able to perceive the moving obstacles in advance with the knowledge of lidar data in unknown dynamic environment. An intelligent planning method is presented by Q target network which makes each step move towards a higher cumulative reward.

6. Test in Actual Environment with ROS Framework

In order to verify the effect of the algorithm in actual environment, we use an omnidirectional mobile robot based on Mecanum Wheels and achieve autonomous navigation with ROS framework [21–23]. The *move_base package* in ROS provides local path planning algorithms for user, that is, dynamic windows method and track-reckoning method. In this paper, DDQN local path planning algorithm is transplanted into ROS in the form of plug-ins, which is encapsulated in the pure virtual function *base_local_planner* of *nov_core* as subclass, namely, *DDQN_local_planner*. The outputs of the deep reinforcement learning algorithm are the translational movements of the agent in eight directions. Then the movements are mapped to actions of moving forward and turning. The agent firstly changes its direction in situ to the corresponding direction before moving and then moves forward 10 cm with a preset speed.

The path planning of the unknown environment is to navigate without using the prior environment SLAM map. Figure 7 is a local map. On this map, only local information is detected by lidar in real-time, where the gray area is unknown. The agent avoids obstacles with local path planning. The starting position is set to be the current position

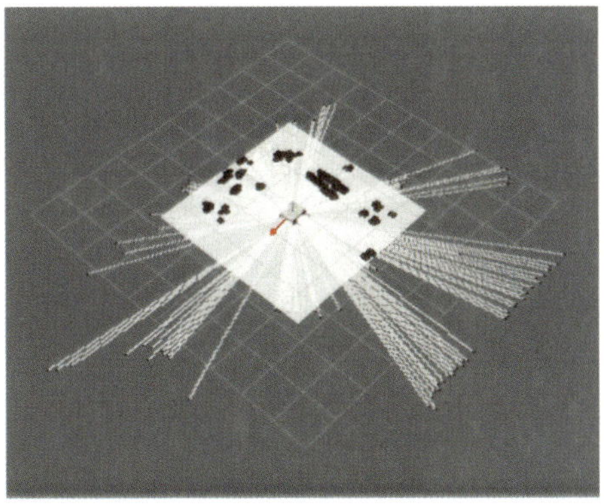

FIGURE 7: A local environment map.

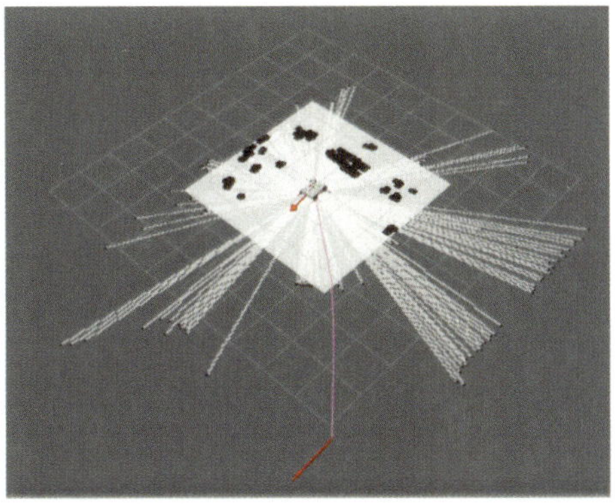

FIGURE 8: The schematic of global path planning in unknown environment.

of agent and the target point is set to be in unknown area. The unknown area between the starting point and the target point is considered as an available area. The schematic of path planning is shown in Figure 8.

The actual layout of the unknown environment is shown in Figure 9. In the figure, the red arrows denote preplanned global paths, which is blocked by obstacle No. 2; thus the agent needs to readjust the path to reach the target position without hitting the obstacles.

The local path planning is completed by the trained DDQN algorithm. The inputs are angles, distance detected by lidar, and local planning points. The output is moving direction. And, meanwhile, the trajectory is smoothed. During the movement, the agent builds map and navigates in real-time. The local path planning effect of the algorithm and generalization ability of the convolutional network are tested in unknown environment. Figure 10 shows the final path. The agent followed the global path but modified its path and successfully avoided the obstacles when it encountered obstacles. A new environmental map was gradually built in the process of moving.

The movement effect of the agent in actual environment is shown in Figure 11. When the agent was in front of the obstacle with a distance of about 30 cm, it changed direction and avoided the obstacle on the left. The agent approached

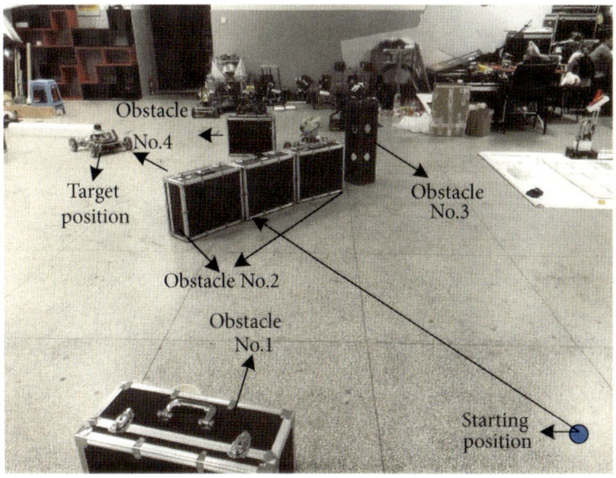

FIGURE 9: The actual layout of the unknown environment.

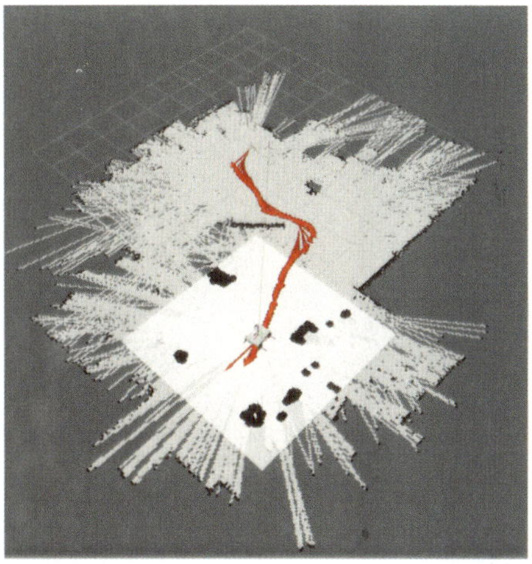

FIGURE 10: Path planning in unknown environment.

the global path again, finally arriving at the target point. The test verifies the feasibility and efficiency of the algorithm in unknown environment.

7. Conclusions

Deep reinforcement learning solves the problem of curse of dimensionality well and is able to process multidimensional inputs. In this paper, we design a specific training method and reward function, which effectively addresses the problem of disconvergence of the algorithm due to the reward sparsity of state space. The experimental results show that DDQN algorithm is capable and flexible for local path planning in unknown dynamic environment with the knowledge of lidar data. Compared with visual navigation tasks, the local lidar information, which can be transplanted into more complex environments without retraining the network parameters, has wider applications and better generalization performance.

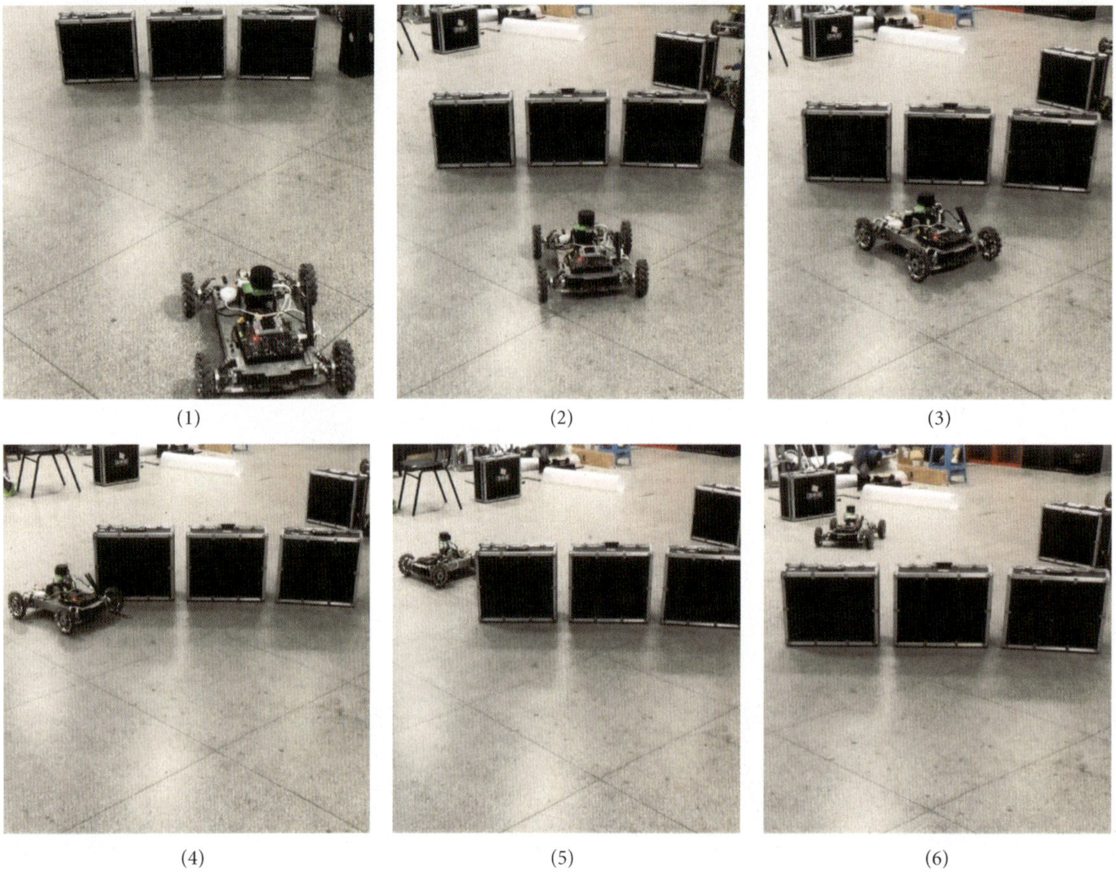

FIGURE 11: The movements of the agent in actual environment.

Conflicts of Interest

The authors declare that there are no conflicts of interest regarding the publication of this paper.

Acknowledgments

This work is supported by National Natural Science Foundation (NNSF) of China under Grant 11472008.

References

[1] R. Sutton and A. Barto, *Reinforcement Learning: An Introduction*, MIT Press, 1998.

[2] S. Legg and M. Hutter, "Universal intelligence: A definition of machine intelligence," *Minds and Machines*, vol. 17, no. 4, pp. 391–444, 2007.

[3] V. Mnih, K. Kavukcuoglu, D. Silver et al., "Playing atari with deep reinforcement learning," https://arxiv.org/abs/1312.5602, 2013.

[4] V. Mnih, K. Kavukcuoglu, D. Silver et al., "Human-level control through deep reinforcement learning," *Nature*, vol. 518, no. 7540, pp. 529–533, 2015.

[5] Y. Bengio, "Learning deep architectures for AI," *Foundations and Trends in Machine Learning*, vol. 2, no. 1, pp. 1–27, 2009.

[6] A. Krizhevsky, I. Sutskever, and G. E. Hinton, "Imagenet classification with deep convolutional neural networks," in *Proceedings of the 26th Annual Conference on Neural Information Processing Systems (NIPS '12)*, pp. 1097–1105, Lake Tahoe, Nev, USA, December 2012.

[7] G. E. Hinton and R. R. Salakhutdinov, "Reducing the dimensionality of data with neural networks," *The American Association for the Advancement of Science: Science*, vol. 313, no. 5786, pp. 504–507, 2006.

[8] D. Silver, J. Schrittwieser, K. Simonyan et al., "Mastering the game of Go without human knowledge," *Nature*, vol. 550, no. 7676, pp. 354–359, 2017.

[9] Z. Tang, K. Shao, D. Zhao, and Y. Zhu, "Recent progress of deep reinforcement learning: from AlphaGo to AlphaGo Zero," *Control Theory & Applications*, vol. 34, no. 12, pp. 1529–1546, 2017.

[10] G. Lample and D. S. Chaplot, "Playing FPS games with deep reinforcement learning," in *Proceedings of the 31st AAAI Conference on Artificial Intelligence, AAAI 2017*, pp. 2140–2146, USA, February 2017.

[11] P. Mirowski, *Learning to navigate in complex environments*, 2016.

[12] Y. Zhu, R. Mottaghi, E. Kolve et al., "Target-driven visual navigation in indoor scenes using deep reinforcement learning," in *Proceedings of the 2017 IEEE International Conference on*

Robotics and Automation, ICRA 2017, pp. 3357–3364, Singapore, June 2017.

[13] J. N. Tsitsiklis and B. Van Roy, "An analysis of temporal-difference learning with function approximation Technical," Tech. Rep. LIDS-P-2322, Laboratory for Information and Decision Systems, Massachusetts Institute of Technology, 1996.

[14] J. L. McClelland, B. L. McNaughton, and R. C. O'Reilly, "Why there are complementary learning systems in the hippocampus and neocortex: insights from the successes and failures of connectionist models of learning and memory," *Psychological Review*, vol. 102, no. 3, pp. 419–457, 1995.

[15] J. O'Neill, B. Pleydell-Bouverie, D. Dupret, and J. Csicsvari, "Play it again: reactivation of waking experience and memory," *Trends in Neurosciences*, vol. 33, no. 5, pp. 220–229, 2010.

[16] L. J. Lin, "Reinforcement learning for robots using neural networks," Tech. Rep. CMU-CS-93-103, Carnegie-Mellon University Pittsburgh PA School of Computer Science, 1993.

[17] H. Van Hasselt, A. Guez, and D. Silver, "Deep Reinforcement Learning with Double Q-Learning," *Association for the Advancement of Artificial Intelligence (AAAI)*, vol. 2, 2016.

[18] Q. Zhang, "Convolutional Neural Networks," in *Proceedings of the 3rd International Conference on Electromechanical Control Technology and Transportation*, pp. 434–439, Chongqing, China, January 2018.

[19] J. Gu, Z. Wang, J. Kuen et al., "Recent advances in convolutional neural networks," *Pattern Recognition*, vol. 77, pp. 354–377, 2018.

[20] T. N. Sainath, A.-R. Mohamed, B. Kingsbury, and B. Ramabhadran, "Deep convolutional neural networks for LVCSR," in *Proceedings of the 38th IEEE International Conference on Acoustics, Speech, and Signal Processing (ICASSP '13)*, pp. 8614–8618, IEEE, Vancouver, Canada, May 2013.

[21] Willow Garage, *Robot Operating System*, Willow Garage, 2015, http://www.ros.org/.

[22] M. Quigley, K. Conley, B. Gerkey et al., "ROS: an open-source Robot Operating System," vol. 3, no. 3.2, ICRA workshop on open source software, May 2009.

[23] M. Cashmore, M. Fox, D. Long et al., "Rosplan: Planning in the robot operating system," in *Proceedings of the 25th International Conference on Automated Planning and Scheduling, (ICAPS '15)*, pp. 333–341, June 2015.

Research on Path Planning Method of Coal Mine Robot to Avoid Obstacle in Gas Distribution Area

Ruiqing Mao[1,2] and Xiliang Ma[1,2]

[1]*Xuzhou Institute of Technology, Xuzhou, Jiangsu 221111, China*
[2]*School of Mechanical and Electrical Engineering, China University of Mining and Technology, Xuzhou, Jiangsu 221116, China*

Correspondence should be addressed to Xiliang Ma; xlma818@sina.com

Academic Editor: Gordon R. Pennock

As the explosion-proof safety level of a coal mine robot has not yet reached the level of intrinsic safety "ia" and it cannot work in a dangerous gas distribution area, therefore, path planning methods for coal mine robot to avoid the dangerous area of gas are necessary. In this paper, to avoid a secondary explosion when the coal mine robot passes through gas hazard zones, a path planning method is proposed with consideration of gas concentration distributions. First, with consideration of gas distribution area and obstacles, MAKLINK method is adopted to describe the working environment network diagram of the coal mine robot. Second, the initial working paths for the coal mine robot are obtained based on Dijkstra algorithm, and then the global optimal working path for the coal mine robot is obtained based on ant colony algorithm. Lastly, experiments are conducted in a roadway after an accident, and results by different path planning methods are compared, which verified the effectiveness of the proposed path planning method.

1. Introduction

Environment exploration (such as detection of methane concentration) is often necessary before coal mining, and a coal mine robot is needed. Similarly, rescue must be carried out after a coal mine accident, and the working environment in the roadway is always very complex; therefore, coal mine robot is also a good choice. In the process of environment detection or rescue, a secondary explosion must be avoided.

In China, the coal mine safe electrical apparatus is Class I, which is strict with the use of mining equipment. Therefore, with regard to coal mine robot, the explosion-proof grade needs to reach the level of intrinsic safety "ia." However, at present, very few apparatuses can reach the level of "ia" [1]. In order to promote the utility of coal mine robot and avoid the danger during the working process, studies on path planning for coal mine robot are of great significance. On the other hand, the existing studies demonstrated that gas in roadways has the regularities of regional distribution and migration [2, 3]. And the authors of this paper have studied the methane concentration prediction method in the previous studies, as seen in [4].

Path planning, which can find out the optimal working path without obstacles, is very important for coal mine robot. Generally, path planning is a nondeterministic polynomial problem with such constraints from much information of environment and obstacles [5]. Artificial Potential Field Method [6], Grid Method [7], and Visibility Graph Method [8] are the common path planning methods. These methods have their own shortages. For example, Artificial Potential Field Method is prone to fall into local optimal point; in complex environment, a surge of storage capacity may arise if Grid Method is adopted; Visibility Graph Method has such shortages as high searching complexity and low efficiency. Pan et al. [9] applied an improved ant colony algorithm to the path planning for mobile robot with peaks on the path considered. Zhang et al. [10] proposed a global path planning method for coal mine rescue robot based on ant colony algorithm. Li et al. [11] studied the global path planning method for mobile robot based on Particle Swarm

Optimization. The complex working environment in roadway after coal mine accident brings about great challenges to the path planning for coal mine robot. And there are few researches on path planning method for coal mine robot with gas concentration distributions considered.

In the paper, to avoid a secondary explosion when the coal mine robot passes through gas hazard zones, path planning method is studied. First, MAKLINK method is adopted to describe the working environment network diagram of the coal mine robot with consideration of gas distribution area and obstacles. Second, with the help of Dijkstra algorithm, the initial working paths for coal mine robot are obtained, and then ant colony algorithm is adopted to optimize the initial working paths, and the global optimal working paths for coal mine robot are obtained. Lastly, experiments are conducted in a roadway after an accident, and results by different path planning methods are compared.

2. Path Planning for Coal Mine Robot

In this paper, with consideration of gas distribution area and obstacles, MAKLINK method is adopted to describe the working environment network graph of coal mine robot. Suppose that the roadway bottom is parallel to XY plane of the space rectangular coordinate system $OXYZ$ and the height of gas concentration distributions and obstacles is parallel to z-axis. The lines linking the nodes, which do not intersect with the obstacles and gas concentration distribution areas, are defined as MAKLINK lines. The convex polygon areas, which are divided by the MAKLINK lines, make up the feasible working space of coal mine robot. Based on the feasible working space, the midpoints of the MAKLINK lines are defined as the possible path nodes $v_1, v_2, \ldots, v_n$. The feasible paths of coal mine robot are formed by the lines that link the starting point, the possible path nodes [12], and the target point.

Based on the GIS system before the mine accident and the modeling of roadway environment after the mine accident, the optimal working path for coal mine robot is obtained by Dijkstra-ant colony algorithm. The optimal working path planning method for coal mine robot is shown in Figure 1.

2.1. Initial Path Planning for Coal Mine Robot. The initial working paths for coal mine robot are obtained based on Dijkstra algorithm. The working environment network graph obtained by MAKLINK method is defined as $G = (V, E)$, where V is the set of all the path nodes which include the starting point and the target point. E is the set of edges which link the path nodes.

The adjacency matrix of the network graph $G = (V, E)$ is defined as X. The value of the element at ith row and jth column can be expressed as follows [13]:

$$X_{i,j} = \begin{cases} H\left(v_i, v_j\right) & v_i \text{ and } v_j \text{ neighboring} \\ \infty & v_i \text{ and } v_j \text{ not neighboring,} \end{cases} \tag{1}$$

where $H(v_i, v_j)$ is the distance between v_i and v_j.

The set of path nodes of coal mine robot is divided into two subsets S and U, and $V = \{S, U\}$, where S is the set of path nodes that have been admitted to be on the shortest path and U is the set of the path nodes that have not.

Based on Dijkstra algorithm, the steps to obtain the initial working paths for coal mine robot are as follows:

(1) Put the starting point St into S and the rest of the path nodes into U. The adjacency matrix X is used to initialize the shortest paths W between St and each point in U.

(2) Select the minimum value W_i in W, and put the corresponding point v_i into S.

(3) Use point v_i as the intermediate point, and update the distances W' between St and each point in U. If the direct distance between St and v_j in U is larger than the sum of the distance between St and v_i and that between v_i and v_j, the distance between St and v_j is updated to be $W_j' = W_i + H(v_i, v_j)$.

(4) Repeat step (2) and step (3), until the set U is empty. The shortest paths between St and each path node are obtained.

2.2. Optimal Path Planning for Coal Mine Robot. As it is known that a safe working path for coal mine robot can be determined through Dijkstra algorithm, due to the inherent shortcomings of Dijkstra algorithm, it is difficult to guarantee that the obtained path is the optimal one. Therefore, to get the optimal working path for coal mine robot, the initial working paths obtained by Dijkstra algorithm are further optimized by ant colony algorithm.

A working path from St to Ds for coal mine robot obtained by Dijkstra algorithm can be expressed as follows:

$$\text{St} \longrightarrow n_1 \longrightarrow n_2 \longrightarrow \cdots \longrightarrow n_l \longrightarrow \text{Ds}, \tag{2}$$

where n_i $(i = 1, 2, \ldots, l)$ is the sequential path nodes.

The initial length of the paths and the distances between the path nodes and the nodes of obstacles or gas concentration distributions should be considered simultaneously when ant colony algorithm is used. The objective function of the optimal working path for coal mine robot is defined as follows [14]:

$$F_{\text{goal}} = H^k + \frac{1}{\min\left({d^k}_{\text{ob}}\right)}, \tag{3}$$

where H^k is the length of the path that is passed by the kth ant; $\min({d^k}_{\text{ob}})$ is the minimum value of the distance between the path nodes and the nodes of obstacles or gas concentration distributions passed by the kth ant.

The edges on the initial path in MAKLINK graph are defined as H_i $(i = 1, 2, \ldots, l)$. The endpoints of H_i are $v_{i,0}$ and $v_{i,1}$. Other path nodes can be expressed as follows:

$$v_i = v_{i,0} + \left(v_{i,1} - v_{i,0}\right) \times k_i, \tag{4}$$

where k_i is a scale factor and $k_i \in [0, 1]$.

For a given set of parameters k_i $(i = 1, 2, \ldots, l)$, a new path can be obtained by the determined new path nodes on each

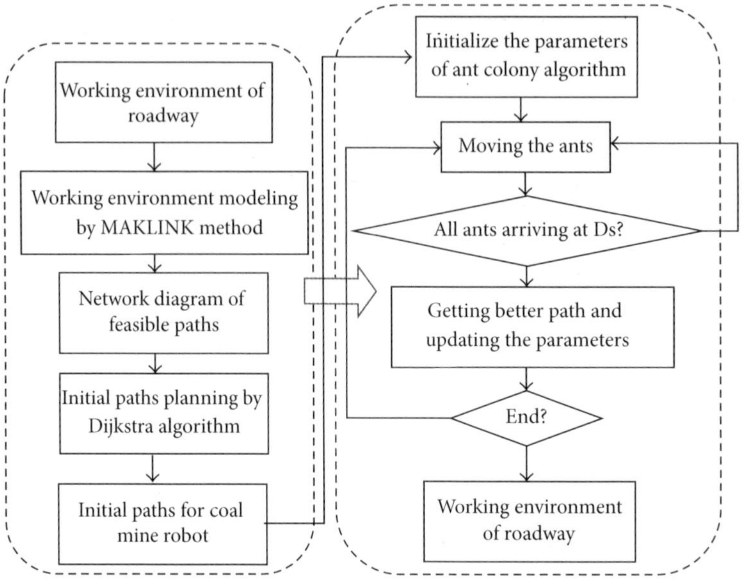

FIGURE 1: Dijkstra-ant colony algorithm for coal mine robot.

H_i. The path with the minimum function value is the optimal one for coal mine robot.

The initial path H_i is split by a fixed-length ΔH. The numbers of the splits of each initial path H_i can be expressed as follows:

$$n(L_i) = \begin{cases} \left\lfloor \dfrac{H_i}{\Delta H} \right\rfloor & \left\lfloor \dfrac{H_i}{\Delta H} \right\rfloor \text{ is an even number} \\ \left\lfloor \dfrac{H_i}{\Delta H} \right\rfloor + 1 & \left\lfloor \dfrac{H_i}{\Delta H} \right\rfloor \text{ is an odd number.} \end{cases} \quad (5)$$

There are $n(H_{i+1}) + 1$ selective paths from each node on edge H_i to the adjacent edge H_{i+1}. Supposing that the initial ant quantity of ant colony algorithm is m, the path from the starting point St to the target point Ds can be expressed as follows:

$$\text{St} \longrightarrow n_{1j} \longrightarrow n_{2j} \longrightarrow \cdots \longrightarrow n_{lj} \longrightarrow \text{Ds}, \quad (6)$$

where n_{lj} represents the jth equal division point of the path node on the edge H_l.

During ant colony algorithm solving process, to avoid falling into a local optimum, a randomly selected parameter q ($q \in [0,1]$) is defined as the selective probability when the kth ant moves from v_i to v_j. The selective probability is compared with a scheduled selectivity constant q_0 ($q_0 \in [0,1]$), and the final selective probability is determined by the compared result and can be expressed as follows:

$$P_{ij}^k = \begin{cases} \max\left(\tau_{i,j}{}^{\varepsilon} \eta_{i,j}{}^{\delta}\right) & q < q_0 \\ \dfrac{\tau_{i,j}{}^{\varepsilon} \eta_{i,j}{}^{\delta}}{\sum_{s \subset \text{allowed}_k} \tau_{i,s}{}^{\varepsilon} \eta_{i,s}{}^{\delta}} & q \geq q_0, \end{cases} \quad (7)$$

where $\tau_{i,j}$ is the pheromone concentration between v_i and v_j; $\eta_{i,j}$ is the inspired value between v_i and v_j, $\eta_{i,j} = 1/d_{i,j}$; $d_{i,j}$ is the distance between v_i and v_j; allowed$_k$ is the next elective nodes set of the kth ant; ε and δ are the relative importance degree of the pheromone concentration and the inspired value, respectively.

After the selective probability of the kth ant which moves from v_i to v_j is obtained, a roulette method is used to determine the right node v_j to which the kth ant moves. As the kth ant reaches v_j, pheromone concentration of the node should be updated as follows:

$$\tau_{i,j} = (1-\rho)\,\tau'_{i,j} + \rho\tau_0, \quad (8)$$

where $\tau_{i,j}$ is the pheromone concentration between v_i and v_j, $\tau'_{i,j}$ is the pheromone concentration between v_i and v_j before update, τ_0 is the initial pheromone concentration, ρ is the volatile coefficient of pheromone concentration, and $\rho \subset [0,1]$.

After all the m ants arrive at the target point Ds, the path with the minimum objective function value is selected, and the pheromone concentration of every node on this path is updated as follows:

$$\tau_{i,j} = (1-\rho)\,\tau'_{i,j} + \rho\Delta\tau_{i,j}, \quad (9)$$

where $\Delta\tau_{i,j}$ is the increment of the pheromone concentration between v_i and v_j on this cycle path, $\Delta\tau_{i,j} = 1/\min(F_{\text{goal}})$, and $\min(F_{\text{goal}})$ is the minimum value of the objective function for the paths that all the m ants have passed.

By using ant colony algorithm, the steps of the working path optimization for the rescue robot can be described as follows [15]:

(1) Get the initial path by Dijkstra algorithm, and initialize parameters of ant colony algorithm.

(2) Calculate the selective probability of each ant, and employ a roulette method to determine which node the ant will move to, and move the ant.

(3) Update the pheromone concentration.

(4) Repeat step (2) and step (3) until all the ants arrive at Ds.

(5) Compute the objective functions F_{igoal} $(i = 1, 2, \ldots, m)$ of each ant, and compare F_{igoal} with the current minimum value $F_{goal\,min}$; if $F_{igoal} < F_{goal\,min}$, let $F_{goal\,min} = F_{igoal}$. Then, the pheromone concentration of the nodes that correspond to F_{igoal} is updated according to (9).

(6) If the loop ends, export the optimal path; otherwise, go to step (2).

The proposed path planning method is suitable for the coal mine robot that works in the roadway of longwall mine.

3. Experiments and Results

3.1. Experimental Conditions. The experiments are carried out in the roadway of Huainan Panyi Mine, whose length is 50 meters and width is 3 meters. In the roadway, gas is distributed regionally and the gas concentration is not over the standard (≤5%). To ensure the safety of the experiments, the area is assumed to be a gas hazard area if the area gas concentration is greater than 2%. Additionally, the length and width of the coal mine robot are 600 mm and 450 mm, respectively. The picture of the coal mine robot is shown in Figure 2. The coal mine robot is equipped with multiparameter gas detector, which can be used to detect methane, carbon dioxide, carbon monoxide, oxygen, wind speed and direction, temperature, humidity, and so on. It can also be used to predict the gas concentration ahead. Laser radar and infrared range sensors are also equipped on the coal mine robot. A sensing space corresponding to the real environment is built through laser perception. The distance between the robot and the obstacle is detected by the infrared distance sensor. The designed walking speed of the coal mine robot is 1 m/s [16]. During the walking process of the coal mine robot, the landform would be evaluated, and if the landform could meet the requirements of walking, the coal mine robot would be walking at the speed of 1 m/s, which will reduce quickly to 0.2 m/s when the obstacle or the gas distribution area is about 3 meters far away. Then, the coal mine robot will go around the obstacle or the gas distribution area at a low speed. In the gas detection process, gas densities are recorded once every 0.5 m in length and width directions, and then gas distribution map is generated according to the gas concentration. Path planning for the coal mine robot is performed under two different circumstances. In one circumstance, it is assumed that the roadway is a flat one; the robot can gain access through the tunnel except the side wall and the dangerous area of the gas. In the other circumstance, it is assumed that other obstacles exist in the roadway, such as stones, mound, and gas danger area.

FIGURE 2: The coal mine robot.

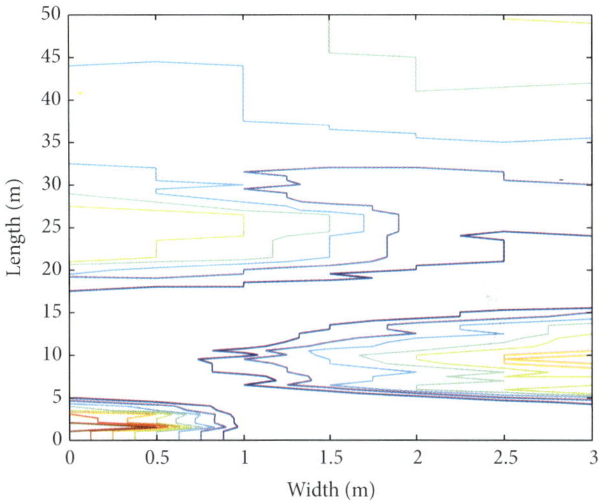

FIGURE 3: Gas concentration distributions.

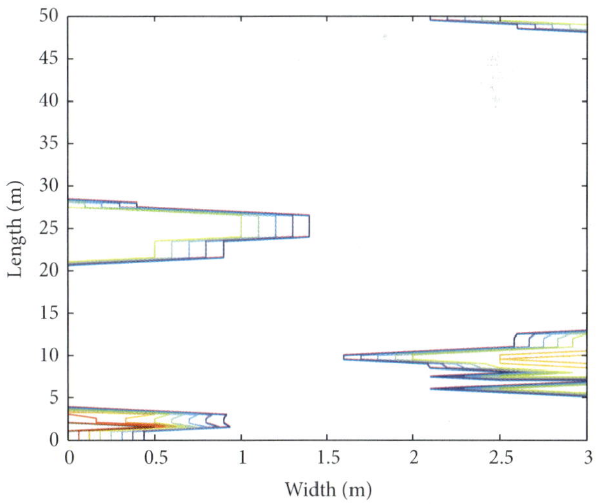

FIGURE 4: Regional distributions of gas prone to explosion.

3.2. Working Environment Modeling for Rescue Robot. The data collected by the coal mine robot is processed, and the gas concentration distributions are shown in Figure 3. Not all regions in Figure 3 meet the conditions of a gas explosion, and, in order to increase the working space of the coal mine robot and avoid a secondary explosion when passing through gas hazard zones, the gas concentration distributions are disposed, which is shown in Figure 4. In Figures 3 and 4, the maximum concentration of the gas exists in the innermost of the closed curve. Although the high concentration of gas

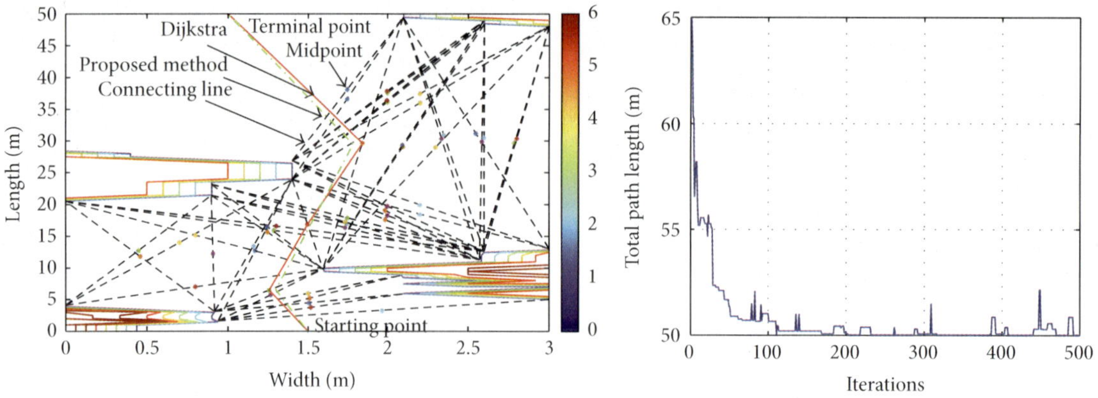

(a) Result of the path planning (b) Variation of the total path length

FIGURE 5: Result of the path planning with only consideration of gas concentration distributions. ∗ denotes the midpoint of the MAKLINK lines.

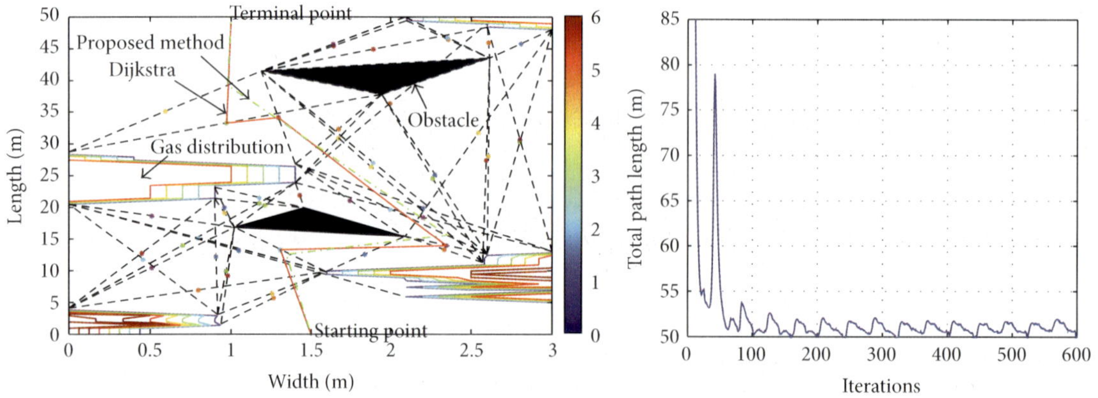

(a) Result of the path planning (b) Variation of the total path length

FIGURE 6: Result of the path planning with consideration of gas concentration distributions and obstacles. ∗ denotes the midpoint of the MAKLINK lines.

may not cause an explosion, the coal mine robot cannot pass through.

The proposed method is applied to the working path planning for the coal mine robot with only the gas concentration area distributions. The coordinate of the starting point and the target point for the rescue robot is (1.5 m, 0) and (1 m, 50 m), respectively. Dijkstra algorithm and Dijkstra-ant colony optimization algorithm are used to solve the optimal path planning of coal mine robot. The parameters of ant colony algorithm are listed as follows: $\varepsilon = 1, \delta = 1.2, q_0 = 0.8$, $\rho = 0.15$, $\tau_0 = 0.0004$, and $m = 30$, and the maximum iteration is 500. The final results of the path planning are shown in Figure 5.

As observed from Figure 5, both the planning working paths by Dijkstra algorithm and the proposed method can avoid the danger gas zones. The curve of the total path length is converged at about 100 times. The total path length by Dijkstra algorithm is 63.39 m, while the one by the proposed method is 60.71 m. Therefore, it can be concluded that the proposed method possesses better performance with only consideration of the gas concentration distributions.

The working environment in the roadway is always very complex after a coal mine accident. The proposed method is also applied to the working path planning for the coal mine robot with consideration of the gas concentration distributions and obstacles simultaneously. The final results of the path planning are shown in Figure 6.

As observed from Figure 6, both the planning working paths by Dijkstra algorithm and the proposed method can avoid the danger gas zones with consideration of the gas concentration distributions and obstacles simultaneously. The curve of the total path length is also converged at about 100 times. The total path length by Dijkstra algorithm is 85.61 m, while the one by the proposed method is 73.95 m. Therefore, it can be concluded that the proposed method possesses better performance.

From Figures 5 and 6, it can be seen that the planning paths through the proposed method are shorter, which is of great significance to the battery powered coal mine robot. Therefore, the proposed path planning method possesses better performance.

4. Conclusions

To avoid a secondary explosion when the coal mine robot passes through gas hazard zones, a path planning method is proposed with consideration of gas concentration distributions and obstacles simultaneously. First, MAKLINK method is adopted to describe the working environment network diagram of the coal mine robot. Second, the initial working paths for the coal mine robot are obtained based on Dijkstra algorithm, and then the global optimal working path for the coal mine robot is obtained based on ant colony algorithm. To verify the effectiveness of the proposed method, lastly, experiments are conducted in a roadway after an accident, and results by different path planning methods are compared. The experiments are divided into two cases; in the first case, it is assumed that the roadway is a flat roadway, and the robot can gain access through the tunnel except the side wall and the dangerous area of the gas; in the other case, obstacles exist in the roadway, such as stones, mound, and gas danger area.

With only consideration of gas concentration distributions, the total path length by Dijkstra algorithm is 63.39 m, while the one by the proposed method is 60.71 m. With consideration of gas concentration distributions and obstacles, the total path length by Dijkstra algorithm is 63.39 m, while the one by the proposed method is 60.71 m. From the results of the path planning with only consideration of gas concentration distributions and the path planning with consideration of gas concentration distributions and obstacles, it can be seen that the planning paths through the proposed method are shorter, which is of great significance to the battery powered coal mine robot. Therefore, the proposed path planning method possesses better performance. The experimental results verified the effectiveness of the proposed path planning method.

Competing Interests

The authors declare that they have no competing interests.

Acknowledgments

This work is supported by a grant of the National 863 Program of China (no. 2012AA041504).

References

[1] D. M. Yang, L. Wang, J. Chen et al., "The basic safety requirements of the coal mine rescue robot," *Coal Mine Safety*, vol. 42, no. 12, pp. 104–107, 2009.

[2] R.-J. Si, R.-Z. Li, L. Wang, and Z.-K. Wu, "Experiment study on the propagation laws of gas and coal dust explosion in coal mine," *Journal of Coal Science & Engineering*, vol. 15, no. 3, pp. 262–266, 2009.

[3] A. P. Sasmito, E. Birgersson, H. C. Ly, and A. S. Mujumdar, "Some approaches to improve ventilation system in underground coal mines environment-a computational fluid dynamic study," *Tunnelling and Underground Space Technology*, vol. 34, no. 1, pp. 82–95, 2013.

[4] X. Ma and H. Zhu, "Gas concentration prediction based on the measured data of a coal mine rescue robot," *Journal of Robotics*, vol. 2016, Article ID 6858970, 10 pages, 2016.

[5] B. Huang and R. Kadall, *Dynamic Modeling, Predictive Control and Performance Monitoring: A Data-Driven Subs-Pace Approach*, Springer, London, UK, 2008.

[6] S. S. Ge and Y. J. Cui, "New potential functions for mobile robot path planning," *IEEE Transactions on Robotics and Automation*, vol. 16, no. 5, pp. 615–620, 2000.

[7] L. Zhu, J.-Z. Fan, J. Zhao, X.-G. Wu, and G. Liu, "Global path planning and local obstacle avoidance of searching robot in mine disasters based on grid method," *Journal of Central South University (Science and Technology)*, vol. 42, no. 11, pp. 3421–3428, 2011.

[8] H.-Q. Yang, X.-G. Xiao, and D. Yao, "A V-graph based global path planning algorithm for mobile robot," *Journal of Shenyang University of Technology*, vol. 31, no. 2, pp. 225–228, 2009.

[9] J. Pan, X.-S. Wang, and Y.-H. Cheng, "Improved ant colony algorithm for mobile robot path planning," *Journal of China University of Mining and Technology*, vol. 41, no. 1, pp. 108–113, 2012.

[10] X. Y. Zhang, X. Y. Zhou, and J. Wei, "Global path planning for coal mine rescue robot," *Journal of Xi'an University of Science and Technology*, vol. 28, no. 2, pp. 323–325, 2008.

[11] Q. Li, Y. M. Xu, D. Z. Zhang, and Y. X. Yin, "Global path planning method for mobile robot based on the particle swarm algorithm," *Journal of University of Science and Technology Beijing*, no. 3, pp. 397–402, 2010.

[12] F. Wang and H.-Y. Wang, "A re-routing path planning method based on Maklink graph and GA algorithm," *Journal of Transportation Systems Engineering and Information Technology*, vol. 14, no. 5, pp. 154–160, 2014.

[13] G.-Z. Tan, H. He, and A. Sloman, "Global optimal path planning for mobile robot based on improved Dijkstra algorithm and ant system algorithm," *Journal of Central South University of Technology*, vol. 13, no. 1, pp. 80–86, 2006.

[14] C. Jin, S. G. Wang, and Y. Y. Yin, "Path planning for multi-robot rescue system under coal mine," *Machine Tool and Hydraulics*, vol. 42, no. 15, pp. 10–14, 2014.

[15] Y. Cen, C. Song, N. Xie, and L. Wang, "Path planning method for mobile robot based on ant colony optimization algorithm," in *Proceedings of the 3rd IEEE Conference on Industrial Electronics and Applications (ICIEA '08)*, pp. 298–301, Singapore, June 2008.

[16] W. Wang, Z. Du, and L. Sun, "Obstacle performance analysis of mine research robot based on terramechanics," in *Proceedings of the IEEE International Conference on Mechatronics and Automation (ICMA '07)*, pp. 1382–1387, IEEE, Harbin, China, August 2007.

Path Planning with Obstacle Avoidance based on Normalized *R*-Functions

Songqiao Tao and Juan Tan

School of Mechanic and Electronic Engineering, Wuhan Technical College of Communications, Wuhan 430065, China

Correspondence should be addressed to Songqiao Tao; taosongqiao@163.com

Academic Editor: Gordon R. Pennock

Existing methods for path planning with obstacle avoidance need to check having the interference between a moving part and an obstacle at iteration and even to calculate their shortest distance in the case of given motion parameters. Besides, the tasks like collision-checking and minimum-distance calculating themselves are complicated and time-consuming. Rigorous mathematical analysis might be a practical way for dealing with the above-mentioned problems. An *R*-function is a real-valued function whose properties are fully determined by corresponding attributes of their parameters, which is usually applied to express a geometrical object. Thus, a signed distance function based on *R*-functions is created to represent whether two objects intervene and their level of intervention or separation. As the signed function is continuous and differentiable, the gradient information of the objective function guides a moving part to avoid its obstacles and to approach its target position rapidly. Therefore, a path planning approach with obstacle avoidance based on normalized *R*-functions is proposed in this paper. A discrete convex hull approach is adopted to solve the problem that *R*-function is inappropriate to represent a geometric object with some curves or surfaces, and pendent points and edges are generated in Boolean operations. Besides, a normalized approach ensures accuracy calculation of signed distance function. Experimental results have shown that the presented approach is a feasible way for path planning with obstacle avoidance.

1. Introduction

In recent years, technologies of path planning with obstacle avoidance have received extensive attention in the fields of robotics, computer simulation, and game animation. In the product design process, path planning with obstacle avoidance is an essential element of the simulations of product assemblies and disassemblies. Path planning with obstacle avoidance is in essence an optimization problem, which usually introduces the evaluation criteria (such as the shortest path or the shortest planning time) to plan an optimal path with obstacle avoidance from a given initial position to the destination in high-interference environments [1].

To avoid obstacles, existing approaches for path planning are necessary to frequently detect whether two objects intervene or to estimate the shortest path between a moving part and an obstacle at the given location. Therefore, there is not a unified form for planning path with obstacle avoidance. Moreover, collision-checking and minimum-distance estimating themselves are complex and time-consuming tasks. To deal with the above-mentioned problems, rigorous mathematical analysis might be a practical way for path planning with obstacle avoidance.

The attributes of an *R*-function are fully determined by corresponding attributes of their argument. Thus, the sign of a real function is fully determined by the sign of its argument. *R*-function is usually applied to represent a geometric object [2]. If g is an *R*-function that expresses a geometric object, we can use a signed distance function $g(x) \leq 0$ to describe interference constraints between this geometric object and a space point x. Thus, $g(x) < 0$, $g(x) = 0$, or $g(x) > 0$, respectively, represents that space point x is the internal, the boundary, or the external of the geometric object.

Based on the above-mentioned analysis, a path planning method with obstacle avoidance based on normalized R-functions is presented in this paper. Firstly, Boolean operations based on *R*-functions are introduced to generate an implicit function for expressing a geometric object. Secondly, a discrete convex hull method is adopted to solve problems that R-function is inappropriate to represent a geometric object with some curves or surfaces, and there are some pendent points and edges in Boolean operations. Then, a normalized method is introduced to transform the implicit function into the *m-th* order normalization function so that the calculation of signed distance function is accurate. Finally, planning paths with obstacle avoidance for a single-point and a polygonal region are conducted to validate the presented approach. Experimental results have shown that the proposed method is a feasible way for path planning with obstacle avoidance.

The rest of this paper is organized as follows. First, related works are introduced in Section 2. Then, the theory of R-function is presented in detail in Section 3. In Section 4, a discrete-convex hull method for generating *CSG* models of geometric objects is introduced. Section 5 adopts the normalization method to transform *g* into the *m-th* order normalization function. And path planning with obstacle avoidance based on the normalization function is given in Section 6. Following this section, experimental results applying the proposed method are presented in Section 7. Finally, the paper ends up with some conclusions in Section 8.

2. Related Works

Path planning with obstacle avoidance has been extensively applied for the robot navigation. For a robot, path planning carries out a series of translational motions and rotational motions from an initial position to the target position while obstacles are avoided in its working environment [3]. Most of path planning methods with obstacle avoidance can be classified into three categories: potential field method (PFM), sampling-based method, and heuristic-based algorithm.

Artificial PFM appeared earlier in [4], which is simple and highly efficient. However, disadvantages like local minima, nonsmooth movement, and oscillations also appear in artificial PFM. To escape from the location of the local minima, Kim [5] presented a new artificial potential function for path planning. Lee et al. [6] proposed new potential function with random force algorithm using PFM for solving the risks of a robot-obstacle collision and local minima trap. To overcome artificial PFM with slow progress and system instability in the presence of obstacles and narrow passages, a modified Newton's approach was applied to potential field [7]. Sfei et al. [8] proposed a new formula of repelling potential for reducing oscillations and avoiding collisions when the target approaches obstacles. To solve problems where a robot only follows the destination while obstacles are rejected in conventional artificial PFMs, Cosio et al. [9] proposed a path planning method with obstacle avoidance based on improved artificial PFM and genetic algorithm. Yin et al. [10] and Zhang et al. [11] adopted modified PFMs to solve path planning problem in dynamic environments.

The advantage of sampling-based path planning methods (SPPMs) has abilities to deal with planning problems in complex and/or time critical real world. At present, most influential SPPMs are probabilistic road-maps (PRMs) and rapidly exploring random trees (RRTs) [12]. Although PRMs and RRTs both randomly sample connecting points from their state space, their graphs connecting points are generated in different ways [13]. More work related to SPPMs can be found in [14]. PRMs have better performance for path planning in high-dimensional state space. As curves or straight lines are used for constructing PRMs, a robot may go anywhere in its free space. Visibility graph [15–17] and Voronoi diagram [18–20] are two typical road-map approaches, which have achieved better results with dramatically various types of roads. RRT algorithms randomly build a space-filling tree to efficiently handle problems with obstacles and differential constraints by searching nonconvex and high-dimensional spaces [21]. Kothari et al. [22] proposed a RRT method to handle path planning problem of unmanned aerial vehicles with uncertainty. The probability of constraint violation and uncertain dynamic obstacles are, respectively, limited by a chance constraint and its extended framework. To further improve search efficiency, Bi-RRT algorithms with bidirectional search have appeared [23, 24]. Two RRTs, respectively, root at the initial and the target state, and algorithms seek states that are "common" for two trees. In [24], a Bi-RRT∗ approach with the intelligence is adopted to handle path planning problem with complex cluttered environments.

A complete overview to heuristic-based algorithms for path planning, which can be found in [3].

3. Theory of *R*-Functions

3.1. Terminology Definition. The sign of an *R*-function is completely determined by the sign of its variable and has nothing to do with the value of its variable. For example, a function $g_1 = x_1 x_2 x_3$ is negative, if the number of its negative variables is odd or positive otherwise. In general, the domain of variable not just has positive or negative but has the form $\geq a$ or $< a$ (a is a real number). In a practical application, the real domain is usually divided into (-∞, -0] and [+0, +∞). Then, the definition of an *R*-function can be expressed as follows.

Definition 1 (*R- function y*). Give a function $S(x)$ with real parameter x, which has the forms

$$S(x) = \begin{cases} 0 & \text{if } x \leq -0 \\ 1 & \text{if } x \geq +0. \end{cases} \quad (1)$$

A real function $y = g(x_1, \ldots, x_n)$ is an *R*-function, if there is a logic function $Y = G(X_1, \ldots, X_n)$ which makes the equation $S[f(x_1, \ldots, x_n)] = G[S(x_1), \ldots, S(x_n)]$ true. Here, the function Y is called *g's* adjoint function. Some *R*-functions have the same adjoint function, such as functions

$$g_1 = x_1 x_2 x_3,$$
$$g_2 = x_1 + x_2 + \sqrt{x_1 x_2 + x_1^2 + x_2^2}. \quad (2)$$

A group of *R*-functions with the same adjoint function is called an *R*-function branch. Key characteristics of an *R*-function are given in the following section (more characteristics can be found in [2]).

(1) The set of R-functions has closure properties under a composition operator. Thus, a function can be obtained by compositing some R-functions, and it is also an R-function.

(2) If g_1 and g_2 are R-functions from the same branch, then the function $g_1 + g_2$ is an R-function belonging to the same branch.

3.2. Sufficiency Completeness. The characteristics of an R-function show that a complex R-function may be generated from simple R-functions. If there is a group of R-functions, which can generate R-functions belonging to any branch by using composition operators, we define this group of R-functions as a function system with sufficiency completeness.

Rvachev [25] presented some useful R-functions with sufficiency completeness. Here, we give a R_a-function system for basic logic operation, such as "negation" ($\neg$), "conjunction" ($\wedge$), or "disjunction" ($\vee$).

$$\neg x = -x;$$
$$x_1 \wedge_a x_2 = \frac{1}{1+a}\left(x_1 + x_2 - \sqrt{x_1^2 + x_2^2 - 2a x_1 x_2}\right); \quad (3)$$
$$x_1 \vee_a x_2 = \frac{1}{1+a}\left(x_1 + x_2 + \sqrt{x_1^2 + x_2^2 - 2a x_1 x_2}\right).$$

Here, $a(x_1, x_2) \in [-1, 1]$. If a=0 or a=1, we call a R_a system an R_0 system or an R_1 system. For function systems R_a^m and R_p, their logical operations can be defined as follows:

$$x_1 \wedge_a^m x_2 = (x_1 \wedge_a x_2)\left(x_1^2 + x_2^2\right)^{m/2};$$
$$x_1 \vee_a x_2 = (x_1 \vee_a x_2)\left(x_1^2 + x_2^2\right)^{m/2}; \quad (4)$$
$$x_1 \wedge_p x_2 = x_1 + x_2 - \left(x_1^{2p} + x_2^{2p}\right)^{1/2p};$$
$$x_1 \vee_p x_2 = x_1 + x_2 + \left(x_1^{2p} + x_2^{2p}\right)^{1/2p}.$$

And their "$\neg$" operations have the same handling rule with R_a-function system.

Although the above-mentioned R_0-function system and R_1-function system have the same logic function, their differential characteristics are various. For instance, R_0-function system is nondifferentiable at the origin of coordinate while other points are differentiable; but R_1-function system is nondifferentiable at the point x_1=x_2 or x_1=-x_2; and R_a^m-function system has m-order differential operators at the origin of coordinate while other points have infinite differential operators. Differential characteristics of R_a-function system have great influence in practical applications. R_a^m-function system has better differential characteristics while the calculation of R_0-function system is simple.

3.3. R-Function Representation for a Geometric Object. Boolean operations (such as "intersection" and "union") between basic primitives can be used to express a *CSG* model [26]. Then, a geometric object B has the following form:

$$B = G\left[B_1, B_2, \ldots, B_m\right]. \quad (5)$$

Here, B_i (i=1, 2, . . ., m) is a basic primitive; and G is a regular Boolean function whose operations have forms $\cap^*$, $\cup^*$, and $-^*$. If we use logical operators $\wedge$, $\vee$, and $\neg$ to displace Boolean operations, G will become a logical function. Combined with the definition of R-function, a geometric object B can be expressed as follows:

$$G\left[S(B_1), S(B_2), \ldots, S(B_m)\right] = 1 \quad (6)$$

If we define the signed distance function $g_i(x)$ for a basic primitive B_i as

$$B_i : g_i(x) \geq 0 \quad (i = 1, 2, \ldots, m), \quad (7)$$

(6) can be rewritten as follows:

$$G\left[S(g_1), S(g_2), \ldots, S(g_m)\right] = 1. \quad (8)$$

Given a continuous function $g(x)$ and its adjoint function $G(X_1, \ldots, X_n)$, a geometric object B in (6) can be expressed as the following form [16]:

$$g(g_1, g_2, \ldots, g_m) \geq 0. \quad (9)$$

In the above formula, g_i (i=1, 2, . . ., m) is a real continuous function while g is an implicit function of a geometric object. In the following, we give an example to illustrate the definition of implicit function g. In Figure 1(a), basic regions B_1 and B_2, respectively, have the following equations:

$$B_1 : g_1(x, y) = 4 - x^2 \geq 0; \quad (10)$$
$$B_2 : g_2(x, y) = 1 - y^2 \geq 0. \quad (11)$$

It can be seen from Figure 1(a), rectangular region B has logic function B=$B_1 \wedge B_2$. Then, using R_0-function system to represent region B has the following implicit function g:

$$g = g_1 \wedge_0 g_2 = g_1 + g_2 - \sqrt{g_1^2 + g_2^2}. \quad (12)$$

If g_1 in (10) and g_2 in (11) are plugged into (12), (2) has

$$g = 5 - x^2 - y^2 - \sqrt{x^4 + y^4 - 8x^2 - 2y^2 + 17}. \quad (13)$$

4. Discrete-Convex Hull Method

In Section 3.3, using R-function to express a geometric object is based on its *CSG* model. However, a geometric object usually has various *CSG* representations. For example, rectangular region B in Figure 1(a) has logic function B=$B_1 \vee B_2$ in Figure 1(b). *CSG* representations usually have various logic functions and implicit functions, but they have the same logic functionality. In fact, it is an intractable task to automatically generate a *CSG* representation for a complex object. Even so, existing algorithms can automatically generate *CSG* representation for polygon regions and polyhedrons. In [27], a polygon region is automatically divided into some Sidechains with the intersection of convex hull vertices. Then, the intersection of the half-space defined by Sidechains can be adopted to express a polygon region while the half-space

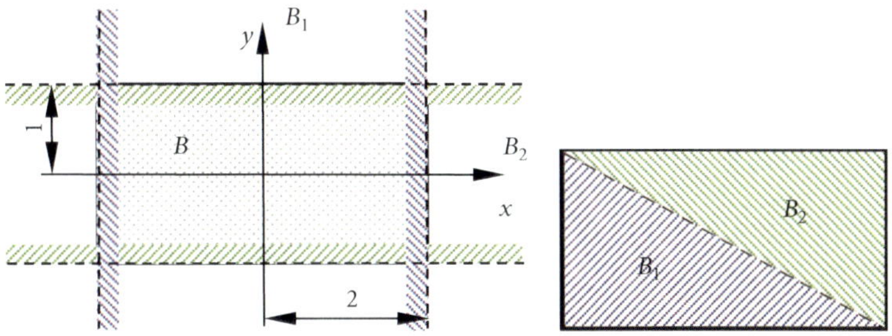

(a) $B = B_1 \wedge B_2$ (b) $B = B_1 \vee B_2$

FIGURE 1: The rectangular region B and its logic function g.

is represented by the intersection or the union of the edges. If the intersection of two edges is a concave, the half-space is the union; otherwise, the half-space is the intersection. For instance, polygonal region P in Figure 2 has logic function $P = C_1 \wedge C_2 \wedge C_3 \wedge C_4$. Here, $C_1 = e_1 \wedge e_2 \vee e_3$, $C_2 = e_4$, $C_3 = e_5 \vee e_6$, and $C_4 = e_7 \wedge e_8$. Besides, convex hull method has been applied to polyhedrons in a three-dimensional space [28, 29].

However, the above-mentioned method is unsuitable for a geometric object with some curves or surfaces. Then, the surface is approximately discretized into planes while the curve is approximately discretized into straight lines. Thus, plane region P in Figure 3(a) has logic function $P = e_1 \wedge e_2 \wedge e_3 \wedge e_4$ (see Figure 3(b)) and $P = e_1 \wedge (e_{21} \wedge e_{22} \wedge e_{23} \wedge e_{24}) \wedge e_3 \wedge e_4$ (see Figure 3(c)). In Figure 3(c), the curve e_2 is discretized into straight lines e_{21}, e_{22}, e_{23}, and e_{24}.

The above-mentioned approach for dealing with a geometric object is called a discrete-convex hull method. If the geometric object is a polygon without a curve or a polyhedron without a surface, the implicit function is accurate; otherwise, the implicit function approximately expresses the geometric object.

Discrete-convex hull method not only automatically generates *CSG* representation for a geometric object but also solves the problem of Point Membership Classification (*PMC*) [30]. *PMC* problem might lead to the implicit function with 0 value for a nongeometric object on its boundary. There are the following two aspects. First, irregular Boolean intersection ∩ for a basic primitive might produce pendent points, pendent edges, or pendent planes (see Figure 4). As the operation $\wedge$ for an *R*-function is irregular, the implicit function of a geometric object may has pendent points on its outside part. Second, rectangular region B has logic function $B = B_1 \vee B_2$ in geometry (see Figure 1(b)). But for mathematical analysis, the operation $\vee$ of an *R*-function cannot ensure that the implicit function has 0 value on B's boundary. As *PMC* exists, the implicit function cannot define the interference constraint between a geometric object and a space point. Fortunately, a discrete-convex hull method avoids the above-mentioned problems. The reasons are that two adjacent lines (planes) of a polygon (polyhedron) just intersect in one point (straight line), and this point (straight line) must belong to the boundary of this polygon (polyhedron). For the operation Boolean union, the discrete-convex hull method will not produce the *PMC* problem.

Therefore, the implicit function of a geometric object based on discrete-convex hull method has 0 value on the object's boundary. And the internal point and external point, respectively, have a value < 0 or > 0. Then, the sign of the function determines whether a space point and a geometric object have the interference. If $g(x)$ represents an implicit function of a geometric object, $g(x) \leq 0$ shows that there is no interference between the space point x and the object.

5. Normalized R-Function

The implicit function based on the above-mentioned approach cannot accurately calculate the signed distance between a space point and an object's boundary. For instance, the function has value of -0.84 at the point (1, 0) for the rectangular region given by the implicit function equation (13) in Section 3. But the minimum signed distance between the point (1, 0) and the boundary of the rectangular region is -1.0 in practices.

Let L be the boundary of a geometric object and p be a space point; the signed distance $\rho(p, L)$ between the point p and the boundary L can be expressed as follows:

$$\rho(p, L) \equiv \inf \rho(p, q), \quad q \in L. \tag{14}$$

where $\rho(p, q)$ is the signed distance between the point p and the point q; if $p \in G$, the signed distance $\rho(p, L)$ is "+" (there is the interference between the point p and the point q); otherwise, it is "-". If all space points satisfy the condition $g(p) = \rho(p, L)$, the function g is called G's normal function. For example, let (a, b) and r, respectively, be the centre and radius of a circle; the normal function g of this circular region has the following form:

$$g = r - \sqrt{(x-a)^2 + (y-b)^2}. \tag{15}$$

As most of geometric objects have some nonsmooth adjacent surfaces, their normal functions usually do not exist. Thus, we introduce the concept of normalized function for a geometric object. A similar normal function $\overline{g}$ with essential

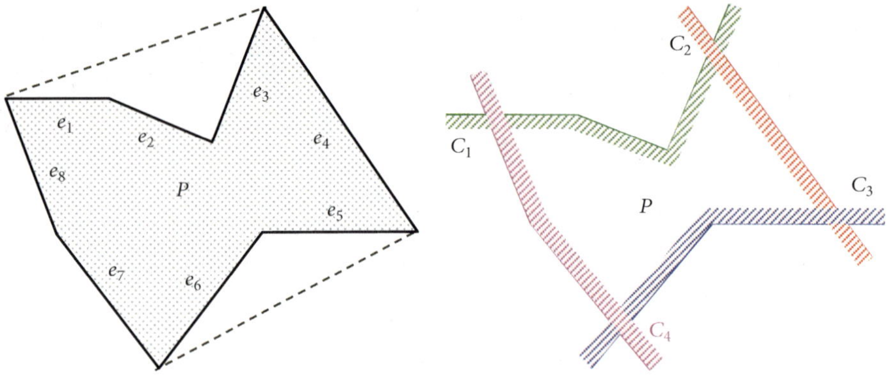

(a) A polygon region P and its convex hull (b) A Sidechain graph for polygon region P

FIGURE 2: Convex hull method for a polygon region.

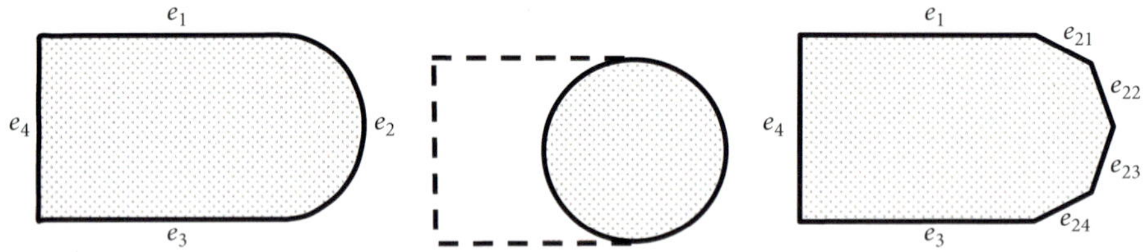

(a) Plane region P (b) $P = e_1 \wedge e_2 \wedge e_3 \wedge e_4$ (c) $P = e_1 \wedge (e_{21} \wedge e_{22} \wedge e_{23} \wedge e_{24}) \wedge e_3 \wedge e_4$

FIGURE 3: Plane region P with a curve is discretized.

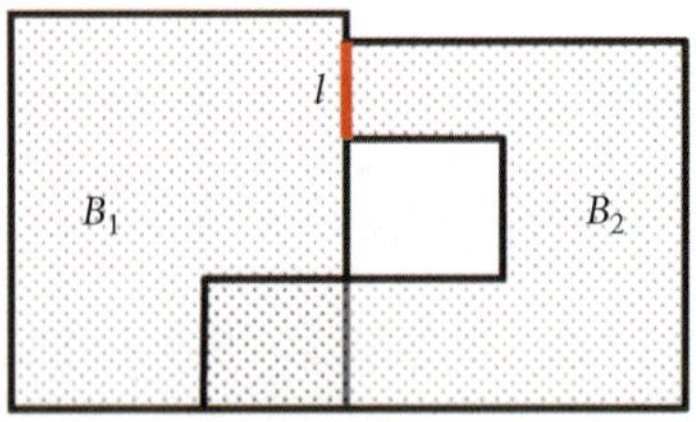

FIGURE 4: Irregular intersection $B_1 \cap B_2$ causes pendent edge l.

differential attributes for a complex object is introduced, which satisfies the following conditions:

$$\begin{aligned} \overline{g}\big|_q &= 0, \\ \frac{\partial \overline{g}}{\partial v}\bigg|_q &= 1, \\ \frac{\partial^k \overline{g}}{\partial v^k}\bigg|_q &= 0 \quad (k = 1, 2, \ldots, m), \end{aligned} \tag{16}$$

where q is a regular point on the boundary L and v is the normal unit vector of the point q. Then, $\overline{g}$ is the m-order normalized function of a geometric object. There are very small differences between normalized function near the boundary and signed distance function. And normalized function far from the boundary is gradually, continuously, and smoothly similar to signed distance function. In general, many R-functions have the above-mentioned normalized properties. For an R-function without normalized properties, the following method can be adopted to normalize.

If the function $g \in C^m$, a point q on the boundary L satisfies the following conditions:

$$\begin{aligned} g\big|_q &= 0, \\ \frac{\partial g}{\partial v}\bigg|_q &> 0. \end{aligned} \tag{17}$$

Then, g can use the following formula to normalize into the first-order form.

$$\overline{g} \equiv g\left(g^2 + \|\nabla g\|^2\right)^{-1/2} \in C^{m-1}. \tag{18}$$

Moreover, higher-order normalized method for the function g can be found in [31]. M-order normalized functions g_1 and g_2 have the same logical operations $g_1 \vee g_2$ and $g_1 \wedge g_2$ with R function. Figure 5 lists the distribution of first-order normalized function (it is also called similar signed distanced field; R_p-function (p=2) system is adopted) for the polygonal region in Figure 2.

The generation of normalized function of a geometric object can be summarized below.

Firstly, a discrete-convex hull method is adopted to generate the implicit function g using the following three steps:

(1) A geometric object is transformed into a polygon (polyhedral). Then, a half-space implicit function that points to the interior of a geometric object is created based on its polygon (polyhedral) and edges (planes).

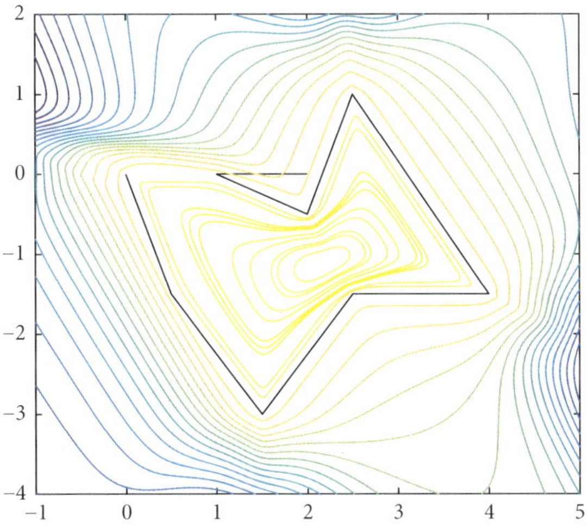

FIGURE 5: The normalized function and its similar signed distanced field.

(2) A convex hull method is introduced to generate *CSG* representation and logic function for this geometric object.

(3) The logic function, the implicit function of an edge (plane), and appropriate *R*-function system are adopted to construct the implicit function g for whole geometric object.

Secondly, the implicit function g is normalized into $\overline{g}$.

6. Path Planning with Obstacle Avoidance

In obstacle environments, path planning with obstacle avoidance is to find an obstacle avoidance path for a moving part from a given initial posture to the final target posture. It is an optimization problem with noninterference between moving parts, noninterference between a moving part and a barrier, or kinematics constraints. For simplicity, we only discuss path planning with a moving part and noninterference between a moving part and a barrier in this paper. For the environments with n obstacles, path planning with obstacle avoidance based on typical optimization can be expressed as follows:

$$\begin{aligned} \min \quad & f(x), \\ s.t. \quad & g_i(x) \le 0 \quad i = 1, 2, \ldots, n. \end{aligned} \tag{19}$$

In the above formula, x is a motion variable and $g_i(x) \le 0$ represents noninterference between a moving part and the *i-th* obstacle and $f(x)$ is an objective function.

For existing path planning approaches with obstacle avoidance based on numerical analysis, most of interference constraints cannot be explicitly defined using the above-mentioned inequality. Thus, two procedures are adopted to solve these problems. First, the interference checking is used to qualitatively determine whether there is interference between a moving part and an obstacle. Second, if it is noninterference, the minimum distance between a moving part and an obstacle and gradient direction of motion parameters are quantitatively calculated, which are used to guide next iteration. Although existing algorithms can deal with them, they usually have heavy computing burden.

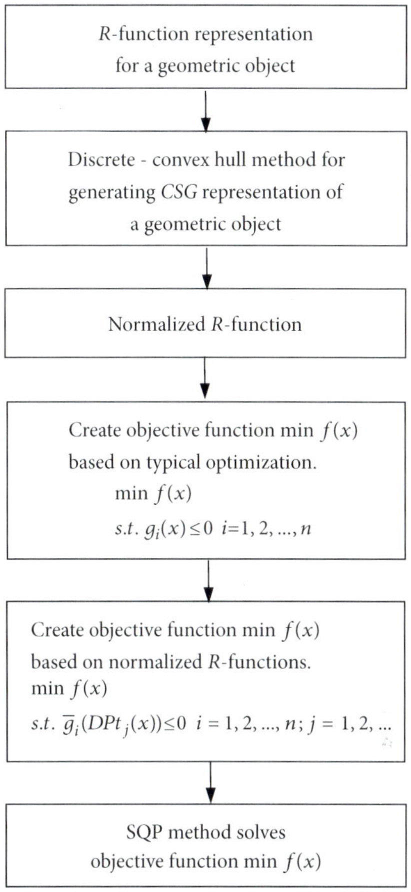

FIGURE 6: The flow diagram of the proposed method.

Fortunately, normalized *R*-function of a geometric object is not only applied to represent interference constraints between two objects, but also there are essential differential properties. Then, (19) can be rewritten as the following forms:

$$\begin{aligned} \min \quad & f(x), \\ s.t. \quad & \overline{g}_i\left(DPt_j(x)\right) \le 0 \\ & (i = 1, 2, \ldots, n;\ j = 1, 2, \ldots). \end{aligned} \tag{20}$$

Here, $\overline{g}_i$ and DPt_j, respectively, express normalized *R*-function of the *i-th* obstacle and a discrete point of moving part *M*'s surface.

For path planning with obstacle avoidance, we must ensure that planning paths have noninterference between a moving part and a barrier. Then, it requires that each iteration point will not default for solving (20). The reason is that interference constraint conditions cannot be strictly ensured for some artificial potential field methods [11]. Therefore, a feasible *SQP* method in [32] is adopted to solve (20). *SQP* is a local optimum algorithm based on the gradient with the superliner convergence.

Based on the above analysis in Sections 3.3, 4, 5, and 6, the flow diagram of the proposed method is given in Figure 6.

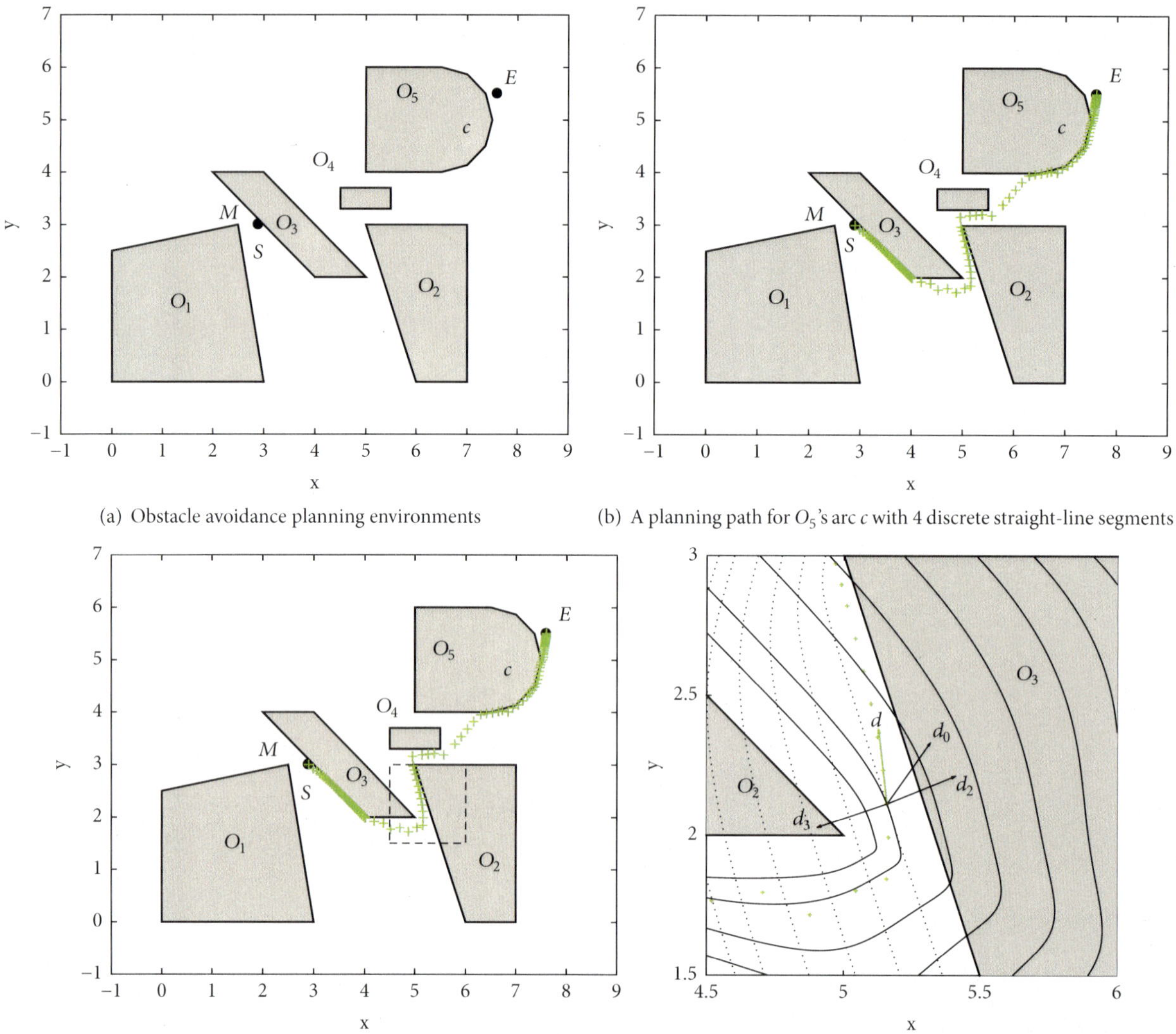

(a) Obstacle avoidance planning environments

(b) A planning path for O_5's arc c with 4 discrete straight-line segments

(c) A planning path for O_5's arc c with 6 discrete straight-line segments

(d) Normalized functions of obstacles affect the iteration direction

FIGURE 7: Planning path with obstacle avoidance for a single-point moving part.

7. Experimental Results

To validate the proposed approach in this paper, some experiments have been performed for path planning with obstacle avoidance on the following three conditions:

(1) The generation of the implicit function is similar in a two-dimensional space and a three-dimensional space, and the representation of the implicit function is relatively complex in a three-dimensional space. For simplicity, examples are conducted in a two-dimensional space.

(2) Compromise differential properties and the computational complexity for a normalized function; an R_p-function system (p=2) is adopted to create the implicit function for an obstacle.

(3) The tests are conducted on a computer with *DELL OPTIPLEX* 760 and Windows *XP* and *Matlab* (R2001a).

First, a single-point moving part M is moved to the target position E from an initial position S through the translation movement in x direction and y direction, and the obstacles $O_1, O_2, \ldots, O_5$ have been avoided (see Figure 7). Besides, the obstacle O_5 has an arc c. Then, it is easy to get normalized function of an obstacle and to create interference constraints for (20).

In Figures 7(b) and 7(c), O_5's arc c is, respectively, discretized into 4 and 6 straight-line segments, and their corresponding planning paths are listed. As there is a higher discrete approximation for O_5's arc c in Figure 7(c), its planning path is more reasonable. In general, the curve of an obstacle has more discrete straight-line segments; normalized function is more complex, which will lead to higher computational complexity. The time is spent on planning paths for O_5's arc c with various discrete approximations is given in Table 1.

Figure 7(d) is partial enlargement of dashed boxes in Figure 7(c), which is listed to further illustrate the role of normalized function in path planning. Red and blue plots in

Table 1: The time is spent on planning paths for O_5's arc c with various discrete approximations.

The number of discrete straight-line segments for arc c	4	6	10	30	100
The time is spent on planning paths(s)	0.6262	0.6434	0.6790	0.92846	1.9188

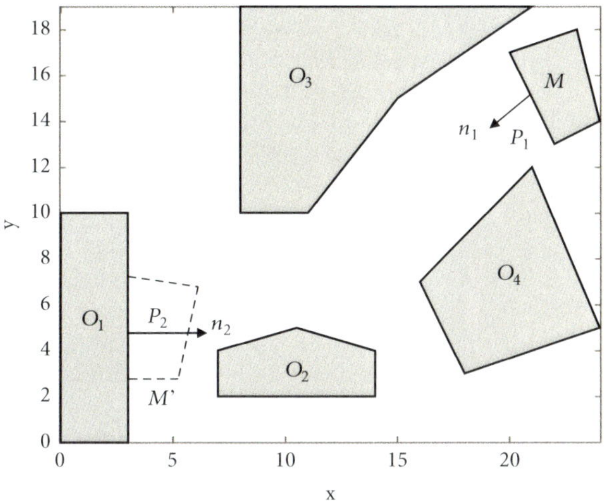

(a) Obstacle avoidance planning environments

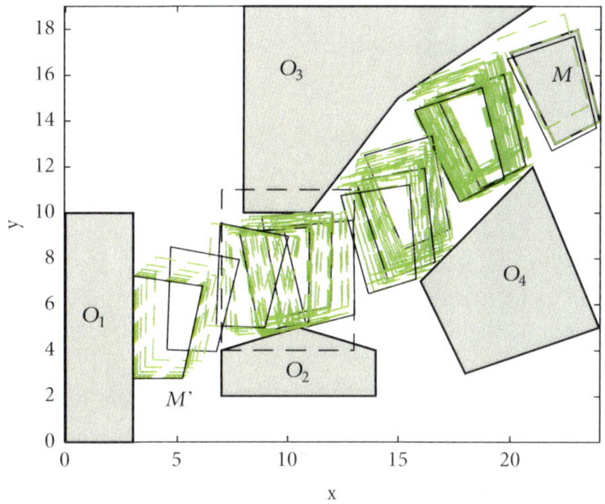

(b) Path planning with obstacle avoidance

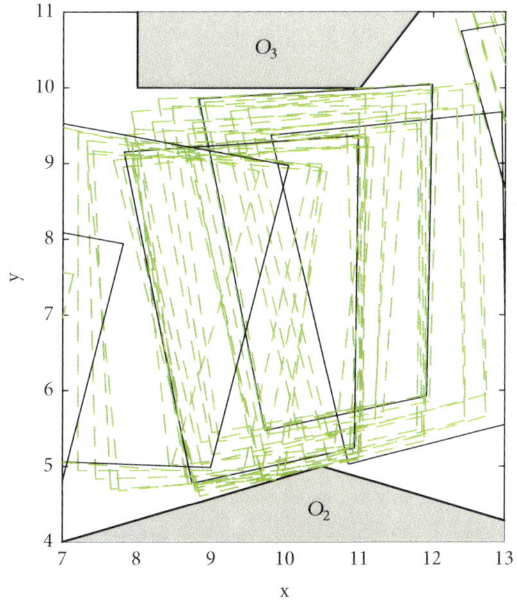

(c) The partial enlargement of dashed boxes in Figure 8(b)

Figure 8: Path planning with obstacle avoidance for a polygonal region moving part.

Figure 7(d), respectively, express outer distance field of obstacles O_2 and O_3. At an iteration point I, d_2 and d_3, respectively, express negative gradient direction of normalized function of obstacles O_2 and O_3, and d_o represents negative gradient direction of the objective function. To avoid obstacles O_2 and O_3, d_2 and d_3 make final search direction be revised to d at iteration point I.

Second, a moving part M is a polygonal region, which has three degrees of freedom (the translation movement in x and y directions, and the rotation motion in z direction), and O_1, O_2, O_3, and O_4 are obstacles (see Figure 8). To generate interference constraints between a moving part and an obstacle, M's boundary is uniformly discretized into 142 points. And M' is the target posture of a moving part. The translation and rotation motions of a moving part M make bold edges of M and the obstacle O_1 are aligned, and their midpoints P_1 and P_2 are coincident in Figure 8(a). Therefore, the objective function for (20) can be expressed as follows:

$$f(x) = W_1 \cos(\langle n_1(x), n_2 \rangle) + W_2 \|P_1(x) - P_2\|^2, \quad (21)$$

where $n_1(x)$ is a function of external normal vector n_1 of M's bold edge for motion parameter x and $n_2 \equiv (1\ 0\ 0)$ expresses external normal vector of O_1's bold edge and $P_1(x)$ represents a function of the midpoint P_1 of M's bold edge for the parameter x and W_1 and W_2 describe weight coefficients. As the cosine value of the vectorial angle is smaller (≥-1 and ≤1), we, respectively, set W_1 =100 and W_2 =1.

Figure 8(b) has listed obstacle avoidance planning paths for a polygonal region M, and dashed lines represent M's posture. To observe clearly, solid lines are adopted to express M's typical postures. And Figure 8(c) is partial enlargement of dashed boxes in Figure 8(b). It can be seen from Figure 8, moving part M successfully avoids obstacles O_2 and O_3 on the basis of posture adjustments. Experimental results have shown that normalized functions of the obstacles can well approximate their signed distanced fields, and their gradient information may guide obstacle avoidance planning paths.

The proposed approach for path planning with obstacle avoidance is based on semianalytic geometry with *R*-functions, which has rigorous mathematical analysis. In general, the characteristics of the algorithm are that the results usually have higher accuracy while the processing speed of the algorithm is relatively slow. In particular, the processing time for various instances has greater diversity.

8. Conclusions

A path planning method with obstacle avoidance based on normalized *R*-functions is presented in this paper. First, an *R*-function is introduced to represent a geometrical object, which ensures that normalized R-function near the boundary of a geometrical object has little difference with signed distance functions, and other parts are continuously and smoothly similar to signed distance functions. Thus, normalized *R*-function can be applied to define interference constraints between a space point x and a geometrical object. As for the generation of interference constraint between two geometrical objects, this needs to plug discrete points of a geometrical object's boundary into normalized R-function of another object. Finally, path planning with obstacle avoidance for a single-point and polygonal region is conducted to validate the presented approach. Experimental results have shown that the proposed method is a feasible way for path planning with obstacle avoidance.

Conflicts of Interest

The authors declare that there are no conflicts of interest regarding the publication of this paper.

Acknowledgments

This work was supported by research team of Wuhan Technical College of Communication (no. CX2018B03).

References

[1] J. C. Latombe, *Robot Motion Planning*, Springer Science & Business Media, 2012.

[2] V. Shapiro, *Theory of R-Functions and Applications: A Primer*, Cornell University, 1991.

[3] T. T. Mac, C. Copot, D. T. Tran, and R. De Keyser, "Heuristic approaches in robot path planning: A survey," *Robotics and Autonomous Systems*, vol. 86, pp. 13–28, 2016.

[4] O. Khatib, "Real-time obstacle avoida nce for manipulators and mobile robots," *International Journal of Robotics Research*, vol. 5, no. 1, pp. 90–98, 1986.

[5] D. H. Kim and S. Shin, "Local path planning using a new artificial potential function composition and its analytical design guidelines," *Advanced Robotics*, vol. 20, no. 1, pp. 115–135, 2006.

[6] J. Lee, Y. Nam, and S. Hong, "New potential functions with random force algorithms using potential field method," *Journal of Intelligent & Robotic Systems*, vol. 66, no. 3, pp. 303–319, 2012.

[7] J. Ren, K. A. McIsaac, and R. V. Patel, "Modified Newton's method applied to potential field-based navigation for mobile robots," *IEEE Transactions on Robotics*, vol. 22, no. 2, pp. 384–391, 2006.

[8] J. Sfeir, M. Saad, and H. Saliah-Hassane, "An improved Artificial Potential Field approach to real-time mobile robot path planning in an unknown environment," in *Proceedings of the 2011 9th IEEE International Symposium on Robotic and Sensors Environments, ROSE 2011*, pp. 208–213, Canada, September 2011.

[9] F. Araınbula Cosío and M. A. Padilla Castañeda, "Autonomous robot navigation using adaptive potential fields," *Mathematical and Computer Modelling*, vol. 40, no. 9-10, pp. 1141–1156, 2004.

[10] L. Yin, Y. Yin, and C.-J. Lin, "A new potential field method for mobile robot path planning in the dynamic environments," *Asian Journal of Control*, vol. 11, no. 2, pp. 214–225, 2009.

[11] Q. Zhang, S. Yue, Q. Yin, and Y. Zha, "Dynamic Obstacle-Avoiding Path Planning for Robots Based on Modified Potential Field Method," in *Intelligent Computing Theories and Technology*, vol. 7996 of *Lecture Notes in Computer Science*, pp. 332–342, Springer Berlin Heidelberg, Berlin, Heidelberg, 2013.

[12] J. Lee, O. Kwon, L. Zhang, and S.-E. Yoon, "A selective retraction-based RRT planner for various environments," *IEEE Transactions on Robotics*, vol. 30, no. 4, pp. 1002–1011, 2014.

[13] S. Karaman and E. Frazzoli, "Sampling-based algorithms for optimal motion planning," *International Journal of Robotics Research*, vol. 30, no. 7, pp. 846–894, 2011.

[14] M. Elbanhawi and M. Simic, "Sampling-based robot motion planning: a review," *IEEE Access*, vol. 2, pp. 56–77, 2014.

[15] T. Siméon, J.-P. Laumond, and C. Nissoux, "Visibility-based probabilistic roadmaps for motion planning," *Advanced Robotics*, vol. 14, no. 6, pp. 477–493, 2000.

[16] J. Kim, M. Kim, and D. Kim, "Variants of the quantized visibility graph for efficient path planning," *Advanced Robotics*, vol. 25, no. 18, pp. 2341–2360, 2011.

[17] M. Zimmermann and C. König, "Integration of a visibility graph based path planning method in the ACT/FHS rotorcraft," *CEAS Aeronautical Journal*, vol. 7, no. 3, pp. 391–403, 2016.

[18] E. Masehian and M. R. Amin-Naseri, "A voronoi diagram-visibility graph-potencial field compound algorith for robot path planning," *Journal of Robotic Systems*, vol. 21, no. 6, pp. 275–300, 2004.

[19] P. Bhattacharya and M. L. Gavrilova, "Roadmap-based path planning - Using the voronoi diagram for a clearance-based shortest path," *IEEE Robotics and Automation Magazine*, vol. 15, no. 2, pp. 58–66, 2008.

[20] Y. V. Pehlivanoglu, "A new vibrational genetic algorithm enhanced with a Voronoi diagram for path planning of autonomous UAV," *Aerospace Science and Technology*, vol. 16, no. 1, pp. 47–55, 2012.

[21] S. M. LaValle, "Rapidly-exploring random trees: A new tool for path planning," Tech. Rep., Computer Science Department, Iowa State University, 1998.

[22] M. Kothari and I. Postlethwaite, "A probabilistically robust path planning algorithm for UAVs using rapidly-exploring random trees," *Journal of Intelligent & Robotic Systems*, vol. 71, no. 2, pp. 231–253, 2013.

[23] S. M. LaValle and J. J. Kuffner Jr., "Randomized kinodynamic planning," *International Journal of Robotics Research*, vol. 20, no. 5, pp. 378–400, 2001.

[24] A. H. Qureshi and Y. Ayaz, "Intelligent bidirectional rapidly-exploring random trees for optimal motion planning in complex cluttered environments," *Robotics and Autonomous Systems*, vol. 68, pp. 1–11, 2015.

[25] V. L. Rvachev, "On the analytical description of some geometric objects," *Reports of Ukrainian Academy of Sciences*, vol. 153, no. 4, pp. 765–767, 1963.

[26] B. Mirtich, "V-Clip: Fast and robust polyhedral collision detection," pp. 177–208.

[27] D. Dobkin, L. Guibas, J. Hershberger, and J. Snoeyink, "An efficient algorithm for finding the CSG representation of a simple polygon," *Algorithmica. An International Journal in Computer Science*, vol. 10, no. 1, pp. 1–23, 1993.

[28] Y. S. Kim and D. J. Wilde, "A convex decomposition using convex hulls and local cause of its non-convergence," *Journal of Mechanical Design*, vol. 114, no. 3, pp. 459–467, 1992.

[29] Y. S. Kim and D. J. Wilde, "A convergent convex decomposition of polyhedral objects," *Journal of Mechanical Design*, vol. 114, no. 3, pp. 468–476, 1992.

[30] V. Shapiro, "Well-formed set representations of solids," *International Journal of Computational Geometry & Applications*, vol. 9, no. 2, pp. 125–150, 1999.

[31] A. Biswas and V. Shapiro, "Approximate distance fields with non-vanishing gradients," *Graphical Models*, vol. 66, no. 3, pp. 133–159, 2004.

[32] Z. Zhu and K. Zhang, "A new SQP method of feasible directions for nonlinear programming," *Applied Mathematics and Computation*, vol. 148, no. 1, pp. 121–134, 2004.

10

FAB: Fast Angular Binary Descriptor for Matching Corner Points in Video Imagery

Mehwish Tahir, Nadia Kanwal, and Fatima Anjum

Department of Computer Science, Lahore College for Women University, Jail Road, Lahore 54000, Pakistan

Correspondence should be addressed to Nadia Kanwal; nadia.kanwal@lcwu.edu.pk

Academic Editor: Farrokh Janabi-Sharifi

Image matching is a fundamental step in several computer vision applications where the requirement is fast, accurate, and robust matching of images in the presence of different transformations. Detection and more importantly description of low-level image features proved to be a more appropriate choice for this purpose, such as edges, corners, or blobs. Modern descriptors use binary values to store neighbourhood information of feature points for matching because binary descriptors are fast to compute and match. This paper proposes a descriptor called Fast Angular Binary (FAB) descriptor that illustrates the neighbourhood of a corner point using a binary vector. It is different from conventional descriptors because of selecting only the useful neighbourhood of corner point instead of the whole circular area of specific radius. The descriptor uses the angle of corner points to reduce the search space and increase the probability of finding an accurate match using binary descriptor. Experiments show that FAB descriptor's performance is good, but the calculation and matching time is significantly less than BRIEF, the best known binary descriptor, and AMIE, a descriptor that uses entropy and average intensities of informative part of a corner point for the description.

1. Introduction

To build an autonomous computing device, computers are required to achieve human vision. This includes detecting and recognizing different objects [1], segmenting objects from their background [2], classifying different scene segments for identification or stitching [3], text classification and identification, and face or hand recognition [4]. Accurate matching of image content is the fundamental part of all these vision based tasks, for which identification and description of salient image regions are the method that is closely related to human vision. Therefore, for the last two or more decades a number of detection and description techniques have been developed. The most popular of these are local image feature extraction and description techniques, which include edges, corners, blobs, or ridges as low-level image features.

Recognizing only distinctive image areas such as corners or blobs is not enough because they cannot be directly matched in other transformed images such as rotated, scaled, or illuminated. However, if we can get sufficient neighbourhood information on a feature, then it becomes easier to identify the same image point in different transformed images. The collection of this neighbourhood information is called description. Although all types of image features are important, corner points are more informative [5]. This is because they carry some inherent information as compared to other features such as edges or blobs. An edge can occur independently anywhere in the image where there is some intensity difference [6]; similarly a blob is a sovereign image point identifying significantly bright or dark area with respect to its neighbourhood [7, 8], which can be matched. However, corner points can only appear when there are two or more intersecting edges. The presence of two or more edges gives additional information such as the presence of an object or existence of two separate planes which can be used for image contents' classification. The literature shows a number of feature descriptors; however, none of them explore this inherent information available with the corner points.

Previously proposed state-of-the-art descriptors use image pixels gradient and its orientation such as Scale Invariant Feature Transform (SIFT) [9], Speeded Up Robust Features (SURF) [10], and Gradient Location Orientation

Histogram (GLOH) and store it as numeric/float descriptor data for matching. These descriptors show very good performance in matching images under transformations, but at high computational cost. However, some other descriptors use binary vectors to store distinctive gradient information of a feature's neighbourhood such as Binary Robust Independent Elementary Features (BRIEF) [11], Oriented Fast and Rotated Brief (ORB) [12], Binary Robust Invariant Scalable Keypoints (BRISK) [13], and Fast Retina Keypoint (FREAK) [14]. Binary descriptors gained more attention due to being less computationally expensive yet considerably accurate. Although numeric descriptors are more accurate than binary descriptors, they need more time to compute and match which make them undesirable for real-time applications. Furthermore, almost all of the aforementioned descriptors describe circular area around feature point for description and, therefore, produce large size descriptors.

This paper presents a descriptor that describes the useful neighbourhood of a corner point: in between two edges and hence named Fast Angular Binary (FAB) descriptor. It combines the strengths of two descriptors BRIEF and a recently proposed numeric descriptor for corner points called Angle, Mean Intensity and Entropy of Informative Arcs (AMIE) [15]. BRIEF takes the circular region around feature point to create binary descriptor and, therefore, cannot produce good matches in dynamically changing backgrounds, whereas AIME selects only the informative area (that lies inside two edges) of a corner point but computes descriptor using basic image intensities and hence produces more false matches. FAB on the other hand combines the distinctiveness of BRIEF and efficiency of AMIE to calculate more capable, robust, and reliable descriptor. Therefore, it is more suitable for identifying objects with dynamically changing backgrounds in real-time. Contrary to BRIEF and AMIE, FAB is capable of accommodating scale and rotational changes while matching frames of a video for tracking, robot navigation, and other video analysis based applications. Furthermore, minimum processing should enable faster processing of data in autonomous robot navigation and other applications.

The rest of the paper is organized as follows: Section 2.1 describes the working principles of BRIEF, AMIE, and FAB. Sections 2.2 and 3 present the matching of FAB descriptor and how it is more efficient and accurate than other descriptors. Performance of newly developed descriptor is compared with BRIEF and AMIE in Section 4.1. Section 4.2 describes the application of matching images using FAB descriptor in different vision applications and lastly Section 5 concludes the paper and suggests some future modification.

2. Material and Methods

2.1. Feature Descriptors. Feature description is a process of creating vectors of neighbourhood information of a feature point. This descriptor is supposed to contain sufficiently unique information to be matched in another transformed image. These transformations include scaling, rotation, illumination, and change in viewpoint also called affine transformation. However, considering a problem of object tracking in video imagery, the amount of image transformation is quite low. For example, at a normal frame rate the scale changes between frames are very small and sometimes negligible. Similarly rotation only occurs due to camera jittering and there is no affine transformation in consecutive video frames. Therefore, for video imagery feature, matching does not need complex computation to make descriptors scale, rotation, or affine invariant which happens in case of SIFT and SURF. Moreover, small and binary descriptor vectors take considerably less time to compute and match and, therefore, are more attractive for real-time applications such as BRIEF and AMIE. The following sections briefly describe their working principles.

2.1.1. BRIEF. It randomly selects two pixels P_i and P_j from a circular neighbourhood of a feature point. The intensities of these two pixels are compared to produce a binary value (0/1) as shown below:

$$\tau(p; x, y) = \begin{cases} 1 & \text{if } P_i(x) < P_j(y); \\ 0 & \text{otherwise.} \end{cases} \quad (1)$$

A 64-bit descriptor vector is produced for each image feature. But due to the binary nature of the descriptor, matching becomes efficient—calculating Exclusive OR, Jackard-Needham, or Dice matrix [16] of two vectors is very fast. However, for BRIEF, the selection of pair of neighbourhood pixels to be compared is an important step. Authors in [11] performed five tests to select the best sample pair and found Isotropic Gaussian distribution to be able to produce good number of matched points as compared to the selection of pixels according to some pattern such as equivalent distance points.

2.1.2. AMIE. AMIE is a numeric descriptor that stores mean intensity and entropy of circular arcs around a corner point. Furthermore, these arcs only cover the informative region around corner point instead of whole circular area. Although the descriptor contains very basic pixels information, it becomes discriminating when calculated only for arc pixels. It produces a vector of length of 16 only that contains angle, mean intensity, and entropy of different arcs and a value to describe the direction of informative region. Although AIME is a numeric descriptor, its matching time is comparable to that of BRIEF due to its small size. Likewise, AMIE is not scale invariant and is not suitable to be used in applications where matching is required between transformed images such as scaled, rotated, or illuminated.

2.1.3. FAB. Fast Angular Binary (FAB) descriptor for corner points is a descriptor that combines the strengths of BRIEF and AMIE discussed above to facilitate matching in real-time with improved accuracy. It calculates the binary vector for five different arcs of different radii similar to AMIE. However, its descriptor is completely different from AMIE because of its variable length vectors depending on corner's angle. So every corner point may have different length descriptor; however, an average length of 60 bits is found for 1000 video frames. This includes binary vector for five different arcs of radii 3,

5, 7, 9, and 11 along with angle and a direction bit. AMIE's descriptor's length is fixed as it stores collective information of each arc such as average intensity and entropy. Detailed description of steps required to calculate FAB descriptor is given below.

Following the detection of corner points in an image using a good detector (Harris and Stephens algorithm [17] is used here), below is the pseudo code followed for descriptor calculation:

```
Read video frames

    for each frame
    {
      (1) convert into grayscale
      (2) detect corner points
      (3) for all corner points
      (4) find edges around the corner
      (5) calculate angle between the two detected
          edges
      (6) find orientation of the angle
      (7) select orientation most related to corner
          point
      (8) calculate descriptor for only selected part of
          the corner
    }
```

The following sections describe these steps in detail.

2.1.4. Angle and Orientation of a Corner Point. Figure 1 shows five circular arcs of radii 3, 5, 7, 9, and 11 around a corner point. These radii are selected to cover the maximum nonoverlapped neighbourhood area. Further, these circular arcs are used to find edge pixels which then decide the angle of a corner point. Each circular arc is scanned to find eigen values (λ_1 and λ_2) for the categorization of image pixels according to Figure 2. Angle of the corner points is calculated by counting number of pixels between two edge pixels at each arc as explained by (2), where N_r is the number of pixels in an arc of radius r while M_r is the number of pixels in the circle of radius r. Consider the following:

$$\theta_r = 2\pi\left(\frac{N_r}{M_r}\right). \quad (2)$$

Although due to rasterization in digital images each arc can give different angle information, it is useful at the same time as it compensates the irregular shapes (nongeometrical shapes such as hands and clothes). The average angle of all five arcs is stored as corner's angle. Furthermore, these edge pixels divide the circular area around corner point into two parts. One may give information about an object (maybe from edge 1 to edge 2 in Figure 1) and the other one may belong to background (from edge 2 to edge 1 in Figure 1).

Once we have angle information next step is to find the orientation of its internal part which should be used for the description purpose. Logically the part which shows more similarity to the corner point should be selected. Therefore the intensities of each part are compared with corner point and the one that shows close average intensity is selected for descriptor calculation and its direction is stored as the orientation of a corner's angle. For simplicity bit value 1 represents clockwise direction and 0 represents anticlockwise direction.

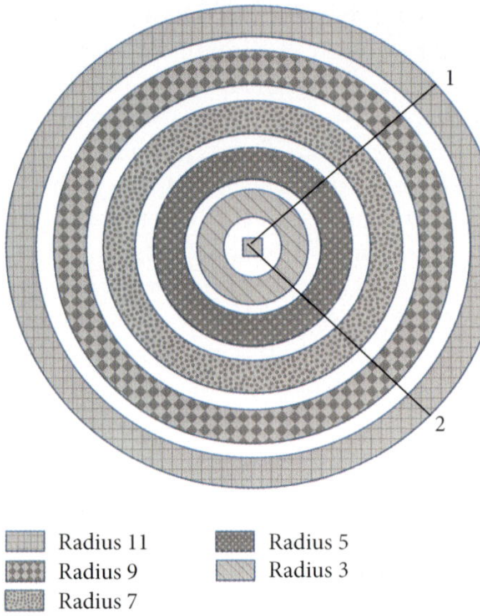

FIGURE 1: Corner point, corresponding edges, and 5 radii around it used to identify internal and external part for description.

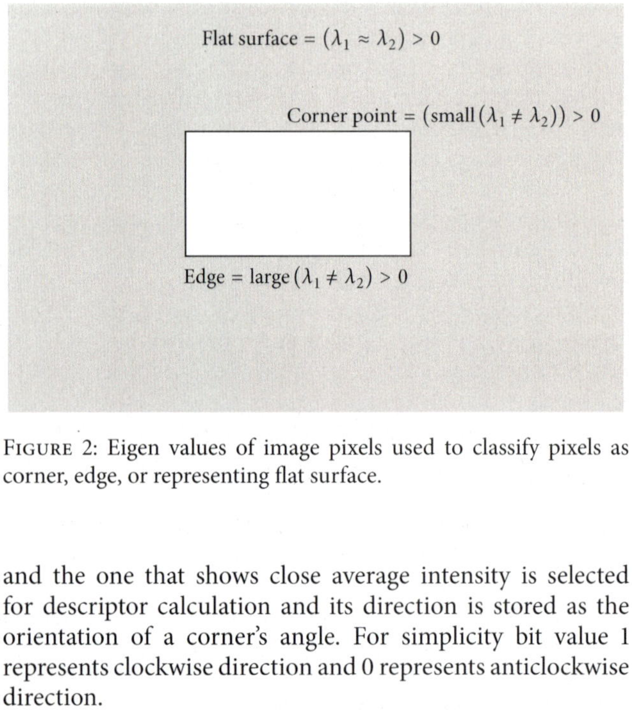

FIGURE 2: Eigen values of image pixels used to classify pixels as corner, edge, or representing flat surface.

2.1.5. Descriptor Calculation. Conventionally, binary descriptors are constructed by comparing either randomly selected pair of pixels or some specified patterns. For example, BRIEF choose pixel pairs based on Isotropic Gaussian distribution while FREAK uses a pattern similar to human eye retina and it claims to give better performance than BRIEF. Beneficially, selecting circular arcs gives patterns similar to FREAK and, therefore, appears to be more effective. Each arc of different radius contributes to the descriptor data. Because of dissimilar radii the number of bits for each arc is different which is why it is called variable length descriptor. Maximum number

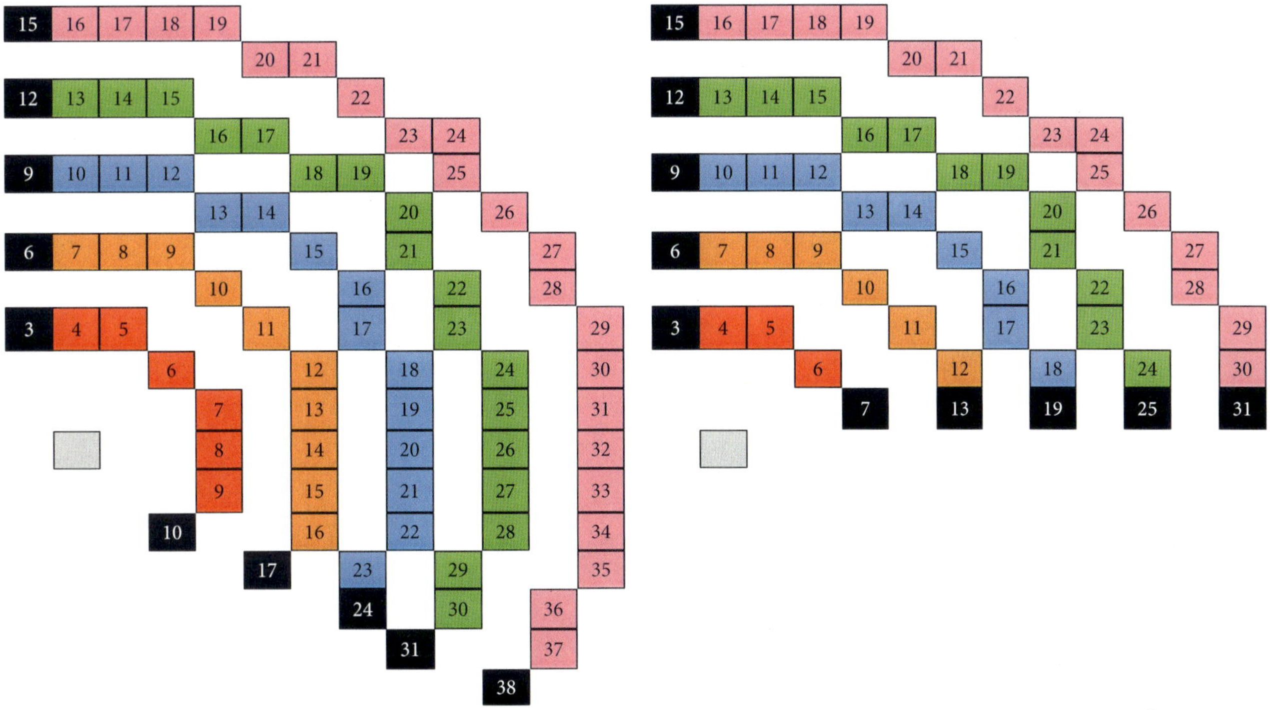

FIGURE 3: Two corner points with different angles and, therefore, different arc lengths for description.

of bits for each arc is equal to the total number of pixels in the circle of radius r, which are used for descriptor calculation, although, rarely, it happens when there is only one edge pixel found on majority of the circular arcs or no edge pixel is found. Describing this kind of detected points can be argued because they should not be detected as corner points and can be considered false responses of corner detection algorithm. However, to increase the amount of information about whole image content we considered them equally important and complete circular arcs are used for the description purpose. In order to build binary descriptor for the internal part of corner point, consecutive pixels at each arc are compared using the following equation:

$$\tau(p; x, y) = \begin{cases} 0 & \text{if abs } (p(x) - p(y)) < t; \\ 1 & \text{otherwise,} \end{cases} \quad (3)$$

where t is a threshold set to 1.5 here. Figure 3 shows two corner points with different angles which usefully resulted in different size of arcs or descriptor. For left side corner point the number of pixels is 6, 10, 14, 18, and 22 for radii 3, 5, 7, 9, and 11, respectively. So the descriptor length will be 71; similarly the descriptor length of right side corner point is 46 including 6 bits of each arc's angle and one bit of orientation bit, smaller than BRIEF or FREAK (64 bits) and SIFT or SURF (128 bits).

Edge pixels are not used in descriptor calculation because of maximum intensity difference from its neighbouring pixels. Therefore, it does not contribute in making descriptor more discriminative.

2.2. Matching Corner Points. Matching is the step that makes FAB most attractive descriptor. As discussed before the FAB descriptor contains numeric and binary values purposefully. Therefore, matching is performed in two steps that are

(i) matching angle and orientation,

(ii) matching binary descriptor.

This two-step matching technique reduces the search space and helps finding optimal solution all the time. Binary descriptor only matched the corner points for which the orientation is the same and the angle is in between some specified range. This further helps achieving scale invariance for FAB descriptor.

Orientation and angle part of the descriptor are matched by finding minimum Euclidean distance of each corner point of reference image from all corner points belonging to test image. So every time a close match is found, binary descriptor is matched using XOR and its hamming distance is stored as matching difference which should be minimum. Hence, at each iteration if the previous match difference is greater than the newly calculated one, current point is selected as a better choice and its difference is stored for future comparisons. This way it is possible that, comparing orientation and angle part of one point with 500 points, the binary descriptor may match only 10 of these 500, a more than significant reduction of unnecessary comparisons. However, it is important to highlight the significance of FAB's variable length. If the two descriptors under comparison have different length, then both descriptors are aligned in such a way that ensures correct matching part. For this the central part of both descriptors is aligned leaving left and right bits of larger descriptor unmatched. Although there seems to be some loss of information, practically it is not the case because of matching corner points with similar angles; maximum of two to four bits are usually unmatched. Further it facilitates matching images under scale transformation making FAB scale invariant. Experimental results shown in next sections prove this argument.

2.2.1. Outliers Removal Using RANSAC. Matches are computed until this step can have some wrong matches. This is because descriptors of a number of points in the image can have similar angles at different locations. Therefore, to discard wrong matches, Random Sample Consensus (RANSAC) method is used. This method finds a homography matrix (a 3 × 3 matrix representing translation and rotation between an image pair) between two images using a randomly selected subset of matched points. Then, it works out support in favour of calculated homography from the rest of the matches and keeps the homography with maximum support. The matched points favouring the selected homography are considered as true matches and also called inliers and the rest as false matches or outliers.

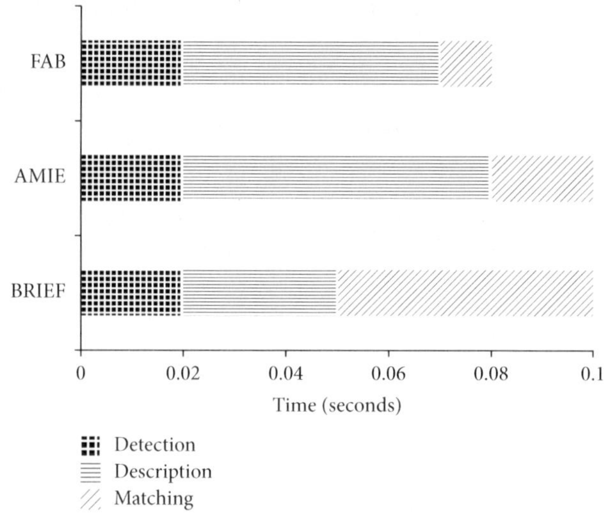

FIGURE 4: Comparison of computation and matching time of FAB with BRIEF and AMIE averaged over 100 images.

3. Computation and Matching Time Comparison

Figure 4 shows average computation time of three descriptors for 100 images containing approximately 50,000 corner points. Detection of corner points is done using Harris and Stephens corner detector for all three descriptors; therefore, its computation time remains the same. Descriptor computation time of FAB and AMIE is slightly more than BRIEF, but still less than 0.2 seconds. However, significantly less matching time of FAB descriptor makes it an overall winner. It is definitely due to angle based matching of binary descriptor, which itself is very fast.

4. Results and Discussion

4.1. Performance Evaluation. Performance of FAB descriptor is compared with BRIEF and AMIE descriptors. This descriptor is mainly developed for matching video frames for applications such as object tracking, template matching, and navigation. Therefore, a number of video results are presented here. As it is not possible to show thousands of video frames as images in the paper, a framework is defined for this purpose called as difference of images (DOI). Computation of DOI is done using the following procedure:

(i) Find homography for pair of images using corresponding matched points.

(ii) Warp the first frame using the computed homography matrix.

(iii) Subtract warped frame from second frame and result will be zero image in case of good matches and a nonzero image otherwise. In other words the sum of pixel intensities DOI will be close to zero if the image is warped using good matched points and greater than zero otherwise.

This framework is applied in all of the performance evaluation tests presented below. For testing a number of videos are captured using handheld camera along with using surveillance video data publicly available from PETS2009 [18]. Furthermore, homography matrix has been previously used and proved to be a better performance analysis metric as compared to repeatability or detection rate [19]. Before discussing these results, it is worth showing that FAB is more scale invariant than AMIE or BRIEF. For this purpose three tests have been performed using two different images and a video sequence, in which images with scale difference are matched and results are shown below.

Figure 5 shows such images and Figure 6 displays the result of FAB descriptors while matching each reference image with corresponding scaled images. Similarly, to show matching results under scale change in video imagery, first frame of a video captured by a handheld camera is matched with subsequent frames as shown in Figure 7. Here, two matched pairs of frames are shown and because video matching results are difficult to display in printed form, a graph is shown in Figure 8 where the sum of all pixel intensities in difference of warped images (DOI) is plotted for all three descriptors. In case of good results the DOI should contain black pixels (with 0 pixels' intensities) for which the sum of whole image will be close to zero in the graph. The results produced by FAB descriptor shown in Figure 8 outperformed all others by producing good matched points and minimum value of DOI (sum of pixel intensities of difference image) for whole video.

4.2. Effectiveness of FAB in Different Vision Applications. Performance evaluation of FAB descriptors for scaled images shows promising results both in still images and in video imagery. Therefore, it should also be tested for different vision

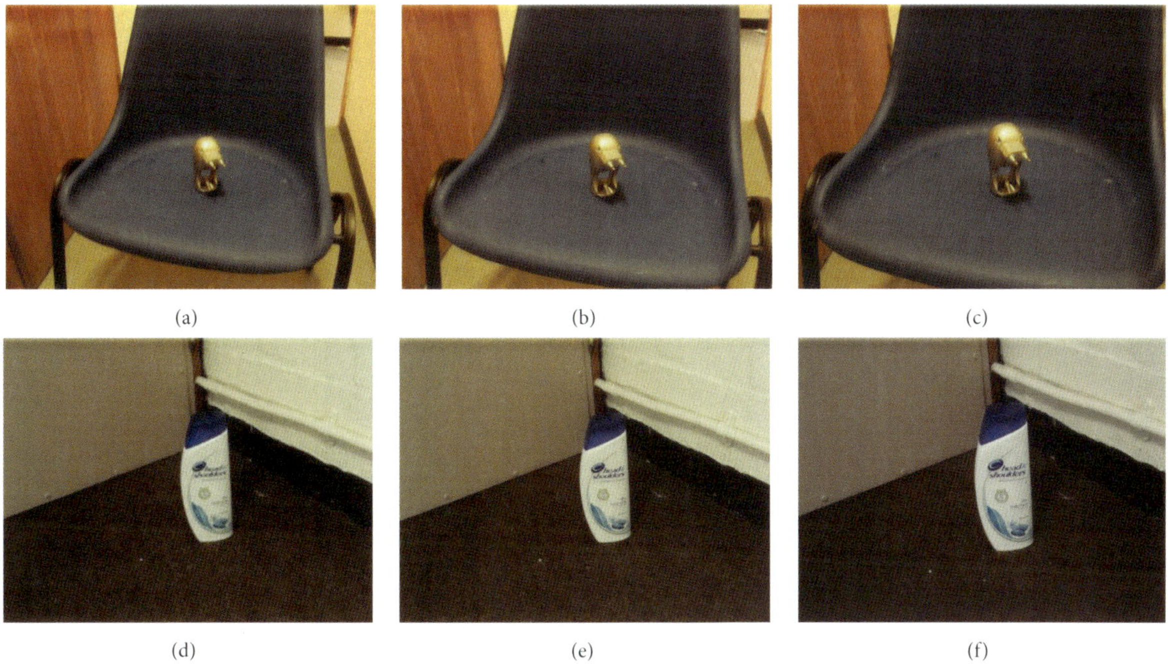

(a) (b) (c)

(d) (e) (f)

FIGURE 5: Images with scale differences. First image in each row is selected as reference image and matched with test images (second and third in the row).

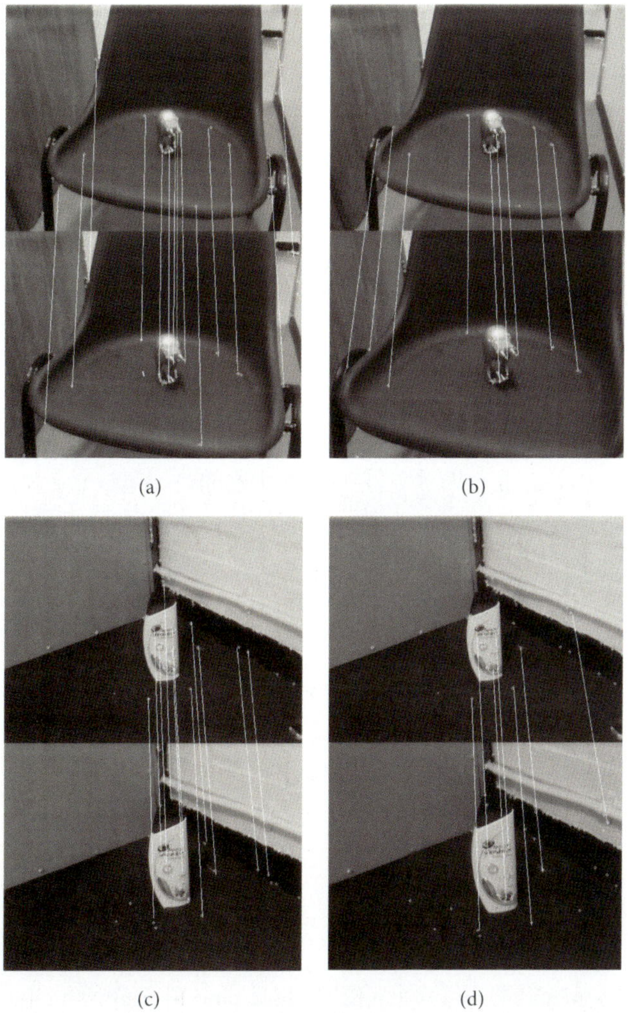

(a) (b)

(c) (d)

FIGURE 6: Scaled image pairs matched using FAB descriptor. Reference image in Figure 5 is matched with corresponding test images.

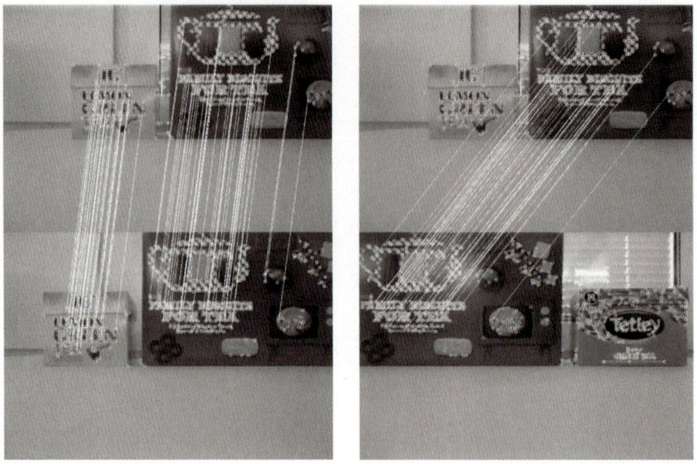

FIGURE 7: Initial video frame is matched with subsequent frames of the video using FAB descriptor to show matching under scale differences. Results are shown for frame #1 matched with frame #150 (a) and frame #205 (b).

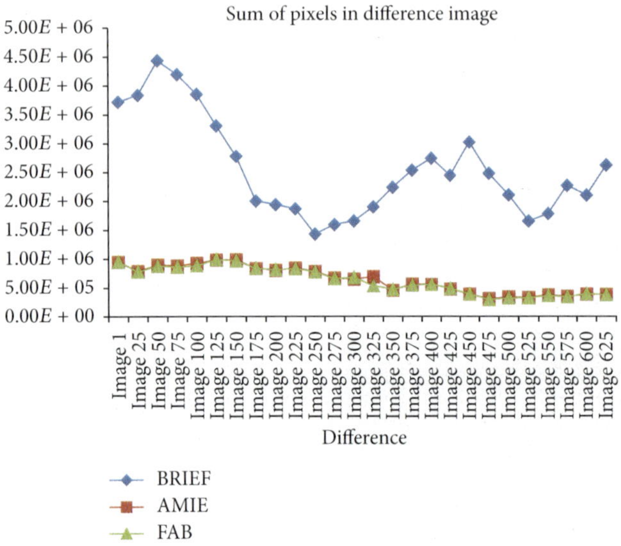

FIGURE 8: DOI analysis of video frames captured using handheld moving camera (as shown in Figure 7); small values show minimum difference in matched images and hence good matching results.

applications such as template matching, motion detection in surveillance videos, and object tracking.

4.2.1. Object Tracking. Object tracking is one of the most popular applications of computer vision. Figure 9 shows two sample frames from a video captured using a handheld camera. In the video a car is moving and its position is tracked by finding corner points witch are matched in subsequent frames using FAB descriptor. Matching results obtained by FAB descriptor to track this moving car are presented in Figure 10. The tracked car's position is marked by a rectangle in the difference images.

4.2.2. Surveillance. Next application, that is, motion detection using surveillance cameras, has great significance in today's world where security is becoming the foremost global and local issue. Vision based surveillance systems are getting more attention these days because of its low cost and an unobstructed use of sensor (camera/s). PETS is a very popular public domain for vision community where a number of surveillance video datasets are available for research purposes. We also took advantage of getting different surveillance videos to test FAB descriptor. Some sample images are shown in Figure 11 while the matching results are shown in Figure 12 where the movement can easily be detected using DOI.

4.2.3. Template Matching. Lastly, template matching is done to show that FAB can work equally good for applications such as vision based engineering, medical, and text classification. A template containing some text is matched in a video sequence. Figure 13 shows the template which is matched in different video frames. Figure 14 shows accurately matched template in different timely spaced video frames. In this figure first two frames are initial part of the video while last two frames are from the middle to end part of the video.

FAB descriptor for corner points proposes improvement in two different dimensions, one using corner points in different vision applications such as tracking, surveillance, and template matching which helps identify useful image content. Secondly, it describes only the useful neighbourhood of corner point using binary vectors which facilitate matching an object in changing background as can be seen in tracking and template matching results shown in Figures 10 and 14. Moreover, these results along with matching results of scaled images shown in Figures 6 and 7 proved FAB to be scale invariant descriptor that is very useful and required property of a fast feature descriptor. Time comparison of FAB with

(a) (b)

FIGURE 9: Sample frames from a video captured using a handheld camera with a moving car.

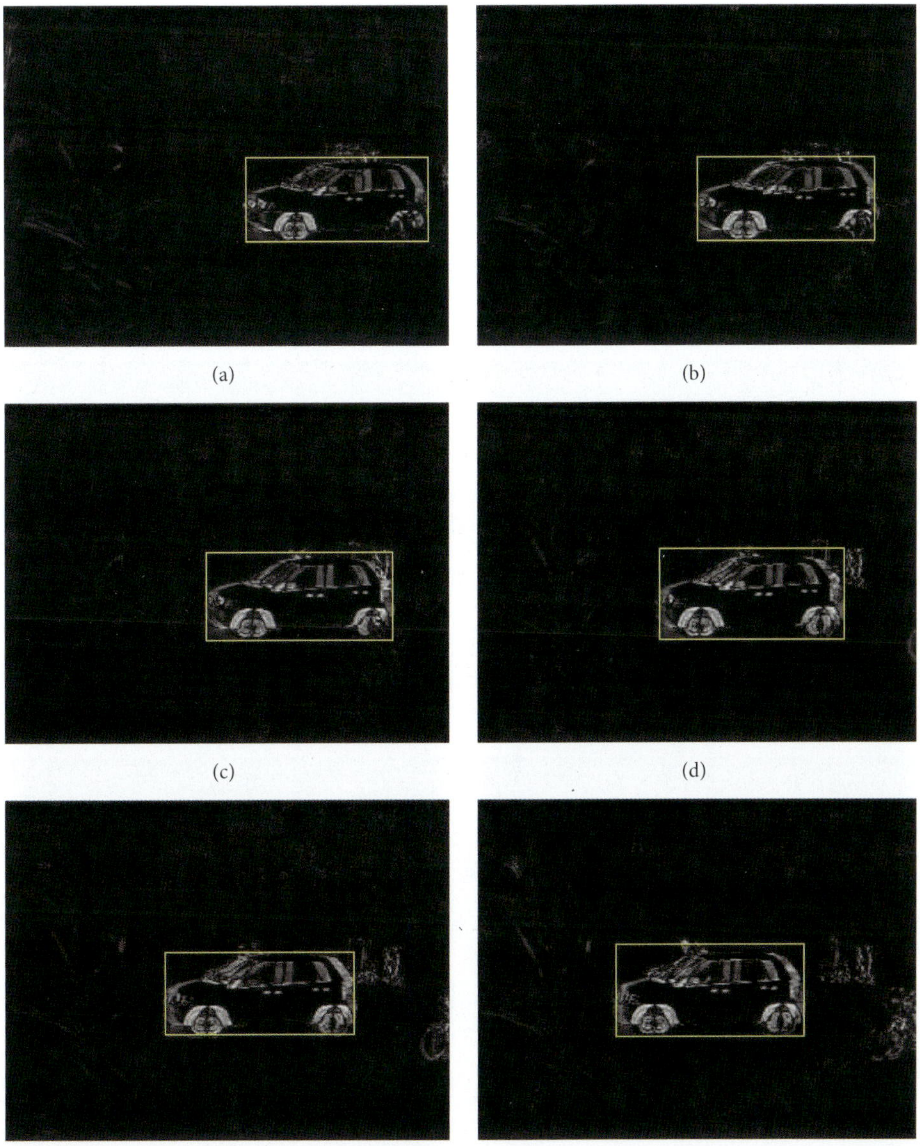

(a) (b) (c) (d) (e) (f)

FIGURE 10: Tracking a moving car using matched corner points by FAB descriptor; DOI are showing the new location of the moving object.

(a) (b)

FIGURE 11: Sample frames from a PETS2009 dataset View_001 captured using a fixed camera.

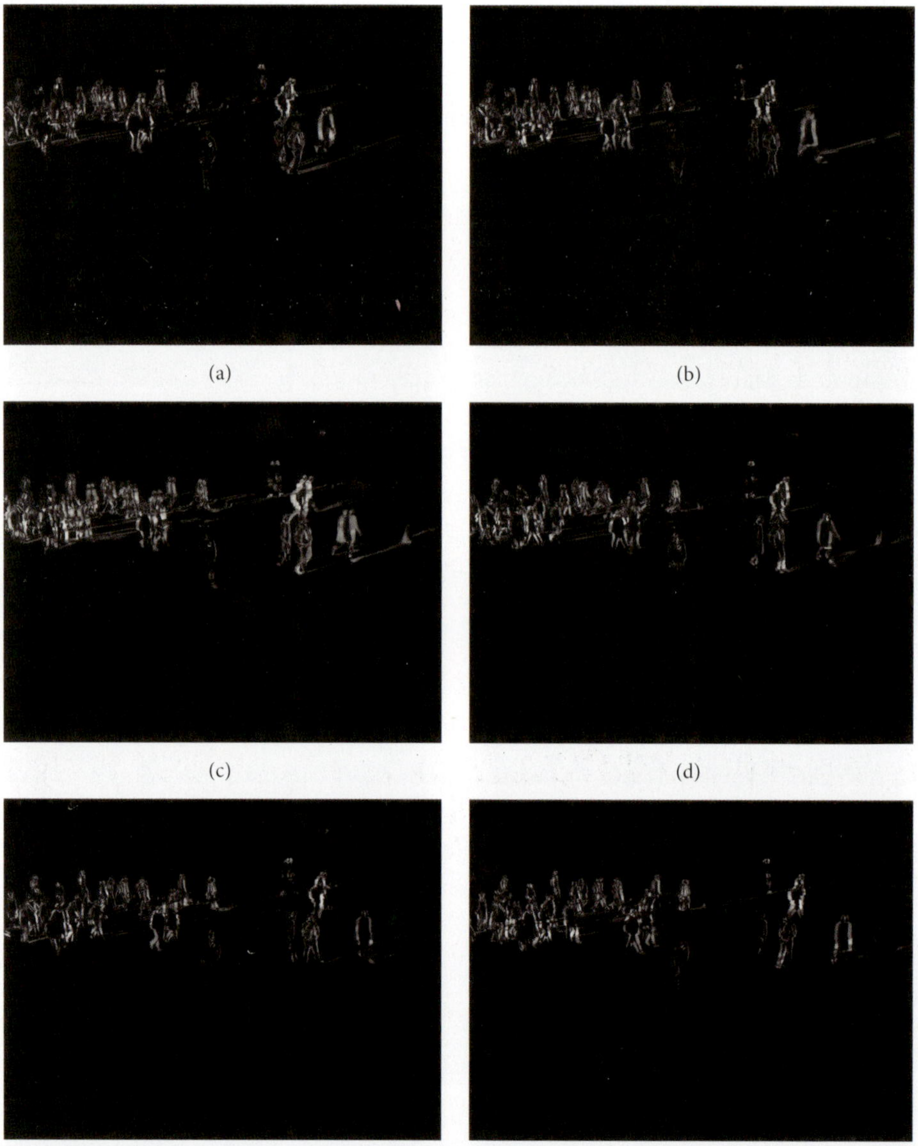

(a) (b) (c) (d) (e) (f)

FIGURE 12: PETS dataset is used for motion detection using surveillance cameras. Dataset used in this application is captured using still camera whereas objects are moving. The white area is showing that multiple persons are passing from this area. This application can be used to track if someone entered a place where entrance is not allowed.

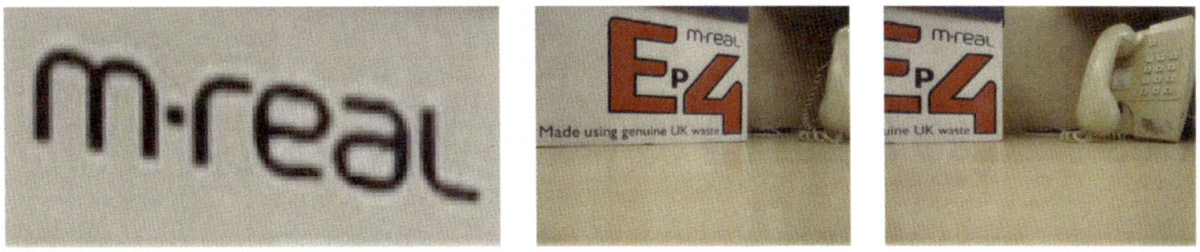

(a) (b) (c)

FIGURE 13: A template and some sample video frames for template matching. A moving handheld camera is used to capture this video in which a text template is matched.

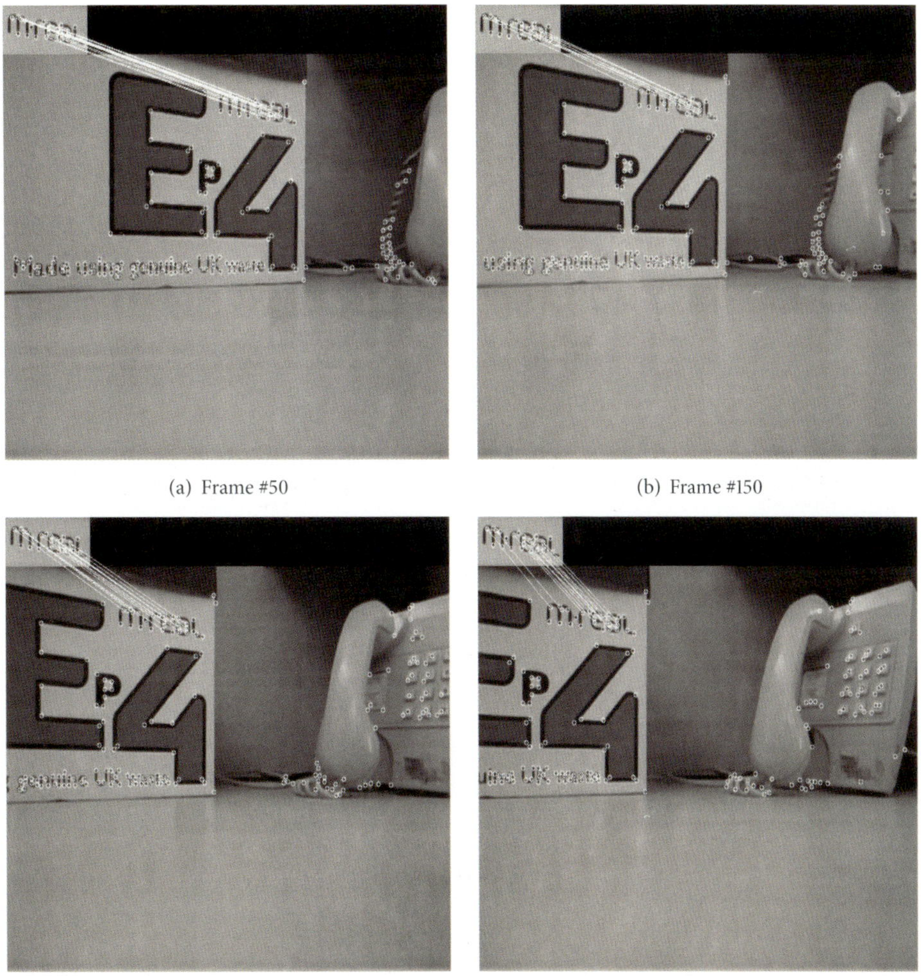

(a) Frame #50 (b) Frame #150

(c) Frame #250 (d) Frame #350

FIGURE 14: Template is successfully matched in whole video; however, some equally spaced frames are shown here.

BRIEF and AIME shows that it can be used in real-time applications with improved accuracy.

5. Conclusion

Efficient and robust feature descriptor is the key to match images accurately. Feature descriptor with less computation time and less storage space is the basic need of real-time applications. This paper presented a new hybrid descriptor called FAB (Fast Angular Binary) descriptor, containing two kinds of feature's neighbourhood information, first angle and orientation of a corner point and the second binary vector representing pixels' intensity comparison for only useful neighbours of corner point. This hybrid information assists in implementing a two-step matching technique which is actually the trivial part of FAB's success. Matching only those corner points which have similar angle and orientation makes search space significantly small and supportive for getting optimal results quickly.

Performance comparison of FAB with other state-of-the-art descriptors proved it to be more efficient, reliable, and accurate for different vision applications. For future modifications, FAB descriptor can loose orientation bit to achieve rotation and affine invariance, but with some added information.

Competing Interests

The authors declare that they have no competing interests.

References

[1] F. Porikli and A. Yilmaz, "Object detection and tracking," in *Video Analytics for Business Intelligence*, vol. 409 of *Studies in Computational Intelligence*, pp. 3–41, Springer, New York, NY, USA, 2012.

[2] P. Kapsalas and S. Kollias, "Affine morphological Shape Stable Boundary Regions (SSBR) for image representation," in *Proceedings of the 18th IEEE International Conference on Image Processing (ICIP '11)*, pp. 3381–3384, Brussels, Belgium, September 2011.

[3] M. Brown and D. G. Lowe, "Automatic panoramic image stitching using invariant features," *International Journal of Computer Vision*, vol. 74, no. 1, pp. 59–73, 2007.

[4] R. R. Igorevich, P. Park, J. Choi, and D. Min, "Two hand gesture recognition using stereo camera," *International Journal of Computer and Electrical Engineering*, vol. 5, no. 1, pp. 69–72, 2013.

[5] N. Kanwal, S. Ehsan, E. Bostanci, and A. F. Clark, "Evaluating the angular sensitivity of corner detectors," in *Proceedings of the IEEE International Conference on Virtual Environments, Human-Computer Interfaces and Measurement Systems (VECIMS '11)*, pp. 28–31, Ottawa, Canada, September 2011.

[6] J. Canny, "A computational approach to edge detection," *IEEE Transactions on Pattern Analysis and Machine Intelligence*, vol. 8, no. 6, pp. 679–698, 1986.

[7] A. J. Danker and A. Rosenfeld, "Blob detection by relaxation," *IEEE Transactions on Pattern Analysis and Machine Intelligence*, vol. 3, no. 1, pp. 79–92, 1981.

[8] S. Hinz, "Fast and subpixel precise blob detection and attribution," in *Proceedings of the IEEE International Conference on Image Processing (ICIP '05)*, vol. 3, pp. 457–460, September 2005.

[9] D. G. Lowe, "Object recognition from local scale-invariant features," in *Proceedings of the 7th IEEE International Conference on Computer Vision (ICCV'99)*, vol. 2, pp. 1150–1157, September 1999.

[10] H. Bay, T. Tuytelaars, and L. Van Gool, "SURF: speeded up robust features," in *Computer Vision—ECCV 2006*, A. Leonardis, H. Bischof, and A. Pinz, Eds., vol. 3951 of *Lecture Notes in Computer Science*, pp. 404–417, Springer, New York, NY, USA, 2006.

[11] M. Calonder, V. Lepetit, C. Strecha, and P. Fua, "BRIEF: binary robust independent elementary features," in *Computer Vision—ECCV 2010: 11th European Conference on Computer Vision, Heraklion, Crete, Greece, September 5-11, 2010, Proceedings, Part IV*, vol. 6314 of *Lecture Notes in Computer Science*, pp. 778–792, Springer, Berlin, Germany, 2010.

[12] E. Rublee, V. Rabaud, K. Konolige, and G. Bradski, "ORB: an efficient alternative to SIFT or SURF," in *Proceedings of the IEEE International Conference on Computer Vision (ICCV '11)*, pp. 2564–2571, Barcelona, Spain, November 2011.

[13] S. Leutenegger, M. Chli, and R. Y. Siegwart, "BRISK: Binary Robust invariant scalable keypoints," in *Proceedings of the IEEE International Conference on Computer Vision (ICCV '11)*, pp. 2548–2555, Barcelona, Spain, November 2011.

[14] A. Alahi, R. Ortiz, and P. Vandergheynst, "FREAK: fast retina keypoint," in *Proceedings of the IEEE Conference on Computer Vision and Pattern Recognition (CVPR '12)*, pp. 510–517, IEEE, Providence, RI, USA, June 2012.

[15] N. Kanwal, E. Bostanci, and A. F. Clark, "Matching corners using the informative arc," *IET Computer Vision*, vol. 8, no. 3, pp. 245–253, 2014.

[16] E. Bostanci, "Is Hamming distance only way for matching binary image feature descriptors?" *Electronics Letters*, vol. 50, no. 11, pp. 806–808, 2014.

[17] C. Harris and M. Stephens, "A combined corner and edge detector," in *Proceedings of the 4th Alvey Vision Conference*, vol. 15, pp. 147–151, Manchester, UK, September 1988.

[18] J. Ferryman and A. Shahrokni, "An overview of the pets 2009 challenge," in *Proceedings of the 11th IEEE International Workshop on Performance Evaluation of Tracking and Surveillance*, Miami, Fla, USA, 2009.

[19] E. Bostanci, N. Kanwal, and A. F. Clark, "Spatial statistics of image features for performance comparison," *IEEE Transactions on Image Processing*, vol. 23, no. 1, pp. 153–162, 2014.

11

Keyframes Global Map Establishing Method for Robot Localization through Content-Based Image Matching

Tianyang Cao,[1,2] Haoyuan Cai,[1] Dongming Fang,[1] Hui Huang,[1] and Chang Liu[1]

[1] *State Key Laboratory of Transducer Technology, Institute of Electronics, Chinese Academy of Sciences, Beijing 100190, China*
[2] *The University of Chinese Academy of Sciences, Beijing 100190, China*

Correspondence should be addressed to Haoyuan Cai; hycai@mail.ie.ac.cn and Chang Liu; tuengineer@qq.com

Academic Editor: Huosheng Hu

Self-localization and mapping are important for indoor mobile robot. We report a robust algorithm for map building and subsequent localization especially suited for indoor floor-cleaning robots. Common methods, for example, SLAM, can easily be kidnapped by colliding or disturbed by similar objects. Therefore, keyframes global map establishing method for robot localization in multiple rooms and corridors is needed. Content-based image matching is the core of this method. It is designed for the situation, by establishing keyframes containing both floor and distorted wall images. Image distortion, caused by robot view angle and movement, is analyzed and deduced. And an image matching solution is presented, consisting of extraction of overlap regions of keyframes extraction and overlap region rebuild through subblocks matching. For improving accuracy, ceiling points detecting and mismatching subblocks checking methods are incorporated. This matching method can process environment video effectively. In experiments, less than 5% frames are extracted as keyframes to build global map, which have large space distance and overlap each other. Through this method, robot can localize itself by matching its real-time vision frames with our keyframes map. Even with many similar objects/background in the environment or kidnapping robot, robot localization is achieved with position RMSE <0.5 m.

1. Introduction

The ideal indoor mobile robot should store a global map of the entire indoor space for self-localization, especially in a building including multiple rooms and corridors [1]. It would be much better if this global map can be built by the robot itself during indoor environment studying and without the help of human [2].

For this research field, SLAM (Simultaneous Localization and Mapping) is the commonly employed [3], especially the V-SLAM (Visual Simultaneous Localization and Mapping) [4, 5]. Many latest variants take advantage of new features or 3D information to build navigation map [6, 7], including ORB SLAM [8], dense SLAM [9, 10], semidense SLAM [11], LSD SLAM [12], and CV-SLAM [13, 14]. The CV-SLAM method is specially designed for indoor robot localization [15] and could make good use of ceiling features as navigation map through an upward-looking camera [16]. This technology has been used widely in the Roomba, a product of iRobot Inc. Similar methods are used in other models made by Dyson, SAMSUNG, and LG.

However, the SLAM method is easy to be disturbed, and the most troublesome problems include kidnap problem and similar objects interference [17]. The kidnap problem occurs frequently when a robot is suddenly involved in a collision, kicked, or intentionally repositioned during operation. The robot cannot localize itself according to video information of previous moments [18]. The similar object interference refers to the fact that a robot is easy to be confused by similar features on similar objects in different places and error fixed itself to the wrong position. An indoor robot that looks mostly at ceilings of indoor space do not detect many features of significance. To make matters worse, there are usually many similar objects in the indoor environment (e.g., air condition outlets, ceiling lamps, and ceiling). Hence established methods are not sufficiently reliable for global indoor positioning.

In order to solve these problems, this paper presents keyframes global map establishing method for robot localization through content-based image matching, involving both ceiling and wall regions. The robot could establish global map by itself and fix its position at any time without the need of its localization information of previous moments and cannot be kidnapped.

Comparing with the common keyframes extraction method (e.g., according to the robot moving time or distance), our keyframes extracted according to image content would be more useful (image content mainly including the objects in the room and their layout and their color distribution). Problems related to losing keyframes or redundancy can be avoided, especially when robot is changing speed or turning suddenly [19, 20].

In this paper, the keyframes global map establishing method is also based on a specially designed content-based image matching method, which can calculate images similarity exactly. The keyframes for each room and corridor can be extracted from robot environment studying video by image content matching. This matching method is designed to avert image distortion interference caused by robot view angle and movement. It is designed based on a distortion model, which is deduced to analyze the distortion features of ceiling and wall region (e.g., wall, door, window, and furniture). According to these features, two image processing methods, including overlap region extraction method and overlap region rebuild method through subblocks matching, are presented. We first *adjust the distortion of vision frames nearly consistent and then calculate the similarity between frames*. In order to adjust distortion more exactly, ceiling points detecting method and mismatching subblocks checking method are incorporated. Through content-based image matching, this paper can take advantage of the information of different objects imaged in different room to extract keyframes for each room from robot vision. *These keyframes have very large space distance, and the whole indoor environment could be described only by 10 s of keyframes*. Keyframes overlap with each other, and their position relationship can be extracted. These keyframes and their global position would compose global map for indoor environment. This content-based image matching method can also be used for robot self-localization. When the robot is moving in rooms or corridors, for the real-time updated robot vision, the most similar keyframe for each robot vision frame can be chosen automatically from this global map and the robot can fix its position precisely.

In the test, the experiment site is composed of two rooms (20-meter-square) and two corridors (11 m). The robot can move automatically in this site to study environment features, and a video about 9.5 min (1710 frames) long had been taken by robot vision during indoor environment studying. Through the map establishing method presented in this paper, 72 keyframes (less than 5% frames) are extracted from this video to build the map of experiment site. The robot can make good use of this map to localize itself in the whole building. It is able to draw robot trajectory exactly, even in the event of kidnapping, with position RMSE less than 0.4 m.

The rest of this paper is organized as follows. Section 2 describes the design of robot vision system and the distortion features of common indoor objects. In Section 3, we discuss content-based image matching method and the keyframes global map establishing method. In Sections 4 and 5, the experiment result and conclusion are presented and discussed.

2. Image Distortion Model and Feature Analysis for Common Indoor Objects

Content-based image matching with both ceilings and walls is one of the best ways for detecting keyframes and building global map, which can extract keyframes from environment video as efficiently as possible to describe the indoor environment and keep the connection between these frames. However, the main problem for content-based image matching is the image distortion. If two frames are taken in different view angle and position, the objects distortions in them are different, and image content similarity between the two frames is difficult to be calculated.

Here we discuss the method for analyzing features of image content distortion through establishing a distortion model. An upward-looking camera (SY012HD wide angle camera, Weixinshijie Technology Co., Ltd, Shenzhen, China) is installed on the wheeled robot (Roomba, R1-L081A, Midea, Suzhou, China) as the robot vision sensor [21], as shown in Figure 1(a), the same as that carrying a CV-SLAM [22, 23].

2.1. The Features of Robot View Angle and Movement for Ceiling and Wall Region. The camera model, which can change the three-dimensional object in the real world to two-dimensional picture, is as follows [24, 25]:

$$z\begin{bmatrix} x \\ y \\ 1 \end{bmatrix} = \begin{bmatrix} \dfrac{1}{dx} & 0 & 0 \\ 0 & \dfrac{1}{dy} & 0 \\ 0 & 0 & 1 \end{bmatrix} \begin{bmatrix} f & 0 & 0 \\ 0 & f & 0 \\ 0 & 0 & 1 \end{bmatrix} \begin{bmatrix} R & T \end{bmatrix} \begin{bmatrix} X \\ Y \\ Z \\ 1 \end{bmatrix}, \quad (1)$$

where

$$R = \begin{bmatrix} \cos R & 0 & -\sin R \\ 0 & 1 & 0 \\ \sin R & 0 & \cos R \end{bmatrix} \cdot \begin{bmatrix} 1 & 0 & 0 \\ 0 & \cos P & \sin P \\ 0 & -\sin P & \cos P \end{bmatrix} \cdot \begin{bmatrix} \cos H & \sin H & 0 \\ -\sin H & \cos H & 0 \\ 0 & 0 & 1 \end{bmatrix} \quad (2)$$

is calculated through robot view angles of roll R, pitch P, and heading H. The term $T = \begin{bmatrix} T_x \\ T_y \\ T_z \end{bmatrix}$ is calculated through robot movements T_x and T_y and ceiling height T_z, (X, Y, Z) are the coordinates of an object point in three-dimensional real world, and (x, y) are its corresponding points in the robot vision. The terms f, dx, and dy are the constant of the

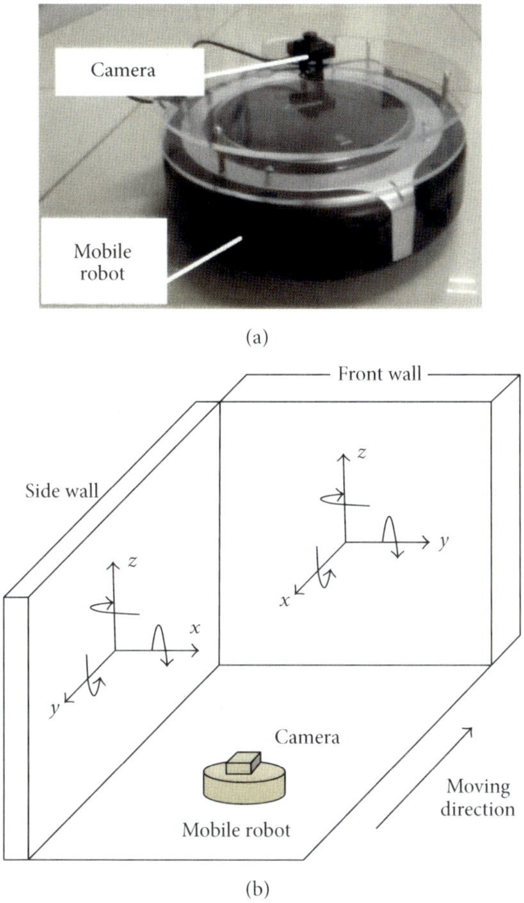

FIGURE 1: The relationship between wall and upward-looking camera. (a) Robot and camera. (b) Relationship between front wall and side wall and camera.

camera: f is the camera focus and dx and dy are the scale factors.

There are two kinds of main indoor objects: ceiling and wall regions. The ceiling is parallel to the floor, and wall regions (e.g., wall, window, door, and furniture) are perpendicular to the floor. For the upward-looking camera on the robot, the view angles for them are different.

When a robot is moving on the floor, its route parallels the ceiling, the heading (given H) is the only robot view angle can be changed, and both roll and pitch are equal to 0.

For the wall region, because they are fixed on the floor, the freedom of robot view angles is also very limited. We analyze each wall separately, by giving each wall an independent coordinate (front/back wall or side wall), as shown in Figure 1(b), with the X-axis perpendicular to the wall, Y-axis parallel to the wall and floor, and Z-axis perpendicular to the floor. The heading H between a wall and the robot is the only view angle which can be changed. As wall is perpendicular to floor and ceiling, it can be seen as a part of ceiling with roll or pitch of 90 deg. For the front wall (or back) of robot, the pitch is 90 deg and roll is 0, as shown in Figure 1(b). For the left wall (or right wall) on the robot, the roll is 90 deg and pitch is 0, as shown in Figure 1(b).

Because of the different view angles, ceiling and wall deform differently in the robot vision. In order to match precisely, this deformation should be revised before image content matching. According to (1), it can also be deduced that robot movement (T_x, T_y) and object position (X, Y, Z) can also contribute to image distortion. Therefore, bringing the robot view angle and movement and object position into (1), the distortion models of ceiling and wall region could be established, and their distortion features can be extracted.

2.2. Ceiling Distortion Model and Feature Extraction. Given the coordinate origin on the ceiling, T_x and T_y show robot movement, and T_z is ceiling height. The heading between ceiling and camera is H, and roll and pitch are 0. Therefore (1) can be transformed as follows:

$$z \begin{bmatrix} x \\ y \\ 1 \end{bmatrix} = \begin{bmatrix} \frac{1}{dx} & 0 & 0 \\ 0 & \frac{1}{dy} & 0 \\ 0 & 0 & 1 \end{bmatrix} \begin{bmatrix} f & 0 & 0 \\ 0 & f & 0 \\ 0 & 0 & 1 \end{bmatrix} \begin{bmatrix} \cos H & \sin H & 0 & T_x \\ -\sin H & \cos H & 0 & T_y \\ 0 & 0 & 1 & T_z \end{bmatrix} \begin{bmatrix} X \\ Y \\ Z \\ 1 \end{bmatrix}. \tag{3}$$

It can be simplified as follows:

$$\begin{aligned} x &= \frac{f}{dx(T_z + Z)} (X \cos H + Y \sin H + T_x), \\ y &= \frac{f}{dy(T_z + Z)} (-X \sin H + Y \cos H + T_y). \end{aligned} \tag{4}$$

The distortion of ceiling only includes rotation and translation, and the shape of ceiling is unchanged. As rotation and translation are affine transform, they can be revised by affine transform too. Through SURF (Speed Up Robust Features), the matched feature points in frames A and B are extracted [26, 27]. Bringing these points coordinate into (5), the heading difference ΔH and movement differences Δt_x and Δt_y between these two frames could be resolved [28]:

$$\begin{pmatrix} x_B \\ y_B \end{pmatrix} = \begin{pmatrix} \cos \Delta H & \sin \Delta H \\ -\sin \Delta H & \cos \Delta H \end{pmatrix} \begin{pmatrix} x_A \\ y_A \end{pmatrix} + \begin{pmatrix} \Delta t_x \\ \Delta t_y \end{pmatrix}. \tag{5}$$

Two frames, A and B, can be adjusted to the same robot view angles and camera shooting position after rotation and translation according to ΔH, Δt_x, and Δt_y. Their overlap regions (containing the same objects in two frames) could be extracted. The image similarity would be obtained by comparing the similarity of their overlap regions.

2.3. Wall Distortion Model and Feature Extraction. For the front wall ($P = 90$, $R = 0$) and side wall ($P = 0$, $R = 90$), the heading difference is 90 deg (or −90 deg), but their image distortions are the same and can be deduced as follows.

Given that the robot heading is H when it is taking frame A, for a point (X, Y, Z) on the front wall, its corresponding point (x_{Aw}, y_{Aw}) on frame A of robot vision is calculated as follows:

$$z\begin{bmatrix} x_{Aw} \\ y_{Aw} \\ 1 \end{bmatrix} = \begin{bmatrix} \frac{1}{dx} & 0 & 0 \\ 0 & \frac{1}{dy} & 0 \\ 0 & 0 & 1 \end{bmatrix}\begin{bmatrix} f & 0 & 0 \\ 0 & f & 0 \\ 0 & 0 & 1 \end{bmatrix}\left[\begin{bmatrix} 1 & 0 & 0 \\ 0 & 0 & 1 \\ 0 & -1 & 0 \end{bmatrix}\begin{bmatrix} \cos H & \sin H & 0 & T_x \\ -\sin H & \cos H & 0 & T_y \\ 0 & 0 & 1 & T_z \end{bmatrix}\right]\begin{bmatrix} X \\ Y \\ Z \\ 1 \end{bmatrix}. \quad (6)$$

And it can be simplified as follows:

$$x_{Aw} = \frac{f}{dx}\frac{X\cos H + Y\sin H + T_x}{X\sin H - Y\cos H + T_z},$$
$$y_{Bw} = \frac{f}{dy}\frac{Z + T_y}{X\sin H - Y\cos H + T_z}. \quad (7)$$

For the point (X_s, Y_s, Z_s) on the side wall, its corresponding point (x_{Aws}, y_{Aws}) on the frame A of robot vision is calculated as follows:

$$z\begin{bmatrix} x_{Aws} \\ y_{Aws} \\ 1 \end{bmatrix} = \begin{bmatrix} \frac{1}{dx} & 0 & 0 \\ 0 & \frac{1}{dy} & 0 \\ 0 & 0 & 1 \end{bmatrix}\begin{bmatrix} f & 0 & 0 \\ 0 & f & 0 \\ 0 & 0 & 1 \end{bmatrix}\left[\begin{bmatrix} 0 & 0 & -1 \\ 0 & 1 & 0 \\ 1 & 0 & 0 \end{bmatrix}\begin{bmatrix} \cos H & \sin H & 0 & T_x \\ -\sin H & \cos H & 0 & T_y \\ 0 & 0 & 1 & T_z \end{bmatrix}\right]\begin{bmatrix} X_s \\ Y_s \\ Z_s \\ 1 \end{bmatrix}. \quad (8)$$

And it can be simplified as

$$x_{Aws} = \frac{f}{dx}\frac{-Z_s + T_x}{X_s\cos H + Y_s\sin H + T_z}, \quad (9)$$

$$y_{Aws} = \frac{f}{dy}\frac{-X_s\sin H + Y_s\cos H + T_y}{X_s\cos H + Y_s\sin H + T_z}. \quad (10)$$

Comparing with (7), it means that the x-axis and y-axis terms are exchanged in (9) and (10). Accordingly, x, y are exchanged, and T_x, T_y are exchanged, too. The distortion feature of front wall in the robot vision is the same as that of side wall, and the distortion revise method for the front wall would be the same as that of side wall.

Take the distortion of front wall for example. Given the robot heading and movement for frame A are H, t_x, t_y for frame B which would be marched with frame A, its parameters are $H + \Delta H$, $t_x + \Delta t_x$, and $t_y + \Delta t_y$. According to (8), the object point (X, Y, Z) on frame B is (x_{Bw}, y_{Bw}) and is calculated as follows:

$$x_{Bw} = \frac{f}{dx} \cdot \frac{X\cos(H+\Delta H) + Y\sin(H+\Delta H) + T_x + \Delta t_x}{X\sin(H+\Delta H) - Y\cos(H+\Delta H) + T_z}, \quad (11)$$

$$y_{Bw} = \frac{f}{dy}\frac{Z + T_y + \Delta t_y}{X\sin(H+\Delta H) - Y\cos(H+\Delta H) + T_z}. \quad (12)$$

For the ceiling and wall region in frames A and B, the deformations of their ceiling images, including rotation and translation, can be adjusted similarly through (5). However, the wall region in frame A, after rotation and translation, would be transformed differently from that of frame B:

$$\begin{pmatrix} x_{AB} \\ y_{AB} \end{pmatrix} = \begin{pmatrix} \cos\Delta H & \sin\Delta H \\ -\sin\Delta H & \cos\Delta H \end{pmatrix}\begin{pmatrix} \frac{f}{dx}\frac{X\cos H + Y\sin H + T_x}{X\sin H - Y\cos H + T_z} \\ \frac{f}{dy}\frac{Z + T_y}{X\sin H - Y\cos H + T_z} \end{pmatrix} + \begin{pmatrix} \Delta t_x \\ \Delta t_y \end{pmatrix}, \quad (13)$$

where (x_{AB}, y_{AB}) is point in frame A processed by (5) and can be simplified as

$$x_{AB} = \frac{f}{dx}\frac{X\cos H\cos\Delta H + Y\sin H\cos\Delta H + T_x\cos\Delta H + Z\sin\Delta H + T_y\sin\Delta H}{X\sin H - Y\cos H + T_z} + \Delta t_x,$$
$$y_{AB} = \frac{f}{dy}\frac{-X\cos H\sin\Delta H - Y\sin H\sin\Delta H + Z\cos\Delta H - T_x\sin\Delta H + T_y\cos\Delta H}{X\sin H - Y\cos H + T_z} + \Delta t_y. \quad (14)$$

If this point is translated by an amount (S_x, S_y),

$$\begin{aligned} S_x &= \cos\Delta H \frac{f}{dx} \frac{-T_x}{X\sin H - Y\cos H + T_z} \\ &\quad + \sin\Delta H \frac{f}{dx} \frac{-X\sin H + Y\cos H - Z - T_y}{X\sin H - Y\cos H + T_z} \\ &\quad + \frac{f}{dx} \frac{\Delta t_x}{X\sin H - Y\cos H + T_z} - \Delta t_x, \\ S_y &= \sin\Delta H \frac{f}{dy} \frac{X\cos H + Y\sin H + T_x}{X\sin H - Y\cos H + T_z} \\ &\quad + \cos\Delta H \frac{f}{dy} \frac{-Z - T_y}{X\sin H - Y\cos H + T_z} \\ &\quad + \frac{f}{dy} \frac{\Delta t_y + Z + T_y}{X\sin H - Y\cos H + T_z} - \Delta t_y. \end{aligned} \tag{15}$$

The transform result is

$$\begin{aligned} x_{AB} &+ S_x \\ &= \frac{f}{dx} \frac{X\cos(H+\Delta H) + Y\sin(H+\Delta H) + T_x + \Delta t_x}{X\sin H - Y\cos H + T_z} \\ y_{AB} &+ S_y = \frac{f}{dy} \frac{Z + T_y + \Delta t_y}{X\sin H - Y\cos H + T_z}. \end{aligned} \tag{16}$$

Comparing with (10) and (11), the transformed frame A is much closer to frame B than the original frame A. The similarity between frames A and B can be calculated more exactly through matching with transformed frame A compared to original frame A.

But there is still some difference between transformed frame A and frame B. On one hand, their denominators are different, which are $X\sin H - Y\cos H + T_z$ and $X\sin(H + \Delta H) - Y\cos(H + \Delta H) + T_z$. On the other hand, some terms in (15) are unknown (e.g., point coordinate (X, Y, Z) in the room space), as the terms (S_x, S_y) cannot be calculated through the two equations. In order to neutralize these disadvantages, overlap region extraction method and subblock matching method would be presented, and (S_x, S_y) can be resolved through subblock matching. The term subblock is defined as follows. An image is divided into many small blocks which are equal sized, and these blocks are named as subblock in this paper. Equation (5) would also be accomplished through overlap region extraction method designed in this paper.

If frames A and B contain the same image content, their transform results would be very similar after being processed by (5) and (15) and very useful for image matching and similarity analysis. For the same ceiling regions in frames A and B, the correlation coefficient between them would be very large. For the same wall regions in two frames, as containing the same objects, the number of correct matching subblocks would be also very large. This paper would combine them together to analyze the similarity between frames A and B.

3. Match Method Design

Content-based image matching method would be designed firstly and then the keyframes global map can be established through this matching method.

3.1. Content-Based Image Matching through Overlap Region Extraction and Subblocks Rebuild. The content-based image matching method is designed according to the features of image distortion and consists of three parts: image overlap region extraction, overlap region rebuild through subblock matching, and, lastly, image content similarity calculating, as shown in Figure 2. For frames A and B, this method would adjust the distortion of frame A similar to that of frame B and then calculate their degree of similarity. In order to adjust more exactly, ceiling points detecting method is designed to improve the accuracy of overlap region extraction, and mismatched subblocks checking method is deigned to rebuilt the overlap region more exactly.

3.1.1. Image Overlap Region Extraction. For frames A and B, which are taken by robot at different positions, only parts of them might overlap. Content-based image matching would focus on analyzing the image similarity in this overlap region.

In order to extract this overlap region, frames A and B would be adjusted to the same view angle and camera shooting position according to their matched feature points extracted through the SURF method. If all the feature points are on the ceiling, the translation Δt_x, Δt_y, and rotation ΔH between the two frames can be calculated directly according to (5). Through rotation and translation, the same objects in frames A and B would be adjusted to the same position, and this image processing progress is shown in Figure 3. The overlap regions can be extracted effectively from these two adjusted frames.

The image procession result is shown in Figure 4. The rotation and translation of frames A and B are shown in Figures 4(c) and 4(d), and the same objects are overlapped, as shown in Figure 4(e). The overlap region mask is the pixels covered by the two image, as shown in Figure 4(f). The overlap regions in frames A and B can be extracted through this mask.

If the feature points were extracted from wall regions, they would interfere with the calculation result of Δt_x, Δt_y, and ΔH. In order to delete the points not on the ceiling (including wall, furniture, window, and door), this paper designs ceiling points detecting method according the feature of ceiling, which are of equal height.

For the ceiling points a and b, given that the origin of coordinate is on the ceiling and ceiling points $Z = 0$, according to (4), the length of their connect lines in frames A and B (for points "a" and "b" in Figures 5(a) and 5(b), their connect line is "line ab") can be calculated as follows:

$$\begin{aligned} d_{A,ab} &= \left((x_{A,a} - x_{A,b})^2 + (y_{A,a} - y_{A,b})^2 \right)^{1/2} \\ &= \left(\left(\frac{f}{dxT_z} \right)^2 ((X_a \cos H + Y_a \sin H + T_{x,A}) \right. \end{aligned}$$

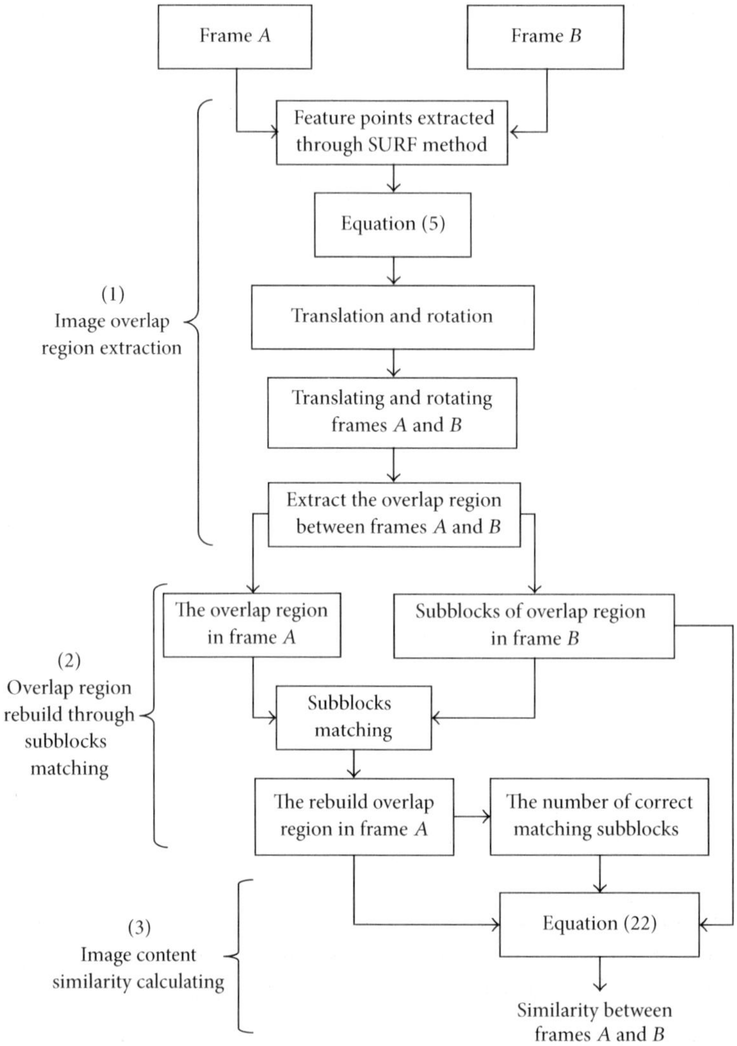

FIGURE 2: The image processing of content-based image matching.

FIGURE 3: The image processing of rotation and translation.

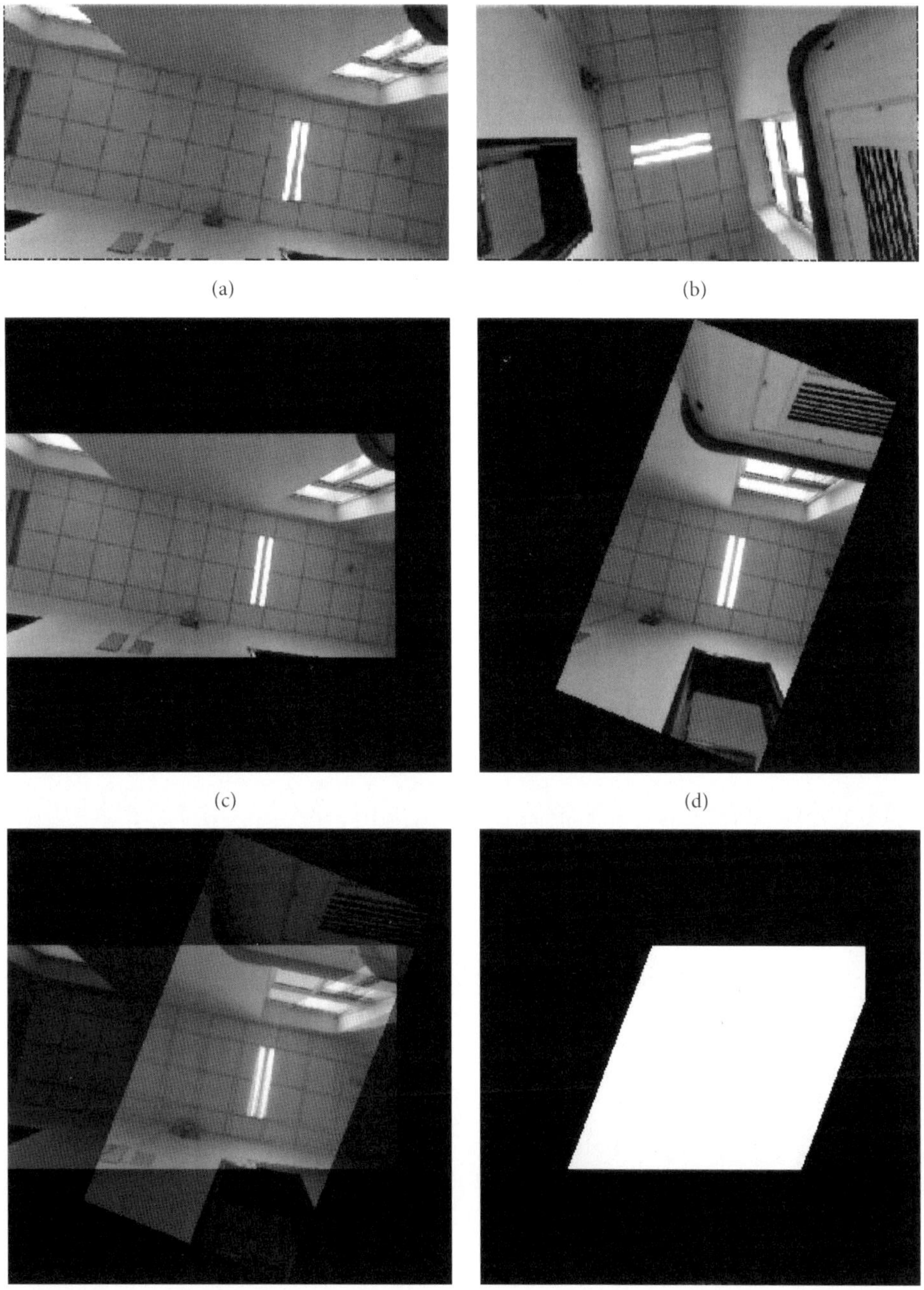

FIGURE 4: The overlap region extraction progress. (a) Frame A, (b) Frame B, (c) the translation result, (d) the rotation result, (e) the overlap between two frames, and (f) the overlap region mask.

$$- \left(X_b \cos H + Y_b \sin H + T_{x,A}\right)\right)^2 - \left(\frac{f}{dyT_y}\right)^2$$

$$\cdot \left(\left(-X_a \sin H + Y_b \cos H + T_{y,A}\right) - \left(-X_a \sin H\right.\right.$$

$$\left.\left.+ Y_b \cos H + T_{y,A}\right)\right)^2\Bigg)^{1/2} = \left(\left(\frac{f}{dxT_z}\right)^2 (X_a \cos H\right.$$

$$+ Y_a \sin H - X_b \cos H - Y_b \sin H)^2 + \left(\frac{f}{dyT_y}\right)^2$$

$$\cdot (-X_a \sin H + Y_b \cos H + X_a \sin H - Y_b$$

$$\left.\cdot \cos H)^2\right)^{1/2},$$

$$d_{B,ab} = \left(\left(x_{B,a} - x_{B,b}\right)^2 + \left(y_{B,a} - y_{B,b}\right)^2\right)^{1/2}$$

$$= \left(\left(\frac{f}{dxT_z}\right)^2 ((X_a \cos (H + \Delta H)\right.$$

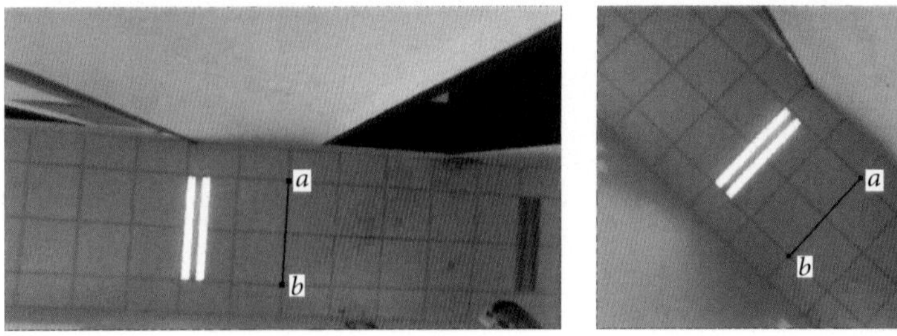

(a) (b)

FIGURE 5: The connect lines on the ceiling. (a) The connect lines on the original ceiling. (b) The connect lines on the ceiling after rotation and translation.

$$\begin{aligned}&+ Y_a \sin(H+\Delta H) + T_{x,B}) - (X_b \cos(H+\Delta H)\\ &+ Y_b \sin(H+\Delta H) + T_{x,B}))^2 - \left(\frac{f}{dyT_y}\right)^2\\ &\cdot\big((-X_a \sin(H+\Delta H) + Y_b \cos(H+\Delta H) + T_{y,B})\\ &- (-X_a \sin(H+\Delta H) + Y_b \cos(H+\Delta H)\\ &+ T_{y,B}))^2\big)^{1/2} = \left(\left(\frac{f}{dxT_z}\right)^2 (X_a \cos H + Y_a \sin H\right.\\ &- X_b \cos H - Y_b \sin H)^2 + \left(\frac{f}{dyT_y}\right)^2 (-X_a \sin H\\ &\left. + Y_b \cos H + X_a \sin H - Y_b \cos H)^2\right)^{1/2} = d_{A,ab},\end{aligned} \tag{17}$$

where (X_a, Y_a, Z_a) and (X_b, Y_b, Z_b) are the coordinates of points a and b in the real world, $(x_{A,a}, y_{A,a})$ and $(x_{A,b}, y_{A,b})$ are the coordinates of points a and b in frame A, and $(x_{B,a}, y_{B,a})$ and $(x_{B,b}, y_{B,b})$ are the coordinates of points a and b in frame B. $(T_{x,A}, T_{y,A})$ and $(T_{x,B}, T_{y,B})$ are the robot movement for frames A and B. As points a and b are equal in height, it can be deduced by (17) that $d_{A,ab} = d_{B,ab}$ and the length of connect lines of the same ceiling points in the two images are unchanged, as shown in Figure 5.

However, if one of the feature points b_w is on the wall (given height is Z_{bw}) and not on the ceiling, its height would differ from that of ceiling (T_z). In this case, the length of the connect line is

$$\begin{aligned}d_{A,abw} &= \left((x_{A,a} - x_{A,bw})^2 + (y_{A,a} - y_{A,bw})^2\right)^{1/2}\\ &= \left(\left(\frac{f}{dxT_z}(X_a \cos H + Y_a \sin H + T_{x,A}) - \frac{f}{dx}\right.\right.\\ &\left.\cdot \frac{X_{bw} \cos H + Y_{bw} \sin H + T_{x,A}}{X_{bw} \sin H - Y_{bw} \cos H + T_z}\right)^2\\ &+ \left(\frac{f}{dyT_z}(-X_a \sin H + Y_{bw} \cos H + T_{y,A}) - \frac{f}{dy}\right.\\ &\left.\left.\cdot \frac{Z_{bw} + T_{y,A}}{X_a \sin H - Y_{bw} \cos H + T_z}\right)\right)^{1/2},\end{aligned} \tag{18}$$

$$\begin{aligned}d_{B,abw} &= \left((x_{B,a} - x_{B,bw})^2 + (y_{B,a} - y_{B,bw})^2\right)^{1/2}\\ &= \left(\left(\frac{f}{dxT_z}(X_a \cos(H+\Delta H) + Y_a \sin(H+\Delta H)\right.\right.\\ &+ T_{x,B}) - \frac{f}{dx}\\ &\left.\cdot \frac{X_{bw} \cos(H+\Delta H) + Y_{bw} \sin(H+\Delta H) + T_{x,B}}{X_{bw} \sin(H+\Delta H) - Y_{bw} \cos(H+\Delta H) + T_z}\right)^2\\ &+ \left(\frac{f}{dyT_z}(-X_a \sin(H+\Delta H) + Y_{bw} \cos(H+\Delta H)\right.\\ &+ T_{y,B}) - \frac{f}{dy}\\ &\left.\left.\cdot \frac{Z_{bw} + T_{y,B}}{X_a \sin(H+\Delta H) - Y_{bw} \cos(H+\Delta H) + T_z}\right)\right)^{1/2}\\ &\neq d_{A,abw},\end{aligned} \tag{19}$$

where $(x_{A,bw}, y_{A,bw})$ and $(x_{B,bw}, y_{B,bw})$ are coordinates of wall point b_w in frames A and B and $d_{A,abw}$ and $d_{B,abw}$ are the length of connect lines ab_w in these two frames, as ΔH is in the denominator of (19), and (18) and (19) are not equal, and $d_{A,abw} \neq d_{B,abw}$. Therefore, the points on the wall can be deleted effectively by comparing the length of connect lines of feature points in the two frames, and the feature points on the ceiling can be reserved. The image process is shown in the Figure 6.

Therefore the points on the ceiling can be extracted by comparing the length of their connect lines in frames A and B. And these points can make the calculating result of (5) more exact and extracting overlap regions more effective. In the overlap regions, as rotation and translation have been

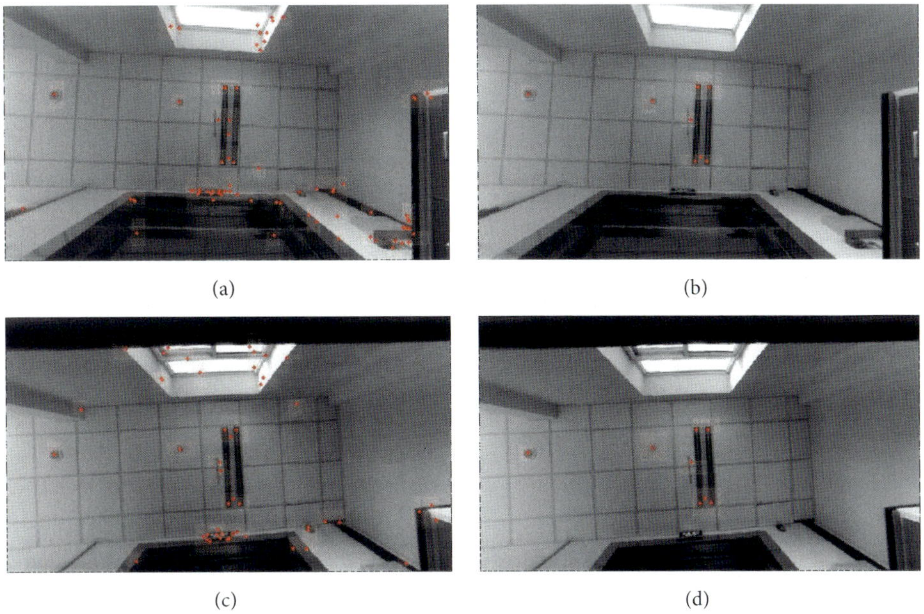

FIGURE 6: The image processing of ceiling feature point detection (red points are the feature points). (a) Original feature points in frame A, (b) ceiling points detecting result in frame A, (c) original feature points in frame B, and (d) ceiling points detecting result in frame B.

completed, the difference of ceiling image between frames A and B caused by robot rotation and translation can be deleted. However, for the wall image distortion difference between these two frames, the distortion adjustment would be completed in the next section.

3.1.2. Overlap Region Rebuild through Subblocks Matching. For the points on the ceiling region, it is difficult to calculate the translation result (S_x, S_y) of each point in the wall directly through (15), as robot cannot measure each point coordinates (X, Y, Z) in the room through its monocular camera. This paper presents subblocks matching method to calculate the translation value.

As the distorted object in the frame is still an entity in the image, the translation values (S_x, S_y) of distorted object points in a small image region are nearly equal. Therefore this region can be translated as smaller units according to the average translation value $(\overline{S_x}, \overline{S_y})$ of this region. The overlap region in frame A can be divided into many small subblocks firstly and then match the overlap region in frame B through SAD method (Sum of Absolute Difference) to get the average translation value $(\overline{S_x}, \overline{S_y})$ of each block. Therefore the distortion of overlap region in frame A can be adjusted similar to that of frame B, as shown in Figure 7.

The SAD matching method is as follows:

$$v(i,j) = \arg \min_{\substack{i \in [1,L] \\ j \in [1,L]}} \sum_{s=1}^{M} \sum_{t=1}^{N} \left| B_L(i+s-1, j+t-1) - A_T(s,t) \right|, \quad (20)$$

where A_T is one of the subblocks in the overlap region in frame A, and size is $M \times N$, and B_L is the overlap region in frame B, and size is $L \times D$. Traversal B_L, the terms (i, j), which can make (20) smallest, is the most suitable rebuild position of A_T in B_L. This (i, j) is the $(\overline{S_x}, \overline{S_y})$ of subblock A_T. Through this method, every subblock in frame A can find its rebuild position and get its translation $(\overline{S_x}, \overline{S_y})$, and the rebuild overlap region of frame A is similar to that of frame B.

Considering that there might be some mismatched subblocks, which are caused by similar objects and mismatched to wrong position, this paper presents mismatched subblocks checking method to check the wrong $(\overline{S_x}, \overline{S_y})$.

As $(\overline{S_x}, \overline{S_y})$ is the average of (S_x, S_y) of all the points in a subblock; the performance of $(\overline{S_x}, \overline{S_y})$ is very similar to that of (S_x, S_y). For (S_x, S_y), (15) can changed into matrix from:

$$\begin{aligned}
\begin{pmatrix} S_x \\ S_y \end{pmatrix} &= \frac{f}{X \sin H - Y \cos H + T_z} \\
&\cdot \begin{pmatrix} \frac{1}{dx} & 0 \\ 0 & \frac{1}{dy} \end{pmatrix} \begin{pmatrix} \cos(-H) & \sin(-H) \\ -\sin(-H) & \cos(-H) \end{pmatrix} \begin{pmatrix} Y \\ X \end{pmatrix} \\
&\cdot \sin \Delta H - \frac{f}{X \sin H - Y \cos H + T_z} \\
&\cdot \begin{pmatrix} \frac{1}{dx} & 0 \\ 0 & \frac{1}{dy} \end{pmatrix} \begin{pmatrix} \cos \Delta H & \sin \Delta H \\ -\sin \Delta H & \cos \Delta H \end{pmatrix} \begin{pmatrix} T_x \\ Z + T_y \end{pmatrix} \\
&+ \frac{f}{X \sin H - Y \cos H + T_z} \\
&\cdot \begin{pmatrix} \frac{1}{dx} & 0 \\ 0 & \frac{1}{dy} \end{pmatrix} \begin{pmatrix} \Delta t_x \\ \Delta t_y + T_y + Z \end{pmatrix} - \begin{pmatrix} \Delta t_x \\ \Delta t_y \end{pmatrix}.
\end{aligned} \quad (21)$$

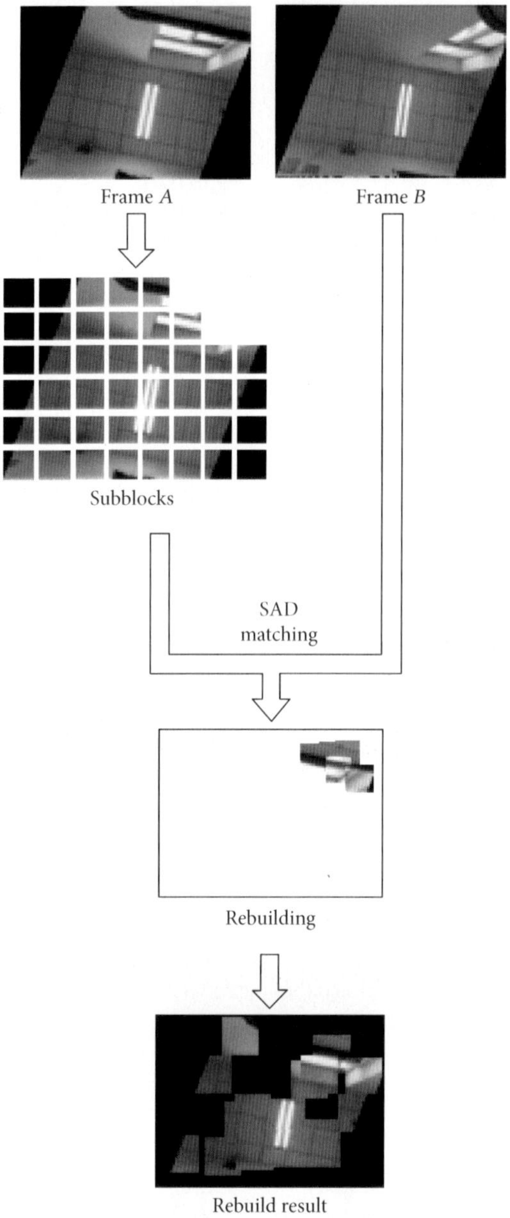

FIGURE 7: The overlap region rebuilding process.

It can be seen from (21) that the translation (S_x, S_y) is constituted of two parts: the projection of heading difference ΔH produced by the first and second term and the translation produced by the third and fourth term. The core of first term is $\begin{pmatrix} \cos(-H) & \sin(-H) \\ -\sin(-H) & \cos(-H) \end{pmatrix}\begin{pmatrix} Y \\ X \end{pmatrix} \sin \Delta H$, and it is the function of $\sin \Delta H$, which is the projection of ΔH. The core of second term is $\begin{pmatrix} \cos \Delta H & \sin \Delta H \\ -\sin \Delta H & \cos \Delta H \end{pmatrix}\begin{pmatrix} T_x \\ Z+T_y \end{pmatrix}$, the same as the rotation part in (5); it shows frame A rotating ΔH during overlap region extraction. So the subtraction result of the first two terms in (21) can extract the projection of heading difference ΔH between frame A and B and excludes the rotation caused by overlap region extraction. The core of third term is $\begin{pmatrix} \Delta t_x \\ \Delta t_y + T_y + Z \end{pmatrix}$; it is the translation difference between frames A and B caused by wall object height and robot movement difference between frames A and B. For the fourth term, the same as the translation part in (5), it shows frame A translation result during overlap region extraction. So the subtraction result of the last two terms in (21) can extract the translation difference between frames A and B and exclude the translation of overlap region extraction.

Therefore the wrong $(\overline{S_x}, \overline{S_y})$ of mismatched subblocks can be picked out according to the rotation difference and translation difference between frames A and B after overlap region extraction, as $(f/(X \sin H - Y \cos H + T_z))\begin{pmatrix} 1/dx & 0 \\ 0 & 1/dy \end{pmatrix}$ is the projection ratio between real world and camera image, and (X, Y, Z) can be substituted by its corresponding subblock coordinate (x, y). The thresholds $(\Delta P_x, \Delta P_y)$ for $(\overline{S_x}, \overline{S_y})$ of subblock in (x, y) can be calculated by affine model:

$$\begin{pmatrix} \Delta P_x \\ \Delta P_y \end{pmatrix} = \begin{pmatrix} \cos \Delta H_T & \sin \Delta H_T \\ -\sin \Delta H_T & \cos \Delta H_T \end{pmatrix}\begin{pmatrix} x \\ y \end{pmatrix} + \begin{pmatrix} \Delta x_T \\ \Delta y_T \end{pmatrix}, \quad (22)$$

where ΔH_T, Δx_T, and Δy_T are the thresholds of rotation difference and translation difference between frames A and B after overlap region extraction. In order to delete the wrong mismatched subblocks effectively, ΔH_T, Δx_T, and Δy_T are smaller than the maximums of rotation difference and translation difference.

If $(\overline{S}_x, \overline{S}_y)$ of a subblock is larger than its threshold $(\Delta P_x, \Delta P_y)$, it can be deduced that this block is mismatched and should be deleted. Through mismatched block detection, the correct content-based image matching result can be gotten. The matching result for two similar frames is shown in Figure 8, and the matching result for the dissimilar frames is shown in Figure 9. This subblock matching method can match similar frames and evaluate their similarity effectively.

3.1.3. Image Content Similarity Calculating. If the content of frames A and B is similar, the number of correct matched subblocks is very great, and the correlation coefficients between rebuild overlap region in frame A and overlap region in frame B are also very large [29, 30], and the similarity S_k between two frames is the product of subblock number and correlation coefficients:

$$S_{AB} = N_k \cdot \frac{\sum_{x,y}\left(C_A(x,y) - \overline{C}_A\right)\left(C_B(x,y) - \overline{C}_B\right)}{\sqrt{\sum_{x,y}\left(C_A(x,y) - \overline{C}_A\right)^2 \cdot \sum_{x,y}\left(C_B(x,y) - \overline{C}_B\right)^2}}, \quad (23)$$

where N_k is the number of matched subblock, $C_A(x, y)$ is the pixel value in rebuild overlap region in frame A, $C_B(x, y)$ is the pixel value in overlap region in frame B, and $\overline{C}_A$ and $\overline{C}_B$ are the average image pixel value.

3.2. Keyframes Global Map Establishing through Content-Based Image Matching. Before establishing map, robot would move in the building automatically to take a video of indoor environment through its camera firstly. And then, through this content-based image matching method, robot can extract keyframes sequence from vision video to build global map of indoor environment by itself.

The first keyframe is the first frame of vision video. And the other keyframes extraction progress is as follows:

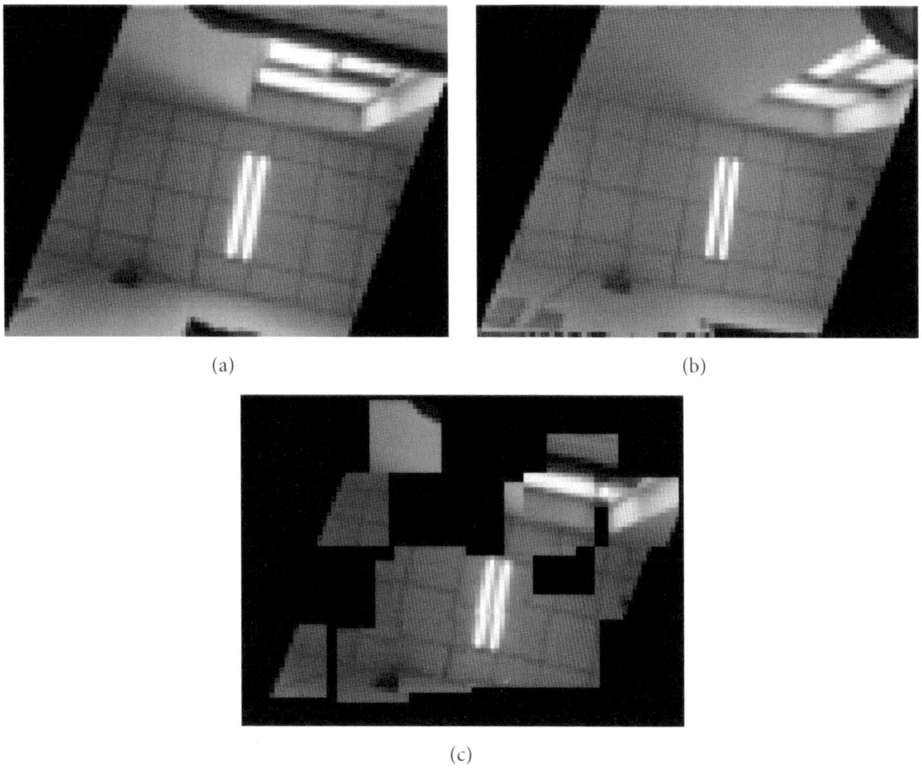

FIGURE 8: The rebuild overlap region by two similar frames. (a) The overlap region in frame A, (b) the overlap region in frame B, and (c) the rebuild overlap region in frame A.

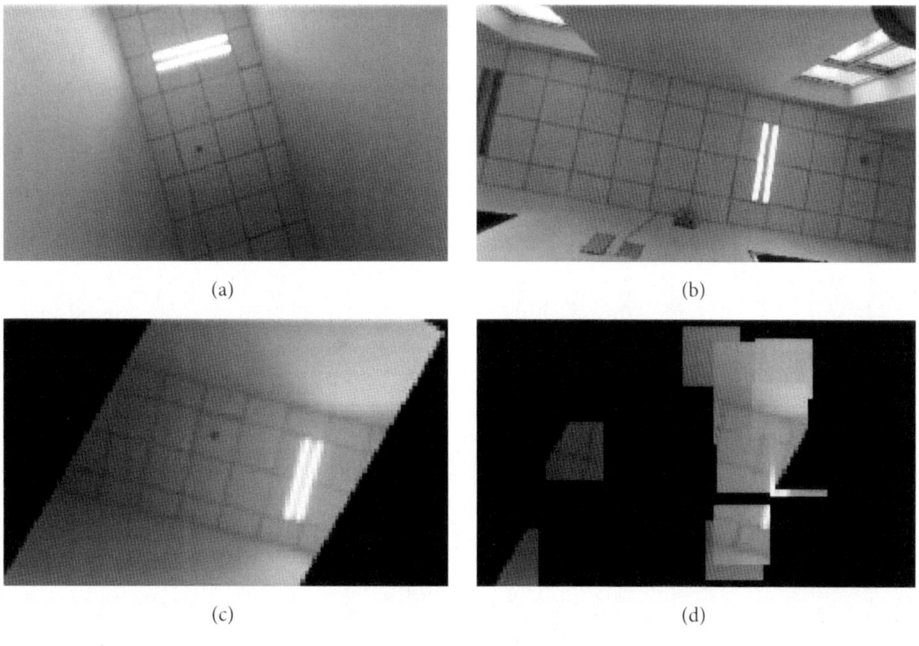

FIGURE 9: The rebuild overlap region by two dissimilar frames. (a) Frame A, (b) frame B, (c) the overlap region in frame A, and (d) the rebuild overlap region in frame A.

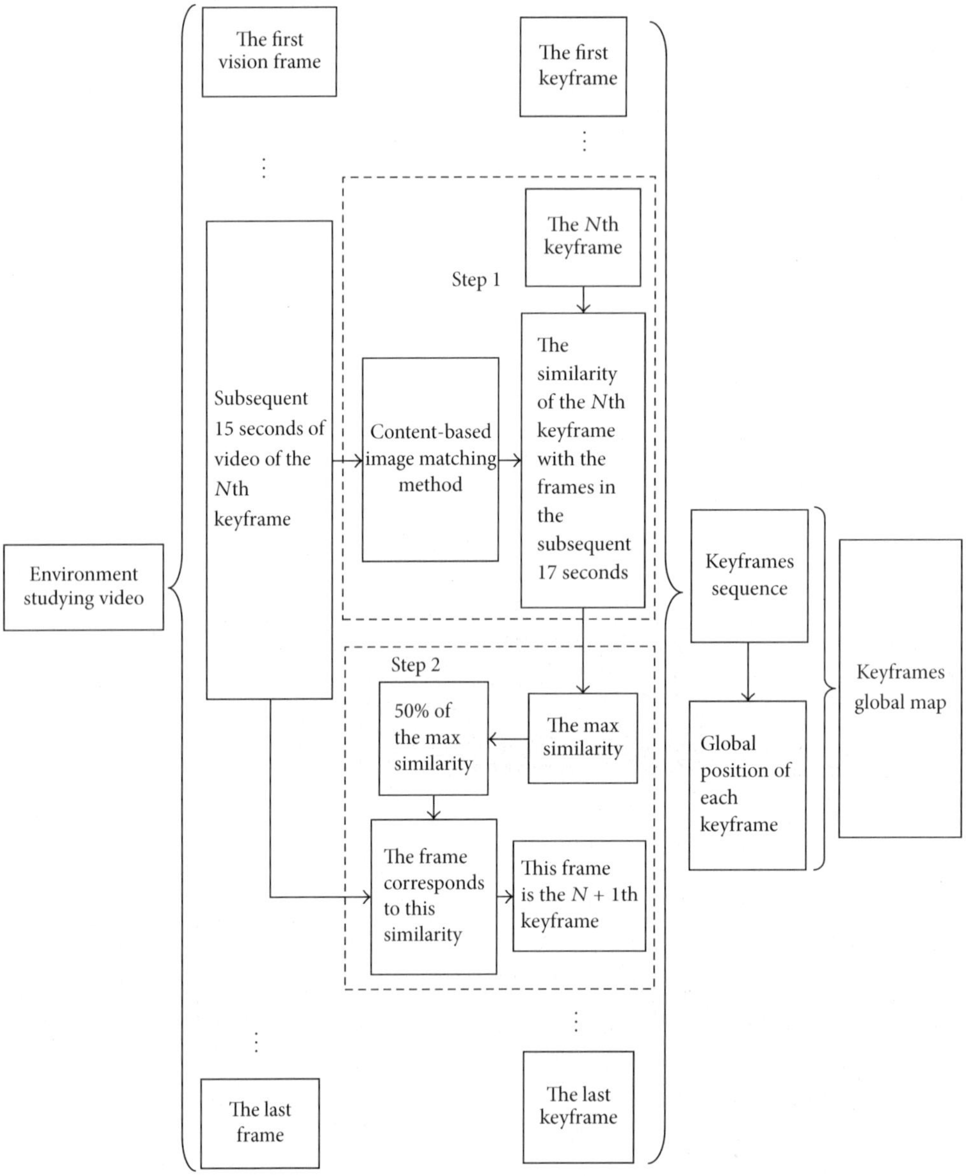

FIGURE 10: The image processing of keyframes global establishing.

Step 1. For the Nth keyframe, its similarity with subsequent 50 frames video (about 17 seconds) is calculated by the robot.

Step 2. The max similarity in the subsequent 50 frames is found out firstly, and it is correspondent to the frame whose space position is nearest to the Nth keyframe. And, then, the frame whose similarity is 50% of the max similarity can be extracted as the $N+1$th keyframe. If all the subsequent 50 frames are larger than 50% of the max similarity, the 50th frame is the $N+1$th keyframe. As there is 50% similarity between the N and $N+1$ keyframes, they can both interval long space distance and overlap each other.

Repeating Steps 1 and 2 and processing all the frames of indoor environment video, keyframes sequence can be extracted, as shown in Figure 10.

This global map would be established by these keyframes, and consists of two parts: keyframes sequence and the global position of each keyframe. As the keyframes can partially overlap each other, extracting the feature points in the overlap region between neighboring keyframes and bringing them into (5), the relative position relationship of these keyframes, including heading difference $H_{N+1,N}$ and position difference $x_{N+1,N}$, $y_{N+1,N}$, would be resolved. Then, bringing these relation positions into (24), the global position of each keyframe can be gotten:

$$\begin{pmatrix} x_{N+1} \\ y_{N+1} \end{pmatrix} = \begin{pmatrix} \cos H_{N+1,N} & \sin H_{N+1,N} \\ -\sin H_{N+1,N} & \cos H_{N+1,N} \end{pmatrix} \begin{pmatrix} x_{N+1,N} \\ y_{N+1,N} \end{pmatrix} + \begin{pmatrix} x_N \\ y_N \end{pmatrix}, \tag{24}$$

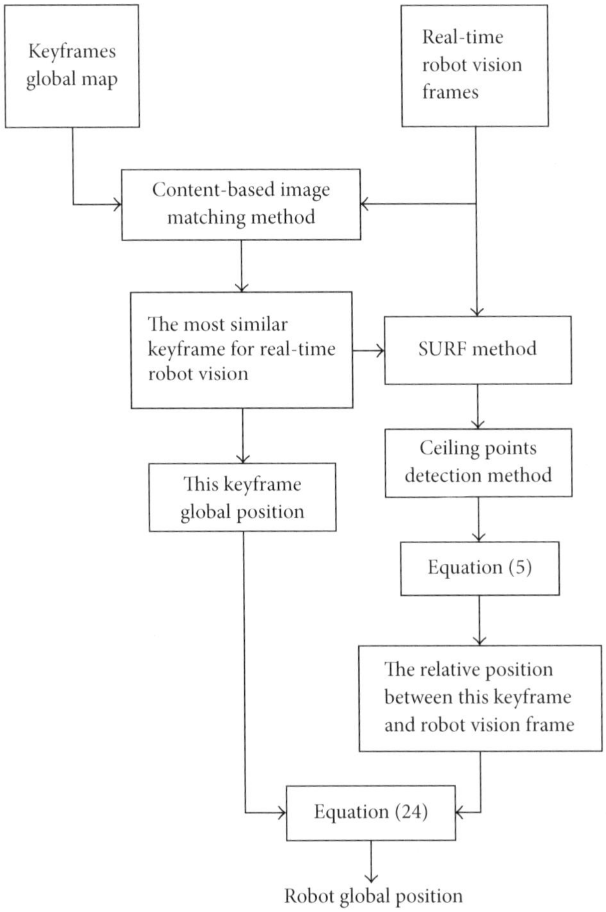

FIGURE 11: The image processing of robot localization.

where (x_N, y_N) and (x_{N+1}, y_{N+1}) are the global position of the Nth and $N+1$th keyframe.

Through this content-based image matching method, the space distance between each keyframe could be very large, and the number of keyframes is very less. Robot would resolve its position quickly by matching with these keyframes. In the experiment, total 1710 frames are taken by the robot during indoor environment studying, and 72 keyframes are extracted by this method.

Through this global map, the robot can localize itself in real-time, and the image processing of robot localization is shown in Figure 11. When robot moving in the indoor environment to render service to human, this content-based image matching method could be used to match real-time robot vision frames with the keyframes sequence in the map and find out the most similar keyframe for each robot vision frame. The same as the global map establishing progress, feature points between robot vision and this keyframe can be extracted, and their relative position can be resolved through (5). And the global position of each robot vision frame, which is also the robot global position, can be resolved through (24), the same as resolving progress of the $N+1$th keyframe global position.

TABLE 1: Position RMSE comparison (m).

	Method in this paper	ORB SLAM	CV-SLAM
Corridor 1 (left)	0.144	0.086	0.123
Corridor 2 (right)	0.159	0.117	0.245
Room 1 (left)	0.468	2.026	0.358
Room 2 (right)	0.483	2.665	0.546

4. Results of Experiments

This keyframes global map establishing method for robot localization through content-based image matching had been tested, including two parts: map building and localization in large indoor environment and the test for robot localization under kidnap condition.

4.1. The Experiment of Map Building and Localization in Large Indoor Environment. The experiment site is composed of two rooms (20-meter-square) and two corridors (11 m). The experiment progress is shown in Figure 12. During robot indoor environment studying, a video about 9.5-minute long (1710 frames) had been taken by robot vision during robot indoor environment studying.

Total 72 keyframes (less than 5%) are extracted by our method from these video frames to build map of experiment site. Their global position resolving result is shown in Figure 13. The environment of the experiment site, including ceiling and wall region (wall, window, door, and furniture), can also be described by this keyframe sequence and their global position, and the position RMSE is less than 0.3 m, as shown in Figure 14.

The robot can make good use of this map to localize itself in the whole building and draw its route in the different room and corridor effectively, as shown in Figure 15.

In order to evaluate the localization precision of this map building method, the corners of tile floor are taken as ground signs, and the air conditioning port and ceiling lamp on the ceiling are taken as ceiling signs. And the localization RMSE between robot localization result and these signs is less than 0.5 m, as shown in Table 1.

Table 1 also shows the comparison between our method and ORB SLAM. The algorithm architecture of our method is similar to that of ORB SLAM, but this paper uses image content matching taking place of feature points matching. Through image content matching, the robot can pick up the keyframe from global map more precisely, which is most similar to its real-time robot vision, and is seldom disturbed by similar objects in the indoor environment. As the experiment site (Figure 14) includes four parts, two corridors and two rooms, the comparison between our method and ORB SLAM is also divided into four parts, as shown in Table 1.

It can be seen from Figure 14 that there are many similar objects in the experiment site, and this would be a serious test for our method and ORB SLAM.

When robot is moving the in the two rooms, the experiment result of our method is better than that of ORB SLAM. This is due to the fact that there are many similar objects in the two rooms, such as the air condition outlet and ceiling

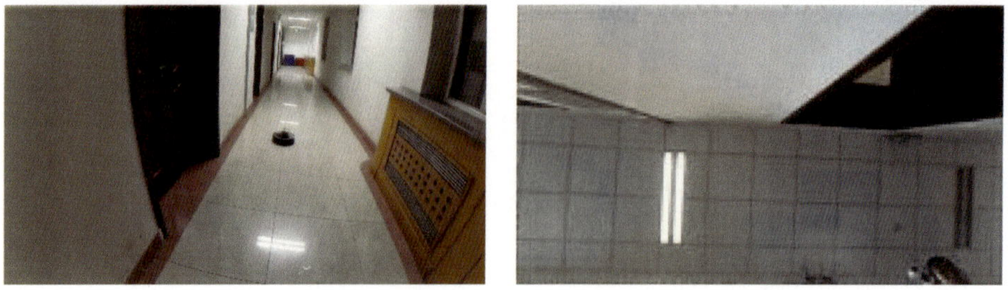

(a) (b)

FIGURE 12: The experiment site. (a) The robot moving in the experiment site. (b) The ceiling in the robot vision.

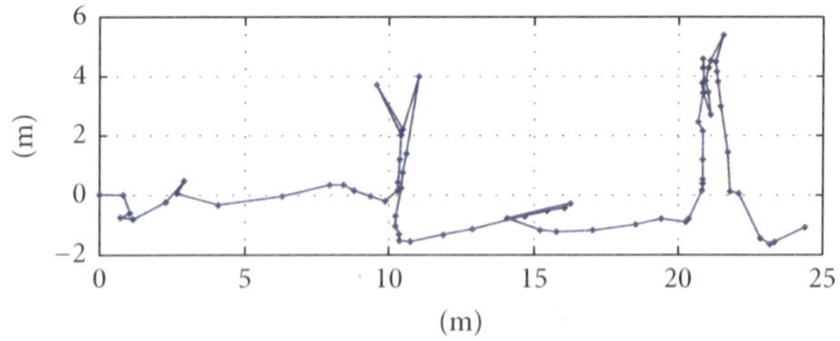

FIGURE 13: The global position relationship of keyframes sequence.

FIGURE 14: Indoor environment described by keyframes sequence.

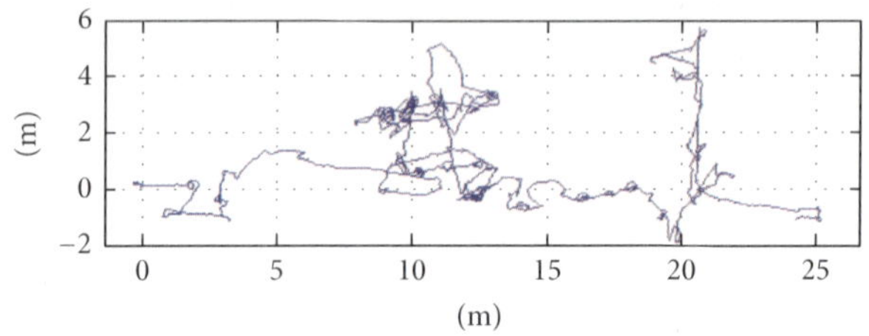

FIGURE 15: The robot route resolved through keyframes global map and content-based image matching.

lamps. The feature points on these similar objects in different rooms are easy to be error matched as the same points. If ORB SLAM needs to match all keyframes in the map, the robot can easily localize itself to wrong room (the distance between the two rooms is 12 m). But our method is able to take advantage of the image of dissimilar objects in different rooms, the interference caused by similar objects can be suppressed, and robot can fix its position precisely in these rooms and less error localized.

While the robot is moving in the two corridors, as when there are fewer similar objects than that of rooms and ORB, the method can extract feature points more effectively, the experiment result of ORB SLAM is better than that of our method.

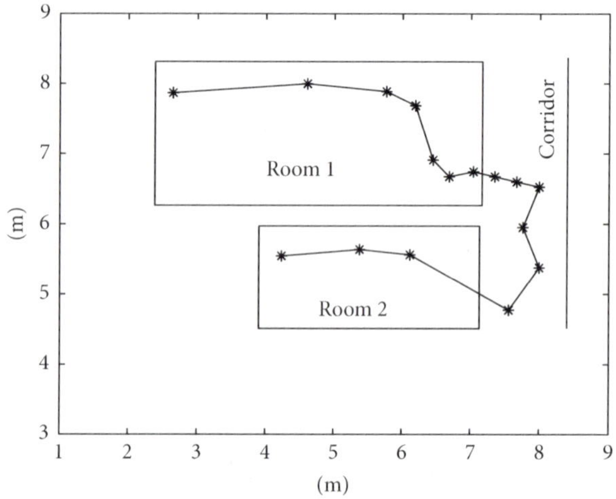

FIGURE 16: The global position relationship of keyframes sequence; "*" is the position of each keyframe.

Table 1 also shows the comparison between our method and CV-SLAM. The difference is not significant under normal condition. However, compared with CV-SLAM method on commercial equipment, our method is immune to kidnapping events, because our method can build global map of indoor environment and make good use of this map to fix robot position at any time.

4.2. The Test for Robot Localization under Kidnap Condition. The kidnap problem for robot self-localization can be solved effectively by our method. In order to test robot in more complex kidnap condition, two adjoining rooms (20-meter-square and 10-meter-square) and the corridor (5 m) are chosen as experiment site. Through our method, 16 frames were extracted as keyframes to build the global map of these rooms and the corridor, with the position relationship of keyframes sequence shown in Figure 16. The keyframes mosaic result is shown in Figure 17. The rooms and small corridor could restrict the field of robot camera, such that the robot cannot fix its position by watching the distant landmarks. It was kidnapped and suddenly put into a far-away place. In the test, the robot was frequently taken from one place to another place (2 or 3 m apart) by experimenters to achieve kidnap. Through matching with global map, the robot was still able to fix its position effectively, especially

FIGURE 17: The global map for kidnap test described by keyframes sequence.

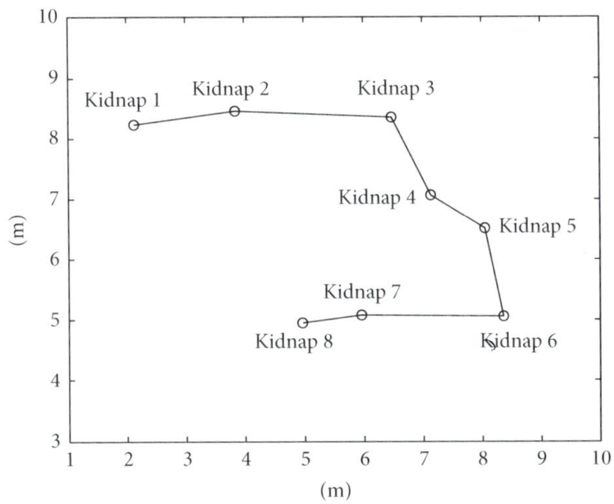

FIGURE 18: The kidnap experiment; eight consecutive kidnaps were executed by experimenters, and sign "o" is the localization result for each kidnap position fix by the robot after landing.

when the robot is suddenly moved from room to the corridor, as shown in Figure 17, and position RMSE is less than 0.4 m.

5. Conclusion

For the issue of robot localization and mapping in the indoor environment, this paper presents keyframes global map establishing method for robot localization through content-based image matching with ability to analyze distortion and overlapping of keyframes. Results show that common problems, such as kidnapping or disturbances by similar objects, can be resolved through the content-based image matching method presented in this paper, which is specially designed for indoor environment. In the test, the keyframes global map can be established by this method and describe the indoor environment effectively. Although there are many similar objects in the experiment site, the robot cannot be kidnapped and can localize itself accurately (Figure 18), with the position RMSE being less than 0.5 m.

Competing Interests

The authors declare that there are no competing interests regarding the publication of this paper.

Acknowledgments

This work is supported by the NSFC (nos. 61271147, 61372052).

References

[1] F. Y. Zhou, X. F. Yuan, Y. Yang et al., "A high precision visual localization sensor and its working methodology for an indoor mobile robot," *Frontiers of Information Technology & Electronic Engineering*, vol. 17, no. 4, pp. 365–374, 2016.

[2] E. Garcia-Fidalgo and A. Ortiz, "Vision-based topological mapping and localization by means of local invariant features and map refinement," *Robotica*, vol. 33, no. 7, pp. 1446–1470, 2015.

[3] J. M. Santos, M. S. Couceiro, D. Portugal, and R. P. Rocha, "A sensor fusion layer to cope with reduced visibility in SLAM," *Journal of Intelligent & Robotic Systems*, vol. 80, no. 3, pp. 401–422, 2015.

[4] S. Jia, K. Wang, and X. Li, "Mobile robot simultaneous localization and mapping based on a monocular camera," *Journal of Robotics*, vol. 2016, Article ID 7630340, 11 pages, 2016.

[5] W. Y. Jeong and K. M. Lee, "CV-SLAM: a new ceiling vision-based slam technique," in *Proceedings of the IEEE/RSJ International Conference on Intelligent Robots and Systems (IROS '05)*, pp. 3195–3200, IEEE, Edmonton, Canada, 2005.

[6] I. Dryanovski, R. G. Valenti, and J. Xiao, "Fast visual odometry and mapping from RGB-D data," in *Proceedings of the IEEE International Conference on Robotics and Automation (ICRA '13)*, pp. 2305–2310, IEEE, Karlsruhe, Germany, May 2013.

[7] Y. Ji, A. Yamashita, and H. Asama, "RGB-D SLAM using vanishing point and door plate information in corridor environment," *Intelligent Service Robotics*, vol. 8, no. 2, pp. 105–114, 2015.

[8] R. Mur-Artal, J. M. M. Montiel, and J. D. Tardos, "ORB-SLAM: a versatile and accurate monocular SLAM system," *IEEE Transactions on Robotics*, vol. 31, no. 5, pp. 1147–1163, 2015.

[9] K. Tateno, F. Tombari, and N. Navab, "Real-time and scalable incremental segmentation on dense SLAM," in *Proceedings of the IEEE/RSJ International Conference on Intelligent Robots and Systems (IROS '15)*, pp. 4465–4472, IEEE, Hamburg, Germany, October 2015.

[10] T. Schöps, J. Enge, and D. Cremers, "Dense planar SLAM," in *Proceedings of the International Symposium on Mixed and Augmented Reality (ISMAR '14)*, pp. 157–164, IEEE, Munich, Germany, September 2014.

[11] T. Schöps, J. Enge, and D. Cremers, "Semi-dense visual odometry for AR on a smartphone," in *Proceedings of the 13th IEEE International Symposium on Mixed and Augmented Reality (ISMAR '14)*, pp. 145–150, IEEE, Munich, Germany, September 2014.

[12] J. Engel, J. Stückler, and D. Cremers, "Large-scale direct SLAM with stereo cameras," in *Proceedings of the IEEE/RSJ International Conference on Intelligent Robots and Systems (IROS '15)*, pp. 1935–1942, Hamburg, Germany, October 2015.

[13] W. Y. Jeong and K. M. Lee, "CV-SLAM: a new ceiling vision-based SLAM technique," in *Proceedings of the IEEE IRS/RSJ*

International Conference on Intelligent Robots and Systems (IROS '05), pp. 3195–3200, IEEE, Alberta, Canada, August 2005.

[14] S. Jo, H. Choi, and E. Kim, "Ceiling vision based SLAM approach using sensor fusion of sonar sensor and monocular camera," in *Proceedings of the 12th International Conference on Control, Automation and Systems (ICCAS '12)*, pp. 1461–1464, Jeju Island, Korea, October 2012.

[15] H. Choi, D. Y. Kim, J. P. Hwang, C.-W. Park, and E. Kim, "Efficient simultaneous localization and mapping based on ceiling-view: ceiling boundary feature map approach," *Advanced Robotics*, vol. 26, no. 5-6, pp. 653–671, 2012.

[16] D. Y. Kim, H. Choi, H. Lee, and E. Kim, "A new cvSLAM exploiting a partially known landmark association," *Advanced Robotics*, vol. 27, no. 14, pp. 1073–1086, 2013.

[17] S. Lee, S. Lee, and S. Baek, "Vision-based kidnap recovery with SLAM for home cleaning robots," *Journal of Intelligent & Robotic Systems*, vol. 67, no. 1, pp. 7–24, 2012.

[18] S. T. Pfister and J. W. Burdick, "Multi-scale point and line range data algorithms for mapping and localization," in *Proceedings of the IEEE International Conference on Robotics and Automation (ICRA '06)*, pp. 1159–1166, IEEE, Orlando, Fla, USA, May 2006.

[19] R. L. Stewart and H. Zhang, "Image similarity from feature-flow for keyframe detection in appearance-based SLAM," in *Proceedings of the IEEE International Conference on Robotics and Biomimetics (ROBIO '11)*, pp. 305–312, Phuket, Thailand, December 2011.

[20] J. Kim and I.-S. Kweon, "Robust feature matching for loop closing and localization," in *Proceedings of the IEEE/RSJ International Conference on Intelligent Robots and Systems (IROS '07)*, pp. 3905–3910, IEEE, San Diego, Calif, USA, November 2007.

[21] P. W. Smith, K. B. Johnson, and M. A. Abidi, "Efficient techniques for wide-angle stereo vision using surface projection mode," in *Proceedings of the IEEE Computer Society Conference on Computer Vision and Pattern Recognition*, pp. 113–118, IEEE, Fort Collins, Colo, USA, 1999.

[22] H. Choi, Y. K. Dong, J. P. Hwang et al., "CV-SLAM using ceiling boundary," in *Proceedings of the 5th IEEE Conference on Industrial Electronics and Applications (ICIEA '10)*, pp. 228–233, IEEE, Taichung, Taiwan, 2010.

[23] P. Chen, Z. Gu, G. Zhang, and H. Liu, "Ceiling vision localization with feature pairs for home service robots," in *Proceedings of the IEEE International Conference on Robotics and Biomimetics (ROBIO '14)*, pp. 2274–2279, IEEE, Bali, Indonesia, December 2014.

[24] Z. Zhang, "A flexible new technique for camera calibration," *IEEE Transactions on Pattern Analysis and Machine Intelligence*, vol. 22, no. 11, pp. 1330–1334, 2000.

[25] K. H. Strobl and M. Lingenauber, "Stepwise calibration of focused plenoptic cameras," *Computer Vision and Image Understanding*, vol. 145, pp. 140–147, 2016.

[26] J. Yuan, Y. L. Huang, F. C. Sun, and T. Tao, "Active exploration using a scheme for autonomous allocation of landmarks," *Robotica*, vol. 32, no. 5, pp. 757–782, 2014.

[27] H. Bay, A. Ess, T. Tuytelaars, and L. Van Gool, "Speeded-Up Robust Features (SURF)," *Computer Vision and Image Understanding*, vol. 110, no. 3, pp. 346–359, 2008.

[28] Z. Li, C. Shao, and Y. M. Liu, "Motion estimation based on axis affine model," in *Proceedings of the IEEE International Conference on Mechatronics and Automation (ICMA '10)*, pp. 572–576, IEEE, Xi'an, China, August 2010.

[29] D. Caballero, T. Antequera, A. Caro, M. L. Duran, and T. Perez-Palacios, "Data mining on MRI-computational texture features to predict sensory characteristics in ham," *Food and Bioprocess Technology*, vol. 9, no. 4, pp. 699–708, 2016.

[30] J.-S. Cho, H.-J. Lee, J.-H. Park, J.-H. Sung, J.-Y. Choi, and K.-D. Moon, "Image analysis to evaluate the browning degree of banana (*Musa* spp.) peel," *Food Chemistry*, vol. 194, pp. 1028–1033, 2016.

12

Generation and Control of Basic Geometric Trajectories for a Robot Manipulator using CompactRIO®

Jorge Luis Aroca Trujillo,[1] Alexander Pérez-Ruiz,[2] and Ruthber Rodriguez Serrezuela[1]

[1] *University Corporation of Huila, Corhuila, Neiva, Colombia*
[2] *Escuela Colombiana de Ingeniería Julio Garavito, Bogotá D.C., Colombia*

Correspondence should be addressed to Ruthber Rodriguez Serrezuela; ruthber.rodriguez@corhuila.edu.co

Academic Editor: L. Fortuna

The utility of a robot manipulator focuses on the ability to locate its end effector in a position with a determined orientation following a specified trajectory. For this, algorithms were used in order to generate and control the movements joints of robot in a synchronized way. The high-level languages to program robots are based on three types of movement: joint interpolation (MOVEJ), linear interpolation (MOVES), and circular arcs (MOVEC), which are used to develop any type of task. In this work, these three movements are implemented in the industrial controller CompactRIO, as part of the reconditioning process of a robot manipulator of five degrees of freedom (5 DOF) whose controller was obsolete. As a result, it will have an interface in LabVIEW where you can view and modify the basic parameters implemented in the industrial controller. In addition, the results of the validation tests of the joint positions and the end effector of the manipulator will be found.

1. Introduction

The robot manipulators have been constituted as an essential part of modern industries. Some of the reasons are the low production cost, the optimization in the processes, their flexibility, and the adaptation to several productive situations. Additionally, the robot manipulator with only small modification in the programming allows industries to adopt this technology in their operations instead of company staff. Generally, the algorithms designed for the monitoring and control of trajectories must be accompanied by the mathematical support that guarantees the good performance of the scheduled tasks. There are techniques with different approaches to control robot manipulator movements, such as neural networks [1], uncoupled control [2], and game theory [3]. However, some control units that currently accompany these manipulators and incorporate these techniques, due to the passage of time, have presented faults or are very outdated, which limits their use.

In this article, we intend to endow a robot manipulator of five degrees of freedom (5 DOF) a new industrial controller (CompactRIO) (Figure 1), which allows designing a modern interface with open control algorithms. This new control unit offers the possibility of using a generic or different industrial controller to factory design for robot manipulators, which are functional, reliable, and user friendly.

In this new control unit, three types of path-generating algorithms are implemented, which provide the positions that a robot manipulator must follow to execute a specific task [4, 5]. The first algorithm develops a joint path (MOVEJ), in which the end effector does not have any relevance. The other two will generate Cartesian trajectories that will draw with the end effector the shape of a line or an arc (MOVES and MOVEC). The programming will be done through LabVIEW and will be implemented in the same control unit. It will interact with the user through a computer connected remotely to the Ethernet port, which will make it easier to visualize, create, and modify the tasks that the user would like the manipulator to replicate. The developed interface will contain the reading of the commands, the mathematical modeling found, the algorithms for generating trajectories, the visualization of variables, and the position control (PID) for motors [6].

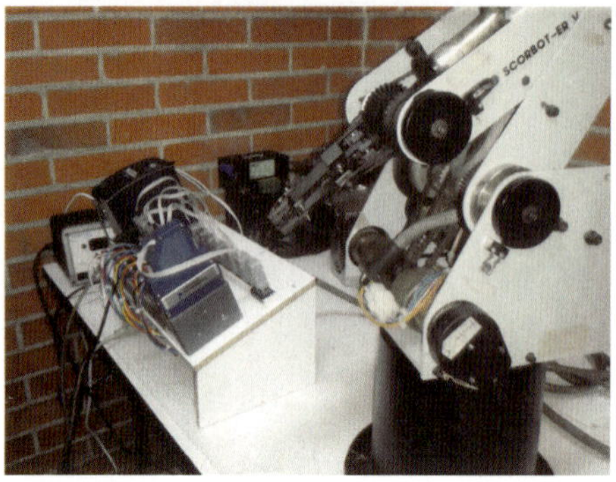

FIGURE 1: Robot manipulator Scorbot ER_4PC.

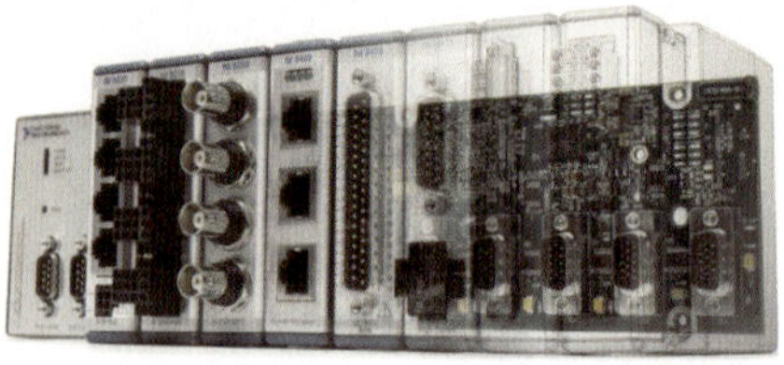

FIGURE 3: Industrial controller, NI CompactRio [retrieved on June 8, 2017, from http://www.ni.com/compactrio/value-controller/esa/].

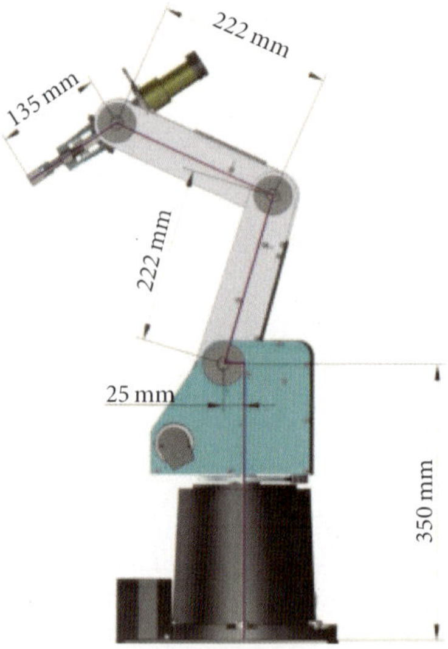

FIGURE 2: Mechanical structure of the Scorbot manipulator [7].

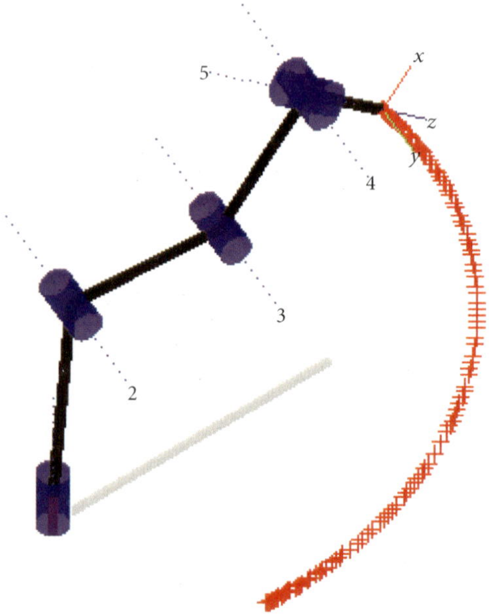

FIGURE 4: Simulation of movement joint.

2. Hardware Description

2.1. Robot Manipulator. The Scorbot manipulator (Figure 2) has an anthropomorphic configuration, composed of a mechanical structure with five joints, DC motors of 12 Volts, rotary encoders coupled to the axes of the motors, and five microswitches to determine the position of home.

The anthropomorphic design provides a broad region of end effector operation [8, 9]. It has a rotation at the base of 310°, shoulder rotation of +130°/−35°, rotation of the elbow of ±130°, tilting of the clamp of ±130°, and rotation of the mechanically unlimited clamp but originally restricted to ±570° electrically (configurable software) [5].

The Scorbot-ER 4PC has several types of transmission to generate the movement in the joints of the robot; these transmissions are composed of gears with different configurations of trains, pulley, and toothed belts. It also has a spindle transmission in charge of opening and closing the clamp.

2.2. Industrial Controller: CompactRio. Manufactured by the company *National Instruments*, it provides reliable and deterministic performance for real-time monitoring and control applications. It has a robust and modular architecture, with a reconfigurable FPGA chassis capable of executing high-speed processes, which is interconnected with the LabVIEW (*Laboratory Virtual Instrumentation Engineering Workbench*) development platform (Figure 3) [10].

3. The Algorithms of Trajectories Implemented Description

Three movements (MOVEJ, MOVES, and MOVEC) will be dealt with, which will be implemented in the industrial controller CompactRio, which can be combined so that the robot can develop a specific task [11].

3.1. Joint Movement: MOVEJ. The use of this movement allows the manipulator to reach an end position, with independent joint movements. This does not ensure a linear trajectory between the initial and final points (Figure 4); that is to say the end effector moves from one point to another by a random path [12].

The coordinates in the end effector workspace can be obtained through the Direct Kinematic Model (DKM) [13–15], using the joint values executed in the MOVEJ command. The pseudo-code can be seen in Algorithm 1.

Data: Actual position: $[\theta_1\ \ \theta_2\ \ \theta_3\ \ \theta_4\ \ \theta_5]$,
desired position: $[\theta_{1d}\ \ \theta_{2d}\ \ \theta_{3d}\ \ \theta_{4d}\ \ \theta_{5d}]$
Result: New position: $q = [\theta_1\ \ \theta_2\ \ \theta_3\ \ \theta_4\ \ \theta_5]$
while $[\theta_1\ \ \theta_2\ \ \theta_3\ \ \theta_4\ \ \theta_5] \neq [\theta_{1d}\ \ \theta_{2d}\ \ \theta_{3d}\ \ \theta_{4d}\ \ \theta_{5d}]$ **do**
    if $\theta_1 \neq \theta_{1d}$ **then**
        $\theta_1 = \theta_1 \pm \Delta\theta_1$;
    end
    if $\theta_2 \neq \theta_{2d}$ **then**
        $\theta_2 = \theta_2 \pm \Delta\theta_2$;
    end
    if $\theta_3 \neq \theta_{3d}$ **then**
        $\theta_3 = \theta_3 \pm \Delta\theta_3$;
    end
    if $\theta_4 \neq \theta_{4d}$ **then**
        $\theta_4 = \theta_4 \pm \Delta\theta_4$;
    end
    if $\theta_5 \neq \theta_{5d}$ **then**
        $\theta_5 = \theta_5 \pm \Delta\theta_5$;
    end
    $q = [q;\ \theta_1\ \ \theta_2\ \ \theta_3\ \ \theta_4\ \ \theta_5]$;
end
New position $q = [\theta_1\ \ \theta_2\ \ \theta_3\ \ \theta_4\ \ \theta_5]$;

ALGORITHM 1: Algorithm for joint movement of MOVEJ.

3.2. Linear Motion: MOVES. If it is desired to arrive at a Cartesian position following a trajectory in a straight line with the end effector, the command MOVES is used. This command requires a vector with six values; the first three are Cartesian variables (X, Y, and Z) of the desired position in the workspace of the origin of the coordinate system of the end effector. The following three refers to your orientation; this orientation is entered with the angles Roll(δ), Pitch(β), and Yaw(γ), also known as angles RPY [16].

In this way, you will have a linear movement as Figure 5 [12], through the path generated from the current point or from start to end point (point entered with the command).

For this command, we will briefly discuss the two transformations that must be performed [17]. The treatment is done differently, mainly because the space of translations is Euclidean, while the space of rotations is not.

Comparing the two movements, translation is the simplest of the two interpolations, since it is expected that a point $P_o \in \mathbb{R}^3$, denoted by three variables (x_o, y_o, and z_o) defined in the Cartesian space, is moved through a translation vector (Δx, Δy, and Δz).

The interpolation of orientation and rotation is done by the mathematical method SLERP (*Spherical Linear Interpolation*) (it is a unitary sphere of $\mathbb{R}^4$) [17–19]. For this, the orientation and rotation must first be represented in four dimensions or quaternions and then interpolated by the method SLERP. This method is to perform the movement at constant speed along a circle, on the surface of a hypersphere (Figure 6) [10], under the assumption that the positions of the trajectories of the end effector are given between the points of P_o and P_f. The effect is a uniform rotation with angular velocity around a fixed axis of rotation, which guarantees

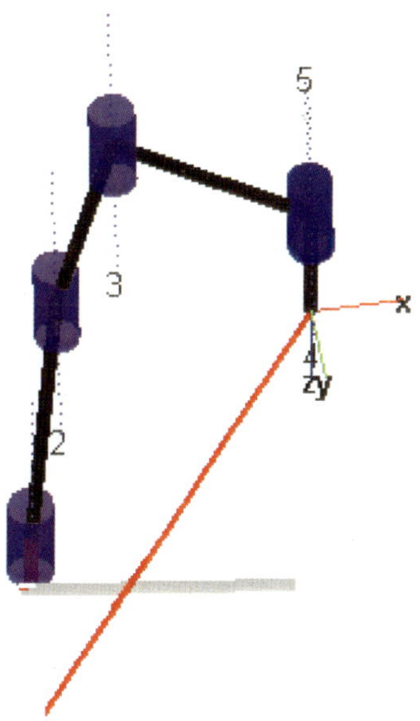

FIGURE 5: Simulation of motion in a straight line.

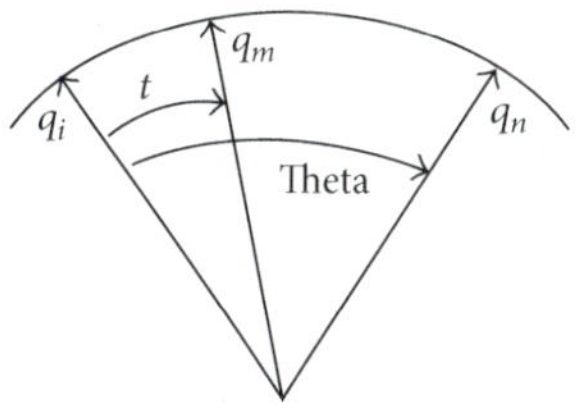

FIGURE 6: Spherical linear interpolation.

Data: Transformations Matrix T_o, T_f
Result: Interpolated positions $[\theta_1 \ \ \theta_2 \ \ \theta_3 \ \ \theta_4 \ \ \theta_5]$
$p = 0 : 1 : (|P_f| - |P_o|)$;
$N = \text{length}(p)$;
$t = 0 : (0.1\,(N/(N-1))) : N$;
$P_f = T_f(1:3, 4)$;
$P_o = T_o(1:3, 4)$;
$R_f = T_f(1:3, 1:3)$;
$R_o = T_o(1:3, 1:3)$;
for $i = 1 : 1 : N$ **do**
  $P_n = P_o + (P_f - P_o) \cdot t(i)$;
  $q_o \leftarrow R_o$;
  $q_1 \leftarrow R_1$;
  $\theta = \arccos(q_o \cdot q_1)$;
  $q_m = q_o \cdot \sin((1 - t(i)) \cdot \Omega)/\sin(\Omega) + q_1 \cdot \sin(t(i) \cdot \Omega)/\sin(\Omega)$;
  $R_n \leftarrow q_m$;
  $T_n = [R_n \ \ P_n; \ 0 \ \ 0 \ \ 0 \ \ 1]$;
  $q = [\text{IKM}(T_n)]$ ⇝ Inverse Kinematic Model
end
Interpolated positions $q = [\theta_1 \ \ \theta_2 \ \ \theta_3 \ \ \theta_4 \ \ \theta_5]$;

ALGORITHM 2: Algorithm for joint movement of MOVES.

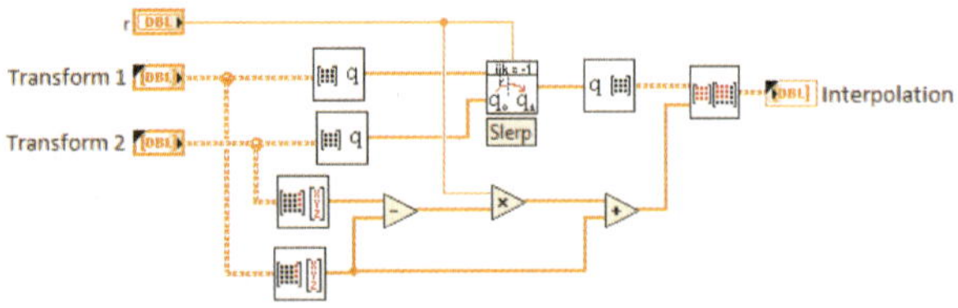

FIGURE 7: Interpolation implemented in *LabVIEW*.

a single route [18, 20]. Equation (1) [21] represents such movement.

$$q_m = \frac{q_o \cdot \sin((1 - t(i)) \cdot \Theta) + q_1 \cdot \sin(t(i) \cdot \Theta)}{\sin(\Theta)}, \quad (1)$$

where q_m is *quaternion* interpolated, q_o is first quaternion, q_1 is second quaternion, Θ is the angle between the vectors q_o and q_1, and t is the spherical interpolation between vectors q_o and q_1, with values between 0 and 1.

With this method, implemented in Algorithm 2, any interpolation of quaternions can be drawn to make a smooth movement of the rotation of an object with complex rotations and the orientation of the same ones. Once the interpolated matrices are obtained, we proceed to change Cartesian coordinates to articular coordinates using the Inverse Kinematic Model (IKM) [22, 23].

Algorithm 2 shows the description of the execution of the MOVES command whose main part of the implementation can be seen in Figure 7.

3.3. MOVEC: Circular Movement. Three linearly independent points are indispensable to do this movement. The first is the value of X, Y, and Z of the Cartesian plane where the end effector is located, the second is the destination point, and the third belongs to the via point. That last one, in addition to helping in defining the path where the end effector should slide, defines the plane where it is going to work, as shown in Figure 8.

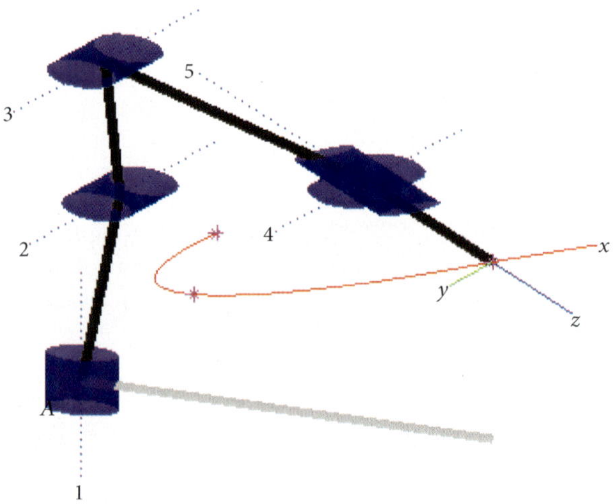

FIGURE 8: Arc and circular movement.

To perform the stroke, we define the orientation and direction of a vector associated with the systems X_1, Y_1, and Z_1. The representation is made by a vector in a frame that rotates angle θ around X, an angle α around Y, and an angle β around Z and moves the distances **a**, **b**, and **c** with respect to the axes x, y, and z; all this transformed to the coordinates of the frame of reference. On the frames X_1, Y_1, and Z_1, the three linearly independent points define the plane that will contain the distribution of the points that make up the curve.

Data: P_f, P_v
Result: Interpolated positions $[\theta_1 \ \ \theta_2 \ \ \theta_3 \ \ \theta_4 \ \ \theta_5]$

Vectors a and $b \mapsto a = (P_f - P_o) \leftrightarrow b = (P_v - P_o)$;
$x = a/\text{norm}(a) \mapsto$ Value in x associated to the system to O_1;
$z = \text{cross}(a, b)$;
$z = z/\text{norm}(z) \mapsto$ Vector normal to the plane or z associated to the system to O_1;
$y = \text{cross}(z, x) \mapsto$ Value in y associated with the system to O_1;
$\Theta = \text{acos}(\text{dot}(a, b)/(\text{norm}(a) * \text{norm}(b))) \mapsto$ Angle between vectors a and b;
$x_{vp} = \text{norm}(b) * \cos(\Theta)$;
$y_{vp} = \text{norm}(b) * \sin(\Theta)$;
$X_{fp} = \text{norm}(a)$;
$m = -x_{vp}/y_{vp}$;
$x_{cp} = P_{fp}/2 \mapsto x$ of the center of the circumference;
$y_{cp} = (y_{vp}/2) + m(x_{cp} - (x_{vp}/2)) \mapsto y$ of the center of the circumference;
$r = \text{norm}([X_{fp} \ \ 0] - [x_{cp} \ \ y_{cp}])$;
$\alpha_1 = -\text{atan2}(y_{cp}, x_{cp}) \Rightarrow \alpha_2 = \pi - \alpha_1$;
$\sigma = \text{linspace}(\alpha_2, \alpha_1, N)$;
$y_s = r * \sin(\sigma) + y_{cp} \mapsto x$ interpolated from the circumference;
$x_s = r * \cos(\sigma) + x_{cp} \mapsto y$ interpolated from the circumference;
$T = [x \ \ y \ \ z \ \ P_o; \ 0 \ \ 0 \ \ 0 \ \ 1]$;
for $i = 1 : N$ **do**
  $P = T * [x_s(i) \ \ y_s(i) \ \ 0 \ \ 1]'$;
  $c = P(1:3) - P_o$;
  $x_n = c/\text{norm}(c)$;
  $y_n = \text{cross}(z, x_n)$;
  $Tc = [x_n \ \ y_n \ \ z \ \ P(1:3); \ 0 \ \ 0 \ \ 0 \ \ 1]$;
  $q = [\text{IKM}(Tc)] \rightsquigarrow$ Inverse Kinematic Model
end
Interpolated positions $q = [\theta_1 \ \ \theta_2 \ \ \theta_3 \ \ \theta_4 \ \ \theta_5]$;

ALGORITHM 3: Algorithm for joint movement of MOVEC.

The circumference in the plane is drawn from the perpendicular bisector of P_oP_v and the perpendicular bisector of P_oP_f, where they will be cut into a single point, which is the center of the circumference, passing through P_o, P_v, and P_f, since the three are equidistant from it (Figure 9). From another point of view, it may be said that it is necessary to find a circumference circumscribed in the triangle formed by the three points, the center of this circumference is obtained through the intersection of the two-side bisectors of the triangle. The problem generates as a result a circumference that tends to infinity.

Similarly if you wish to complete the drawn arc to form a complete circle, it is necessary to generate another movement on the same plane, which has, as its starting point, the end point of the first and the end point, as the starting point of the first command. The diameter is defined by the distance between the start and end points and in turn must be equal to the distance between the two-way points. If this proportion is not met, it is also possible to obtain an ellipsoidal trace whose distance between its height and width can vary depending on the distances of the initial and final points with respect to the distances of the way points (or vice versa). The operations that they use to implement the MOVEC command are defined in Algorithm 3.

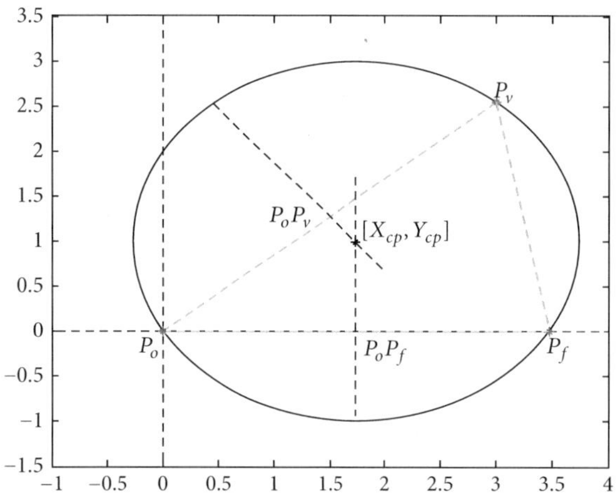

FIGURE 9: Construction of the circle, using the perpendicular bisectors P_oP_v and P_oP_f.

4. Scheme Implemented

The implementation of the algorithms of trajectories for the robot manipulator required the implementation of the

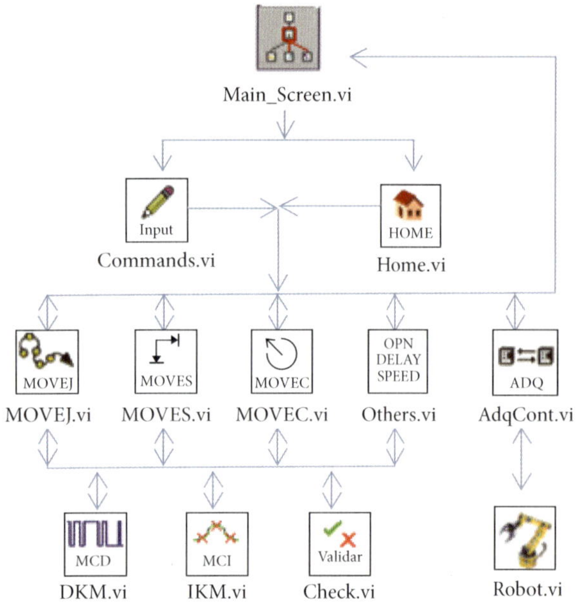

FIGURE 10: SubVI scheme for robot arm control.

graphical interface that incorporates the acquisition and control of variables supplied by the industrial controller, the input of parameters for the generation of the movements, and the algorithms generators of trajectories, among others. Next, a brief description will be given on the most relevant components that are part of the work done.

4.1. General Scheme of the Interface Implemented. To integrate the mechanical system of the robot and the industrial controller CompactRIO, to the environment of LabVIEW, the programming and interconnection scheme of the figure was developed in Figure 10, which is programmed in the industrial controller [24].

The main VI of the system is completely designed and intended to interact with the user. To do this, it receives and displays the articulator positions of the manipulator, together with the Cartesian positions of the end effector. In addition, it sends the commands that are to be executed to the two blocks responsible for interpreting them (Commands.vi and Home.vi).

The VI Home is just responsible for providing the articular positions to bring the manipulator to the initial configuration (Home). On the other hand, the VI Commands organizes and sends the other orders to the VIs in charge of executing each of the algorithms (MOVEJ, MOVES, and MOVEC), the desired speed with which to execute each robot movement (SPEED), the time delay between commands (DELAY), and the opening and closing of the fingers of the clamp (OPEN and CLOSE).

As the names indicate, the MOVEJ, MOVES, and MOVEC VIs, with the help of the DKM, IKM, and validate blocks, are in charge of creating the algorithms generating articular, linear, and circular trajectories, respectively. These three first blocks give the point to point the route that must be made to comply with the command, the acquisition block,

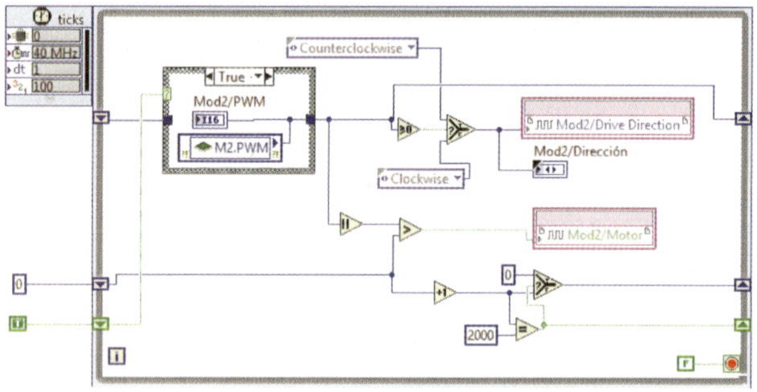

FIGURE 11: PWM signal control block.

and the control of variables (Adpcont.vi). The other VI is destined to interpret the commands SPEED, DELAY, OPEN, and CLOSE to later execute them in the manipulator.

4.2. Robot, CompactRio, and Computer Communication. NI 9503 FPGA modules will be responsible for the acquisition of the signals generated by the encoders, through the entrances: phase A+ and phase B+. In addition, it will be responsible for energizing the motors that move each joint, through the PWM that each module counts. The signals from the switches, used to calibrate the positions, are read in the inputs 1, 2, 3, 4, and 5 of the NI 9403 digital input module.

These six modules, five to control the motors and one for the calibration switches, are connected to the industrial controller CompactRio, which, in turn, communicates with the computer through a network connection that owns the chassis.

4.2.1. Acquisition and Control of Signals. The purpose of this VI is to provide the program with variable reading and writing, such as the direction of rotation of the motors, number of pulses generated by the *encoders* increment, currents delivered to each motor, PWMs, stop, reset, and the activation of the push-buttons. The latter is obtained through the configuration of the NI-9403 module and the other variables are done through the NI-9505 modules.

To obtain the PWM signal with which the speed of each of the motors will be controlled, a block is created in a time frame, which is executed at a clock frequency of 40 Mhz (internal frequency of the CompactRio), as the picture shows in Figure 11.

In addition, the NI-9505 module delivers signals A and B of the encoders that are 90 degrees out of phase, so that the diagram in Figure 12 performs the counting of the pulses. Here, the spin direction is detected, based on the relationship phase between the signals, and then add or subtract them, according to the case.

With this count, a small operation is carried out to find the angle that the motor has turned.

$$\theta_{\text{motor}} = \frac{2\pi}{\text{Resolution\_of the\_encoder}} \#\text{pulses}. \quad (2)$$

The way the manipulator is constructed, with θ_3 and θ_4, presents an angle different from that measured on the axis of the motors that move them. In other words, these angles

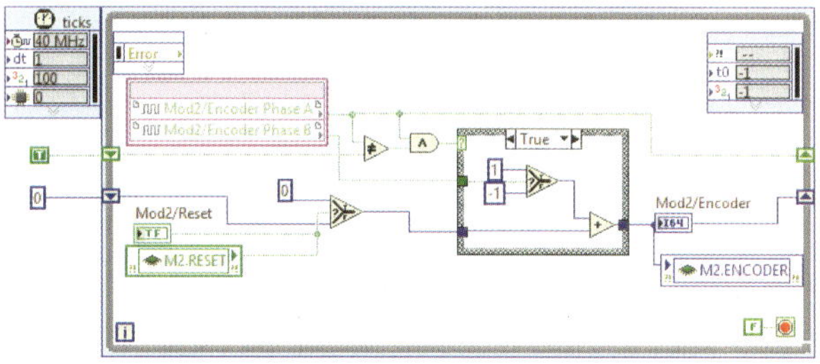

Figure 12: Structure used to read signals from encoders.

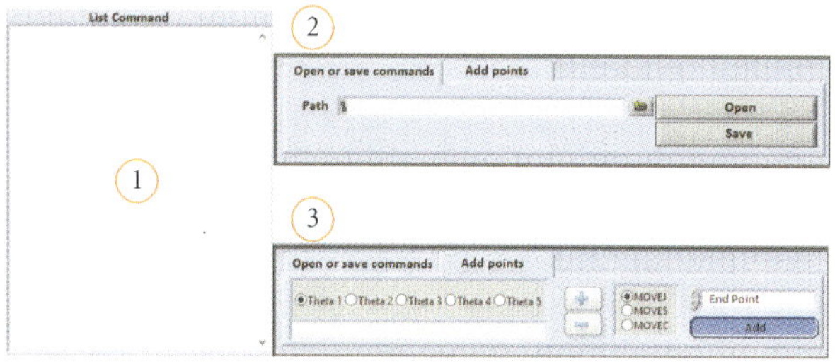

Figure 14: Enter orders to execute.

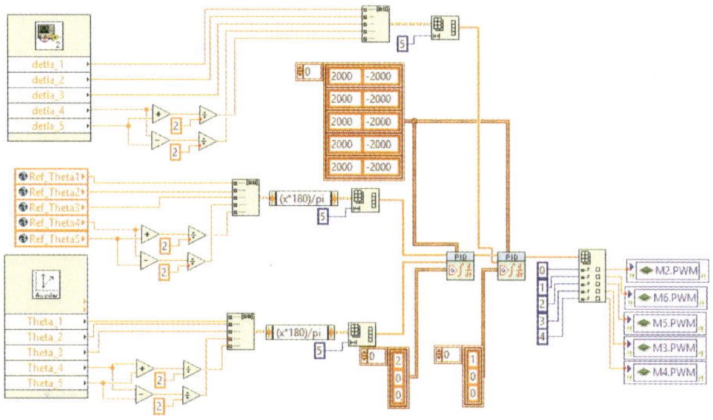

Figure 13: Decoupled control loop diagram.

Table 1: Commands allowed.

Command	Operation
HOME	Press button
DELAY	DELAY [t]
MOVEJ	MOVEJ $\left[\theta_{1d}\ \ \theta_{2d}\ \ \theta_{3d}\ \ \theta_{4d}\ \ \theta_{5d}\right]$
MOVES	MOVES $\left[P_{xd}\ \ P_{yd}\ \ P_{zd}\ \ \delta_d\ \ \beta_d\ \ \gamma_d\right]$
MOVEC	MOVEC $\left[P_{xf}\ \ P_{yf}\ \ P_{zf}\ \ P_{xv}\ \ P_{yv}\ \ P_{zv}\right]$
SPEED	SPEED [%]
OPEN	OPEN
CLOSE	CLOSE

are affected by the previous angles; this indicates that, besides taking into account the angles of the motors, the angles that precede the one that needs to be found must be added or subtracted, in such a way that

$$\begin{aligned}\theta_3 &= \theta_2 + \theta_{\text{motor3}},\\ \theta_4 &= \theta_2 + \theta_3 + \theta_{\text{motor4}} + \theta_{\text{motor5}},\\ \theta_5 &= \theta_{\text{motor4}} - \theta_{\text{motor5}},\end{aligned} \tag{3}$$

where $\theta_{\text{motor4}} + \theta_{\text{motor5}}$ generate the wrist movement that moves the end effector up and down. Additionally, the difference of these angles ($\theta_{\text{motor4}} - \theta_{\text{motor5}}$) produces the rotary movement of the tool; the latter is not affected by other angles, as is the case of θ_3 and θ_4.

The implemented control scheme was carried out through a simple control, which is based on the movement of the joints independently or decoupled (Figure 13), which together with the path-generating algorithms coordinates the movements articular for Cartesian trajectories.

The first control block is in charge of following the reference position given by the algorithms generators of trajectories; the second control block allows reaching the changes of speed that must make the manipulator guarantee the correct articular and Cartesian positions in different instants of time.

The PID controllers are limited to generate a maximum and minimum PWM of ±2000, which corresponds to the input of the DC motor at ±12 volts.

4.2.2. Enter Commands. The development of this VI allows entering the commands for the robot manipulator to perform a task. There are three options: the first is to write directly the command to be used in the command console of the main screen of the program; the second way is to import to the command console the flat file where you have the commands to execute; the last one requires moving the manipulator from grade to grade at each joint until the desired position is reached, then adding the point with the desired operation to the command console (Figure 14).

For correct use of the manual mode (option three), it is necessary to select the joint you want to move and then increase or decrease the angle with the plus (+) or minus (−) buttons. Once the robot is in the desired position, you can choose the operation to perform pressing the add button, which is added to the list. For MOVEJ and MOVES commands, only the end point needs to be stored, but for MOVEC it is necessary to store two points; one will be the end point and the other will be the way point.

The commands allowed and the correct way to enter them are mentioned in Table 1.

The angles θ_{1d}, θ_{2d}, θ_{3d}, θ_{4d}, θ_{5d}, δ_d, β_d, and γ_d are expressed in degrees, even though the execution of the program is done with radians, and the coordinates P are given in millimeters.

4.2.3. HOME Button. This command has the function of calibrating the angles of each joint and bringing the robot manipulator to an initial or reference position (home).

4.2.4. SPEED Command. This order provides the speed change with which the robot manipulator performs the movements, and the SPPED[x] command is entered, where x is the integer value between 0 and 100 given in percent, where 100% supplies the maximum joint velocity of 0.218 [rad/s].

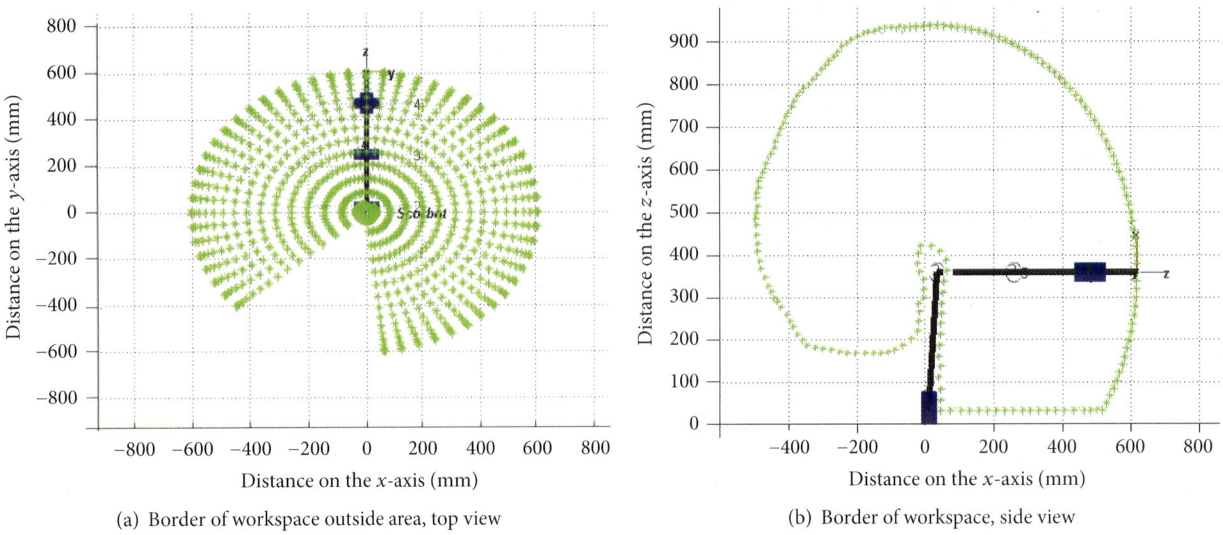

FIGURE 15: Scorbot workspace found in simulation.

4.2.5. DELAY Command. This command forces the robot to remain in its last position during the seconds that it is commanded. For that, the block time delay of LabVIEW is used.

4.2.6. OPEN and CLOSE Commands. With these commands, the manipulator can open and close the clamp, which makes it possible to hold and release objects.

4.2.7. MOVEJ, MOVES, and MOVEC Commands. These commands have the function of generating the interpolations so that the manipulator can trace desired trajectories, as previously exposed.

5. Obtained Results

5.1. Workspace. The effective working space is directly related to the length of the links, the mechanical limitations, and the configuration that the robot has. It can be found mathematically, through the equations describing the position of the end effector along with the mechanical limitations it possesses [25].

Physically, the workspace indicates that it is able to reach these points without exceeding the joint limits. This is essential to be able to propose the movements that are required to develop a task.

Due to the configuration that the Scorbot manipulator has (anthropomorphic), the movement described by the end effector is realized in the three-dimensional space.

The graphs corresponding to the workspace of the robot manipulator Scorbot-ER, with the new restrictions implemented by software in this work, are shown in Figures 15(a) and 15(b). In such graphs, some of the successions of points of the Cartesian coordinates found from the articular positions of the robot are presented. In this way, it gives an idea of the points, in the Cartesian system, that can be reached. For the manipulator to be controlled, most of the positions can be got that are within a radius less than or equal to 604 mm.

5.2. Performance of Joint Movement. An analysis of the result of the kinematic model is performed, based on the MOVEJ command. For that, a known position and a desired joint position are used. For the test, the starting point is chosen as $\theta_0 = [0, 0, 0, -90, 0]$, and the desired joint position is designated, $\theta_d = [90, -10, -45, 0, 0]$, as the point of arrival or end, where angles are given in degrees. With these values, an interpolation of the initial angles to the desired angles is generated (Figure 4).

Afterwards, a comparison of the data of the movement made by the manipulator and those obtained in the simulation is made. Figures 16(a), 16(b), 16(c), 16(d), and 16(e) perform such joint movements generated by the robot, where the red stroke represents the expected values and the blue stroke the articular path made by the manipulator.

5.3. Performance of Linear Motion in the Cartesian Space. A straight stroke is the result of applying the MOVES command (Figure 5), as detailed during the development of this document. The example used to analyze the performance of the movement is performed with the initial coordinates, $[-28.6, -351.04, 64.24, 3.13, 0, -1]$, to the desired point of $[280, -4, 305, 90, 0, -6.3]$, where distances are given in millimeters and angles in degrees.

The desired movement for the end effector is seen in Figures 17(a), 17(b), and 17(c), which describes the path taken in the coordinate axes x, y and z, where the red stroke represents the expected stroke and the blue the actual behavior of the manipulator.

5.4. Cartesian Movement Performance: Circular. Like the previous move, in this command, a planned figure is described

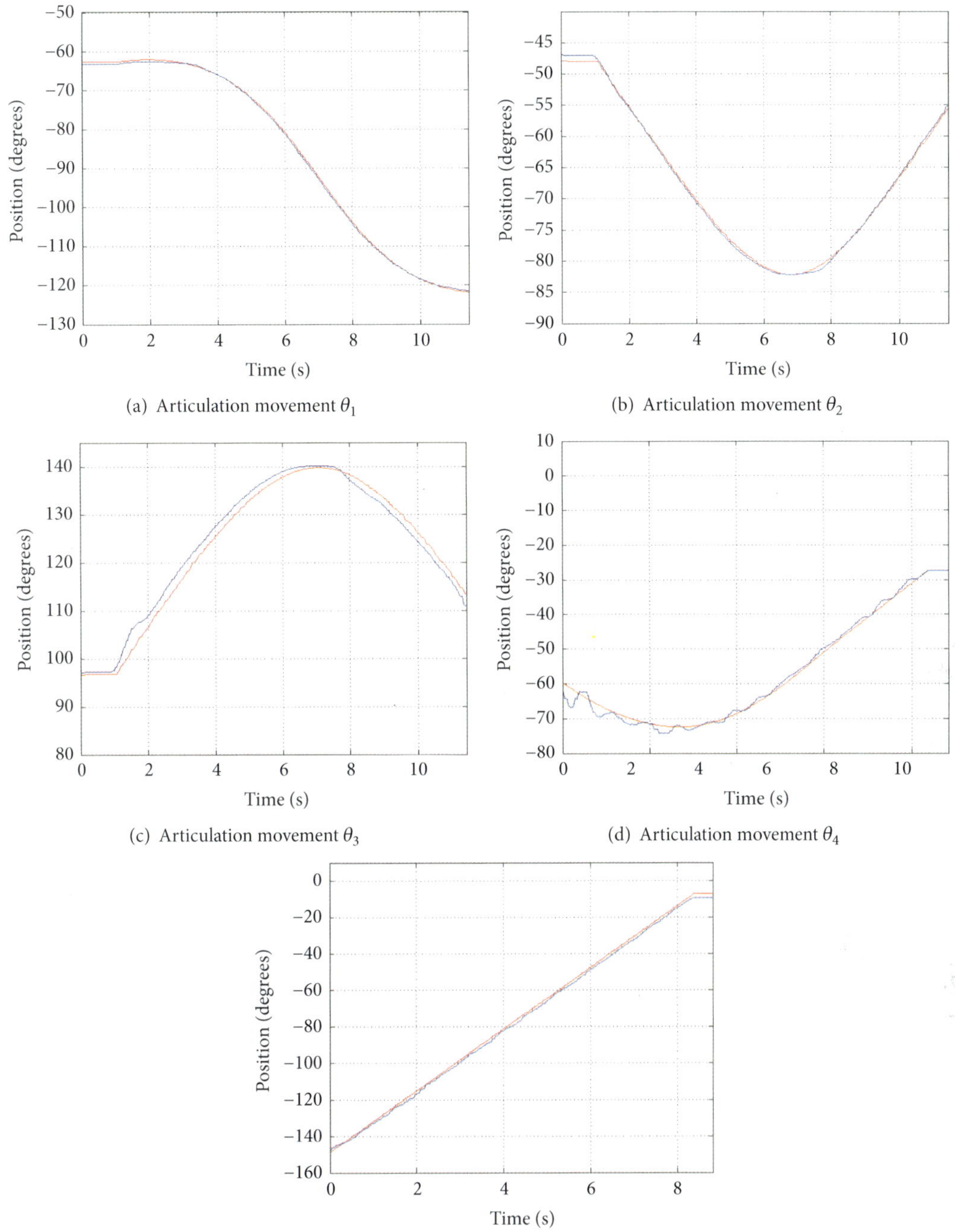

(a) Articulation movement θ_1

(b) Articulation movement θ_2

(c) Articulation movement θ_3

(d) Articulation movement θ_4

(e) Articulation movement θ_5

FIGURE 16: Joint movement of the MOVEJ command.

on the Cartesian plane. This movement realizes a circular stroke realized by the end effector during its trajectory. The starting point was Cartesian coordinates X_o = [145, −276, 185], an obligatory point of passage (way point) X_v = [25, −170, 222], and a point of arrival X_f = [−140, −235, 190].

Consequently, the movements of the coordinate axes shown in Figure 18 present some errors of position of the end effector, due to the response time that has the manipulator, comparing it with the simulation (red line).

6. Conclusion

Having programmed the path generator algorithms and position control system in the CompactRio controller allows the manipulator to continue the task it is developing when there is a connection failure between the computer and the controller.

Three path algorithms are implemented, MOVEJ, MOVES, and MOVEC, which provide a sequence of points in the joint space to generate the different movements

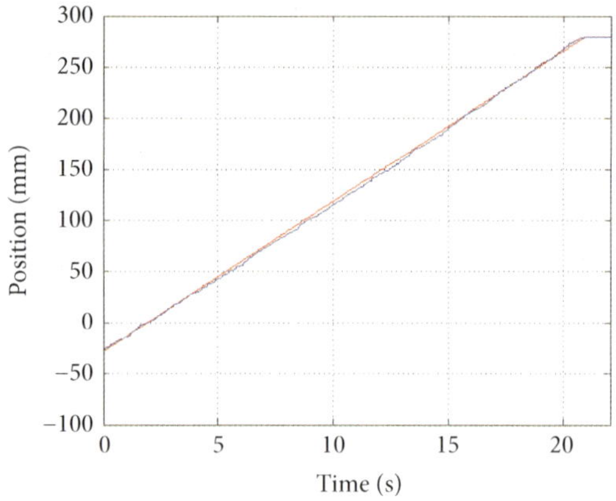

(a) Shaft movement X

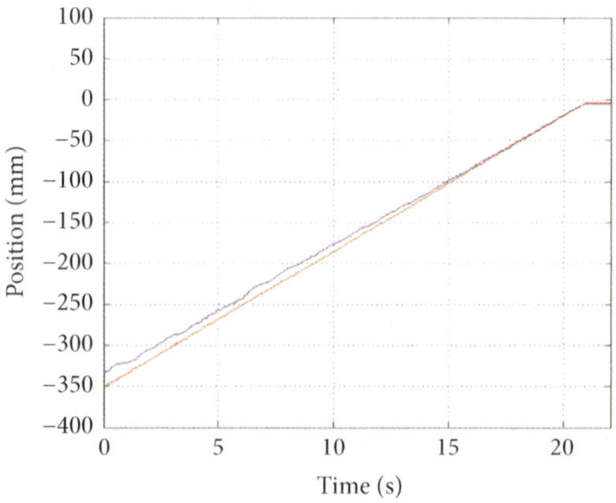

(b) Shaft movement Y

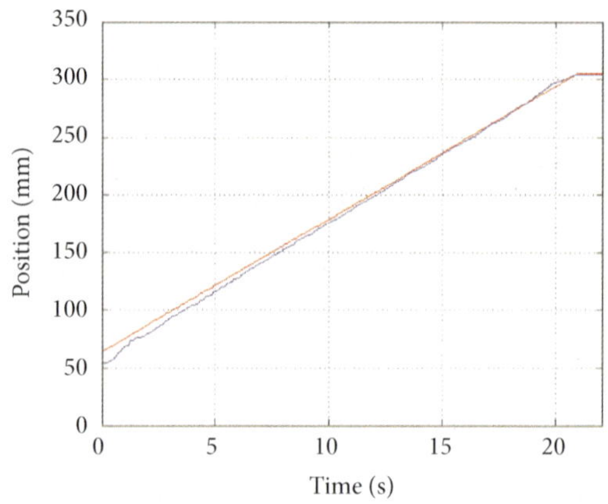

(c) Shaft movement Z

Figure 17: Cartesian movement of the MOVES command.

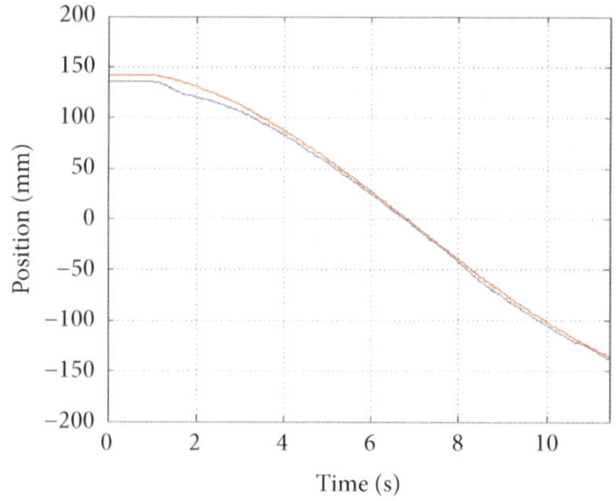

(a) Shaft movement X

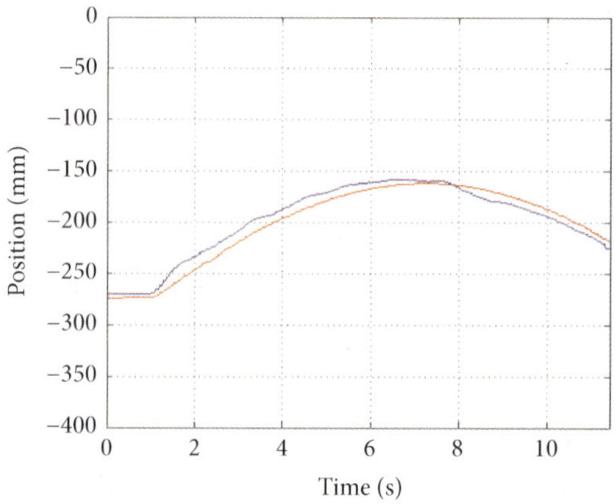

(b) Shaft movement Y

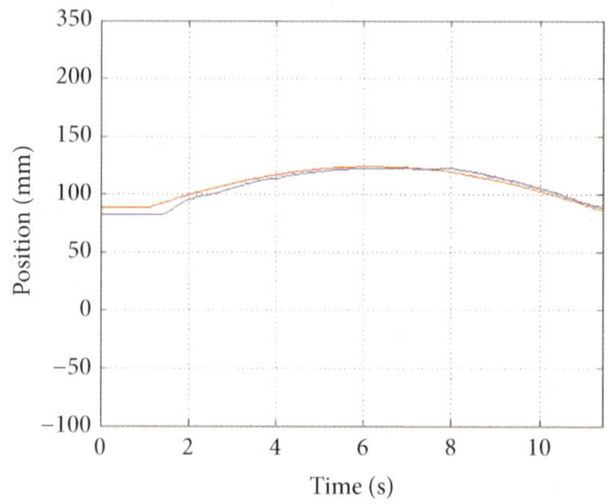

(c) Shaft movement Z

Figure 18: Cartesian movement of the MOVEC command.

(articular, linear, and circular). For the interpolations of trajectories in Cartesian space (MOVES and MOVEC), it is necessary to calculate the IKM to find the articular values of each interpolated Cartesian point, which implies that it requires more processing in these commands, compared to MOVEJ.

It is possible to determine the workspace presented by the robot manipulator Scorbot ER_4PC, taking into consideration the mechanical and electrical limitations. It means that the maximum work space was founded, restricting the manipulator's movements due to the positions that are not allowed for this type of anthropomorphic robot.

The use of rigorous mathematical models allows us to obtain the algorithms generators of trajectories, which supply articular and Cartesian coordinates for the movement of the manipulator. In the latter case, shown in Figures 17 and 18, the real coordinates of the end effector show oscillations on the desired coordinates. These results are acceptable as they are within the subjection tolerance of the tool used for testing (clamp with radius of 32.5 mm).

Conflicts of Interest

The authors declare that there are no conflicts of interest regarding the publication of this paper.

References

[1] W. Xu, L. Dongsheng, and W. Mingming, "Complete calibration of industrial robot with limited parameters and neural network," in *Proceedings of the IEEE 4th International Symposium on Robotics and Intelligent Sensors (IRIS '16)*, pp. 103–108, Tokyo, Japan, December 2016.

[2] F. Petit and A. Albu-Schäffer, "Cartesian impedance control for a variable stiffness robot arm," in *Proceedings of the IEEE/RSJ International Conference on Intelligent Robots and Systems (IROS '11)*, pp. 4180–4186, San Francisco, Claif, USA, September 2011.

[3] P. Arena, S. Fazzino, L. Fortuna, and P. Maniscalco, "Game theory and non-linear dynamics: The Parrondo Paradox case study," *Chaos, Solitons & Fractals*, vol. 17, no. 2-3, pp. 545–555, 2003.

[4] M. S. Kazemi and M. J. Dominguez, "Simulation and evaluation of neuro-controllers applied in a SCORBOT," in *Proceedings of the IEEE International Conference on Automatica (ICA-ACCA '16)*, Curico, Chile, October 2016.

[5] E. Zurek Varela and R. López Beltrán, "Carga y descarga automática de una fresadora de control numérico utilizando un robot Scorbot-ER 4pc," *Revista Científica Ingeniería y Desarrollo*, vol. 7, pp. 113–119, 2011.

[6] V. A. Deshpande and P. M. George, "Analytical Solution for Inverse Kinematics of SCORBOT-ER-Vplus Robot," *International Journal of Emerging Technology and Advanced Engineering*, vol. 2, no. 3, 2012.

[7] A. González Echeverri, *Análisis cinemático y dinámico del robot Scorbot-ER Vplus para la nueva configuración en una base deslizante [Bachelor's thesis]*, Universidad Tecnológica de Pereira, Pereira, Colombia, 2014.

[8] E. Robotec, *Scorbot er-4pc: User's Manual*, 1982.

[9] A. Elfasakhany, E. Yanez, K. Baylon, and R. Salgado, "Design and development of a competitive low-cost robot arm with four degrees of freedom," *Modern Mechanical Engineering*, vol. 1, no. 2, article 47, 2011.

[10] National Instruments, The CompactRIO Platform, endless Capabilities, Unrivaled Performance, 2014, http://www.ni.com/compactrio/.

[11] H. C. Fang, S. K. Ong, and A. Y. C. Nee, "Interactive robot trajectory planning and simulation using augmented reality," *Robotics and Computer-Integrated Manufacturing*, vol. 28, no. 2, pp. 227–237, 2012.

[12] P. Corke, *Robotics, Vision and Control*, Springer, Berlin, Germany, 2011.

[13] A. Hemami, "Kinematics of two-arm robots," *IEEE Journal on Robotics and Automation*, vol. 2, no. 4, pp. 225–228, 1986.

[14] A. A. Mohammed and M. Sunar, "Kinematics modeling of a 4-DOF robotic arm," in *Proceedings of the International Conference on Control, Automation and Robotics (ICCAR '15)*, pp. 87–91, Singapore, May 2015.

[15] J. H. C. Rojas, R. R. Serrezuela, J. A. Q. López, and K. L. R. Perdomo, "LQR hybrid approach control of a robotic arm two degrees of freedom," *International Journal of Applied Engineering Research*, vol. 11, no. 17, pp. 9221–9228, 2016.

[16] C. S. G. Lee, "Robot arm kinematics, dynamics, and control," *Computer*, vol. 15, no. 12, pp. 62–80, 1982.

[17] A. Cayley, *An Elementary Treatise on Elliptic Functions*, Deighton Bell & Co, Cambridge, UK, 1876.

[18] T. Barrera, A. Hast, and E. Bengtsson, "Incremental spherical linear interpolation," in *Proceedings of the Annual SIGRAD Conference. Special Theme-Environmental Visualization*, pp. 7–10, Linköping University Electronic Press, November 2004.

[19] V. E. Kremer, *Quaternions and SLERP*, University of Saarbrucken, Department for Computer Science Seminar Character Animation, Saarbrücken, German, 2008.

[20] J. S. Ahn, W. J. Chung, and S. S. Park, Application of Quaternion Interpolation (SLERP) to the Orientation Control of 6-Axis Articulated Robot using LabVIEW® and RecurDyn®.

[21] K. Shoemake, "Animating rotation with quaternion curves," in *Proceedings of the 12th Annual Conference on Computer Graphics and Interactive Techniques (SIGGRAPH '85)*, pp. 245–254, San Francisco, Calif, USA, July 1985.

[22] R. R. Kumar and P. Chand, "Inverse kinematics solution for trajectory tracking using artificial neural networks for SCORBOT ER-4u," in *Proceedings of the 6th International Conference on Automation, Robotics and Applications (ICARA '15)*, pp. 364–369, Queenstown, New Zealand, February 2015.

[23] R. R. Serrezuela, A. F. C. Chavarro, M. A. T. Cardozo, A. L. Toquica, and L. F. O. Martinez, Kinematic modelling of a robotic arm manipulator using matlab, 2006.

[24] Y. Angal and A. Gade, "LabVIEW controlled robot for object handling using NI myRIO," in *Proceedings of the IEEE International Conference on Advances in Electronics, Communication and Computer Technology (ICAECCT '16)*, Pune, India, December 2016.

[25] M. M. Ali, H. Liu, N. Stoll, and K. Thurow, "Kinematic analysis of 6-DOF arms for H20 mobile robots and labware manipulation for transportation in life science labs," *Journal of Automation Mobile Robotics and Intelligent Systems*, vol. 10, no. 4, pp. 40–52, 2016.

13

Stair Climbing Control for 4-DOF Tracked Vehicle based on Internal Sensors

Daisuke Endo,[1] Atsushi Watanabe,[1] and Keiji Nagatani[2]

[1] *Graduate School of Engineering, Tohoku University, Aramaki-aza Aoba 468-1, Aoba-ku, Sendai 980-0845, Japan*
[2] *New Industry Creation Hatchery Center, Tohoku University, Aramaki-aza Aoba 6-6-10, Aoba-ku, Sendai 980-8579, Japan*

Correspondence should be addressed to Daisuke Endo; endo@frl.mech.tohoku.ac.jp

Academic Editor: Gordon R. Pennock

In search-and-rescue missions, multi-degrees-of-freedom (DOF) tracked robots that are equipped with subtracks are commonly used. These types of robots have superior locomotion performance on rough terrain. However, in teleoperated missions, the performance of tracked robots depends largely on the operators' ability to control every subtrack appropriately. Therefore, an autonomous traversal function can significantly help in the teleoperation of such robots. In this paper, we propose a planning and control method for 4-DOF tracked robots climbing up/down known stairs automatically based on internal sensors. Experimental results obtained using mockup stairs verify the effectiveness of the proposed method.

1. Introduction

In search-and-rescue missions, it is widely known that multi-degrees-of-freedom (DOF) tracked robots that are equipped with subtracks are very useful. Quince [1] is a typical example of such a multi-DOF tracked robot. It explored buildings that were affected by the meltdown accident in the Fukushima Daiichi nuclear power plant [2]. Such a multi-DOF tracked robot has main tracks that constitute a skid-steered mobile base, and each one rotates independently. In addition, it is equipped with subtracks that connect to the main body with rotary joints at the front and rear of the main tracks. Further, it has high traversability on rough terrain in spite of its relatively simple mechanism.

For search-and-rescue missions, a tracked vehicle is sometimes required to traverse stairs to go to another floor and expand the reachable area. However, stairs represent typical uneven ground and are likely to cause some trouble for ground vehicles, even a multi-DOF tracked robot. In particular, if its subtracks are operated inadequately, it can cause fatal failure modes, such as tipping over. Moreover, in teleoperated missions, the operation of the subtracks is a difficult task, and this can easily cause operation errors. Therefore, it requires sophisticated operator skills.

To solve this problem, various related studies have been carried out on tracked robots [3–12]. Okada et al. realized a shared autonomous system to control subtracks adaptively by mounting laser range sensors on the multi-DOF tracked robot to detect the ground shape under the robot body [3]. Li et al. developed an autonomous system to traverse stairs by using kinect sensor [5]. Zhang et al. realized climbing control including the detection of stairs, using a multi-DOF tracked robot equipped with a laser range finder, sonar, and a proximity distance sensor [9]. As mentioned in each paper, these are effective systems. However, each approach requires dedicated external sensors to detect the shape of the stairs. On the other hand, depending on the search target, there are cases where the shape of the stairs is known or where it can be detected using external sensors that were originally installed to achieve automatic navigation or environmental map construction. In such cases, it should be possible to reduce the number of dedicated external sensors that detect the shape of the road surface, which is required by the conventional method. As other related research examples, Ohno et al. proposed a semiautonomous approach to make the subtracks remain in contact with the unknown shape ground by realizing compliant joints that are based on joint-torque information [12]. However, this approach requires a current sensor on each subtrack to obtain the joint torque, and the sensor information is affected by the friction of its transmissions. To reduce the effect of the friction, the reduction ratio of the gear should be small. However, the su-track power that is required to achieve control on uneven terrain is limited. In addition, Nan et al. proposed and verified control while ascending stairs by using a crawler-type robot without any additional sensors, but this method applies to the specific robot used in their study. Therefore, it is not considered suitable for general applications. The above conventional approaches are primarily for unknown natural rough terrain. On the other hand, there are some cases where the shape of the stairs is known, and in such simple cases, the robot does not require external sensors. Therefore, in this paper, we propose an autonomous motion for a 4-DOF tracked robot for climbing up/down a known flight of stairs based on the rotational velocity of the main tracks, the rotational position of the subtracks, and IMU (Inertial Measurement Unit) information, without any other external and internal sensors. Besides, some conventional multi-DOF tracked robots, such as Quince, are 6-DOF and have four subtracks, one each at the front-left, front-right, rear-left, and rear-right of the main tracks. However, in the case when it is undergoing a typical climbing up/down motion on stairs, the left and right subtracks are not operated independently but are synchronized. Therefore, our proposal is also useful for 6-DOF tracked robots. In addition, some verification tests are reported to confirm the validity of the proposed method.

As a supplement, we have already reported a part (only climbing up the stairs) related to this theme [13]. This article has comprehensive contents including climbing up and climbing down stairs in addition to [13].

2. Problem Description

In this study, we aim to realize a motion planning and control for a 4-DOF tracked robot to ascend and descend on various types of stairs by utilizing only the rotational information of each track and the attitude of the robot body. To proceed this study, we assumed the target robots and target environments, as shown in the following subsections. These assumptions are common for many applications such as autonomous inspection in known buildings or search-and-rescue, and therefore the results of this research can be used widely and generally.

2.1. Target Robot. In this study, we employ a 4-DOF tracked robot that consists of two main tracks to enable skid steering. It has two pairs of two synchronous subtracks. One pair is on both sides of the front, and the other is on both sides of the rear. The rotation range of each subtrack exceeds $\pm 90^\circ$ with respect to the front-rear direction of the main tracks. For joint angles, both the front and rear are defined as positive in case directed to upward toward main track. We also assume that the shape of the robot and the parameters of the position of the centroid of the robot are known. In addition, we assume that the position of the centroid of the robot is set such that it fits within the range that is directly above the contacting ground of the main track. Further, this position is in the state where the robot is grounded only by the main crawler onto the horizontal plane for any joint angle of the front and rear subtracks. In addition, there are multiple grousers on the periphery of each track that serve to improve the friction between the track and the ground. These should be sufficient to prevent the robot from slipping. We assume that the number of rotations of the main tracks and the subtracks are obtained by internal sensors, such as rotary encoders, which enable basic velocity/position control and odometry. The robot's pose is obtained by an IMU that contains acceleration sensors and gyroscopes. The definition of each specification parameter is described in Figure 1(i). Δ_f and Δ_r are the angles formed by the outer edges with respect to the symmetry line when the front and rear subtracks are viewed from the side of the robot, respectively. They are represented by the following:

$$\Delta_f = \arctan\left(\frac{r_f - r_m}{l_f}\right)$$
$$\Delta_r = \arctan\left(\frac{r_r - r_m}{l_r}\right). \tag{1}$$

2.2. Target Environment. In this paper, we deal with an environment consisting of two different horizontal floors that are connected by stairs, with all steps in parallel. The distance (pitch) p and inclination θ_s between two adjacent edges of the stairs are constant.

We assume that all steps have a horizontal plane, and the existence of a vertical plane is arbitrary. p, θ_s, and the number of steps n are known. The width of the stairs is sufficiently larger than that of the robot. In addition, we assume that the relationship between p and the maximum length of the flat area under the robot $L = l_f + l_m + l_r$ is $L \geq 2p$. This assumption means that the robot can traverse the stairs while maintaining an inclination of the robot that is equal to that of the stairs. In addition, the conditions of (p, θ_s) are limited in that falling backward (described later) does not occur in the case where the robot stretches all subtracks completely straight. This assumption physically means that the robot is at least able to traverse the stairs.

3. Definitions of Motion Phases and Failure Modes

When the tracked robot climbs up the stairs, there are three motion phases (Figure 2). On the other hand, there are four motion phases in the case involving climbing down the stairs (Figure 3). In both cases (climbing up and climbing down), the failure mode that most easily occurs differs based on the motion phase. In this section, we define failure modes, motion phases, and their relationships.

3.1. Definition of the Three Phases of the Climbing-Up Motion. First, we divide the stair climbing-up motion into three phases as follows:

(i) Pitch-up ascend phase: the phase starts when the robot touches the lowest step of the stairs and lasts

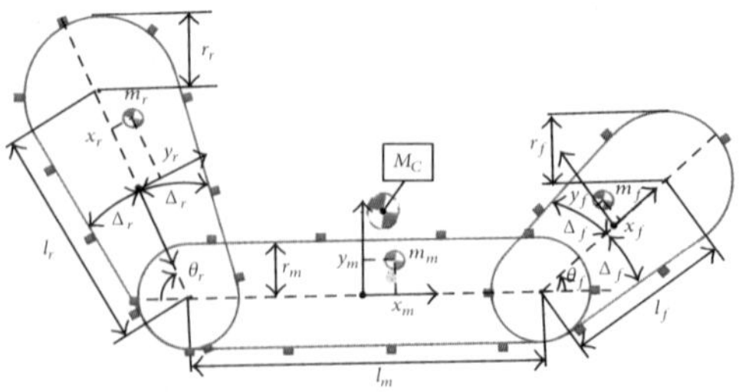

m_f: mass of front subtracks
m_m: mass of main tracks
m_r: mass of rear subtracks
M_C: centroid of the robot
θ_f: joint angle of front subtracks
θ_r: joint angle of rear subtracks
l_f: length of front sub tracks
l_m: length of main tracks
l_r: length of rear subtracks
r_f: tip track radius of front subtracks
r_m: track radius of main tracks
r_r: tip track radius of rear subtracks
x_f, y_f: x- and y-coordinate of front subtrack's centroid
x_m, y_m: x- and y-coordinate of main track's centroid
x_r, y_r: x- and y-coordinate of rear subtrack's centroid

(i)

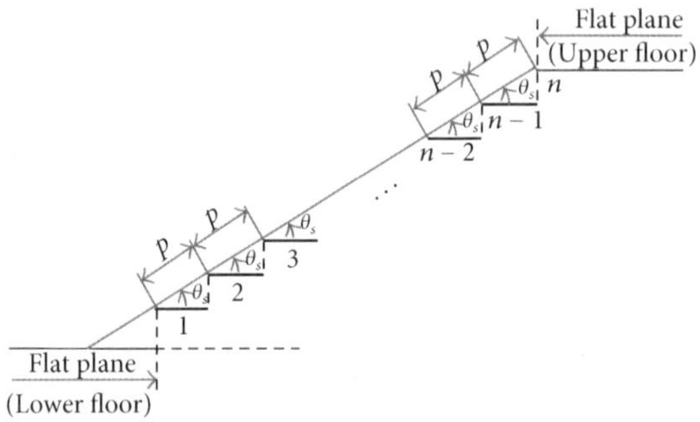

(ii)

FIGURE 1: Assumed shape of the robot and the stairs.

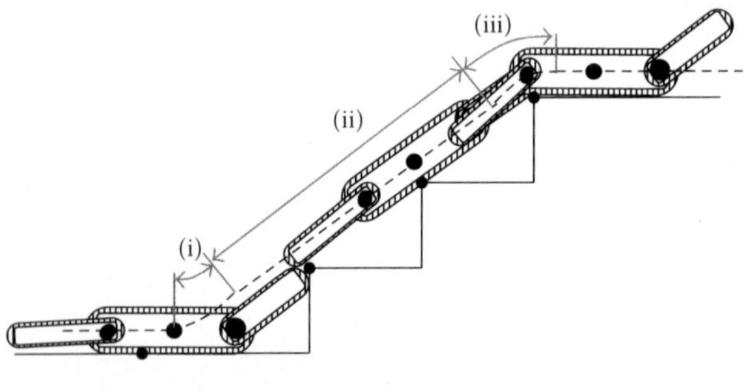

--- Trajectory of a representative position of the robot

FIGURE 2: Three motion phases when climbing up.

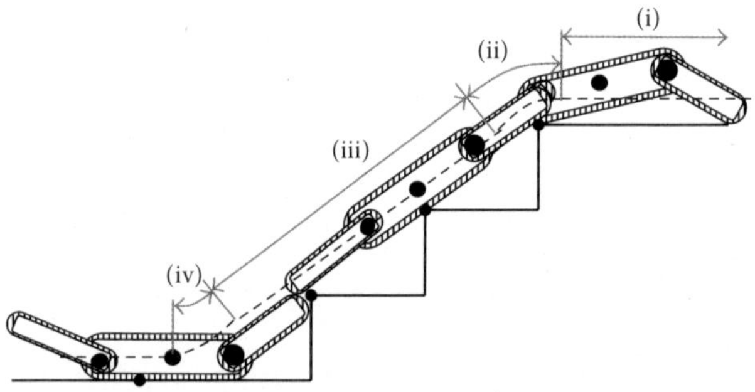

--- Trajectory of a representative position of the robot

FIGURE 3: Four motion phases when climbing down.

until the inclination of the robot, θ_p, corresponds to that of the stairs, θ_s (Figure 2(i)).

(ii) Normal climbing-up phase: this is the phase between (i) and (iii). During this phase, the inclination angle of the robot θ_p is equal to that of the stairs θ_s (Figure 2(ii)).

(iii) Pitch-down ascend phase: this is the phase during which the inclination angle of the robot starts to decrease, which leads to the robot successfully reaching the new level (Figure 2(iii)).

3.2. Definition of the Four Phases of the Climbing-Down Motion. Secondly, we divide the stair climbing down motion into four phases as follows:

(i) Descend-ready phase: an operation state for detecting the relative position between the robot and the highest step.

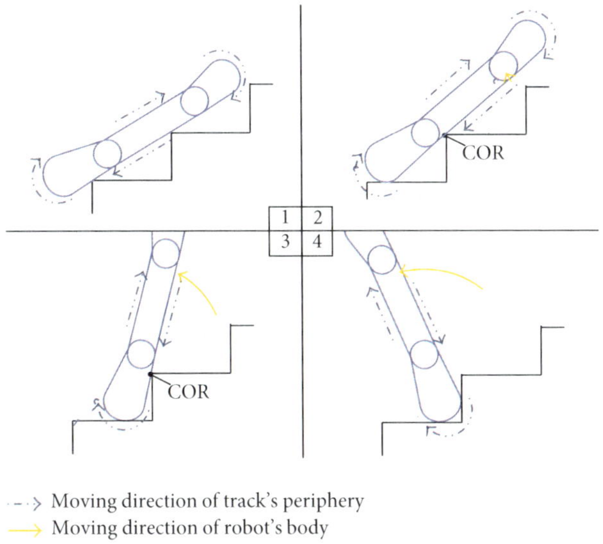

Figure 4: Falling backward.

(ii) Pitch-up descend phase: after the main tracks touch the highest step, the operation state until the pitch angle of the robot body θ_p substantially matches the inclination of the stairs θ_s.

(iii) Normal climbing-down phase: this is the phase between (ii) and (iv). During this phase, the inclination angle of the robot, θ_p, is equal to that of the stairs, θ_s (Figure 3(iii)).

(iv) Pitch-down descend phase: the phase from leaving the lowest step until the pitch angle of the robot body, θ_p, evenly matches the lower floor

3.3. Definition of Failure Modes. While the robot traverses the stairs, it is necessary to prevent the following four failure modes:

(1) Slipping: the friction between the tracks and the steps is not sufficient to move in the direction of the main tracks.

(2) Falling backward: the robot body rotates backward and downward around the lowermost point of contact with the stairs (flips over backwards, as shown in Figure 4).

(3) Direction error: orientation of the robot deviates from the perpendicular direction to each edge of the stairs. Although this deviation is not a failure, as a result of this, falling sideward (tips sideways, as shown in Figure 5) or contact with side wall restricting the width of the stairs or slipping sideward (the frictional force decreases and slides down in the side direction of the robot) can happen. In this paper, as a generic name of these failure modes, it is written as direction error.

(4) Excessive shock: the contact shock exceeds the acceptable level for the robot (or surrounding environment) when it falls and touches the surrounding environment (as shown in Figure 6).

4. Relationships between Motion Phases and Failure Modes

The tendency of each failure mode is different for each motion phase described above. In this section, we explain the relationships with each failure mode that is likely to occur in each motion phase.

4.1. Case Involving Climbing Up

4.1.1. Pitch-Up Ascend Phase. Slipping tends to occur owing to an insufficient frictional force. On the other hand, because the pitch of the robot body is zero or is sufficiently small, it is difficult for it to fall backward and sideward compared to other motion phases. In addition, with respect to the excessive shock, it occurs if the robot body slips after being lifted in the pitch direction. However, this can be prevented by preventing it from slipping. Therefore, for this phase, it is desirable to prioritize the prevention of sliding.

4.1.2. Normal Climbing-Up Phase. According to the assumption described in Section 2, friction is sufficiently generated with respect to the staircase during the normal climbing-up phase. For this reason, it is difficult to slip. On the other hand, because the inclination of the robot body is large, falling backward and sideward are likely to occur. In addition, excessive shock is unlikely to occur, with the exception of those caused by falling backward and sideward. In addition, even if it occurs, its influence tends to be reduced. Therefore, in this phase, it is preferable to prevent falling backward and sideward.

4.1.3. Pitch-Down Ascend Phase. As well as the normal climbing-up phase, slipping is unlikely. In addition, once falling backward and sideward can be prevented in the normal climbing-up phases, it is difficult for slippage to occur, even in this phase. On the other hand, if the subtracks cannot be properly operated, immediately after the centroid of the robot body passes through the highest edge of the step, it will be pulled by gravity and will fall while rotating around the edge, causing excessive shock when landing on the upper floor. Therefore, in this phase, it is desirable to prioritize the prevention of excessive shocks.

4.2. Case Involving Climbing Down

4.2.1. Descend-Ready Phase. Because this phase is primarily an operation to move on the upper even plane, it is difficult for any failure mode described in Section 3.2. However, depending on the position of the centroid of the robot and the length of the subtracks, it can fall over when the subtracks are grounded and the robot body is lifted.

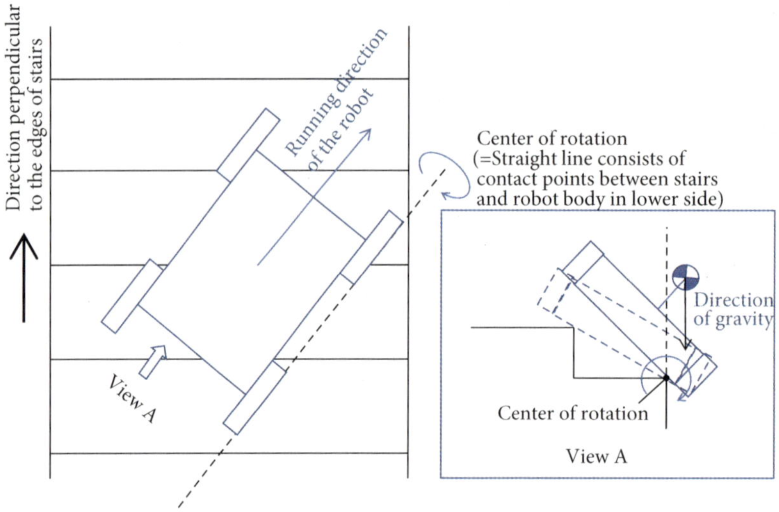

FIGURE 5: Falling sideward caused by direction error.

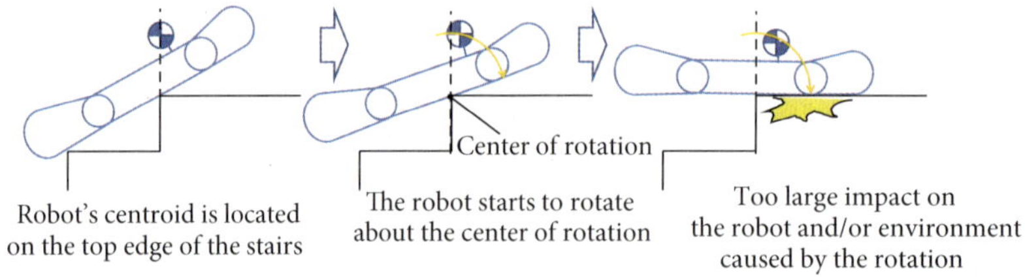

FIGURE 6: Excessive shock.

4.2.2. Pitch-Up Descend Phase. Because the grouser of the crawler belt comes into contact with the highest edge of the step in an oblique direction during this phase, friction is sufficiently generated (described later). Therefore, it is difficult for slippages to occur. In addition, because the pitch of the robot body is zero or sufficiently small, it is difficult to fall backward and sideward compared to the normal climbing-down phase described later. On the other hand, if it fails to properly operate the subtracks, it is pulled by gravity immediately after the centroid of the robot passes the edge of the highest step, falling while rotating, and an excessive shock occurs at the time of landing. Therefore, in this phase, it is also desirable to prioritize the prevention of excessive shocks.

4.2.3. Normal Climbing-Down Phase. This is the same as the normal climbing-up phase. Therefore, in this phase, it is preferable to prevent falling backward and sideward.

4.2.4. Pitch-Down Descend Phase. In addition to the pitch-up ascend phase, slipping is likely to occur. However, because the destination after slipping down is the lower floor on which the robot is going to advance, the sliding itself is not fatal. However, it is possible that the shock at the time of landing as a result of slipping may be excessive. On the other hand, it is difficult for falling backward and sideward to occur in this phase as long as they can be prevented during the normal climbing-down phase grade. Moreover, if the front subtracks cannot be properly operated, it falls from the lowest contacting edge immediately after the front subtracks leave, and an excessive shock is generated at the time of landing. Therefore, in this phase, it is desirable to give priority to preventing any excessive shocks, including those caused by sliding (Figure 7).

The relationship between each motion phase and failure mode described in this section is summarized in Table 2.

5. Operation Strategy

For the three phases involving climbing up the stair and the four phases involving climbing down, the failure modes that are likely to occur and which should be considered are different in each phase. In this section, we describe the operation strategy to prevent failure modes that are likely to occur in each phase. In addition, when the robot straightens the subtracks and the contact plane composed of the main tracks and the subtracks is in a horizontal state, the position of the entire centroid of the robot is generally not at the center of the contact plane and is either at the front or rear. In this paper, the side at which the centroid position is biased is defined as the front direction of the robot. Considering the situation with normal climbing up or down, it is desirable that the centroid be biased toward the upper side in order for the robot to operate without falling backward. This is common in ascending and descending cases. Therefore, it is desirable that the ascending operation be performed by advancing (translational speed is positive), and

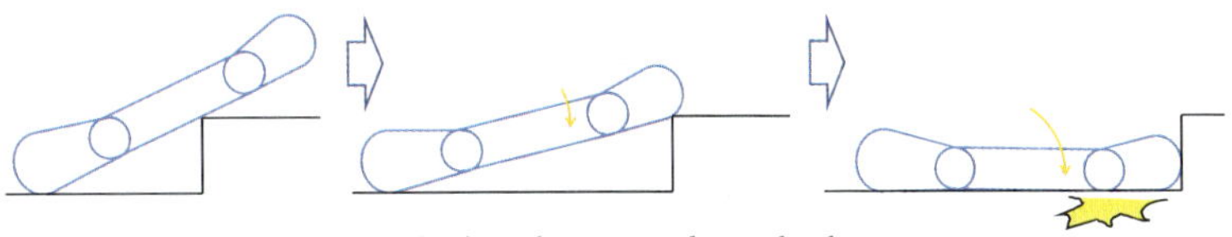

FIGURE 7: Excessive shock can be generated in pitch-down descend phase.

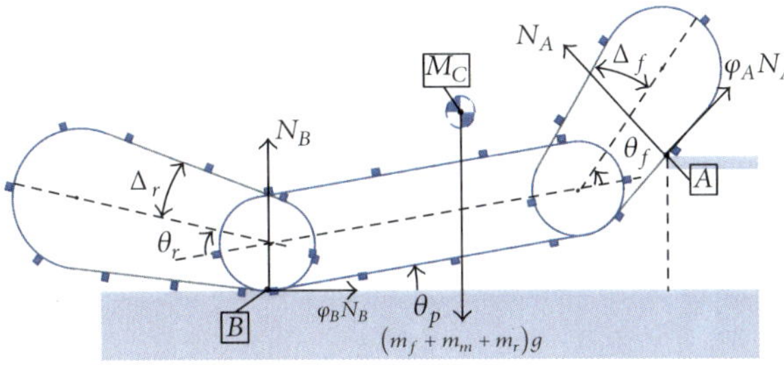

A: forward contact point
B: backward contact point
N_A: normal reaction affecting contact point A
N_B: normal reaction affecting contact point B
φ_A: coefficient of friction at contact point A
φ_B: coefficient of friction at contact point B

FIGURE 8: Physical model in the pitch-up phase.

the descending operation by reversing (translational speed is negative). In this paper, we assume that the climbing up is a forward movement and that the climbing down is a backward movement. Therefore, note that the side closer to the lower floor is the rear of the robot, and the side closer to the upper floor is the front, without depending on the difference between the climbing up and down movements.

5.1. Slip Prevention in Pitch-Up Ascend Phase. In the pitch-up ascend phase, the operation to prevent slipping should be given priority. Thus, the motion of the robot should be predetermined in advance to prevent the slipping mode. For this reason, it is desirable that the reference joint angles of the subtracks θ_f and θ_r be set such that they increase friction. According to Y. Liu and G. Liu, when a tracked robot with grousers makes contact at the peak of a step, as shown in Figure 8, the equivalent friction coefficient (the value calculated from the force affecting it in the peripheral direction divided by the force affecting it in the normal direction) at the contact point A φ_A can be calculated by [14]

$$\varphi_A = \frac{\sin\left(\theta_f + \theta_p - \Delta_f\right) + \mu\cos\left(\theta_f + \theta_p - \Delta_f\right)}{\cos\left(\theta_f + \theta_p - \Delta_f\right) - \mu\sin\left(\theta_f + \theta_p - \Delta_f\right)}, \qquad (2)$$

where μ is the friction coefficient between the materials of the grousers and the materials of the stairs. Therefore, the friction of the front subtrack is maximized when the denominator on the right side of (2) is equal to zero. Hence, the slipping mode can be minimized when the reference angle of the front subtrack θ_f^{ref} is determined as

$$\theta_f^{\text{ref}} = \arctan\left(\frac{1}{\mu}\right) - \theta_p + \Delta_f. \qquad (3)$$

Incidentally, for this condition, φ_A mathematically equals infinity. This means that the finite force affects the contact point A even if the normal force equals zero by the actual hooking of the grousers. In addition, in the case where the robot lifts up its main body from the horizontal state, the friction coefficient required to lift the body becomes smaller as the front contacting point becomes physically closer to the rear contacting point [15].

On the other hand, to prevent the falling backward mode, it is desirable that the rear subtracks remain very close to the ground to prevent it from falling backward. Based on the above, we can conclude that the reference joint angle of the rear subtrack θ_r^{ref} is determined as follows:

$$\theta_r^{\text{ref}} = \theta_p + \Delta_r + \delta, \qquad (4)$$

where δ is the margin of the rear subtrack that does not make contact with the lower floor, even if there is any error or delay of control.

5.2. Prevention of Falling Backward and Sideward in Normal Climbing-Up Phase. In the normal climbing phase, two falling modes, falling backward and direction error, easily occur. It is possible to prevent the falling backward mode when the subtracks are operated properly. On the other hand, it is difficult to prevent the falling sideward mode in the case where the robot's orientation differs from the vertical direction of the edges of the stairs, even if the subtracks are operated properly. In the following subsections, we describe control methods to prevent the above two failure modes.

5.2.1. Prevention of Falling Backward Using Subtracks. The falling backward mode occurs in the case where the zero-moment point (namely, the acceleration projection point of the centroid to the support polygon) [16] of the robot moves downward and moves out of the range of the support polygon. Thus, when the robot climbs up the stairs, it is desirable that the centroid be located as low and frontward as possible and also extend the support polygon backward. Based on the above, it is effective for the front subtrack to be extended frontward, and the rear subtrack is extended backward so that all of the contact points are located in a single flat plane. Hence, the reference joint angles of the subtracks in the normal climbing-up phase are described by

$$\theta_f^{\text{ref}} = \Delta_f$$
$$\theta_r^{\text{ref}} = \Delta_r. \tag{5}$$

5.2.2. Prevention of Direction Error Using Main Tracks. To prevent the robot from direction error, it is desirable that the orientation of the robot matches the vertical direction of the stairs as closely as possible. Therefore, the main tracks need to be controlled based on the control law of (6) using ϕ, which is the angle between the orientation of the robot and the direction of the stairs:

$$\Omega^{\text{ref}} = -k_\phi \phi, \tag{6}$$

where Ω^{ref} is the target angular velocity of the robot body and k_ϕ is the control gain. The problem to be overcome involves the detection of ϕ. In our assumption, the robot does not have any external sensors. Furthermore, in the case where the left and right main tracks experience unequal degrees of slipping on the stairs, and where the yaw angle of the robot obtained by a gyroscope experiences some drift, it causes an increase in the estimation error of ϕ when the robot relies only on a dead-reckoning method such as odometry or its gyroscope. Therefore, to solve this problem, we propose a method to calculate ϕ based on the roll information in the gravity direction. Let $\vec{y}$ be a unit vector pointing to the left of the robot body in the robot coordinate system, and let $\vec{g}$ be a unit vector in the gravity direction (Figure 9).

$$\vec{y} = (-\cos\phi, \sin\phi\cos\theta_s, \sin\phi\sin\theta_s),$$
$$\vec{g} = (0, 0, -1), \tag{7}$$

As $\vec{y}\cdot\vec{g} = \cos(\pi/2 + \theta_R)$, where θ_R is the roll angle of the robot body with respect to the gravity direction, we derive

$$\phi = \arcsin\left(\frac{\sin\theta_R}{\sin\theta_s}\right). \tag{8}$$

Equation (8) indicates that we can obtain ϕ from the inclination of the stairs θ_s and the roll angle of the robot body θ_R. θ_s is also known, and θ_R can be detected using only an internal sensor, IMU, without drift. Besides, the switch of control law from the pitch-up phase to the normal climbing phase is conducted when the pitch angle of the robot body θ_p comes sufficiently close to the inclination of the stairs, θ_s. More specifically, the control law is changed when the condition

$$\theta_p \geq \theta_s - e_p \tag{9}$$

is established. Here, $\pm e_p$ describes the detection error range of the pitch angle of the robot.

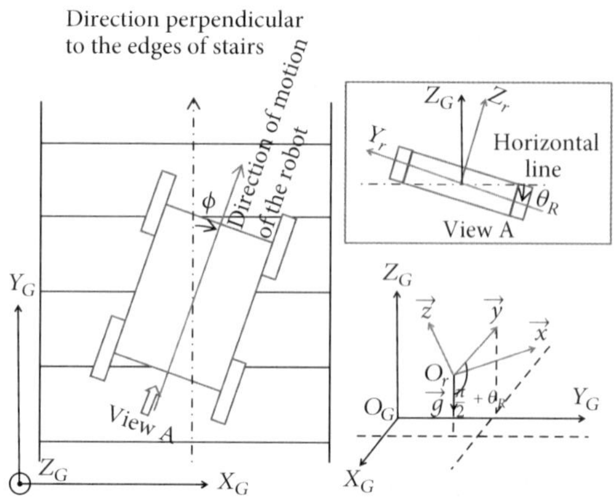

FIGURE 9: Vertical climbing direction and orientation of the robot.

5.3. Shock Mitigation in Pitch-Down Ascend Phase. In the pitch-down ascend phase, the robot should be very careful to mitigate the shock that is encountered between the robot body and the ground when it touches down on the upper floor. For shock mitigation, the ideal of a soft, landing motion makes the height of free fall equal to zero. This motion can be created theoretically by causing the front subtrack to make contact with the upper floor just before the centroid projection point of the robot body passes the peak of the final step. This can be realized only when the position of the robot is estimated without error. However, it is difficult for the robot to maintain the accuracy of the position estimation at high quality on stairs on which it tends to experience slipping or oscillation. In the case where the front subtracks contact the upper floor with the robot's position estimation error, it may hit the peak of the final step or the upper floor with some force. As a result, θ_p increases, and in the worst case, the falling backward mode occurs. To avoid the above situation, in this paper, we propose a method that absorbs the position estimation error for practical applications. Specifically, the robot operates in a motion such that its height of free fall does not exceed the maximum acceptable height $h_{\max}$ for the robot and the surrounding environment when the robot touches down on the upper floor.

Figure 10(a) indicates a state in which the joint angles of the rear subtracks are equal to zero, and the zero-moment point matches the peak of the final step. This is the state-transition configuration for which the robot starts to fall in a

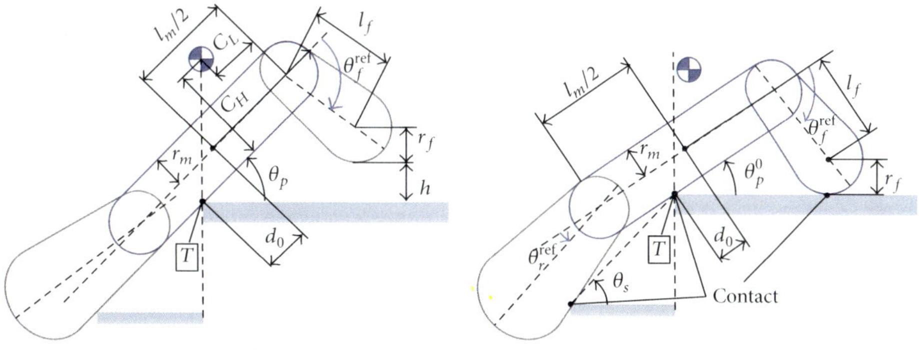

(a) θ_f^{ref} (b) θ_r^{ref}

FIGURE 10: Target angle of subtracks in pitch-down mode.

forward direction, and h indicates the height of free fall. Based on the geometric condition of Figure 10(a), the reference joint angle of the front subtrack should be θ_f^{ref}, as described in (10) below, to make the height of free fall equal to the intended h less than h_{max}.

$$\theta_f^{ref} = \arcsin\left(\frac{(d_0 + l_m/2)\sin\theta_s + r_m\cos\theta_s - r_f - h}{l_f}\right) - \theta_s, \tag{10}$$

where d_0 is the offset distance in the front-back direction of the robot between the peak of the final step and the center of the main tracks, as shown in Figure 10(a). Specifically, it is expressed by the following equation:

$$d_0 = C_H \tan\theta_p + C_L - \frac{l_m}{2}. \tag{11}$$

On the other hand, the joints of the rear subtracks should be controlled downward slightly, to support the robot body falling forward smoothly, as shown in Figure 10(b). We can determine a suitable angle of the rear subtrack by using the configuration of the robot in the figure: the three contact points do not generate the internal force required to grasp the stairs.

Depending on the geometric condition, the reference joint angle of the rear subtrack θ_r^{ref} can be calculated by

$$\theta_r^{ref} = -\arctan\left(\frac{p\sin\theta_s - (l_m/2 - d_0)\sin\theta_p^0}{p\cos\theta_s - (l_m/2 - d_0)\cos\theta_p^0}\right) + \theta_p^0 + \Delta_r, \tag{12}$$

where θ_p^0 is the pitch angle of the robot body when the front subtrack makes contact with the upper floor after the robot falls down in the forward direction. Thus, it satisfies

$$\left(\frac{l_m}{2} + d_0\right)\sin\theta_p^0 + r_m\cos\theta_p^0 + l_f\sin\left(\theta_f + \theta_p^0\right) - r_f = 0. \tag{13}$$

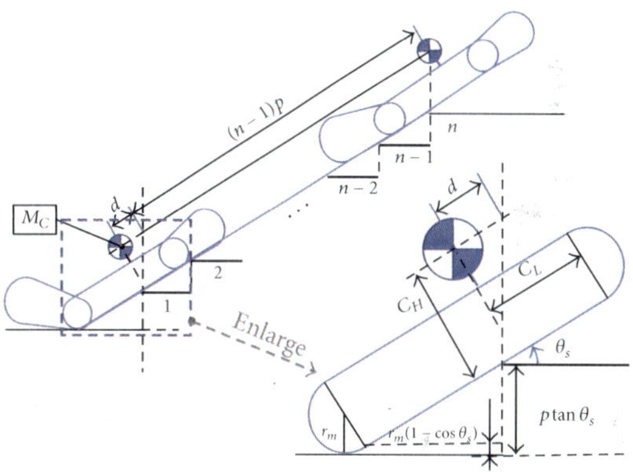

FIGURE 11: Distance to move in the normal climbing phase.

In this study, we assumed that the robot has no external sensors. Therefore, the phase transition from the normal climbing phase to the pitch-down phase is conducted when its running distance D in the normal climbing phase exceeds a fixed value, as shown in the following equation:

$$D \geq d + (n-1)\,p, \tag{14}$$

where n is the number of steps. d is a dimension that is indicated in Figure 11, which according to the geometric condition is determined by

$$d = C_L \tan\theta_s + \frac{p\tan\theta_s - r_m(1-\cos\theta_s)}{\sin\theta_s} - l_m - C_H. \tag{15}$$

5.4. Position Detection of Highest Step in Descend-Ready Phase. It is desirable for the robot to accurately perceive the relative position with the stairs in order to smoothly and safely perform the operation of the climbing down after the pitch-up descend phase. However, the information about the shape of the stairs that can be perceived by the robot on the upper floor applies only to the top edge of the step. Therefore, in order to detect this using a robot that has no external sensors, we let the robot perform the following operation. First, bring the rear subtrack into contact with the ground, and tilt the robot body forward slightly. More specifically, the

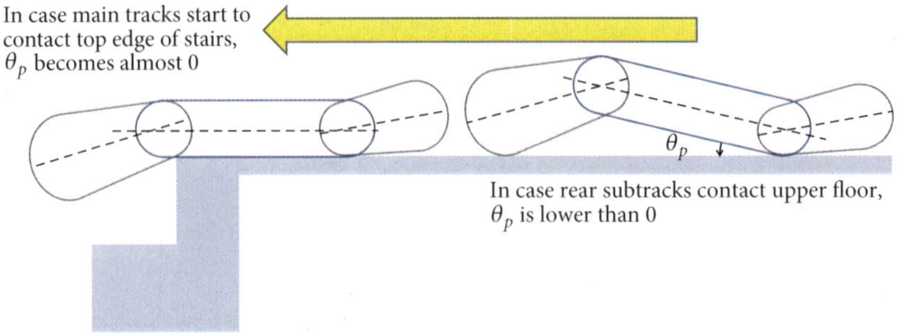

FIGURE 12: Position detection when the robot starts climbing down.

pitch angle of the robot body is set to an angle that exceeds the detection error range of the pitch angle e_p, and the stability margin [17] is sufficient. We define this angle as θ_{de}. On the other hand, with respect to the front subtrack, the position is determined so that it stays very close to the upper horizontal floor such that the forward tilting can be performed smoothly and safely. In summary, the target joint angle of the subtracks can be expressed as the following equations:

$$\theta_f^{\text{ref}} = -\theta_p + \Delta_p + \delta, \tag{16}$$

$$\theta_r^{\text{ref}} = \theta_{\text{de}}. \tag{17}$$

As described above, if the robot moves backward to the stair while controlling the subtracks, the robot can perceive that it is in the relative position shown in Figure 12 when the pitch angle of the robot body becomes horizontal.

5.5. Shock Mitigation in Pitch-Up Descend Phase. In the pitch-up descend phase, it is very important to mitigate the shock when the body collapses to the stairs immediately after the centroid of the robot passes through the top edge of the stairs. In order to decrease this shock, θ_p should be made equal to θ_s as much as possible before the position of the centroid exceeds the edge of the highest step. Based on this idea, the target joint angle of each subtrack is determined as follows. First, the front subtrack is controlled so that the inclination of the robot body θ_p matches the inclination of the stairs θ_s as much as possible. Specifically, the feedback control described in the following equation is performed:

$$\theta_f^{\text{ref}} = K_P\left(\theta_s - \theta_p\right) + K_I \int \left(\theta_s - \theta_p\right) dt. \tag{18}$$

Equation (18) implies PI control of the front subtrack so that the pitch angle of the robot body θ_p matches the stair inclination θ_s. However, based on this control method, it is necessary for it to be limited in $\theta_f^{\text{ref}} \le \pi/2 + \theta_p$ as a countermeasure for the case where the length of the front subtrack is insufficient to make the robot body inclination equal to the inclination of the stairs (see Figure 13).

On the other hand, with respect to the rear subtrack, it needs to be controlled based on the following equation so that the inclination of the robot body coincides with the inclination of the stairs after the centroid position crosses the edge of the highest step:

$$\theta_r^{\text{ref}} = \theta_s - \theta_p + \Delta_r. \tag{19}$$

Further, switching from the immediately preceding descend-ready phase to the pitch-up descend phase can be performed when the inclination angle of the robot body θ_p becomes sufficiently horizontal. Specifically, switching should be performed when the following equation (20) is satisfied using the detection error in the pitch angle direction e_p.

$$\theta_p \ge -e_p. \tag{20}$$

5.6. Prevention of Falling Backward and Sideward in Normal Climbing-Down Phase. It is better to perform the same control as the normal climbing-up phase. That is, in order to prevent falling backward, the robot should extend the front and rear subtracks horizontally, and to prevent direction error, the main crawler should be controlled so that the robot is directed perpendicular to the edges of the stairs based on the roll-angle information. It is reasonable that switching from the immediately preceding pitch-up descend phase is performed when it is believed that the position of the centroid of the robot has moved downward from the edge of the highest step, as well as when the pitch angle of the robot body θ_p matches the inclination of the stairs θ_s. Here, we describe the joint angle of the front subtrack in this situation. Based on Figure 14, the following equation holds when the ZMP is positioned at the vertex of the uppermost stage of the staircase:

$$\begin{aligned} x - C_L \cos\theta_s - \left(C_H - r_m\right)\sin\theta_s &= 0, \\ \therefore x = C_L \cos\theta_s + \left(C_H - r_m\right)\sin\theta_s&, \end{aligned} \tag{21}$$

where x is the horizontal distance between the center of the sprocket in front of the main tracks and the top of the staircase when ZMP is coincident with the top edge of the stair. In addition, let y be the distance in the vertical direction between the sprocket center and the upper floor in front of the main track in the same state, and when θ_p matches θ_s, it is described as

$$y = x \tan\theta_s + \frac{r_m}{\cos\theta_s}. \tag{22}$$

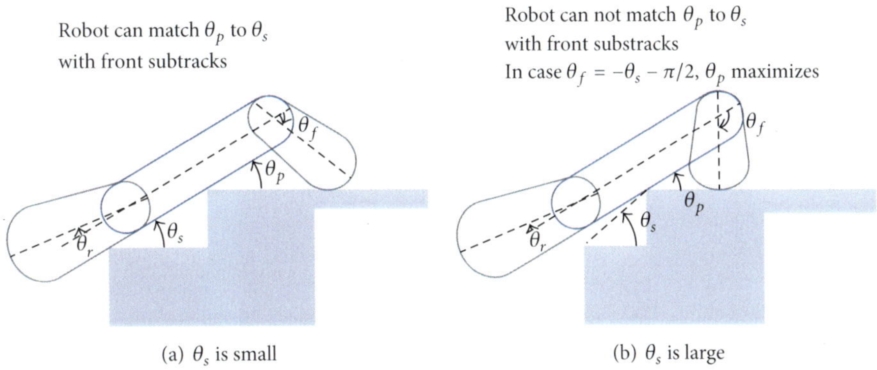

FIGURE 13: Pitch-up descend.

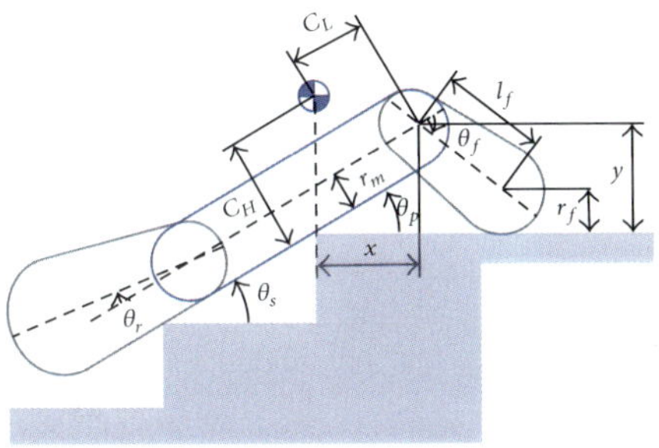

FIGURE 14: State to switch to normal climbing-down phase.

Moreover,

$$y = r + l_f \sin(\theta_f - \theta_s). \tag{23}$$

Therefore, we can derive the following equation:

$$\theta_f = -\theta_s + \arcsin\left(\frac{r_f - x\tan\theta_s - r_m/\cos\theta_s}{l_f}\right). \tag{24}$$

From the above, it is reasonable to switch when both (25) and (26) are satisfied.

$$\theta_f > -\theta_s + \arcsin\left(\frac{r_f - x\tan\theta_s - r_m/\cos\theta_s}{l_f}\right), \tag{25}$$

$$\theta_p > \theta_s - e_p. \tag{26}$$

Here, when $l_f < |r_f - x\tan\theta_s - r_m/\cos\theta_s|$, the arcsin term in (25) cannot be defined. This means that when the pitch angle of the robot body coincides with the staircase, and the position of the centroid exceeds the highest edge of the stairs, the front subcrawler is too short to touch the upper floor. In this case, it is reasonable to ignore the condition of (25).

5.7. Preventing Sliding Down and Impact Mitigation in the Pitch-Down Descend Phase. In order to avoid the shock shown in Figure 7, it is necessary to set the front subtrack to the upper direction than the step inclination and to set the fall height to 0 before leaving. Alternately, if the length of the front subtrack is insufficient to touch the upper floor, it is desirable to make it as small as possible. In addition, it is necessary to consider shocks that can occur due to slipping. Here, as described in Section 5.1, the joint angle of the front subtrack that most prevents slipping is expressed by (3), which means that it is always directed upward than the inclination of the robot body in most cases. In this study, we focus on the prevention of slippage, and we assume that the falling height can be reduced to zero because the front subtrack is sufficiently long, or even if a drop of a certain height occurs, the impact will be allowable. According to this assumption, the operational concept of each subtrack is similar to that described in Section 5.1, and it is reasonable to perform control based on (3) and (4).

Next, we consider a reasonable condition for switching the control method from the last normal climbing-down phase. It is an option to switch when the pitch angle of the robot body becomes smaller than the inclination angle of the stairs. However, switching the control method at this instant increases the pitch angle of the robot after the grounding point to the lower floor moves from the rear subtrack to the main track and generates the risk of falling backward. In this study, we assumed that the backward-falling motion does not occur when the inclination of the robot body is less than the inclination of the stairs. For this reason, it is safe to switch the control method after ensuring that the inclination of the robot body does not exceed the step inclination upon changing the control method. The state before and after switching at this moment is shown in Figure 15. Based on geometric conditions,

$$\begin{aligned}\sin\theta_s &= \frac{p\sin\theta_s - r_m(1-\cos\theta_s)}{z}\\ \therefore z &= p - \frac{(1-\cos\theta_s)}{\sin\theta_s}r_m\end{aligned} \tag{27}$$

is established. Moreover, we also derived the following equation:

$$\begin{aligned}\sin\theta_p &= \frac{p\sin\theta_s - r_r(1-\cos\theta_p)}{l_r\cos\Delta_r + z}\\ \therefore \theta_p &= \arcsin\left(\frac{p\sin\theta_s - r_r}{\sqrt{(l_r\cos\Delta_r + z)^2 + r_r^2}}\right)\\ &\quad + \arctan\left(\frac{r_r}{l_r\cos\Delta_r + z}\right).\end{aligned} \tag{28}$$

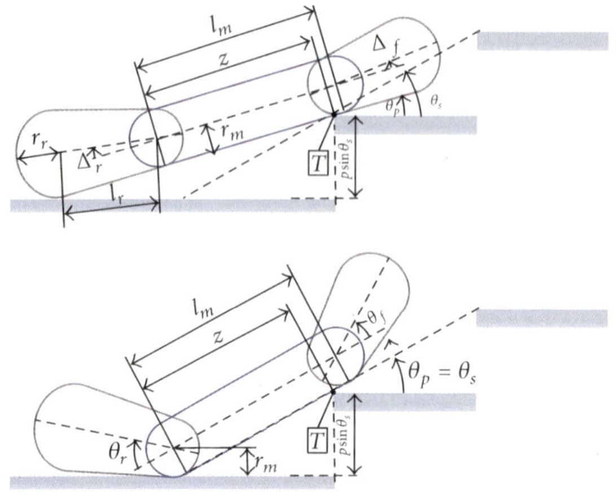

FIGURE 15: State before and after switching the control method.

From the above, it is considered safe to switch when the pitch angle of the robot body θ_s satisfies the following equation:

$$\theta_p < \arcsin\left(\frac{p\sin\theta_s - r_r}{\sqrt{(l_r\cos\Delta_r + z)^2 + r_r^2}}\right) + \arctan\left(\frac{r_r}{l_r\cos\Delta_r + z}\right) - e_p. \quad (29)$$

6. Verification Test

We verified our proposed method described in the previous section by conducting tests with a tracked robot on mockup stairs. In this section, we describe the procedures and results, and we discuss the verification tests.

6.1. Test Equipment

6.1.1. Tracked Robot. We used the tracked robot Kenaf [18] in our laboratory for verification tests (Figure 16). Kenaf is a 6-DOF tracked robot that has two main tracks for traversal, as well as four subtracks, which are located on both sides at the front and rear of the robot and which can be controlled independently. However, the left and right subtracks are synchronized in the verification tests. Table 1 indicates the parameters of Kenaf. In addition, this robot is equipped with an RT-USB-9AXIS-00, which is a 9-axis IMU sensor module manufactured by RT Co., Ltd., in the main track part, and uses the posture-estimation filter proposed by Madgwick et al. [19]. Therefore, the robot can detect the pitch and roll angle with respect to the gravity direction.

In addition, the robot has a 9-axis IMU sensor module RT-USB-9AXIS-00 made by RT Co., Ltd., on the main tracks to detect the pitch and roll angles of the robot with respect to the gravity direction.

6.1.2. Mockup Stairs. In order to confirm whether the proposed method described in Section 5 can be applied even when the inclination angle θ_s and the pitch p of the stair are changed, we used a self-made mockup stair, as shown in Figure 17. The stair can change both the pitch p between two adjacent edges of the stairs and its inclination θ_s. θ_s can be changed by up to 60°, and p can be changed to any value up to 2400 mm.

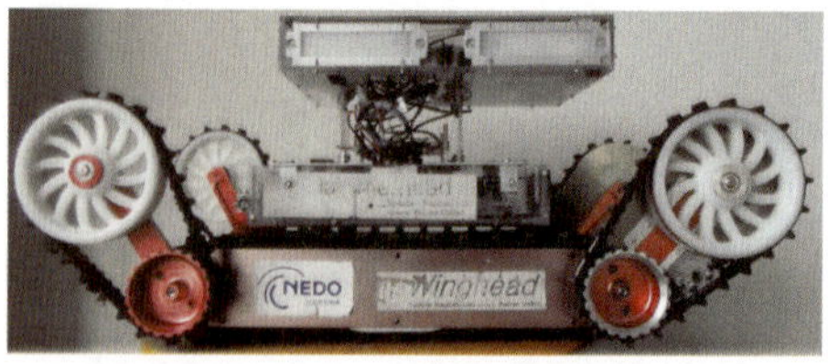

FIGURE 16: Tracked robot used in tests.

FIGURE 17: Changeable mockup stairs.

TABLE 1: Specification of the robot.

Parameter	Symbol	Value
Length of flat area in front ST	l_f	155 [mm]
Length of flat area in MT	l_m	470 [mm]
Length of flat area in rear ST	l_r	155 [mm]
Radius of circular part in front ST	r_f	84 [mm]
Radius of circular part in MT	r_m	56 [mm]
Radius of circular part in rear ST	r_r	84 [mm]
Mass of front ST	m_f	2.1 [kg]
Mass of MT	m_m	22.8 [kg]
Mass of rear ST	m_r	1.6 [kg]
Centroid of x-coordinate of front ST	x_f	15 [mm]
Centroid of y-coordinate of front ST	y_f	15 [mm]
Centroid of x-coordinate of MT	x_m	0 [mm]
Centroid of y-coordinate of MT	y_m	45 [mm]
Centroid of x-coordinate of rear ST	x_r	−10 [mm]
Centroid of y-coordinate of rear ST	y_r	16 [mm]

MT: main tracks; ST: subtracks.

6.2. Test Procedure. We implemented the proposed method on Kenaf, and we conducted traversal tests on the mockup stairs with a safety rope in front of the robot. As the initial condition, the robot is made to face the step, and it started

TABLE 2: Relationship between motion phases and failure modes.

Motion phases (MP)		Failure modes (FM) I slipping	II falling backward	III direction error	IV excessive shock
Ascend	(i) Entering	Easy to occur and critical	Hard to occur		Does not happen if the slipping is prevented
	(ii) Normal climbing	(i) Hard to occur (ii) Not critical because of grousers (=Assumption)	Easy to occur and critical		(i) Hard to occur (ii) Not critical
	(iii) Leaving		Hard to occur if it was prevented in normal climbing		Easy to occur and critical
Descend	(i) Ready		Hard to occur		
	(ii) Entering	Hard to occur	Easy to occur and critical	Hard to occur	Easy to occur and critical
	(iii) Normal climbing	(i) Hard to occur (ii) Not critical because of grousers	Easy to occur and critical		(i) Hard to occur (ii) Not critical
	(iv) Leaving	(i) Easy to occur (ii) Not critical except for excessive shock	Hard to occur if it was prevented in normal climbing		Easy to occur and critical

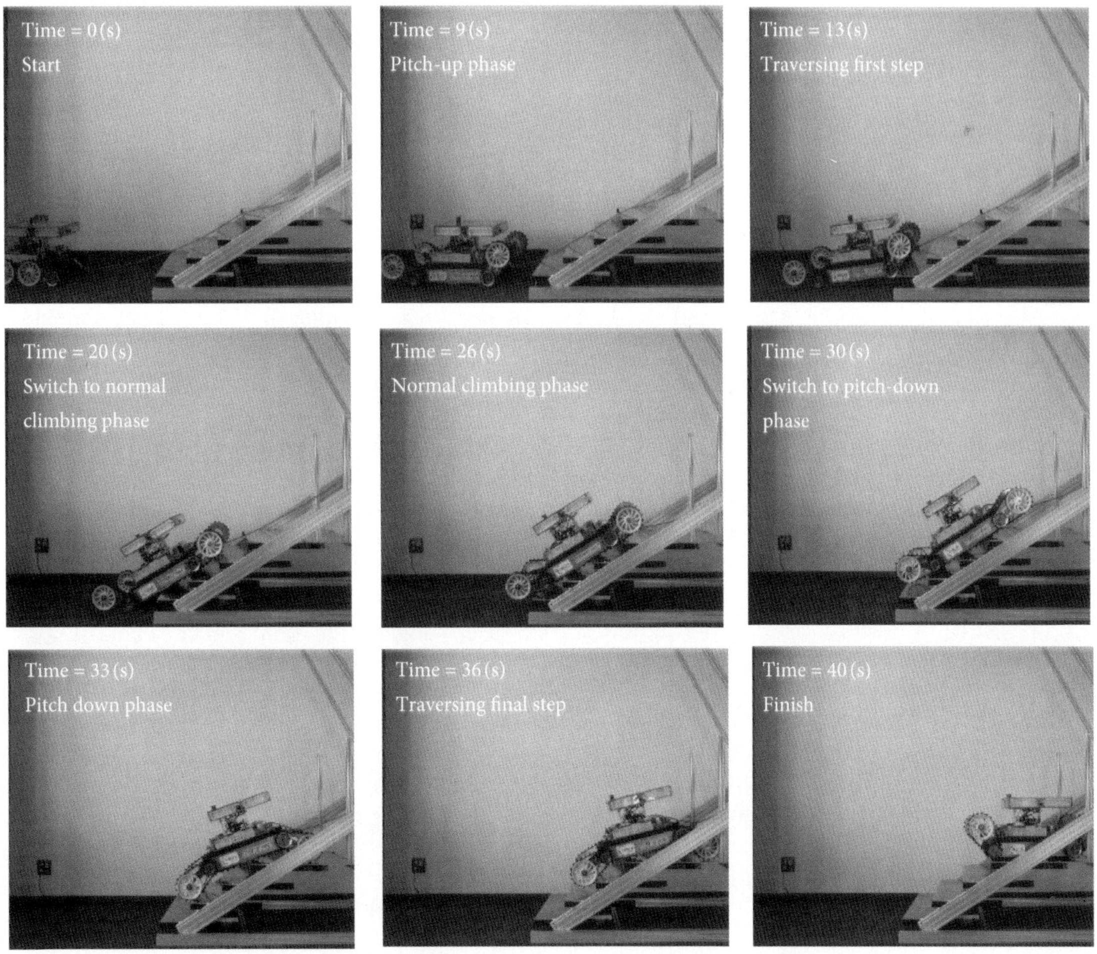

FIGURE 18: Climbing-up behavior for case $\theta_s = 30^{\circ}$, $p = 200$ mm.

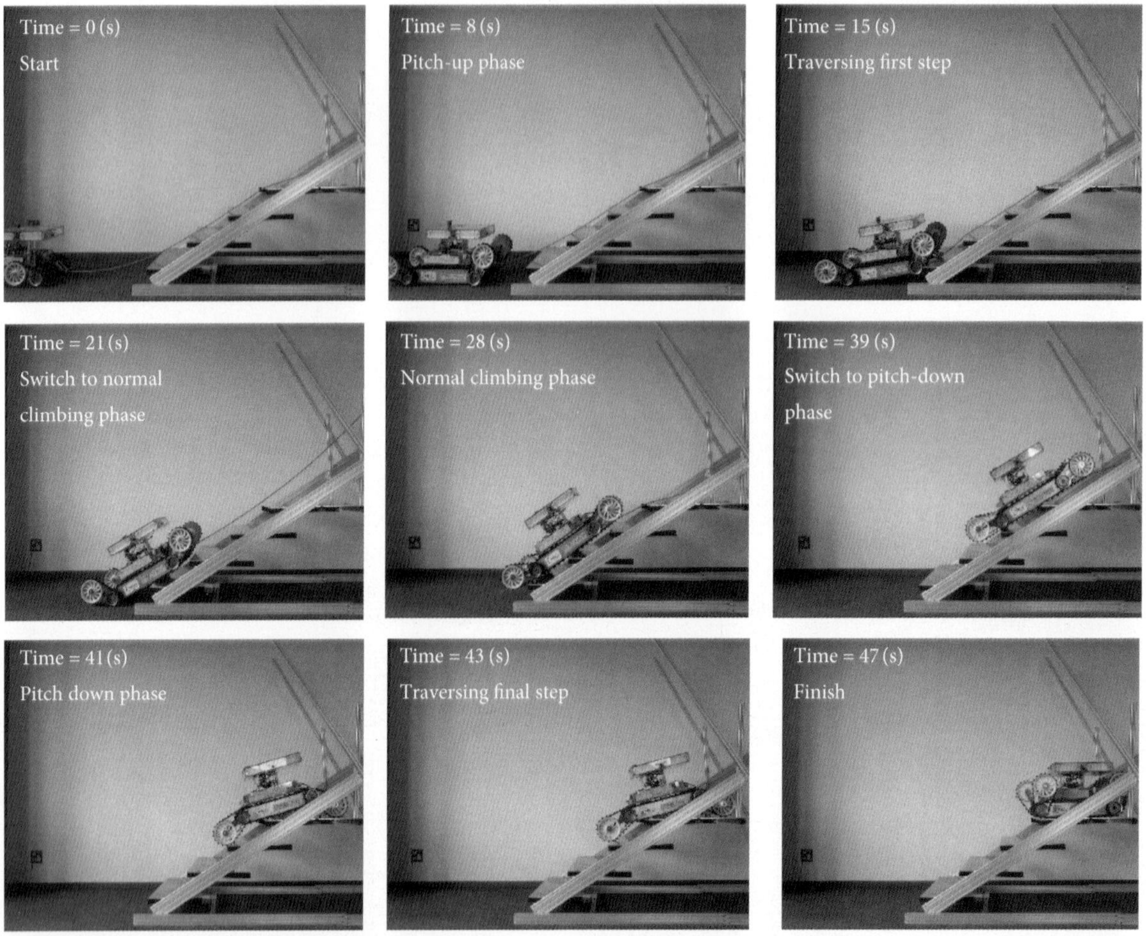

FIGURE 19: Climbing-up behavior for case $\theta_s = 30^\circ$, $p = 300$ mm.

with the front and rear subtracks folded inward ($\theta_f = \theta_r = 165.2[^\circ]$).

The running speed of the main tracks was set at 5 cm/s and −5 cm/s for the ascending and descending tests, respectively. For (2) and (3), we set the friction coefficient to $\mu = 0.74[—]$ based on our preliminary test. The margin of the rear subtrack required to prevent contact with the lower floor was set as $\delta = 6^\circ$. The height of the free-fall margin h was set at 50 mm for the condition $\theta_s = 30^\circ$ and 80 mm for $\theta_s = 45^\circ$. In addition, in the descend-ready phase, the target joint angle of the backward crawler θ_{de} was set to -36°. According to the results of some preliminary experiments, the detection error of the pitch angle of the robot body was $e_p = 2.9^\circ$.

On the other hand, based on the above parameters, we conducted verification tests under three different conditions, (θ_s, p) = (30°, 200 mm), (30°, 300 mm), and (45°, 200 mm), and we performed five trials for each condition. The number of steps was constantly found to be $n = 4$.

In these tests, the pitch angle θ_p and the roll angle θ_R of the robot body, the joint angles θ_f, θ_r, and the operation mode were recorded as log information.

6.3. Test Results. Kenaf traversed the mockup stairs in all conditions without any failure modes. As a representative result in each condition, continuous photographs of the behavior of the robot are shown in Figures 18–23, and the data profiles obtained during these experiments are shown in Figures 24–29.

Based on the continuous photographs in Figures 18–23, we can confirm that the robot can safely climb up and down the stairs based on the proposed motion strategy. Here, we consider the transition of the state of the robot for each mode from the data profile shown in Figures 24–26, which is the test result of climbing up. First, in the pitch-up ascend phase, the pitch angle of the robot remained almost horizontal until it touched the lowest step, so the front and rear subtracks also kept almost constant angles according to (3) and (4). After the front subtrack makes contact with the lowest step, the pitch angle of the robot gradually increases, and along with this, the joint angle of the front subtrack is decreased and the rear subtrack is increased accordingly; therefore, we can see that the pitch angle of the main body is made to smoothly coincide with the stair-inclination angle. After that, when the pitch angle of the main body roughly matches the inclination angle of the stairs, it switches to the normal climbing-up ascend phase, and each joint angle of the front and rear subtracks is kept constant based on (5). In this way, it is possible to climb up stairs while maintaining the pitch angle of the robot body

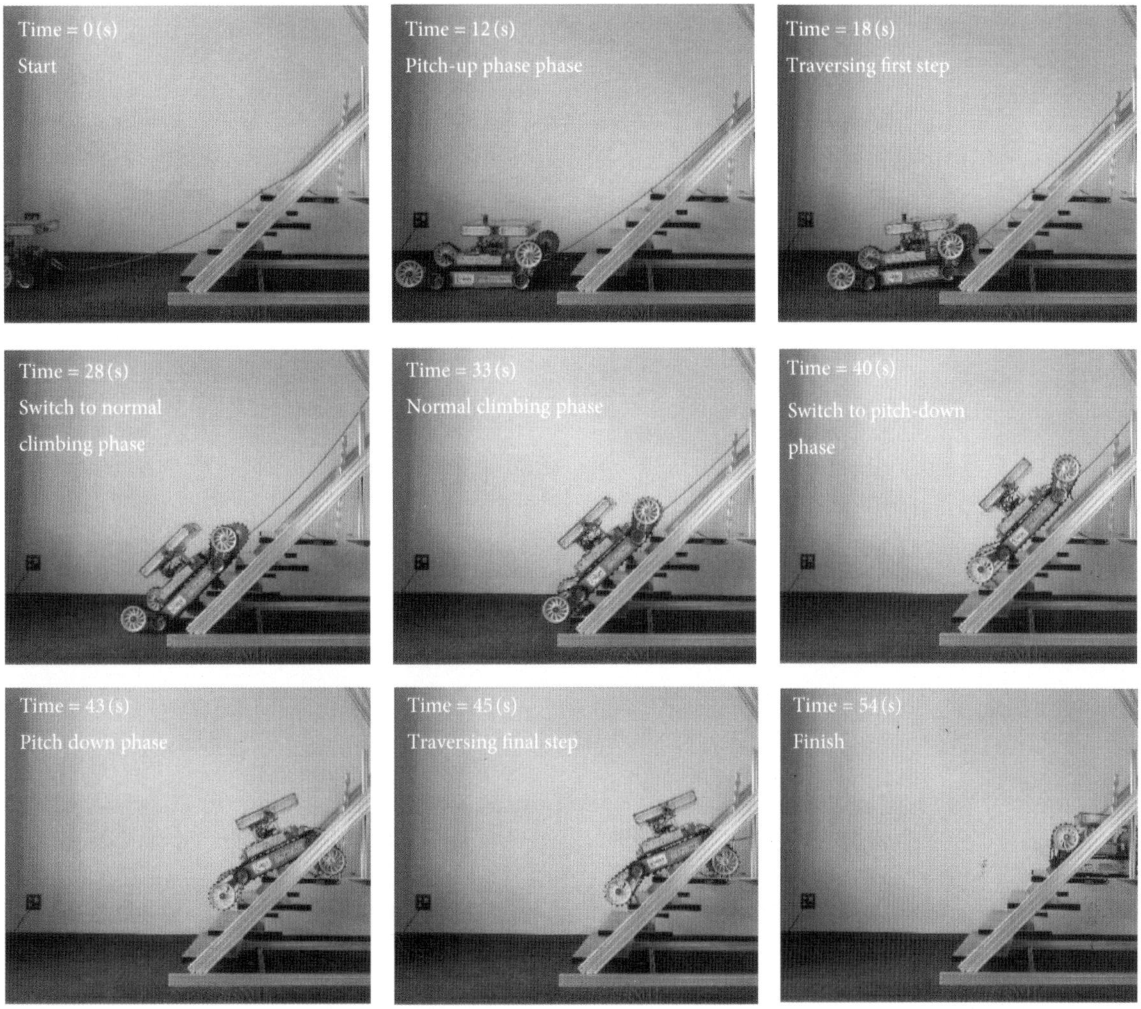

FIGURE 20: Climbing-up behavior for case $\theta_s = 45^\circ$, $p = 200$ mm.

almost at the inclination angle of the stairs. In this normal climbing-up phase, when the traveling distance satisfies (14), the transition to the pitch-down ascend phase is made, and the front and rear subtracks are determined based on (10) and (12). Then, it is possible to move the centroid on the upper floor and to decrease the pitch angle of the robot body.

On the other hand, we should consider the data profile shown in Figures 27–29, which is the test result of climbing down. In the first descend-ready phase, the robot went back keeping with its pitch angle of the robot body which was slightly decreased by controlling the joint angles of the front and rear subtracks based on (16) and (17). As a result, we see that condition (20) was established at approximately the same time that the rear end of the main track made contact with the highest edge of the stairs, and the transition to the subsequent pitch-up descend phase was made. In the pitch-up descend phase, the joint angles of the front and rear subtracks are controlled based on (18) and (19), respectively, and the pitch angle of the robot body can be made to coincide with the target inclination, θ_s. After switching to the normal climbing-down phase, the robot continued to proceed in a backward direction until conditions (25) and (26) were established. Then, it switched to the pitch-down descend phase. In this phase, the subtracks performed the same control as with the pitch-up ascend phase, and it successfully transitioned the pitch angle of the robot body horizontally.

Consequently, we applied the proposed method to various cases for different configurations of stairs.

7. Conclusion

In this paper, we described a method to make a 4-DOF tracked robot climb up and down a known flight of stairs autonomously based on internal sensors only. We classified both motion phases and four failure modes while the robot climbed up and down the stairs, and we then proposed an effective motion in each phase to prevent failure modes. Moreover, the transitions between phases were also realized based mainly on the inclination sensor of the robot body. As a result of the verification tests, we confirmed that a 4-DOF tracked robot climbed up and down the stairs without any failure modes based on our proposed method. From the above, the objective of this research that is automatic floor change for the 4-DOF tracked robots through stairs was

FIGURE 21: Climbing-down behavior for case $\theta_s = 30^\circ$, $p = 200$ mm.

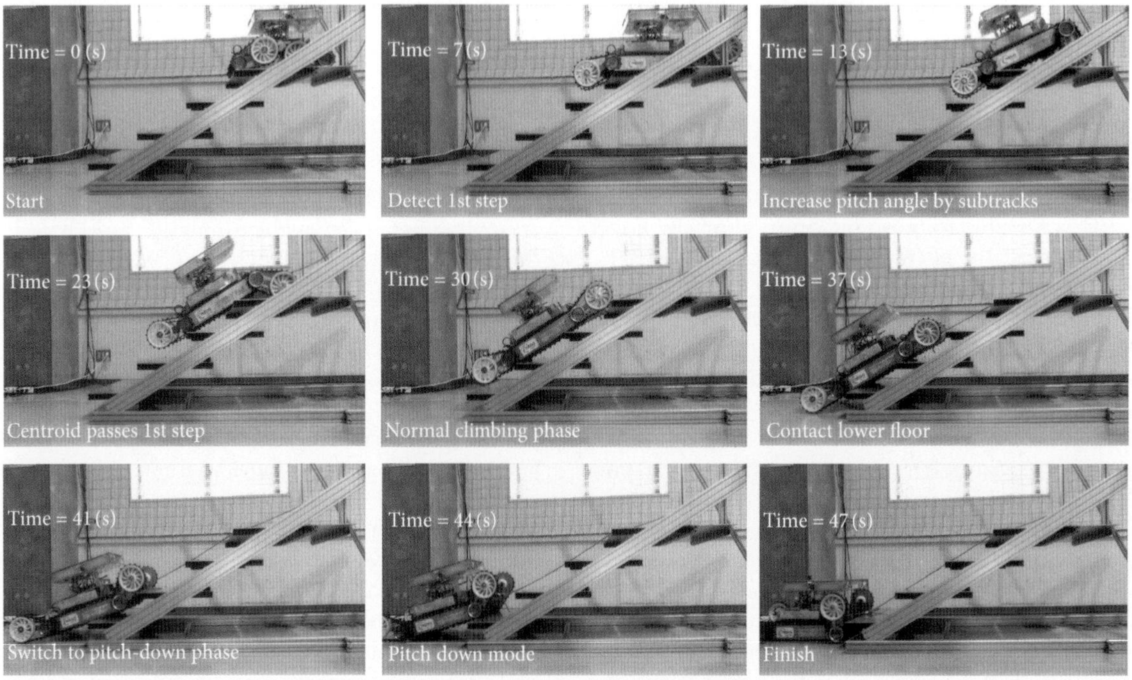

FIGURE 22: Climbing-down behavior for case $\theta_s = 30^\circ$, $p = 300$ mm.

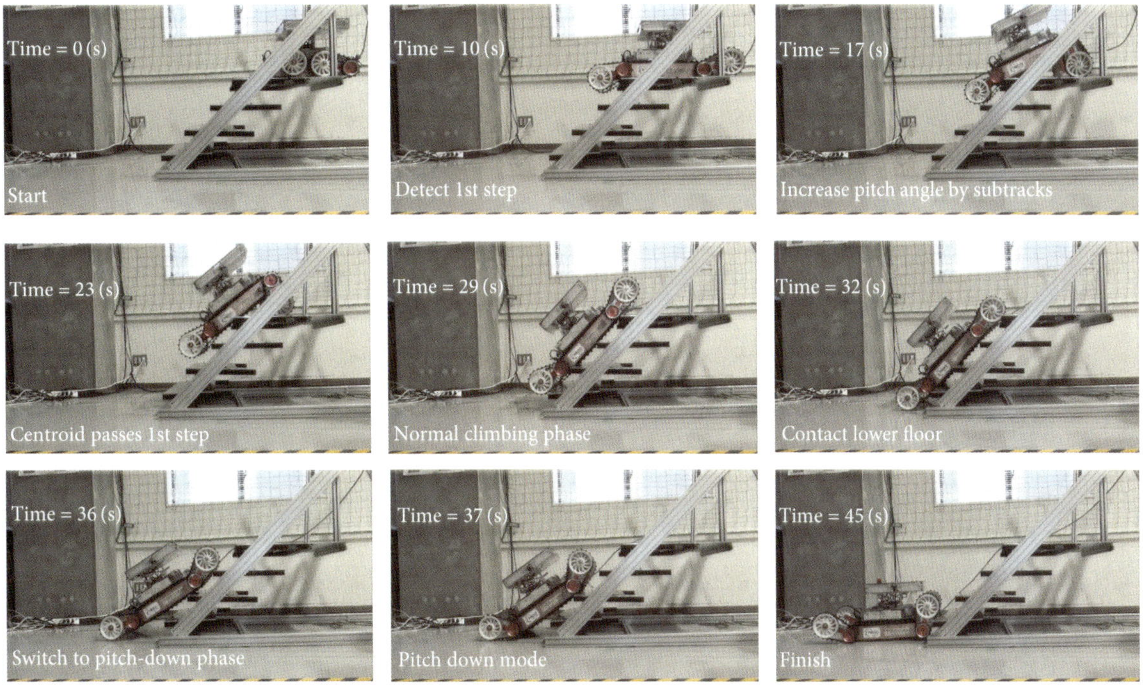

FIGURE 23: Climbing-down behavior for case $\theta_s = 45^\circ$, $p = 200$ mm.

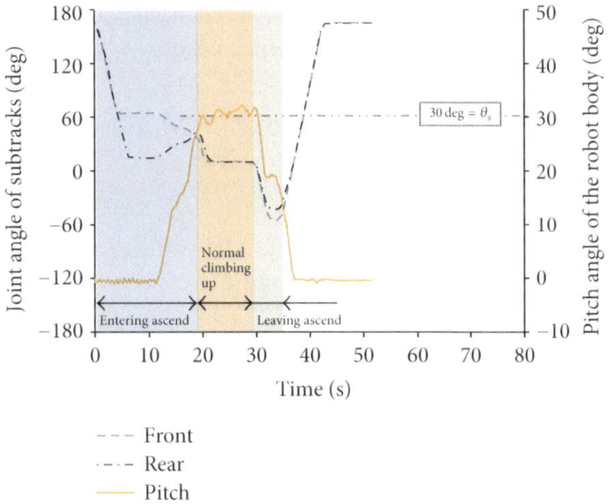

FIGURE 24: Data profile for θ_f, θ_r, θ_p ($\theta_s = 30^\circ$, $p = 200$ mm, ascend).

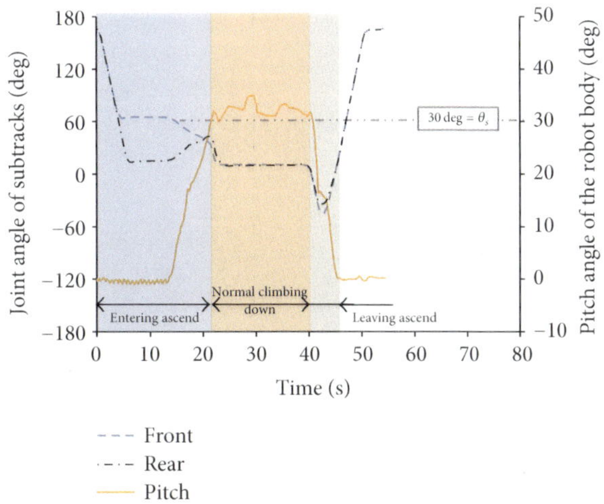

FIGURE 25: Data profile for θ_f, θ_r, θ_p ($\theta_s = 30^\circ$, $p = 300$ mm, ascend).

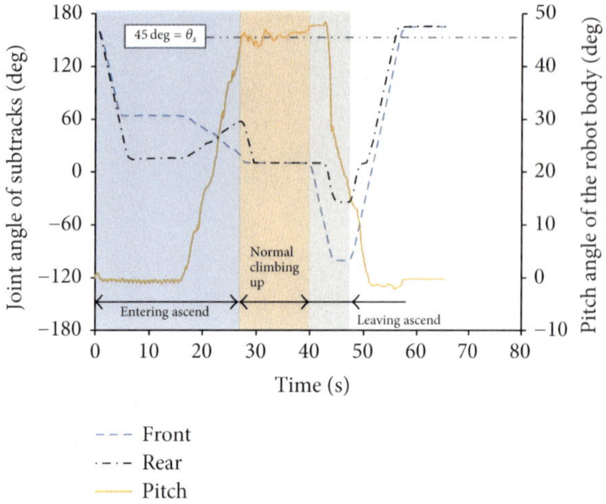

FIGURE 26: Data profile for $\theta_f, \theta_r, \theta_p$ ($\theta_s = 45^\circ$, $p = 200$ mm, ascend).

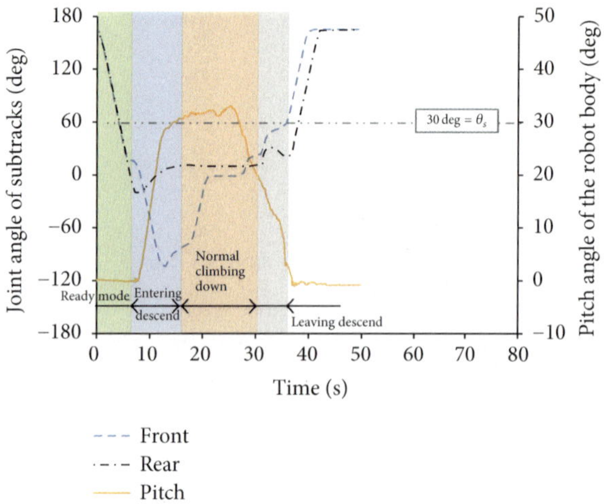

FIGURE 27: Data profile for $\theta_f, \theta_r, \theta_p$ ($\theta_s = 30^\circ$, $p = 200$ mm, descend).

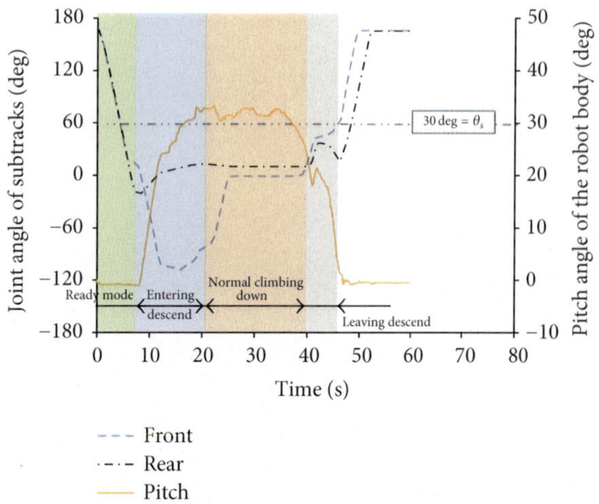

FIGURE 28: Data profile for $\theta_f, \theta_r, \theta_p$ ($\theta_s = 30^\circ$, $p = 300$ mm, descend).

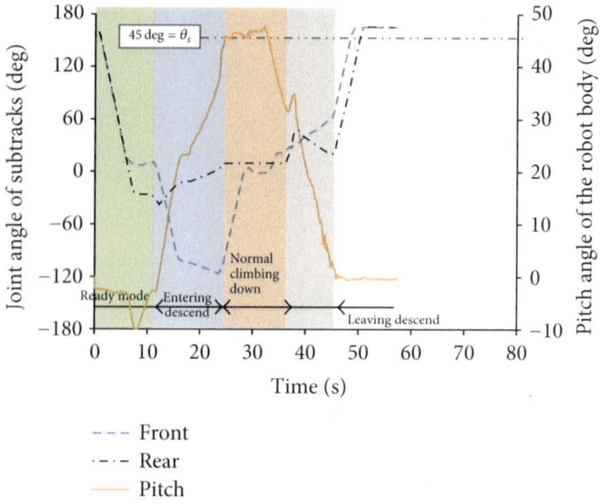

FIGURE 29: Data profile for $\theta_f, \theta_r, \theta_p$ ($\theta_s = 45^\circ$, $p = 200$ mm, descend).

achieved. Moreover, the method described in this paper can be applied to many robots sharing the assumptions described in Section 2, and it can be expected to be widely generalized.

Conflicts of Interest

The authors declare that they have no conflicts of interest.

Acknowledgments

This work was supported in part by a Grant-in-Aid for JSPS Fellows (no. 16J2554). Finally, the authors would like to thank Editage (https://www.editage.jp) for the English language editing services that were provided.

References

[1] T. Yoshida, K. Nagatani, S. Tadokoro, T. Nishimura, and E. Koyanagi, "Improvements to the rescue robot quince toward future indoor surveillance missions in the Fukushima Daiichi nuclear power plant," in *Proceedings of the 8th International Conference on Field and Service Robotics*, 2012.

[2] K. Nagatani, S. Kiribayashi, Y. Okada et al., "Redesign of rescue mobile robot Quince," in *Proceedings of the 9th IEEE International Symposium on Safety, Security, and Rescue Robotics (SSRR '11)*, pp. 13–18, Kyoto, Japan, November 2011.

[3] Y. Okada, K. Nagatani, K. Yoshida, T. Yoshida, and E. Koyanagi, "Shared autonomy system for tracked vehicles to traverse rough terrain based on continuous three-dimensional terrain scanning," in *Proceedings of the 23rd IEEE/RSJ International Conference on Intelligent Robots and Systems (IROS '10)*, pp. 357–362, Taipei, Taiwan, October 2010.

[4] A. I. Mourikis, N. Trawny, S. I. Roumeliotis, D. M. Helmick, and L. Matthies, "Autonomous stair climbing for tracked vehicles," *International Journal of Robotics Research*, vol. 26, no. 7, pp. 737–758, 2007.

[5] I.-H. Li, W.-Y. Wang, and C.-K. Tseng, "A kinect-sensor-based tracked robot for exploring and climbing stairs," *International Journal of Advanced Robotic Systems*, vol. 11, no. 1, article A80, 2014.

[6] A. Kalantari, E. Mihankhah, and S. A. A. Moosavian, "Safe autonomous stair climbing for a tracked mobile robot using a kinematics based controller," in *Proceedings of the IEEE/ASME International Conference on Advanced Intelligent Mechatronics (AIM '09)*, pp. 1891–1896, Singapore, July 2009.

[7] E. Mihankhah, A. Kalantari, E. Aboosaeedan, H. D. Taghirad, and S. A. A. Moosavian, "Autonomous staircase detection and stair climbing for a tracked mobile robot using fuzzy controller," in *Proceedings of the IEEE International Conference on Robotics and Biomimetics (ROBIO '08)*, pp. 1980–1985, IEEE, Bangkok, Thailand, February 2009.

[8] N. Li, S. Ma, B. Li, M. Wang, and Y. Wang, "An online stair-climbing control method for a transformable tracked robot," in *Proceedings of the 2012 IEEE International Conference on Robotics and Automation*, pp. 923–929, St Paul, Minn, USA, May 2012.

[9] Q. Zhang, S. S. Ge, and P. Y. Tao, "Autonomous stair climbing for mobile tracked robot," in *Proceedings of the 9th IEEE International Symposium on Safety, Security, and Rescue Robotics (SSRR '11)*, pp. 92–98, Kyoto, Japan, November 2011.

[10] J. D. Martens and W. S. Newman, "Stabilization of a mobile robot climbing stairs," in *Proceedings of the 1994 IEEE International Conference on Robotics and Automation*, pp. 2501–2507, San Diego, Calif, USA, May 1994.

[11] P. Ben-Tzvi, S. Ito, and A. A. Goldenberg, "Autonomous stair climbing with reconfigurable tracked mobile robot," in *Proceedings of the IEEE International Workshop on Robotic and Sensor Environments (ROSE '07)*, pp. 104–109, Ottawa, Canada, October 2007.

[12] K. Ohno, S. Morimura, S. Tadokoro, E. Koyanagi, and T. Yoshida, "Semi-autonomous control system of rescue crawler robot having flippers for getting over unknown-steps," in *Proceedings of the IEEE/RSJ International Conference on Intelligent Robots and Systems (IROS '07)*, pp. 3012–3018, San Diego, Claif, USA, November 2007.

[13] D. Endo, A. Watanabe, and K. Nagatani, "Stair climbing control of 4-degrees-of-freedom tracked vehicle based on internal sensors," in *Proceedings of the 14th International Symposium on Safety, Security and Rescue Robotics (SSRR '16)*, pp. 112–117, Lausanne, Switzerland, October 2016.

[14] Y. Liu and G. Liu, "Track-stair interaction analysis and online tipover prediction for a self-reconfigurable tracked mobile robot climbing stairs," *IEEE/ASME Transactions on Mechatronics*, vol. 14, no. 5, pp. 528–538, 2009.

[15] J. Liu, Y. Wang, S. Ma, and B. Li, "Analysis of stairs-climbing ability for a tracked reconfigurable modular robot," in *Proceedings of the 2005 IEEE International Workshop on Safety, Security and Rescue Robotics*, pp. 36–41, Kobe, Japan, June 2005.

[16] M. Vukobratovic, B. Branislav, and D. Surdilovic, "Zero-moment point—proper interpretation and new applications," in *Proceedings of the IEEE-RAS International Conference on Humanoid Robots*, pp. 237–244, 2001.

[17] Y. Kan, "Normalized energy stability margin: generalized stability criterion for walking vehicles," in *Proceedings of CRAWLER '98*, pp. 71–76, 1998.

[18] T. Yoshida, E. Koyanagi, S. Tadokoro et al., "A high mobility 6-crawler mobile robot "Kenaf"," in *Proceedings of the 4th International Workshop on Synthetic Simulation and Robotics to Mitigate Earthquake Disaster (SRMED '07)*, p. 38, Atlanta, Ga, USA, July 2007.

[19] S. O. H. Madgwick, A. J. L. Harrison, and R. Vaidyanathan, "Estimation of IMU and MARG orientation using a gradient descent algorithm," in *Proceedings of the IEEE International Conference on Rehabilitation Robotics (ICORR '11)*, pp. 1–7, Zurich, Switzerland, July 2011.

14

A Simple Outdoor Environment Obstacle Detection Method based on Information Fusion of Depth and Infrared

Yaguang Zhu, Baomin Yi, and Tong Guo

Key Laboratory of Road Construction Technology and Equipment of MOE, Chang'an University, Xi'an 710064, China

Correspondence should be addressed to Yaguang Zhu; zhuyaguang@chd.edu.cn

Academic Editor: Shahram Payandeh

In allusion to the existing low recognition rate and robustness problem in obstacle detection; a simple but effective obstacle detection algorithm of information fusion in the depth and infrared is put forward. The scenario is segmented by the mean-shift algorithm and the pixel gradient of foreground is calculated. After pretreatment of edge detection and morphological operation, the depth information and infrared information are fused. The characteristics of depth map and infrared image in edge detection are used for the raised method, the false rate of detection is reduced, and detection precision is improved. Since the depth map and infrared image are not affected by natural sunlight, the influence on obstacle recognition due to the factors such as light intensity and shadow is effectively reduced and the robustness of the algorithm is also improved. Experiments indicate that the detection algorithm of information fusion can accurately identify the small obstacle in the view and the accuracy of obstacle recognition will not be affected by light. Hence, this method has great significance for mobile robot or intelligent vehicles on obstacle detection in outdoor environment.

1. Introduction

Detection of the obstacle has a wide application in smart robots, intelligent vehicles, and autonomous agricultural vehicles. Among them, obstacle detection based on visual image information [1, 2] is mainly used in vision system of automatic driving vehicle or robot, The system completes some special functions similar to the human eye and provides reliable and accurate environmental information for path planning. Hence, the visual navigation system must be able to detect obstacles quickly and accurately and has better robustness.

The sensors used for traditional obstacle detection mainly include laser radar sensors, ultrasonic sensors, infrared sensors, and visual equipment [3]. The basic principle of laser radar sensors is that the distance information can be obtained by calculating the time interval between emission and receiving of laser [4]. Intelligent vehicle named Stanley [5] from Stanford University installs a lot of laser radar sensors at the top of the car, which is used for judging whether the area is an obstacle or an area that can be passed through by 2D grid based on Markov model and height difference between adjacent points. However, the size of the laser radar sensor is relatively large, the cost is very high, and the installation and use are also more complex. In the study of detecting obstacles by using ultrasonic sensors, Figueroa and Lamancusa [6] proposed a method to detect obstacles by using the relationship between the phase delay and the peak value of the echo and then combined with echo time, which had higher detection accuracy. Furthermore, Hua et al. [7] proposed a method to detect phase position by modulating double-sideband amplitude, in which excitation signal of ultrasonic pulse was modulated by using a low frequency signal and adopting appropriate modulation data. This method not only guarantees the resolution of the phase detection obstacle, but also expands the scope of the obstacle detection to a certain extent. But the orientation of the ultrasonic sensors is poorer during the detection. When used in moving or complex conditions such as wild environment, it will not perform well. The principle of infrared sensor is infrared thermal imaging technology [8], which is a wavelength conversion technology. It is the transformation of infrared radiation to the visible graphics. Infrared thermal imaging system has a certain ability to penetrate and it can distinguish right from wrong and has strong anti-interference ability. According to the characteristics of infrared imaging, the infrared sensor has better adaptability and stronger direction. But it is very easy to be affected by the color of the object, and it cannot provide distance information. In recent years, with the development of computer technology, it has become more and more popular to use binocular camera to obtain the depth information of object and scene. However, the disadvantage of this method is that it is easy to be affected by light [9, 10]. Zhang and Zhu [11] proposed a method to segment image based on binocular vision, but it was very easy to be affected by light and has low robustness. Although the use of multiple sensors can avoid the shortage of a single sensor, it is difficult to achieve the integration of information and to ensure the synchronization between the various sensors.

Aiming at the problems of poor robustness and low detection accuracy, a fast and accurate detection method based on depth information and infrared information is proposed. Kinect V2.0 sensor is used in this paper, which can obtain depth information and infrared information at the same time. The main processes of obstacle detection in this paper are as follows. Firstly, the areas to be detected are segmented by the mean-shift algorithm. Secondly, gradient calculation is carried out to highlight the edge information of the object. Finally, the obstacle can be detected by using edge detection or other methods and fusing infrared and depth information. The method proposed in the paper can solve the problems of the accuracy and robustness of obstacle detection of mobile robot and intelligent vehicle in complex environment.

2. Problem Descriptions

Usually, obstacles need to be detected in both indoor and outdoor environment. The only difference is that the outdoor obstacles may be more complex and diverse. Two different scenes collected by Kinect are shown in Figures 1 and 2. It can be seen from the two images that the edge information of contact part between the obstacle and ground in depth map is not very clear. The edge information of the upper part of the obstacle is more obvious. But it is opposite in infrared image. Using the characters of infrared image information, the depth map can be improved. Therefore, if depth information and infrared information are fused together effectively, the detected obstacle boundaries can be segmented from the visual scene accurately. Besides, it will enhance the robustness and can guarantee high recognition rate because the depth information and infrared information are not affected by natural light.

Recently, most of the obstacle detection algorithms based on single vision adopt motion information of the different images [12]. However, due to the lack of additional information, it makes the obstacle produce abnormal values. But it can be solved by fusion of infrared image and depth map. Most obstacles information in the depth map is uniform, because the depth information of the same obstacle has similar characteristics and the depth differences of connection parts between obstacle and upholder are small. Hence, the depth information can be used to remove abnormal values in detection areas as background or some areas beyond the scope of the detection area. If only the depth image is used for obstacle detection and the obstacle is on the edge of the identified range, the depth value of obstacle is similar to the value of the ground when obstacle is smaller. So the ground and obstacle are easily identified as one object, and then errors may occur. Moreover, even though the objects have beyond the setting detection distance, with the good reflection, Kinect camera is still able to get its depth information and be able to detect it. That makes the identified obstacles exceede the scope of the requirements. On the other hand, if only infrared image is used for detection, many wrong areas will be detected because of the characters of infrared information. The colors of areas are diversified and most of these errors are relatively close to the camera and at the same time further obstacles will be ignored. The detection results of infrared images are shown in Figure 3(a).

According to the analysis above, we find that the wrong detection areas of depth map and the wrong detection areas of infrared image are not coincident. So the infrared image can be used for removing the wrong detection areas of depth map. The detection results after the fusion of depth map and infrared image are shown in Figure 3(b), in which the obstacle is a can. Therefore, in order to improve precision and performance, the depth map and the infrared image should be fused to detect the obstacles.

3. Fusion Method

Kinect V2.0 sensor is adopted in this paper, which can collect depth map and infrared image with 512 × 424 resolutions and 30 fps. Meanwhile, it can also collect color image with 1920 × 1080 resolutions and 30 fps. The effective range is 0.5 m–8 m [13]. Compared with other sensors, Kinect V2.0 can collect depth information and infrared and color information at the same time. It effectively avoids the problem of uncoordinated caused by using different sensors. In addition, it has lots of advantages; for example, this sensor has small volume and can obtain accuracy information of image. Besides, the most important is that it has strong adaptability to complex environment. Here, the method proposed by Zhang et al. [14, 15] is adopted to calibrate Kinect camera before the detection of obstacles. The thick red line in Figure 4 represents the vertical wall. Compared with pictures before calibration, the distortion caused by the camera has been corrected obviously.

According to the advantages of Kinect V2.0, obstacle detection algorithm based on information fusion of depth and infrared is proposed. The foreground of depth map is intercepted by mean-shift algorithm firstly, and then the edge information is extracted and processed by the fusion of depth map and infrared image, which is shown in Figure 5. The detection range is 0.5 m–4 m. Edge detection based on depth image or infrared image alone has its own characteristics. Gradient features of depth information can be obtained in obstacle detection with depth image alone. However, since the edge information of the areas segmented by mean-shift algorithm is more prominent, they also can be detected. Similarly, in obstacle detection with infrared image alone, the edge information of the front object in the image taken

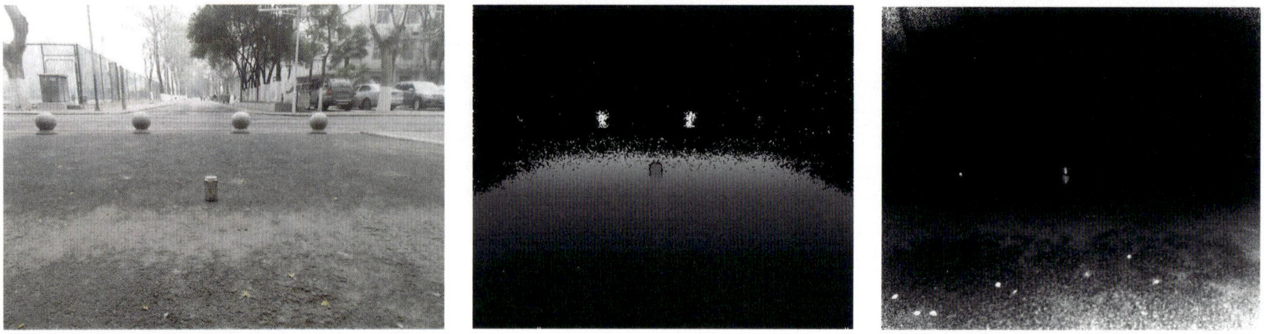

(a) Color image (b) Depth image (c) Infrared image

FIGURE 1: The road scene.

(a) Color image (b) Depth image (c) Infrared image

FIGURE 2: The grassland scene.

by the Kinect camera can be extracted according to the characteristics of infrared image, while the edge information of the rear object cannot be detected. Finally, a perfect detection result can be obtained as long as combining these two features.

In the depth image obtained by Kinect, it is an effective method to segment foreground of detection area via mean-shift algorithm. It has not received much attention when the mean-shift theory was proposed by Fukunaga and Hostetler [16], until the kernel function was proposed by Cheng [17]. The use scope of the mean-shift algorithm was expanded, and it is only beginning to attract more attention. Comaniciu et al. [18, 19] successfully applied the mean-shift algorithm to image processing and got good results in smoothness and segmentation of image. The kernel function defined by Cheng is that: if a function $G : X \rightarrow R$ has a profile function $g : [0, +\infty] \rightarrow R$, namely,

$$G(x) = g\left(\|x\|^2\right) \quad (1)$$

and should meet the following: (1) g is not negative; (2) g is not increasing, namely if $a < b$, then $g(a) \geq g(b)$; (3) g is piecewise continuous, and $\int_0^{\infty} g(r)\,dr < \infty$, then $G(x)$ is a kernel function, in which x is a point in space, represented by column vector and R represents real number field.

According to kernel function, the mean-shift formula can be described as follows [16]:

$$M\left(x_j\right) = \frac{\sum_{i=1}^{n} G_H\left(x_i - x_j\right) w\left(x_i\right) x_i}{\sum_{i=1}^{n} G_H\left(x_i - x_j\right) w\left(x_i\right)} - x_j, \quad (2)$$

in which x_j is the depth of the pixel to be processed, x_i is the depth of the sampled pixel point, $G(x)$ is an unit kernel function, $w(x_i)$ is a weight of sampling point x_i, $G_H(x_i - x_j) = |H|^{-1/2} G(H^{-1/2}(x_i - x_j))$, and H is bandwidth matrix that is usually expressed by a positive definite symmetric matrix with $d \times d$. When this function is used, the error value η is set firstly. The mean-shift algorithm is a process of loop calculation; make

$$m_h\left(x_j\right) = \frac{\sum_{i=1}^{n} G_H\left(x_i - x_j\right) w\left(x_i\right) x_i}{\sum_{i=1}^{n} G_H\left(x_i - x_j\right) w\left(x_i\right)}. \quad (3)$$

Step 1. Calculate $m_h(x_j)$.

Step 2. Substitute the value of $m_h(x_j)$ into x.

Step 3. If $\|m_h(x_j) - x_j\| < \eta$, as for the location of pixel point x_j: location identifier $p_{x_j}(x, y) = 1$, the cycle will terminate. At that time, for any pixel point x_j, we have

$$P_{x_j} = (x, y) \in A_k \quad (4)$$

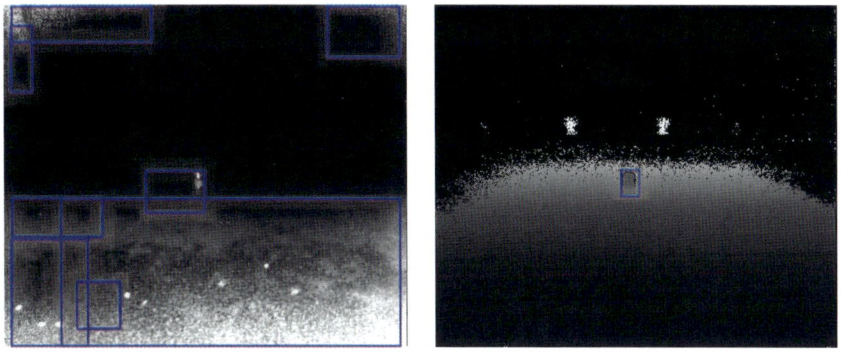

(a) The result of infrared detection (b) The result of fusion

FIGURE 3: An example of single obstacle.

(a) Color image (b) Depth image before calibration (c) Depth map after correction distortion

FIGURE 4: Depth image with calibration.

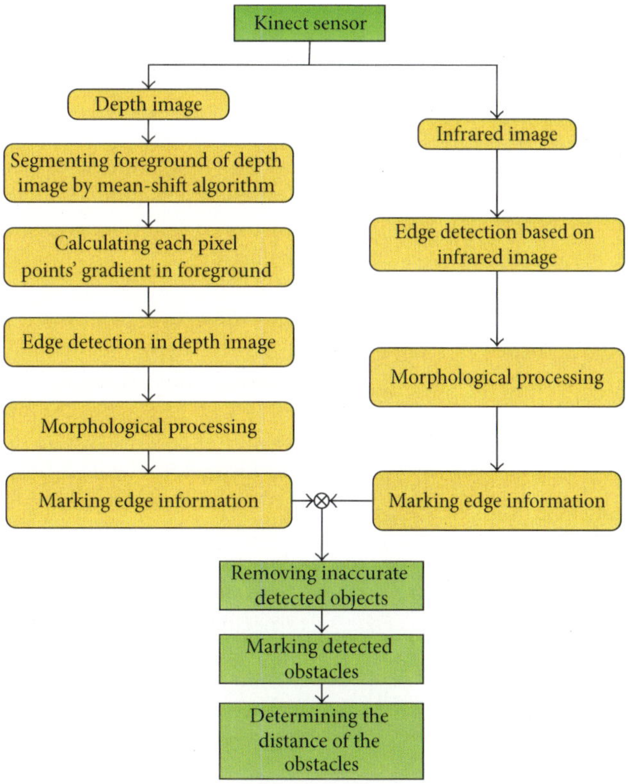

FIGURE 5: Block diagram of detection process.

(a) Color image (b) Depth image (c) Foreground image

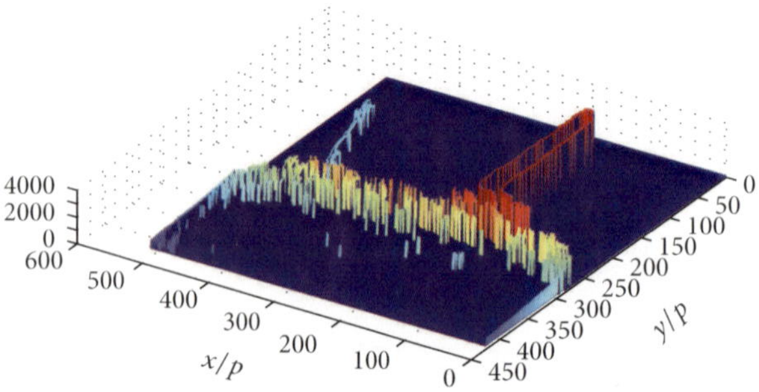

(d) Foreground gradient

FIGURE 6: Foreground information image.

in which A_k is region for segmentation at this time, $p_{x_j}(x, y)$ is a pixel coordinate of x_j in image. Otherwise, the pixel point x_{j+1} will be taken as a new starting point of space, and then the first step will restart.

We can get the expected foreground as long as the $A_1, A_2, A_3, \ldots, A_k$ which satisfied the requirement are combined together. That is,

$$\{A_1, A_2, A_3, \ldots, A_m\} = A. \tag{5}$$

in which $m < k$, A is foreground area. The result of foreground obtained by mean-shift algorithm is shown in Figure 6(c). The foreground is a region near the camera, and the obstacles on the front will be segmented into the foreground overall. The purpose of calculating gradient of each pixel point in the foreground is to highlight the edge information of the obstacle according to the characteristic of gradient. In the regions of continuous depth value, the gradient value will be small, and the change will be continuous. But in the edge of the object, the gradient value will change dramatically, and the edge information is more obvious. In the depth image I, when a pixel point is $(x, y) \in A$ and the depth value of it is represented by $I(x, y)$, the convolution gradient operators of the point in x direction and y direction are as follows [20]:

$$\begin{aligned} T_x(x, y) &= [-1 \;\; 0 \;\; 1] \\ &\quad * [I(x-1, y) \;\; I(x, y) \;\; I(x+1, y)]^T \\ T_y(x, y) &= [-1 \;\; 0 \;\; 1]^T \\ &\quad * [I(x, y-1) \;\; I(x, y) \;\; I(x, y+1)]. \end{aligned} \tag{6}$$

So, the gradient value of pixel point (x, y) is

$$T(x, y) = \sqrt{T_x(x, y)^2 + T_y(x, y)^2}. \tag{7}$$

The new gradient image is obtained after gradient calculation of foreground, which is shown in Figure 6(d). After that, we use edge detection algorithm to detect the gradient image in Figure 6(d). The Canny edge detectors are easier to improve the performance of edge detection algorithms. Similarly, the edge detection of infrared image can be carried out by this way. The value of most areas in the image is uniform. The results of detection indicate that it can completely avoid the effects of light, shadow, and other factors. After obstacles detection in the two images, morphology operations are performed. In detail, dilation operations are carried out, and then erosion operations, finally, merging adjacent points and removing isolated points. A' represents a foreground that has achieved gradient calculation and edge detection. S_1 represents a structural element for dilation. S_2 represents a structural element for erosion. (x, y) represents a location of pixel point. Expanding A' by S_1, we have

$$\begin{aligned} A' &= A \\ (A' \oplus S_1)_{(x,y)} &= \max\{A'(x - x', y - y') \\ &\quad + S_1(x', y') \mid (x', y') \in D_{S_1}\} \end{aligned} \tag{8}$$

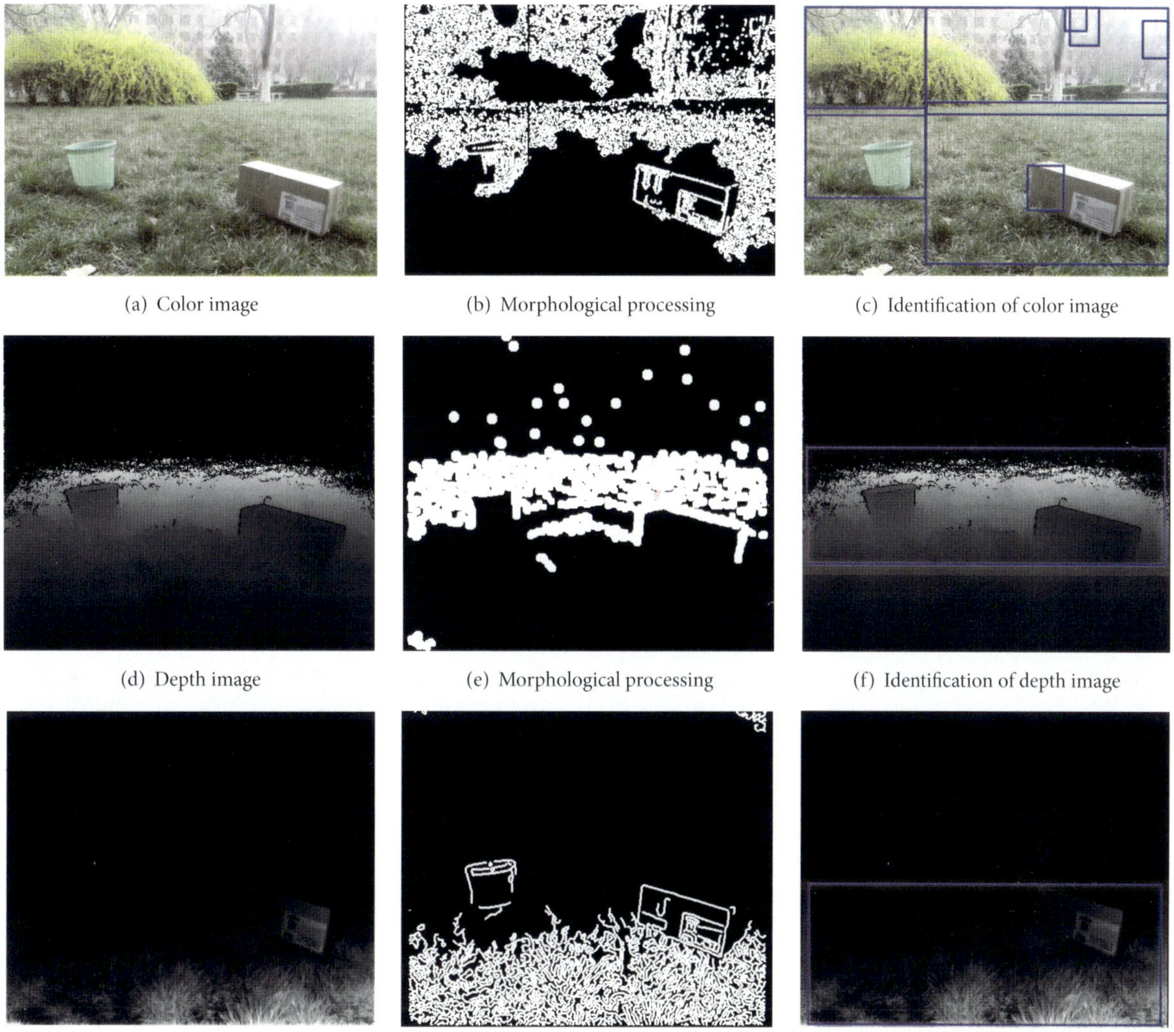

(a) Color image (b) Morphological processing (c) Identification of color image

(d) Depth image (e) Morphological processing (f) Identification of depth image

(g) Infrared image (h) Morphological processing (i) Identification of infrared image

FIGURE 7: The results of using single image identification.

in which D_{S_1} is a domain of S_1. For corrosion A' by S_2, we have

$$\left(A'\Theta S_2\right)_{(x,y)} = \min\left\{A'\left(x-x', y-y'\right) - S_2\left(x', y'\right) \mid \left(x', y'\right) \in D_{S_2}\right\} \quad (9)$$

in which D_{S_2} is a domain of S_2.

Then, we label the results of morphological operations and combine the detection result of two kinds of images by intersection calculation. In this way, the wrong results of detection can be removed. So far, the whole process of obstacle detection has been achieved. The value of the center point of the obstacle region is regarded as the depth value of the obstacle. P_d and P_i, respectively, represent pixel value of depth image and infrared image after labeling at the location of (x, y). p represents pixel value of these points after mixture:

$$p_d(x,y), p_i(x,y) = 1$$

$$p_d(x,y), p_i(x,y) = 0$$

$$p = p_d \otimes p_i. \quad (10)$$

When the obstacle is detected, the value of P_d and P_i is 1; if not, it is 0. If the values of P_d and P_i are 1 at the same time, the value of p is 1. Finally, the identified obstacles are marked by the blue box.

4. Experiments and Result

4.1. Single Image Identification. The results of obstacle detection via a single image on the grass by Kinect sensor are shown in Figure 7. In the picture, the size of carton is 25 cm × 45 cm, and the distance from the sensor is 1 m. The size of the paper baskets is 16 cm × 25 cm, and the distance from the sensor is 1.5 m. The angle of inclination of Kinect sensor is 30°. Since the yellow flowers are far away from the sensor and have over 8 m in color image, their information cannot be collected in depth image and infrared image. We can clearly conclude from the pictures that any results from single image cannot accurately identify the obstacles. In color image, due to the abundant color information and color change, the edge information is really distinct. As shown in Figure 7(b), the edge information of foreground and background both can be detected, but only obstacles in foreground are demanded. So the color image is not suitable for our research. Besides, the obtainment of color map is also easily affected by the illumination condition. When the depth map is used alone, the area edge and obstacle will be easily identified as one object if the obstacle is close to the edge of segmented foreground, which is shown in Figure 7(f). Meanwhile, the edge information is not detected in front regions without obstacles and the error information is less in rear regions, which is shown in Figure 7(e). However, when the infrared image is used alone, the edge information of front obstacles can be detected effectively. But the edge information of farther grass can also be detected and the information of the rear object is ignored, which is shown in Figure 7(h). In a word, if only single image is obtained for detection, it will produce many errors. Accurate obstacles identification cannot be achieved and neither does the robustness.

4.2. Experiments and Results Analysis. However, the method of this paper is to fuse image (e) and image (h) in Figure 7 reasonably. It can not only identify the position of obstacles accurately but also avoid the effect of illumination condition effectively. The experimental result of the obstacle detection using fusion of depth and infrared information is shown in Figure 8. The experimental parameters of edge detection and morphology operation are shown in Table 1, in which σ_R and σ_D, respectively, represent the value of Canny operator in infrared image and depth image. More edge information is detected when the value is small, but when the value is too small, the rough road is also considered as an obstacle. When the value is too large, the edge information of obstacles cannot be detected, so the value should be selected properly. According to our tests, the best value is 3. R_R, R_D, and R_F, respectively, represent expansion radius in three kinds of images. When the value is large, the expansion of the edge information is more obvious. But when the value is too large, it will make the area of the obstacle detection larger than the actual area. On the contrary, if it is too small, one obstacle will be recognized as multiple obstacles. The values in Table 1 will be selected after several tests. N_R, N_D, and N_F, respectively, represent expansion neighborhood in three kinds of images. Their values have little effect on the experimental results so they are 4 in this experiment. The experimental results show that the parameters used above can identify the obstacles accurately.

After foreground segmentation, gradient calculation, edge detection, and morphology operation in depth image, the general outline of obstacle will be conducted, but sometimes the contact surface between the ground and the obstacle cannot be segmented. The foreground edge segmented by mean-shift algorithm will coincide with the edge of obstacles. Besides, there is some other messy identification information. In infrared image, the edge information detected from front objects is abundant. After morphology operation, most areas are identified as obstacles. However, after combining Figures 8(c) and 8(e) by intersecting, we can get Figure 8(f). From Figure 8(f), the location of obstacles can be easily identified and marked. The result is shown in Figure 8(g). After experimental verification, the accuracy rate of the method reached 98%, over detection rate is 1.1%, missing rate is 0.6% according to 50 groups images each with 1–3 objects as obstacles. The accuracy of the obstacles identification will not be affected by the illumination. The method of obstacle detection with fusion of depth and infrared information is more superior to the method with single image information. In addition, it has a strong practicality.

Table 1: Operator parameters.

Operator	Value
Canny operator σ	
σ_R	3
σ_D	3
Inflation factors of infrared image	
R_R	3
N_R	4
Inflation factors of depth image	
R_D	2
N_D	4
Inflation factors of fusion image	
R_F	3
N_F	4

The results of obstacle detection are shown in Table 2. It can be concluded that the distance between the obstacle and the camera is the main factor to the results. Take the can as an example: when the actual distance is 1 m, the error between the measured distance and actual distance is 0.011 m. When the distance reaches 3 m, the error reaches 0.029 m. Because the surface of obstacle is not smooth and sometimes the obstacle is a cluster of objects, the depth value is not constant. The limitation of the camera also contributes to that result. However, such accuracy is relatively high for outdoor robots, smart cars, and so forth. We also find that the smaller obstacle will produce higher measurement accuracy in the scene of road. The reason is that the edge information of small obstacle performs better during the fusion of depth image and infrared image. When the obstacle is large, it will affect the edge information of image and the accuracy of measurement, due to the variant colors in different parts. According to the data from the table, the proposed method can detect the obstacle accurately, and it can be applied to practical problems and holds a significant position for its research value.

5. Conclusions

A simple obstacle detection method based on fusion of depth and infrared image is proposed in this paper. The feature of scene and the information of image are analyzed. According to the character of depth and infrared image in traditional

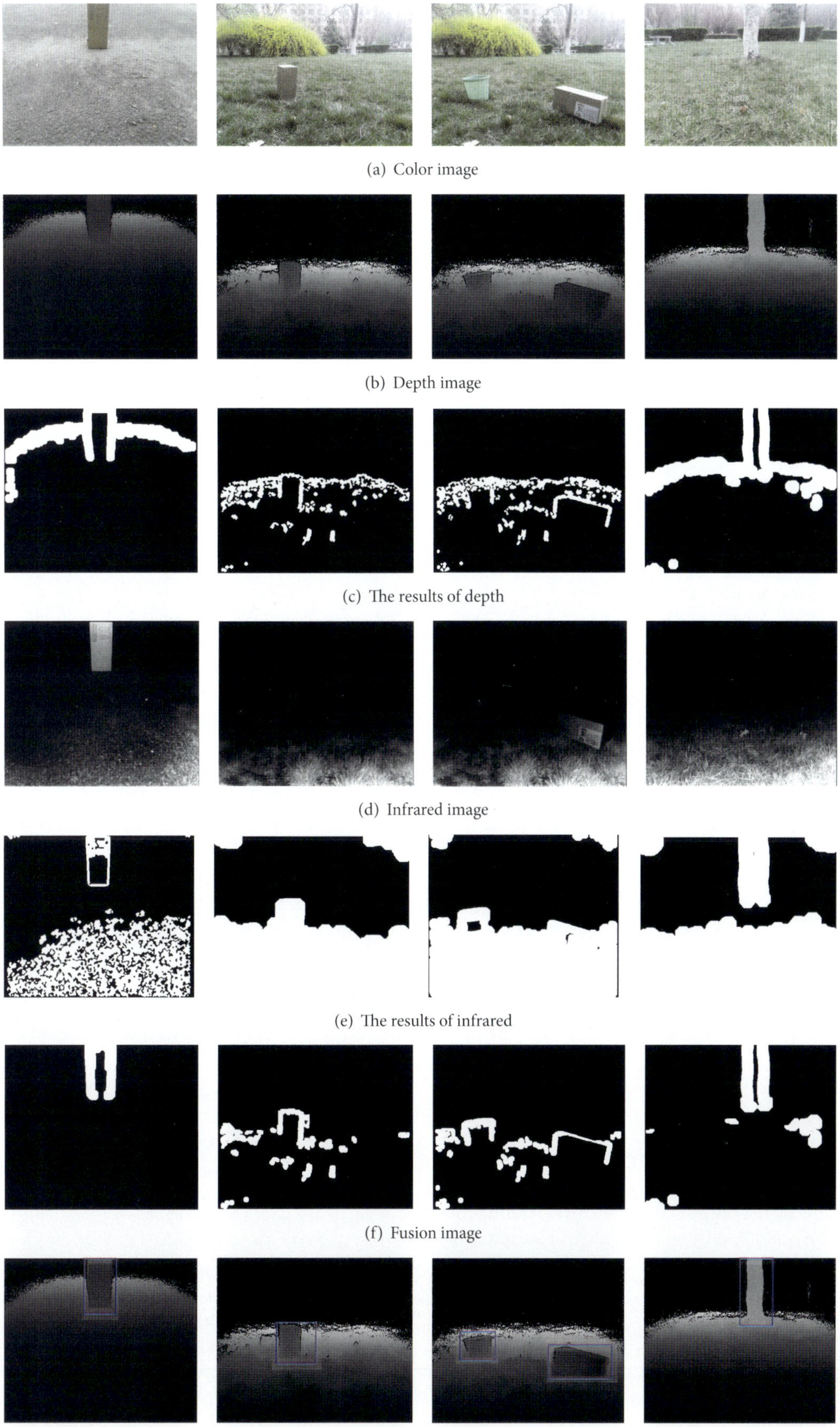

(a) Color image

(b) Depth image

(c) The results of depth

(d) Infrared image

(e) The results of infrared

(f) Fusion image

(g) Detection result

FIGURE 8: Results of obstacle detection.

Table 2: Experimental parameters and results.

The type of scene	Obstacle	Actual distance/m	Actual size Wide × high/cm	Actual geometric center (x, y)	Measured distance/m	Measured size Wide × high/cm	Measured geometric center (x, y)
Road	Can	3	13 × 6	(253, 240)	2.971	12.6 × 6.2	(247, 238)
		1.5		(285, 240)	1.521	12.9 × 5.8	(293, 243)
		1		(323, 248)	1.011	13.4 × 6.2	(325, 253)
	Car	3	160 × 130	(247, 275)	2.914	142.3 × 123.4	(257, 279)
	Carton	1	45 × 25	(62, 246)	1.013	45.2 × 22.6	(67, 255)
Grassland	Carton	1.5	45 × 25	(213, 192)	1.485	45.8 × 26.1	(209, 199)
	Tree	2.2	14 × 200	(72, 306)	2.221	15.1 × 189.2	(84, 309)
	Tree and flowers	3	14 × 210	(132, 341)	2.965	13.5 × 293.1	(146, 344)
		—	—	—	—	—	—
		2.5	68 × 170	(142, 58)	2.481	64.3 × 143.2	(132, 60)
	Carton and paper basket	1	45 × 25	(263, 348)	1.016	44.1 × 26.1	(268, 350)
		—	—	—	—	—	—
		1.4	16 × 25	(231, 126)	1.379	18.1 × 22.3	(223, 121)

edge detection, a field obstacle detection algorithm based on visual information fusion is proposed, which is used for detecting the obstacle in the near area of the sensor (the foreground mentioned before). It overcomes the defects that the color image is easily influenced by the factors such as light and shadow, and it also overcomes the limitation of the single vision algorithm. Finally, the algorithm effectively removes the error detection area caused by using single vision algorithm. The experiments show that the method can identify the obstacle area accurately and it is not easy to be affected by the factors of light condition and intensity. Furthermore, better robustness is achieved. Hence, it can provide reliable data for intelligent robot.

Competing Interests

The authors declare that they have no competing interests.

Acknowledgments

The authors acknowledge the National Natural Science Foundation of China (no. 51605039), the Natural Science Basic Research Plan in Shaanxi Province of China (no. 2016JQ6066), the China Postdoctoral Science Foundation (no. 2016M592728), and Fundamental Research Funds for the Central Universities (no. 310825151041).

References

[1] G. B. Vitor, D. A. Lima, A. C. Victorino, and J. V. Ferreira, "A 2D/3D vision based approach applied to road detection in urban environments," in *Proceedings of the IEEE Intelligent Vehicles Symposium (IEEE IV '13)*, pp. 952–957, Glod Coast City, Australia, June 2013.

[2] A. Benzerrouk, L. Adouane, and P. Martinet, "Obstacle avoidance controller generating attainable set-points for the navigation of multi-robot system," in *Proceedings of the IEEE Intelligent Vehicles Symposium (IV '13)*, pp. 487–492, Glod Coast City, Australia, June 2013.

[3] A. Discant, A. Rogozan, C. Rusu, and A. Bensrhair, "Sensors for obstacle detection—a survey," in *Proceedings of the 30th International Spring Seminar on Electronics Technology*, pp. 100–105, Cluj-Napoca, Romania, May 2007.

[4] S. Wang and Z.-Y. Xiang, "Detecting obstacles in vegetation by multi-spectral fusion," *Journal of Zhejiang University*, vol. 49, no. 11, pp. 2223–2229, 2015.

[5] S. Thrun, M. Montemerlo, H. Dahlkamp et al., "Stanley: the robot that won the DARPA Grand Challenge," *Journal of Field Robotics*, vol. 23, no. 9, pp. 661–692, 2006.

[6] J. F. Figueroa and J. S. Lamancusa, "A method for accurate detection of time of arrival: analysis and design of an ultrasonic ranging system," *Journal of the Acoustical Society of America*, vol. 91, no. 1, pp. 486–494, 1992.

[7] H. Hua, Y. Wang, and D. Yan, "A low-cost dynamic range-finding device based on amplitude-modulated continuous ultrasonic wave," *IEEE Transactions on Instrumentation and Measurement*, vol. 51, no. 2, pp. 362–367, 2002.

[8] K. Oki and K. Omasa, "A technique for mapping thermal infrared radiation variation within land cover," *IEEE Transactions on Geoscience and Remote Sensing*, vol. 41, no. 6, pp. 1521–1524, 2003.

[9] X.-M. Xiao, H.-M. Hu, Z.-X. Cai, and L. Wang, "Fast obstacle diction using adaptive segmentation and stereo vision," *Application Research of Computers*, vol. 24, no. 9, pp. 182–184, 2007.

[10] S. He, Z. Liu, and J. Shi, "Obstacle diction of indoor robots based on monocular vision," *Journal of Computer Applications*, vol. 32, no. 9, pp. 2556–2559, 2012.

[11] H. Zhang and J. Zhu, "Research strategy based on binocular vision obstacle avoidance algorithm," *Hangzhou University of Electronic Science and Technology*, vol. 33, no. 4, pp. 31–34, 2013.

[12] J. Choi, D. Kim, H. Yoo, and K. Sohn, "Rear obstacle detection system based on depth from kinect," in *Proceedings of the 15th International IEEE Conference on Intelligent Transportation Systems (ITSC '12)*, pp. 98–101, Anchorage, Alaska, USA, September 2012.

[13] Microsoft Kinect for Windows [OL], http://www.k4w.cn/news/1.html.

[14] Z. Zhang, "A flexible new technique for camera calibration," *IEEE Transactions on Pattern Analysis and Machine Intelligence*, vol. 22, no. 11, pp. 1330–1334, 2000.

[15] D. Chi, Y. Wang, L. Ning, J. Yi, and L. S. University, "Experimental research of camera 13 calibration based on ZHANG's method," *Journal of Chinese Agricultural Mechanization*, vol. 36, no. 2, pp. 287–289, 2015.

[16] K. Fukunaga and L. D. Hostetler, "The estimation of the gradient of a density function, with applications in pattern recognition," *IEEE Transactions on Information Theory*, vol. 21, no. 1, pp. 32–40, 1975.

[17] Y. Cheng, "Mean shift, mode seeking, and clustering," *IEEE Transactions on Pattern Analysis and Machine Intelligence*, vol. 17, no. 8, pp. 790–799, 1995.

[18] D. Comaniciu and P. Meer, "Mean shift: a robust approach toward feature space analysis," *IEEE Transactions on Pattern Analysis and Machine Intelligence*, vol. 24, no. 5, pp. 603–619, 2002.

[19] D. Comaniciu, V. Ramesh, and P. Meer, "Real-time tracking of non-rigid objects using mean shift," in *Proceedings of the IEEE Conference on Computer Vision and Pattern Recognition (CVPR '00)*, vol. 2, pp. 142–149, June 2000.

[20] N. Dalal and B. Triggs, "Histograms of oriented gradients for human detection," in *Proceedings of the IEEE Conference on Computer Vision and Pattern Recognition (CVPR '05)*, vol. 1, pp. 886–893, 2005.

15

Hands-Free Maneuvers of Robotic Vehicles via Human Intentions Understanding using Wearable Sensing

Weitian Wang, Rui Li, Longxiang Guo, Z. Max Diekel, and Yunyi Jia

Department of Automotive Engineering, Clemson University, Greenville, SC 29607, USA

Correspondence should be addressed to Yunyi Jia; yunyij@clemson.edu

Academic Editor: L. Fortuna

Intelligent robotic vehicles are more and more fully automated, without steering wheels, gas/brake pedals, or gearshifts. However, allowing the human driver to step in and maneuver the robotic vehicle under specific driving requirements is a necessary issue that should be considered. To this end, we propose a wearable-sensing-based hands-free maneuver intention understanding approach to assist the human to naturally operate the robotic vehicle without physical contact. The human intentions are interpreted and modeled based on the fuzzy control using the forearm postures and muscle activities information detected by a wearable sensory system, which incorporates electromyography (EMG) sensors and inertial measurement unit (IMU). Based on the maneuver intention understanding model, the human can flexibly, intuitively, and conveniently control diverse vehicle maneuvers only using his intention expressions. This approach was implemented by a series of experiments in the practical situations on a lab-based 1/10 robotic vehicle research platform. Experimental results and evaluations demonstrated that, by taking advantage of the nonphysical contact and natural handleability of this approach, the robotic vehicle was successfully and effectively maneuvered to finish the driving tasks with considerable accuracy and robustness in human-robotic vehicle interaction.

1. Introduction

With the improvements of computational resources and manufacturing capacities, there is a rapid and steady development in intelligent robotic vehicles technology in recent decades [1, 2]. These robotic vehicles have numerous advantages over traditional vehicles to solve traffic problems such as traffic jams and traffic accident which are caused by the driver's error or negligence [3]. Besides, the intelligent robotic vehicles are more and more likely to eliminate gas/brake pedals, gearshifts, and steering wheels. Google has built a two-seater prototype intelligent vehicle sans the steering wheel or pedals. Even without vehicle controls available to the human driver, this prototype is able to safely maneuver around obstacles via the built-in sensors and the software system [4]. Besides, aiming at ride-hailing and ride-sharing fleets, Ford will build a fully autonomous robotic vehicle without a steering wheel or pedals by 2021 [5].

However, in the intelligent robotic vehicle driving process, the human driver or passenger usually has some specific driving requirements such as accelerating in a straight road, stopping for some emergencies, or turning in a temporary direction. Consequently, how to maneuver the driving modes according to the human special intentions is a necessary issue that should be considered in the vehicle design.

During the human-robotic vehicle interaction process, it is significant for the vehicle to understand the human intentions or behaviors in order to achieve different vehicle maneuvers. Several related works have been conducted in recent years.

By using human motions, the car's speed, and the distance between the car and the intersection, Ohashi et al. proposed a model using case-based learning to construct an experimental system for human driver's intentions understanding [6]. In [7], the research team recognized a set of continuous driver intentions by observing the easily accessible vehicles and environment signals such as pedals or global vehicle positions. Based on the playback system and machine learning, Oliver and Pentland presented a dynamical graphical framework to model and recognize driver's behaviors at a tactical level that focused on how contextual information impacted the driver's performance [8]. Researchers in [9] developed a driver behavior recognition approach by characterizing and detecting driving maneuvers and then modeled and recognized the driver's behaviors in different situations. From the accessible vehicle onboard sensors, Berndt and Dietmayer investigated a method to infer the driver's intentions to leave the lane or other maneuvers. In this work they expected to help drivers predict trajectories or assess risks [10].

However, these recognition and understanding methods for human intentions are too complex to implement in practice. Additionally, we usually cannot get much recorded information from the vehicle embedded system since there are less traditional operation devices in the future robotic vehicles.

Using the gesture to represent human intentions for the robotic vehicle maneuver is a practical and interesting work that attracts a lot of attention. Operating robotic vehicles via human gestures will help the human take his/her hands off the current operation habits to reduce the contributions of the negligence and error which may cause vehicle collisions [11]. Ionescu et al. developed efficient human-vehicle interaction through a smart and real-time depth camera operating in the near infrared spectrum. The acquired depth information was processed for the human gesture detection and recognition to interpret the driving intentions to control the vehicle [12]. Researchers in [13] established the communication by gestures between the human and an intelligent wheelchair through a webcam and sensors. By using an array of cameras which outputted information with an instantaneous state, Kramer and Underkoffler acquired the images of human gestures and then designed a controller that automatically extracted and detected the gesture from the gesture data for the vehicle maneuver [14]. In [15], to enable the human and the vehicle to communicate and work together, Fong et al. used sensor fusion and computer vision to recognize the remote environment and improve the situation awareness. Then they created easily used remote driving tools for the vehicles. Researchers in [16] employed a Leap Motion to detect the gesture data and extracted seven independent instructions for the autonomous vehicle maneuver.

Although there are several vision-based approaches, vision-based driving intention recognition and understanding highly depends on the working surroundings. Its performance is easily interfered by the complex and dynamic background such as the crowded urban settings. Furthermore, the vision system usually requires the human to be within some certain areas in order to capture the motion information, which significantly constrains the activities and working ranges of the humans.

Therefore, with the extensive development and employment of robotic vehicles, researchers expect that humans and vehicles could collaborate seamlessly in different driving situations. Developing a simply configured, naturally operable, and highly robust human intention understanding approach for human-robotic vehicle collaboration is a very necessary issue.

To this end, different from existing approaches of using vehicle built-in devices or vision systems, we propose a wearable-sensing-based maneuver intention understanding approach using a wearable sensory system [17–19] to assist the human in the maneuvering of robotic vehicle without physical contact. This interaction method does not restrain the human's hand to be physically involved in the driving task and can be applied in the complicated human-robotic vehicle interactions.

The major contributions of this work include the following: (1) We propose a natural wearable sensing solution to assist human drivers to maneuver robotic vehicles without traditional operation devices in specific driving situations, which is more robust than existing approaches. (2) We develop a driving intention understanding approach using fuzzy control and human motions information, including forearm postures and muscle activities, which are captured by the wearable sensory system.

2. System Framework

The system framework, which is designed for the human to use the wearable-sensing-based maneuver intention understanding approach to operate the robotic vehicle, is shown in Figure 1. This system contains three layers: the data layer, the decision layer, and the execution layer.

When the human intends to change the robotic vehicle's driving modes, his intentions expressed by forearm postures and muscle activities are detected and calculated via a wearable sensory system, as presented in Figure 2. After being collected, the expression information is preprocessed and fused together to output useful information in the data layer. Then the processed information, including the human hand's rotation angles and arm muscles electromyography (EMG) signals, is sent to the decision layer by means of wireless communication devices in real time.

After that, in the decision layer, the acquired information is further processed to generate intention instructions based on the intention understanding model. Simultaneously, the instruction outputs motivate the vehicle motion planning algorithms by calling the corresponding driving mode function. In order to ensure the vehicle to execute accurately, both intention instructions and algorithm outputs are utilized to make motion planning decisions.

In the execution layer, the vehicle driving commands are generated based on the decision layer outputs for the vehicle to plan motions in the real world workspace. Meanwhile, the vehicle execution states are sent back to the decision layer to alert the motion planning algorithms if the driving intention is accepted. The motion planning algorithms will output the decision again if the driving intention failed to be accepted.

3. Maneuver Intention Representation and Data Acquisition

3.1. Maneuver Intention Representation. The human maneuver intentions [20], including brake, turn, and acceleration, are pretty common in daily driving. These intentions can be reflected and represented by lots of manners, such as body movements and natural languages. Because there are no traditional manual operation devices in the future intelligent robotic vehicles, the normal driving manners are not available in these cases. In this research, in order to make it practical

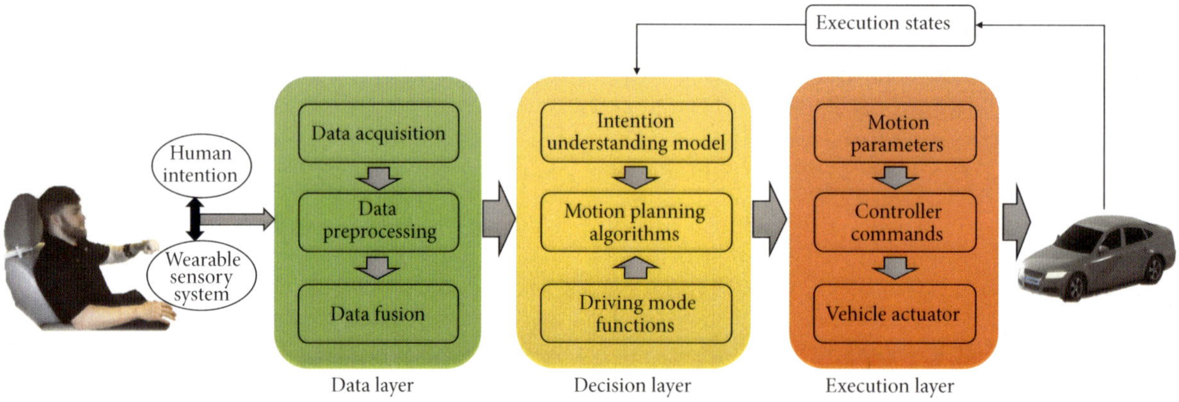

FIGURE 1: The framework of wearable-sensing-based driving intention understanding for the robotic vehicle maneuvers.

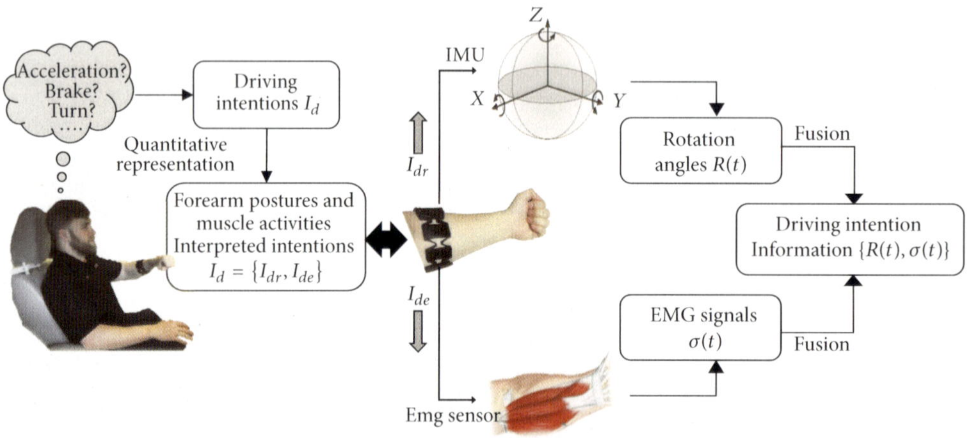

FIGURE 2: The driving intentions represented by the forearm postures and muscle activities information.

and natural, we utilize the human forearm postures and muscle activities to represent these maneuver intentions. As shown in Figure 2, the intention information usually contains forearm rotation angles and EMG signals. Therefore, the maneuver intentions can be described as

$$I_d = \{I_{dr}, I_{de}\}, \tag{1}$$

where I_{dr} denotes the maneuver intention interpreted by forearm rotations; I_{de} denotes the maneuver intention interpreted by EMG signals.

3.2. Wearable Sensory System. We employ a wearable sensory system for human-robotic vehicle interaction to acquire the human forearm postures and muscle activities information in the maneuver process. The sensory system that we choose is Myo [21], which can be worn at the driver's forearm and integrates with an inertial measurement unit (IMU) [22–24] and eight EMG sensors [25–27]. The IMU chip contains an onboard digital motion processor (DMP) and MPU-9150 module which consists of a 3-axis accelerometer, a 3-axis gyroscope, and a 3-axis magnetometer. The detected information from the IMU and EMG sensors is preprocessed by a microcontroller unit (MCU) with a 32 bit ARM architecture 72 MHz Cortex M4 CPU core. All the raw and calculated data are made available through a first-in-first-out (FIFO) buffer that is read by the MCU over the communication bus. The Bluetooth Low Energy (BLE) module on the mainboard is used for external communication between Myo and the client controller [28].

The working principle of the information acquisition by this wearable sensory system is presented in Figure 2. The human forearm postures will be tracked and recorded by the IMU. This data includes acceleration and angular velocity information which can be fused to describe the forearm motions and rotation angles. When the human performs maneuver intentions, the electrical skeletal muscle activities from his forearm will be measured by the EMG sensors. This EMG information can be extracted to estimate human's finger motions such as wave-in, finger-spread, and fist.

3.3. Data Acquisition and Processing. When the human performs his maneuver intentions, as shown in Figure 2, his forearm postures can be quantified by the IMU outputs which contain the 3-axis acceleration information and the 3-axis angular velocity information about forearm motions. Furthermore, these data can be fused together into quaternions

$$q = [q_0, q_1, q_2, q_3]^T, \tag{2}$$

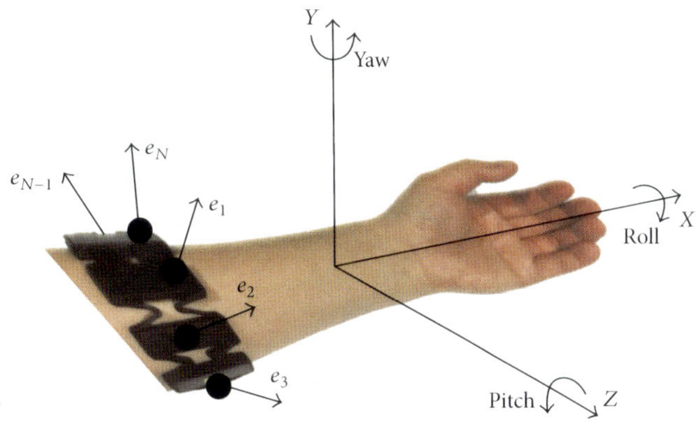

FIGURE 3: EMG signals and spatial rotations from the human driver's forearm.

where $|q|^2 = q_0{}^2 + q_1{}^2 + q_2{}^2 + q_3{}^2 = 1$. The sample frequency of the IMU is 50 Hz in our work.

In order to calculate the forearm postures, Euler angles [29] are utilized to parameterize the forearm spatial rotations in the 3D work space. The Roll-Pitch-Yaw Euler angles can be represented by

$$R(t) = [\phi(t), \theta(t), \psi(t)]^T, \tag{3}$$

where t denotes the IMU sampling time, ϕ is the Roll rotation about the x-axis, θ is the Yaw rotation about the y-axis, and ψ is the Pitch rotation about the z-axis. As presented in Figure 3, Euler angles are able to visually describe the forearm rotation movements in the hand-over process.

Moreover, the Euler angles are able to be calculated by the quaternions as

$$R = \begin{bmatrix} \phi \\ \theta \\ \psi \end{bmatrix} = \begin{bmatrix} \arctan \dfrac{2q_0q_1 + 2q_2q_3}{1 - 2q_1^2 - 2q_2^2} \\ \arcsin(2q_0q_2 - 2q_1q_3) \\ \arctan \dfrac{2q_1q_2 + 2q_0q_3}{1 - 2q_2^2 - 2q_3^2} \end{bmatrix}. \tag{4}$$

Therefore, the driver's intentions I_{dr} interpreted by the arm rotation in (1) can be represented as

$$I_{dr} = \{I_{dr}(x_r) \mid x_r \in R(t)\}. \tag{5}$$

Simultaneously, the human finger motions can be calculated based on the EMG signals which are collected from the human forearm's muscle activities. The EMG data acquired by the wearable sensory system can be described as

$$E(t) = [e_1(t), e_2(t), \ldots, e_n(t)]^T, \tag{6}$$

where t is the sampling time of the EMG sensor, $e(t)$ is the output of each EMG sensor, and n is the number of EMG channels on the wearable sensory system which is 8 in our work. We sample these EMG signals at a frequency of 200 Hz.

The raw EMG signal is a set of discrete points with positive and negative components. Along with the finger activities, the electric potentials generated by muscle cells have a distinct effect on the dispersion of the EMG signal. Therefore, to take advantage of the EMG data accurately, we adopt the standard deviation (SD) σ of the EMG data to extract the characteristics from the finger activities. The standard deviation could reflect the muscle activities observably. In the human-robotic vehicle interaction, the standard deviation can be calculated by

$$\sigma_i = \sqrt{\frac{1}{K}\sum_{k=1}^{K}\left(e_i(k) - \frac{1}{K}\sum_{k=1}^{K} e_i(k)\right)^2}, \tag{7}$$

where $e_i(k)$, $k = 1, 2, \ldots, K$ is a set of EMG signals and K is the window size for determining the number of EMG data to be employed to calculate the stand derivation. We select $K = 150$ in this study.

Moreover, the maneuver intentions I_{de} interpreted by the finger motions in (1) can be represented by

$$I_{de} = \{I_{de}(x_e) \mid x_e \in \sigma(t)\}. \tag{8}$$

According to (5) and (8), the maneuver intention can be represented as

$$I_d = \{I_{dr}(x_r), I_{de}(x_e) \mid x_r \in R(t),\ x_e \in \sigma(t)\}. \tag{9}$$

From the above, it can be concluded that, during the robotic vehicle maneuver process, $R(t)$ and $\sigma(t)$ are dynamically programmed and updated via the human forearm postures and muscle activities. Therefore, the maneuver intentions I_d will be interpreted and updated in real time.

4. Maneuver Intention Understanding Using Fuzzy Control

In this section, based on the wearable sensing information and the maneuver intention representation, we build an intention understanding model using the fuzzy control.

4.1. Maneuver Intention Fuzzification. Before developing the fuzzy controller [30], we should define the fuzzy set and domain of discourse using the wearable sensing information, which contains forearm postures and muscle activities. In this

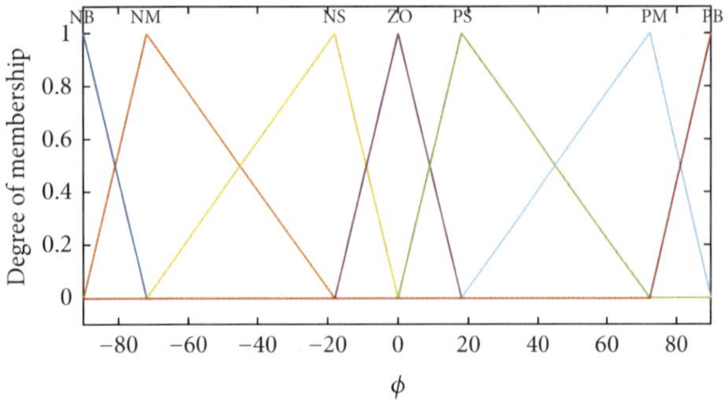

FIGURE 4: The membership function of the Roll angle ϕ.

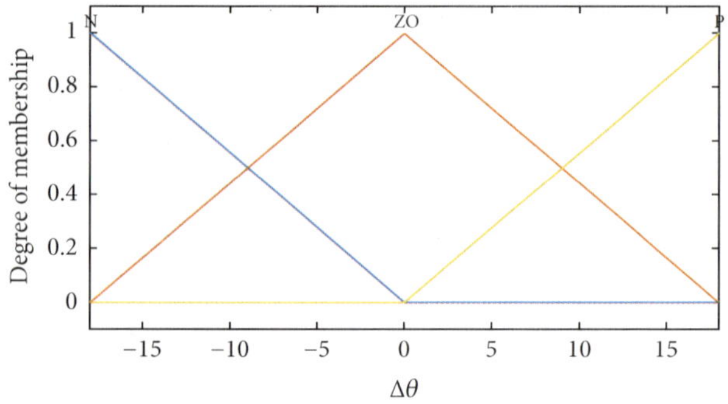

FIGURE 5: The membership function of the Yaw angle $\Delta\theta$.

work, we find it is difficult to distinguish various intentions by directly employing the raw standard derivations of all EMG signals, while the average of them presents clear differences. Hence, we utilize $\overline{\sigma} = \sum_{i=1}^{8} \sigma_i/8$ to denote the muscle activities from the driving intention. Additionally, in order to distinguish steering modes in the robotic vehicle maneuver, we utilized $\Delta\theta = \theta(t+1)-\theta(t)$ to work as an input for the fuzzy controller. Therefore, combining the forearm rotation angles and EMG signals together, we deploy the fuzzy controller with four inputs $(\phi, \Delta\theta, \psi, \overline{\sigma})$ and one output (I_d).

Moreover, the fuzzy sets of the inputs and output are defined as follows:

(i) The Roll angle ϕ : {NB, NM, NS, ZO, PS, PM, PB}

(ii) The Yaw angle $\Delta\theta$: $\{N, \text{ZO}, P\}$

(iii) The Pitch angle ψ : {NB, NM, NS, ZO, PS, PM, PB}

(iv) The EMG signal $\overline{\sigma}$: {ZO, PS, PM, BD, PB}

(v) Driving intentions I_d : {AC, DC, ST, TL, TR, BD},

where "AC," "DC," "ST," "TL," "TR," and "BD" denote the maneuver intentions of "*Acceleration*," "*Deceleration*," "*Stop*," "*TurnLeft*," "*TurnRight*," and "*BackwardDriving*," respectively. Additionally, "NB," "NM," "NS," "ZO," "PS," "PM," and "PB" denote negative big, negative middle, negative small, zero, positive big, positive middle, positive small, and separately, which are employed to represent degree of membership in the maneuver intention understanding.

4.2. Membership Functions. According to the fused wearable sensing information and driving operations in the vehicle maneuver, we define the domains of discourse of the inputs and output as follows:

(i) The Roll angle ϕ : $[-90^\circ, 90^\circ]$

(ii) The Yaw angle $\Delta\theta$: $[-18^\circ, 18^\circ]$

(iii) The Pitch angle ψ : $[-90^\circ, 90^\circ]$

(iv) The EMG signal $\overline{\sigma}$: $[0, 90]$

(v) Driving intentions I_d : $\{-2, -1, 0, 1, 2, 3\}$.

In this study, we ask 5 subjects, who have different hands and muscular tensions when maneuvering the robotic vehicle, to detect the forearm rotation angles and EMG

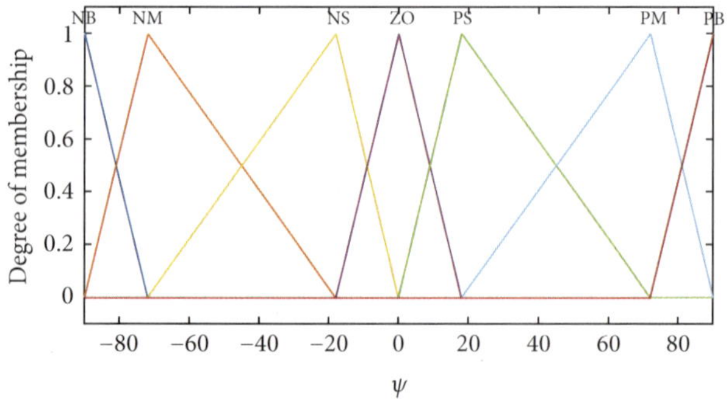

FIGURE 6: The membership function of the Pitch angle ψ.

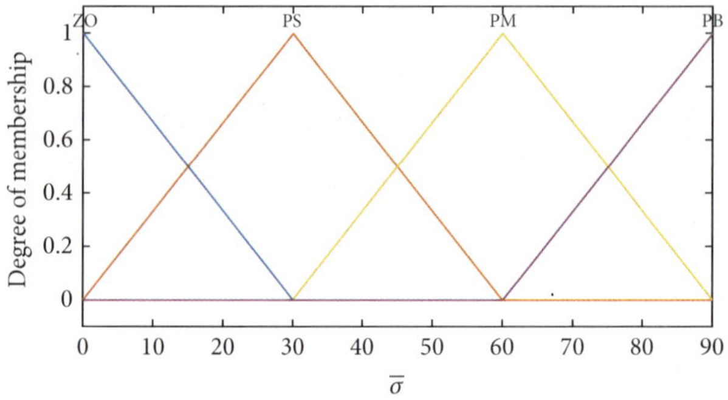

FIGURE 7: The membership function of the EMG signal $\overline{\sigma}$.

signals. Each subject performs 20 times for each maneuver intention. The triangular membership function [31] is employed for each input and output in the fuzzy controller. The membership functions we design are shown in Figures 4~8. It can be seen that, during the maneuver process, the degree of membership is varied with the domain of discourse of each input or output correspondingly.

4.3. Maneuver Intention Understanding. Since the fuzzy controller in this work is configured with four inputs and one output, the fuzzy rules cannot be presented by the traditional rule-table. However, we can use the "IF-THEN" statements [32] to describe the valid fuzzy rules we utilize in the robotic vehicle maneuver. As shown in Table 1, the "IF" parts are antecedents and the "THEN" parts are consequents.

For the maneuver intention understanding, we employ "AND" as the fuzzy operator in each rule and "OR" as the

Table 1: The valid fuzzy rules we utilize in the robotic vehicle maneuver.

IF	ϕ	ZO	ZO	ZO	ZO	ZO	PS, PM, PB
	$\Delta\theta$	N, ZO, P	N, ZO, P	N, ZO, P	P	N	N, ZO, P
	ψ	ZO	ZO	PS, PM, PB	ZO	ZO	ZO
	$\overline{\sigma}$	PM, PB	PS	ZO	ZO	ZO	ZO
THEN	I_d	AC	DC	ST	TL	TR	BD

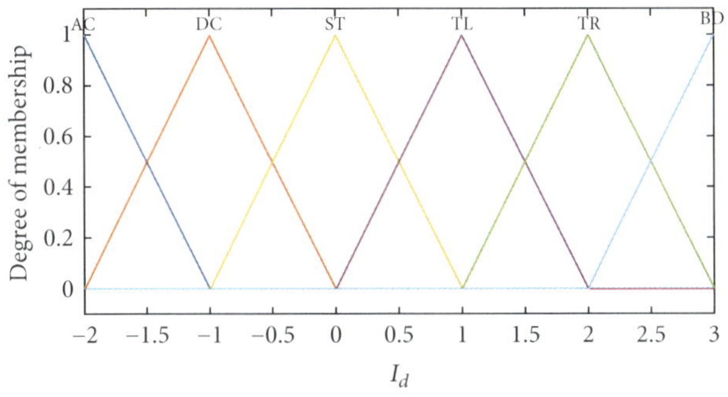

Figure 8: The membership function of the driving Intentions I_d.

fuzzy operator among different rules. As presented in Table 1, for the maneuver intention of "Acceleration," we utilize the aggregate degree of membership as its fuzzy decision, which can be calculated by

$$\begin{aligned}\mu(I_{\mathrm{AC}}) = &\left(\mathrm{ZO}_\phi \cap N_{\Delta\theta} \cap \mathrm{ZO}_\psi \cap \mathrm{PM}_{\overline{\sigma}}\right)\\ &\cup\left(\mathrm{ZO}_\phi \cap \mathrm{ZO}_{\Delta\theta} \cap \mathrm{ZO}_\psi \cap \mathrm{PM}_{\overline{\sigma}}\right)\\ &\cup\left(\mathrm{ZO}_\phi \cap P_{\Delta\theta} \cap \mathrm{ZO}_\psi \cap \mathrm{PM}_{\overline{\sigma}}\right)\\ &\cup\left(\mathrm{ZO}_\phi \cap N_{\Delta\theta} \cap \mathrm{ZO}_\psi \cap \mathrm{PB}_{\overline{\sigma}}\right)\\ &\cup\left(\mathrm{ZO}_\phi \cap \mathrm{ZO}_{\Delta\theta} \cap \mathrm{ZO}_\psi \cap \mathrm{PB}_{\overline{\sigma}}\right)\\ &\cup\left(\mathrm{ZO}_\phi \cap P_{\Delta\theta} \cap \mathrm{ZO}_\psi \cap \mathrm{PB}_{\overline{\sigma}}\right).\end{aligned} \tag{10}$$

Similarly, the fuzzy decisions of other maneuver intentions can be calculated based on Table 1. Afterwards, we employ the Middle of Maximum (MOM) [33] as the defuzzification approach to calculate the corresponding maneuver intention. Furthermore, the maneuver intention understanding result can be expressed as

$$I_d = \mathrm{round}\left[\arg\,\mathrm{mom}\,\mu\left(I_x\right)\right], \tag{11}$$

where $I_x \in \{\mathrm{AC}, \mathrm{DC}, \mathrm{ST}, \mathrm{TL}, \mathrm{TR}, \mathrm{BD}\}$, $\mu(I_x)$ denotes the outputted fuzzy decision, and "round" means that the output is rounded to the nearest integer.

Therefore, based on (11), the maneuver intention can be understood when the human operates the robotic vehicle under specific requirements.

5. Experimental Results and Analysis

5.1. Experimental Platform. The developed approach in this work is implemented on a lab research autonomous robotic vehicle of the 1/10-Scale Vehicle Research Platform (1/10-SVRP). The 1/10-SVRP consists of five 1/10-scale autonomous robotic vehicles, a human manual driving interface, and a 1/10-scale driving environment including an ultra-wideband based indoor GPS system, traffic lights, road signs, and various lane setups. As shown in Figure 9, the wearable sensory system described above is worn around human's arm during the human-robotic vehicle interaction. The sensory information detected by IMU and EMG sensors is sent to the control system in real time via a pair of Bluetooth devices. Once the controller generates new commands, these signals are sent to the vehicle motor drivers to plan the goal motions.

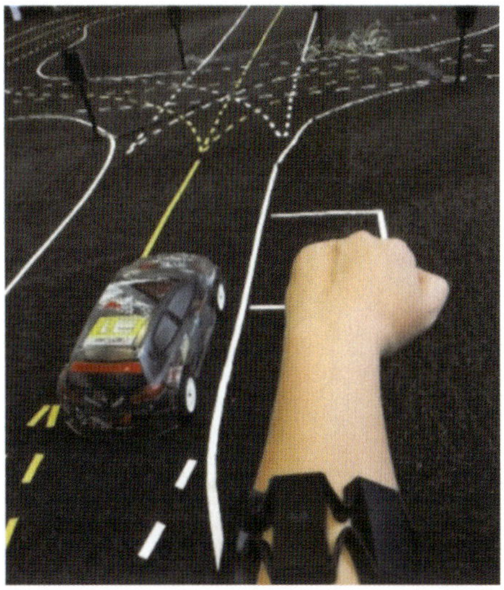

Figure 9: The human is employing the wearable sensing to maneuver the robotic vehicle on the 1/10-Scale Vehicle Research Platform.

In the robotic vehicle maneuver process, the velocity we employed for the robotic vehicle is expressed by

$$V(t) = k_\sigma \overline{\sigma(t)}, \tag{12}$$

where k_σ is the EMG control factor; $\overline{\sigma(t)}$ is determined by the clench and release of the human fist.

The steering angles in the robotic vehicle turning operation are calculated with the following function:

$$\theta_T(t) = k_\theta \theta(t) + l_\theta, \tag{13}$$

where k_θ is the steering angle coefficient; l_θ works as the offset to adjust the initial angle.

5.2. Maneuver Intention Understanding Verification. In this section, we test the maneuver intention understanding

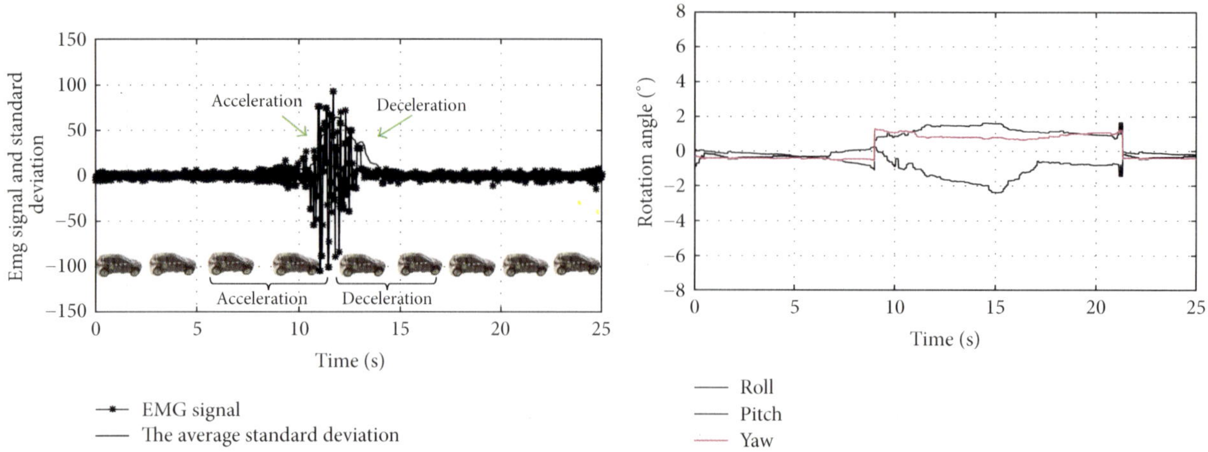

(a) The vehicle movements along with the EMG signals (b) The forearm rotation angles

FIGURE 10: The variation of the human EMG signals and rotation angles in the acceleration and deceleration operation.

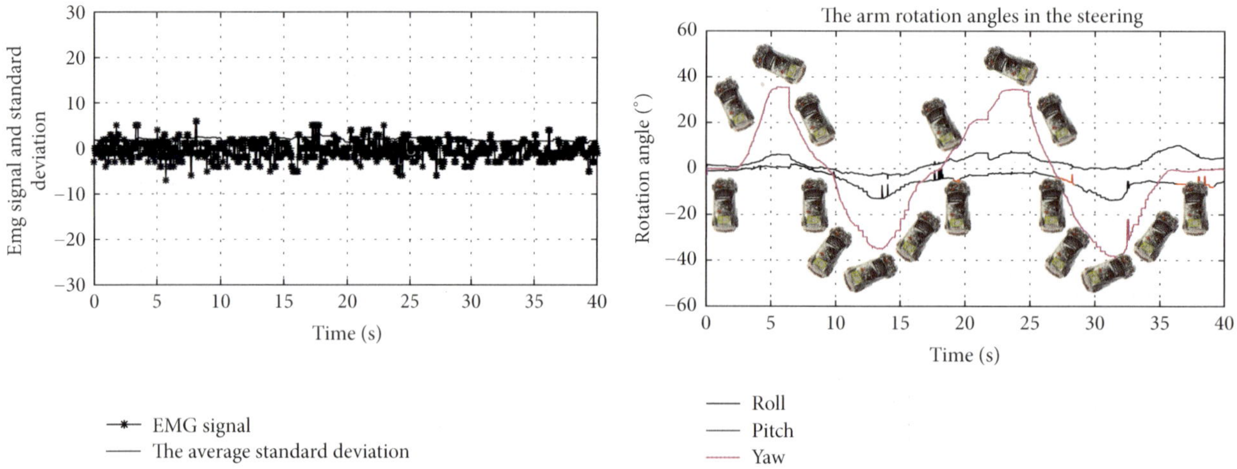

(a) The EMG signals (b) The vehicle movements along with the rotation angles

FIGURE 11: The variation of the human EMG signals and rotation angles in the steering operation.

approach via the forward (acceleration and deceleration) and steering (turning left and turning right) driving modes in practical situations and present the verification results.

5.2.1. Forward Driving. When the human expects the robotic vehicle to speed up or speed down in the forward driving, according to the fuzzy rules, he should present specific finger motions and keep the rotation angles constrained in the required intervals at the same time. As depicted in Figure 10, the human forearm rotations and finger activities properly meet the required rules. Meanwhile, the vehicle's movement procedure in Figure 10(a) shows that the vehicle is accelerated and decelerated correspondingly along with the variation of the EMG signals. Consequently, it can be seen that the maneuver intention understanding approach correctly follows the human intentions to execute the accelerating and decelerating driving.

5.2.2. Steering. When the human wants the robotic vehicle to turn left or right, in accordance with the steering rules of "TL" and "TR," he should control the Yaw angle properly and keep the Roll angle and Pitch angle in the designed constraints simultaneously. The rotation angles information and EMG signals are presented in Figure 11. It can be seen that the robotic vehicle properly performs the steering directions along with the variation of the rotation angles, which indicates that the proposed approach exactly understands the maneuver intentions in the steering operation.

5.3. Accuracy Evaluation. In this section, we conduct understanding accuracy evaluation and compare the results with the work in [16], which utilized a Leap Motion to acquire the human behaviors' information for the vehicle maneuver.

We employ the wearable sensory system to perform all designed maneuver intentions to operate the robotic

TABLE 2: Driving intentions understanding accuracy.

Driving intentions	Testing data	Successful	Failed	Recognition accuracy
AC	40	39	1	97.5%
DC	40	38	2	95%
ST	40	38	2	95%
TL	40	36	4	90%
TR	40	35	5	87.5%
BD	40	38	2	95%

TABLE 3: The comparison of our approach and the previous work.

Works	Interaction interface	Accuracy	SD-E
Our approach	Wearable sensor	93.33%	3.76
The work in [16]	Leap Motion	77.4%	4.52

vehicle without considering the fixed route. Each intention is operated 40 times based on the understanding model. The understanding accuracy results are presented in Table 2. It can be seen that the proposed understanding approach is able to effectively and sensitively identify all the maneuver intentions in the human-robotic vehicle interaction. However, for some intentions understanding such as "TR" and "TL," they present relatively fair accuracy than others. To solve it, we can design much better fuzzy rules through practical trials to improve the accuracy of these maneuver intentions' understanding.

In addition, the average understanding accuracy of this study is about 93.33% which is higher than the work in [16]. Furthermore, from [16] we can calculate that the standard deviation of all errors (SD-E) is about 4.52, which is higher than 3.76 in our work. Therefore, it can be concluded that our approach is more stable in the maneuver intentions understanding. The result comparisons are shown in Table 3.

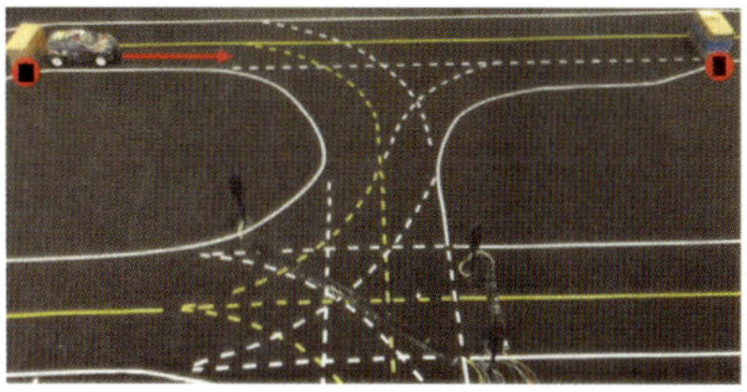

FIGURE 12: The lane tracking task.

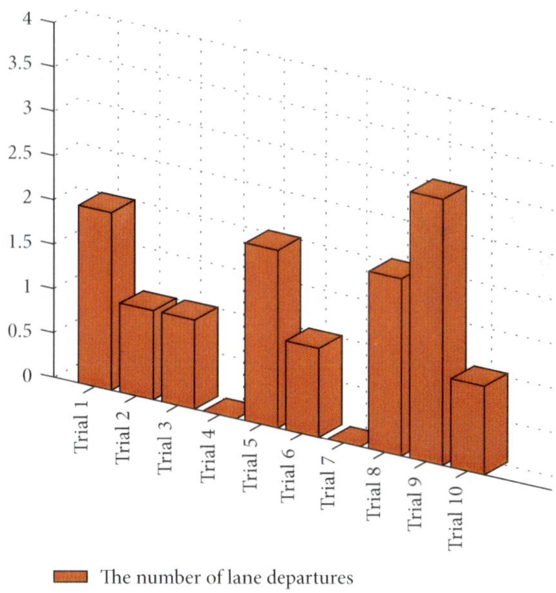

FIGURE 13: The times of lane departure of the two accelerating methods in the lane tracking task.

5.4. Robustness Evaluation. Based on the research platform, we design two tasks to evaluate the robustness of the maneuver intention understanding model in some common driving scenes, such as driving straight on the lane, turning in the intersection, and turning for obstacle avoidance.

5.4.1. Lane Tracking. When the human maneuvers a robotic vehicle in the straight lane, the straightness of driving is very significant to the traffic. Meanwhile serpentine driving usually results in a fine or even a terrible accident. Therefore, the straight driving test is conducted based on the intention understanding model.

We ask 5 individuals with valid driving licenses and considerable driving experiences to maneuver the vehicle one by one for two loops. Each individual operates one straight driving process in each loop. Therefore, we can get 10 driving records from the experiment. As shown in Figure 12, the vehicle is driven forward from A to B.

According to the maneuver results, the occurrences of lane departure in the straight driving are shown in Figure 13. The numbers of lane departures of these ten driving records are "2," "1," "1," "0," "2," "1," "0," "1," "3," and "1," respectively. The average of the number of lane departures is 1.20, which suggests that the maneuver intention understanding approach presents a robust stability and adaptability for different individuals in the straight driving situations.

5.4.2. Obstacle Avoidance. To evaluate the performance of flexibility of the maneuver intention understanding model, some hybrid driving modes to avoid obstacles are allocated to the robotic vehicle in the second task. As presented in Figure 14, the vehicle is driven from A to B. During this process, the vehicle must cross the intersection and avoid colliding with obstacles on the road. The experiment is conducted using the same method by 5 different individuals as task 1.

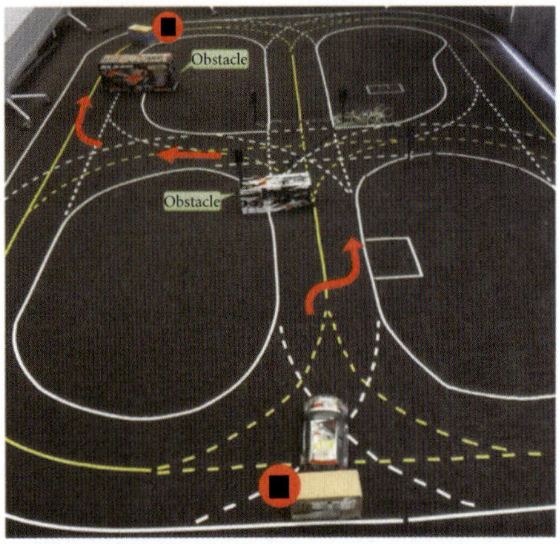

FIGURE 14: The obstacle avoidance task.

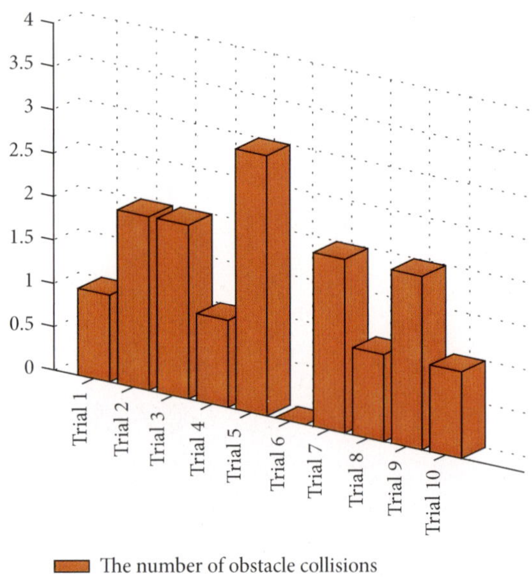

FIGURE 15: The times of obstacle collision of the two accelerating methods in the second task.

Based on the driving records, the numbers of obstacle collisions are "1," "2," "1," "1," "3," "0," "1," "3," "2," and "1," respectively. As shown in Figure 15, the average of the number of obstacle collisions is 1.50. From the results, it can be observed that the maneuver intention understanding approach presents a robust flexibility for the hybrid driving modes in the complex road setting. Comparing to task 1, the standard deviation of the numbers of obstacle collisions (0.97) is higher than that of the numbers of lane departures (0.92), which reveals that the intention understanding approach in hybrid driving modes shows a relatively fair robustness. One of the key reasons is that different fuzzy rules for the intentions present diverse understanding accuracies, in which some of them will impact the overall robustness. Additionally, it is easy for divers to feel nervous in the complicated driving surroundings which can result in obstacle collisions. However, these problems above could be overcome by optimizing fuzzy rules in the proposed approach and taking more practice for the human.

From the above, it is shown that the vehicle maneuvers are successfully and effectively performed by using the maneuver intention understanding approach. Notably, experimental results and evaluations of this approach demonstrate that by taking advantage of the natural wearable sensing information the human driver can maneuver the vehicle only using forearm postures and muscle activities in a much easier and more stable manner with considerable accuracy and robustness.

6. Conclusions

A novel and practical wearable-sensing-based maneuver intention understanding approach was proposed to assist the human driver to naturally operate the robotic vehicles without physical contact. The wearable sensory device can be naturally applied in the complicated human-vehicle interactions without restraining the human's hand to be physically involved in the driving task. First, when the human driver performed his maneuver intentions, the wearable sensory system information which included forearm postures and muscle activities was recorded and updated in real time. Additionally, after getting the parameterized intention information, we developed a maneuver intention understanding approach using the fuzzy control. Afterwards, based on the proposed approach, we conducted a set of experiments on our vehicle research platform. Experimental results and evaluations demonstrated that by taking advantage of the nonphysical contact and natural handleability of this approach the robotic vehicle was successfully and effectively maneuvered to accomplish the driving tasks with considerable accuracy and robustness in human-robotic vehicle interaction.

In human-vehicle interaction, the driver's unconscious gestures and involuntary movements may cause unstable detection and interpretation of the driver's intentions. Therefore, future works will be conducted on integrating multiple kinds of sensing information, such as human gaze information and natural language, as triggers to avoid the false-positive intention understanding. Additionally, with looking forward to extending the applications of our approach in more complicated situations, future works will also be conducted on integrating radar sensing information as the input of the fuzzy control to improve the intention understanding accuracy to avoid potential collisions.

Conflicts of Interest

The authors declare that they have no conflicts of interest.

References

[1] D. L. Fisher, M. Lohrenz, D. Moore, E. D. Nadler, and J. K. Pollard, "Humans and Intelligent Vehicles: The Hope, the Help, and the Harm," *IEEE Transactions on Intelligent Vehicles*, vol. 1, no. 1, pp. 56–67, 2016.

[2] B. Paden, M. Cap, S. Z. Yong, D. Yershov, and E. Frazzoli, "A Survey of Motion Planning and Control Techniques for Self-Driving Urban Vehicles," *IEEE Transactions on Intelligent Vehicles*, vol. 1, no. 1, pp. 33–55, 2016.

[3] L. Li, W. Huang, Y. Liu, N. Zheng, and F. Wang, "Intelligence testing for autonomous vehicles: a new approach," *IEEE Transactions on Intelligent Vehicles*, vol. 1, no. 2, pp. 158–166, 2016.

[4] https://www.cnet.com/news/google-unveils-self-driving-car-sans-steering-wheel.

[5] http://www.theverge.com/2016/8/16/12504300/ford-autonomous-car-ride-sharing-2021.

[6] K. Ohashi, T. Yamaguchi, and I. Tamai, "Humane automotive system using driver intention recognition," in *Proceedings of the SICE Annual Conference 2004*, pp. 2363–2366, jpn, August 2004.

[7] H. Berndt, J. Emmert, and K. Dietmayer, "Continuous driver intention recognition with Hidden Markov Models," in *Proceedings of the 11th International IEEE Conference on Intelligent Transportation Systems, ITSC 2008*, pp. 1189–1194, China, December 2008.

[8] N. Oliver and A. P. Pentland, "Driver behavior recognition and prediction in a SmartCar," in *Proceedings of the Enhanced and Synthetic Vision 2000*, pp. 280–290, April 2000.

[9] N. Kuge, T. Yamamura, O. Shimoyama, and A. Liu, "A driver behavior recognition method based on a driver model framework," *SAE Technical Papers*, 2000.

[10] H. Berndt and K. Dietmayer, "Driver intention inference with vehicle onboard sensors," in *Proceedings of the 2009 IEEE International Conference on Vehicular Electronics and Safety, ICVES 2009*, pp. 102–107, India, November 2009.

[11] C. A. Pickering, K. J. Burnham, and M. J. Richardson, "A research study of hand gesture recognition technologies and applications for human vehicle interaction," in *Proceedings of the 3rd Institution of Engineering and Technology Conference*, Automotive Electronics, pp. 1–15, on, 2007.

[12] B. Ionescu, V. Suse, C. Gadea et al., "Using a NIR camera for car gesture control," *IEEE Latin America Transactions*, vol. 12, no. 3, pp. 520–523, 2014.

[13] S. P. Kang, G. Rodnay, M. Tordon, and J. Katupitiya, "A hand gesture based virtual interface for wheelchair control," in *Proceedings of the 2003 IEEE/ASME International Conference on Advanced Intelligent Mechatronics, AIM 2003*, pp. 778–783, Japan, July 2003.

[14] K. H. Kramer and J. S. Underkoffler, "Gesture-Based Control System For Vehicle Interfaces," in *Google Patents*, 2009.

[15] T. Fong, C. Thorpe, and C. Baur, "Advanced interfaces for vehicle teleoperation: Collaborative control, sensor fusion displays, and remote driving tools," *Autonomous Robots*, vol. 11, no. 1, pp. 77–85, 2001.

[16] U. E. Manawadu, M. Kamezaki, M. Ishikawa, T. Kawano, and S. Sugano, "A hand gesture based driver-vehicle interface to control lateral and longitudinal motions of an autonomous vehicle," in *Proceedings of the 2016 IEEE International Conference on Systems, Man, and Cybernetics, SMC 2016*, pp. 1785–1790, Hungary, October 2016.

[17] A. Nag, S. C. Mukhopadhyay, and J. Kosel, "Wearable Flexible Sensors: A Review," *IEEE Sensors Journal*, vol. 17, no. 13, pp. 3949–3960, 2017.

[18] I. H. Lopez-Nava and A. Munoz-Melendez, "Wearable Inertial Sensors for Human Motion Analysis: A Review," *IEEE Sensors Journal*, vol. 16, no. 22, pp. 7821–7834, 2016.

[19] M. Janidarmian, A. Roshan Fekr, K. Radecka, and Z. Zilic, "Multi-Objective Hierarchical Classification Using Wearable Sensors in a Health Application," *IEEE Sensors Journal*, vol. 17, no. 5, pp. 1421–1433, 2017.

[20] S. Lefèvre, C. Laugier, and J. Ibañez-Guzmán, "Exploiting map information for driver intention estimation at road intersections," in *Proceedings of the 2011 IEEE Intelligent Vehicles Symposium, IV'11*, pp. 583–588, Germany, June 2011.

[21] https://www.myo.com/.

[22] G. Loianno, C. Brunner, G. McGrath, and V. Kumar, "Estimation, Control, and Planning for Aggressive Flight With a Small Quadrotor With a Single Camera and IMU," *IEEE Robotics and Automation Letters*, vol. 2, no. 2, pp. 404–411, 2017.

[23] H. Ahmed and M. Tahir, "Improving the Accuracy of Human Body Orientation Estimation With Wearable IMU Sensors," *IEEE Transactions on Instrumentation and Measurement*, vol. 66, no. 3, pp. 535–542, 2017.

[24] J. Wahlstrom, I. Skog, P. Handel, and A. Nehorai, "IMU-Based Smartphone-to-Vehicle Positioning," *IEEE Transactions on Intelligent Vehicles*, vol. 1, no. 2, pp. 139–147, 2016.

[25] N. D. Bunt, J. C. Moreno, P. Müller, T. Seel, and T. Schauer, "Online Monitoring of Muscle Activity During Walking for Bio-feedback and for Observing the Effects of Transcutaneous Electrical Stimulation," *Biosystems and Biorobotics*, vol. 15, pp. 705–709, 2017.

[26] S. H. Roy, G. De Luca, M. Cheng, A. Johansson, L. D. Gilmore, and C. J. De Luca, "Electro-mechanical stability of surface EMG sensors," *Medical & Biological Engineering & Computing*, vol. 45, pp. 447–457, 2007.

[27] C. J. De Luca, M. Kuznetsov, L. D. Gilmore, and S. H. Roy, "Inter-electrode spacing of surface EMG sensors: Reduction of crosstalk contamination during voluntary contractions," *Journal of Biomechanics*, vol. 45, no. 3, pp. 555–561, 2012.

[28] K. Nymoen, M. R. Haugen, and A. R. Jensenius, "Mumyo–evaluating and exploring the myo armband for musical interaction," 2015.

[29] J. Diebel, "Representing attitude: Euler angles, unit quaternions, and rotation vectors," *Matrix*, vol. 58, pp. 1–35, 2006.

[30] K. Tanaka and H. O. Wang, *Fuzzy control systems design and analysis: a linear matrix inequality approach*, John Wiley and Sons, 2004.

[31] J. S. Kim and K.-S. Whang, "A tolerance approach to the fuzzy goal programming problems with unbalanced triangular membership function," *European Journal of Operational Research*, vol. 107, no. 3, pp. 614–624, 1998.

[32] H. Ishibuchi, K. Nozaki, N. Yamamoto, and H. Tanaka, "Selecting fuzzy if-then rules for classification problems using genetic algorithms," *IEEE Transactions on Fuzzy Systems*, vol. 3, no. 3, pp. 260–270, 1995.

[33] H. Hellendoorn and C. Thomas, "Defuzzification in fuzzy controllers," *Journal of Intelligent & Fuzzy Systems: Applications in Engineering and Technology*, vol. 1, no. 2, pp. 109–123, 1993.

16

A Crowd Avoidance Method using Circular Avoidance Path for Robust Person Following

Kohei Morishita,[1] Yutaka Hiroi,[1] and Akinori Ito[2]

[1]*Osaka Institute of Technology, 5-16-1 Omiya, Asahi-ku, Osaka 535-8585, Japan*
[2]*Tohoku University, 6-6-5 Aramaki Aza Aoba, Aoba-ku, Sendai 980-8579, Japan*

Correspondence should be addressed to Akinori Ito; aito@spcom.ecei.tohoku.ac.jp

Academic Editor: Tao Liu

A life-support service robot must avoid both static and dynamic obstacles for working in a real environment. Here, a static obstacle means an obstacle that does not move, and a dynamic obstacle is the one that moves. Assuming the robot is following a target person, we discuss how the robot avoids a crowd through which the target person passes and arrives at the target position. The purpose of this paper is to propose a crowd avoidance method that makes a robot to be able to avoid both static and dynamic obstacles. The method uses the surface points of the obstacles to form an avoidance region, and the robot moves along the edge of the region. We conducted experiments assuming various situations such that the robot was blocked, there was a wide gap in the crowd, or a person in the crowd yielded for the robot to pass through. As an experimental result, it was confirmed the robot could avoid the crowd even when the obstacles were aligned in an "inverted wedge" shape.

1. Introduction

Many research works have been conducted for developing life-support service robots [1]. Here, the robots we are focusing on include autonomous mobile robots that conduct tasks such as bringing a beverage to the user based on the user's request [2, 3]. A life-support service robot differs from industrial robots because the life-support service robots must work under an environment where humans and the robots coexist. For this purpose, a life-support service robot must have several abilities such as manipulation [4, 5], vision [6, 7], and mobility [8, 9]. We especially focus on the mobility in this paper. The two technologies, the path planning, and the path following are important for realizing the autonomous navigation of the robot. Many research works have been done for those issues so far [10–17]. The path planning and the following behavior are either based on the prepared map of the environment or based on the on-the-fly measurement of the environment. We focus on the latter case so that the robot can move around the new and dynamic environment where the obstacles (mainly humans) move. Our interest in this paper is the robot navigation based on human-following behavior, where the robot moves in an environment by following a specific person. Since the robot's final goal is to follow the preceding person without colliding any obstacles, the robot does not need to have a static map of the environment. Rather, the robot needs to measure the person to follow and the obstacles (walls, objects, other persons, etc.) continuously and determine the robot's trajectory quickly and robustly.

The person-following behavior of the mobile robot has been actively researched [18]. We have developed the life-support service robot ASAHI [19], which helps to carry baggage items in a place such as a shopping mall or a hospital. The person-following behavior is essential for this task. We developed a method to track the person using a single laser range finder (LRF) [20], which enabled robust person-following behavior in an environment where other persons exist. We also developed an obstacle avoidance method while following a person using an LRF. This method can track the target person even when the target is temporally lost. However, the person tracking fails when the target person is completely lost in a crowd. Therefore, we aim to develop a

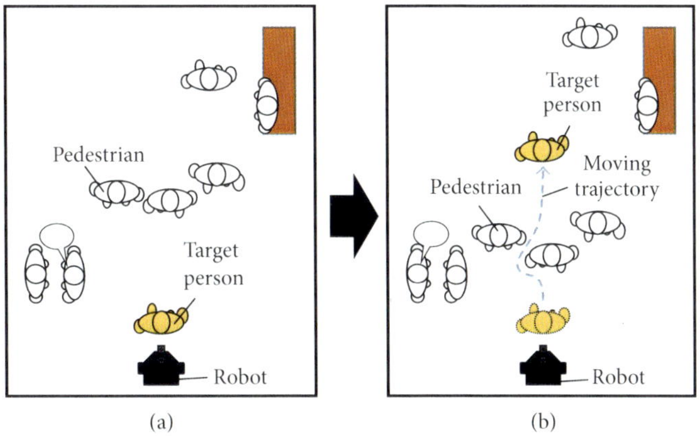

FIGURE 1: Supposed environment.

novel method that enables continuing tracking of the target person even when he/she is lost in the crowd. As a part of the development, this paper describes the crowd avoidance method.

2. Related Works

A number of methods for obstacle avoidance have been proposed so far [10, 12, 13, 21–24]. Usually, the purpose of the path planning and dynamic obstacle avoidance is to navigate a robot between the start point and the goal point, which are known beforehand on the internal map, while avoiding the obstacles that are not on the map [11]. On the other hand, the purpose of our robot is not to navigate the robot to the fixed goal but to follow the person, which can be considered as a "moving goal."

The potential-based path planning method [23] is a widely used technique for path planning under an environment with obstacles. However, there are several problems with the potential-based method. First, most of the techniques are suitable for the static environment, where the obstacles do not move [12]. Cao et al. proposed an obstacle avoidance method for the dynamic environment based on the potential method [13]. Their method assumed that position and velocity of an individual obstacle can be obtained, but it is not always true in a crowded environment where it is sometimes difficult to identify individual obstacle because two or more persons are moving or staying as clusters. Another problem is that the robot sometimes fails to reach the target point when the attractive and repulsive forces balance [24].

Another reason for the difficulty to apply the potential-based method is that the positions of the obstacles are not known to the robot before measuring them. In this paper, the robot is assumed to measure the environment including the target person and obstacles using the LRF and behaves only based on the results of the measurement. To calculate the potential accurately, we need to consider the effect of all obstacles including the ones which cannot be measured directly from the LRF. Therefore, it is impossible to use the accurate potential under the current assumption.

3. The Crowd Avoidance Algorithm

3.1. Overview. The potential method is a general framework, where we can consider any path given an appropriate potential field. However, as stated in the previous section, it is not always easy to calculate an appropriate field under the situation assumed in this paper, where the target person is completely lost, there are clusters of obstacles around the robot, and the obstacles are moving. Thus we made a simple assumption on the situation and solved this problem using a simpler method rather than calculating the potential field around the robot.

Let us consider a robot following the target person in an environment where there are many other persons, such as a shopping mall, as shown in Figure 1. Then the target person passes through the blocking persons as Figure 1(b). If there is a gap in the crowd, the robot can go through the gap. However, if there is no gap to pass through, the robot should avoid the crowd.

Suppose the target person walks through the crowd, and the robot is blocked by the crowd. Figure 2 shows the overview of the assumed situation and the proposed algorithm. We set the "avoidance region" around the obstacles, where it is assured that the robot does not collide with the obstacle if the robot's center is on or outside of the edge of the region. When the robot judges that it have completely lost the target person and the path is blocked by the crowd, the robot makes the hypothetical target line toward the robot's direction (which faces the position where the target person is finally measured) and sets the tentative target on the target line beyond the crowd. The distance between the robot's current position and the tentative target is determined based on the average walking speed of the target person. In the later experiment, the distance is fixed to 4000 mm. Next, the robot moves along the hypothetical target line toward the crowd, measuring the distance between the robot and the nearest point of the obstacle. If the distance between the robot and the obstacle becomes shorter than the predefined threshold, the robot begins to move from the hypothetical target line to the outer edge of the avoidance region. Since the edge of the avoidance region is assured to be far enough from the

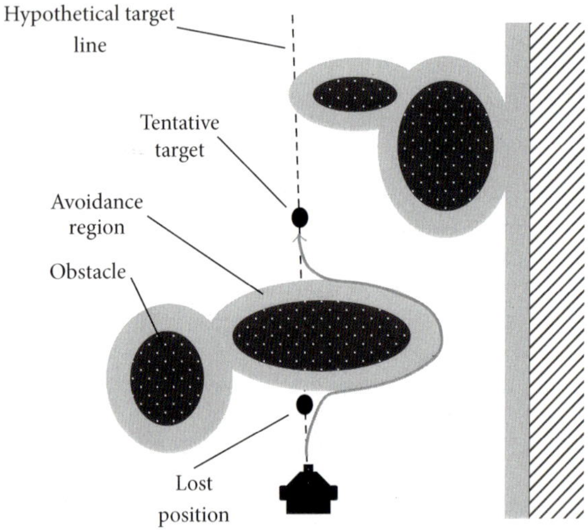

FIGURE 2: Overview of the algorithm.

obstacle, the robot can avoid the obstacle by moving along the edge.

The overview of the crowd avoidance algorithm is as follows.

(1) Follow the target person.

(2) Recognize that the robot has lost the target person.

(3) Generate the hypothetical target line and the tentative target point.

(4) Generate the avoidance region and move along the hypothetical target line.

(5) Measure the surface points of the obstacle and determine the nearest point.

(6) If the distance to the nearest point is longer than the threshold, go to (4).

(7) Measure the nearest surface point of the obstacle.

(8) Follow the circle that centers the nearest point.

(9) If the distance to the hypothetical target line is longer than the threshold, go to (7).

(10) Follow the hypothetical target line until arriving at the tentative target point.

(11) If the distance between the robot and the tentative target is shorter than the threshold, stop the robot.

(12) Go to (10).

As this method uses the "avoidance region" and the robot moves along the region, the path is always at almost the same distance from the obstacles even when the robot can take more distance from them. However, it is assured that the robot can avoid the obstacles keeping a sufficient distance using the proposed algorithm. The proposed algorithm is very easy to implement because it uses only local information (i.e., information of the nearest obstacle) as opposed to the potential-based method.

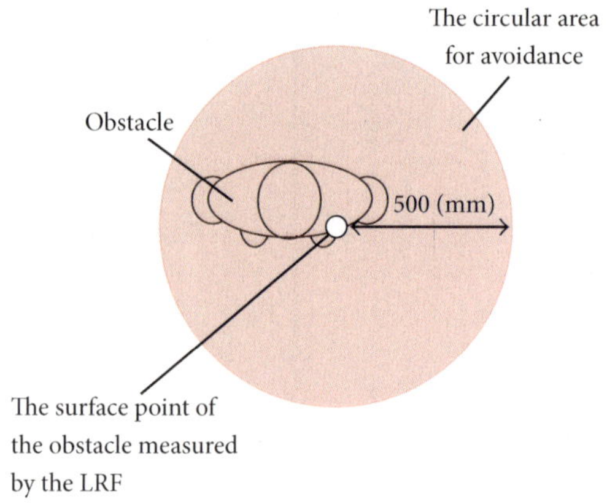

(a) The circular area for avoidance

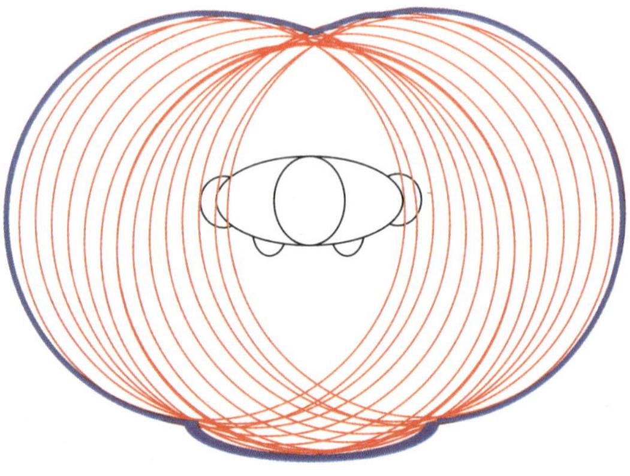

(b) The avoidance region

FIGURE 3: Generation of the avoidance region.

3.2. The Avoidance Region. First, we describe the avoidance region. The avoidance region is an overlapped region of the circular areas, and the center of one circle is the surface point of an obstacle measured by the LRF. The radius of the circle is 500 mm, which is determined considering the size of the robot. Figure 3(a) shows an example of the circular area. As the measurement by the LRF is done for all of the surface points of the obstacle, we have many circular areas that correspond to all the surface points, and the union of all the circular areas is the avoidance region. Obviously, the outer edge of the avoidance region is the envelope of the circular areas. Figure 3(b) shows an example of the avoidance region.

3.3. Crowd Avoidance Using the Avoidance Region. In this section, the detail of the crowd avoidance method is described. First, assume the situation where the target person passed through the crowd, as shown in Figure 4(a). At this point, the robot loses the target person because measurement of the target person is blocked by the crowd. Then the robot assumes that the target exists beyond the hypothetical target line, which is generated by extending the line segment between the center of the robot and the last position where the target person was measured. Then the robot follows the line (Figure 4(b)) and moves toward the crowd. When the distance between the robot and the nearest point of the crowd becomes less than 700 mm, the robot starts to follow the

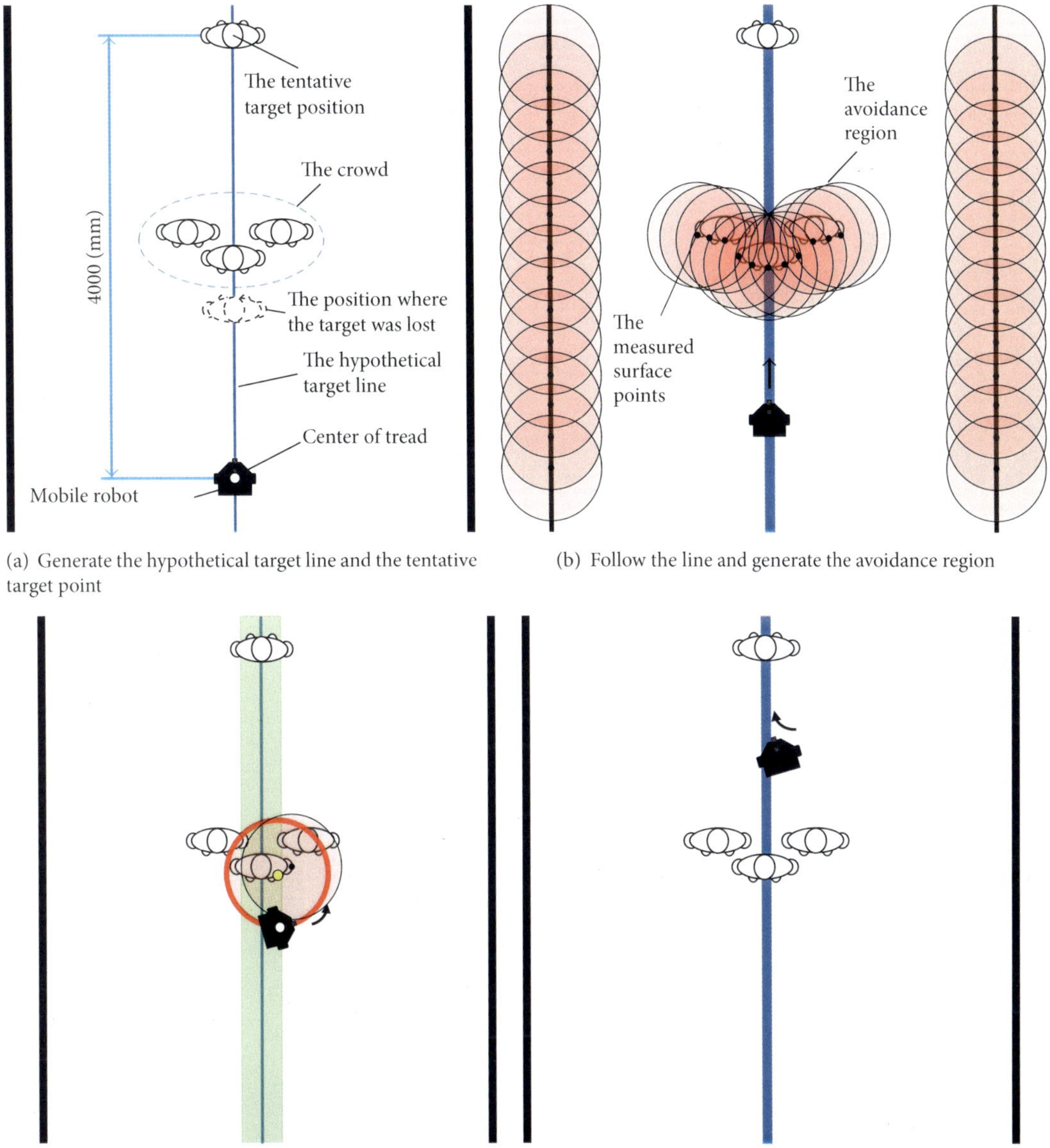

(a) Generate the hypothetical target line and the tentative target point

(b) Follow the line and generate the avoidance region

(c) Change the trajectory to the circle around the nearest surface point

(d) Follow the line again

FIGURE 4: The crowd avoidance using the avoidance region.

piece of the circle (Figure 4(c)). The threshold value 700 mm was determined empirically. Then the robot moves along the edge of the avoidance region and goes behind the crowd. If the robot enters the point near the line (the green region of Figure 4(c)), then the robot starts to follow the line (Figure 4(d)). Here the robot can avoid the crowd when the crowd is moving because the center point of the circle to follow (i.e., the nearest surface point of the obstacle) is always updated when following the circle. The details of the methods to follow a line and a circle are described in the following sections.

3.4. Following a Line. Next, the method to follow a line is described [20]. Figure 5 shows the line-following method.

Let $\Delta\theta$ [deg] be the angular difference between the robot's moving direction and the line to follow, as shown in Figure 5. Let L_1 [mm] be the distance between the center position of the robot (x_{odo}, y_{odo}) and the line to follow. Then the velocities of the left and right wheels, V_L and V_R, are controlled so that L_1 becomes zero.

Let the equation of the line to follow be $y = ax + b$. Then L_1 is calculated as follows.

$$L_1 = \frac{y_{odo} - a \cdot x_{odo} - b}{\sqrt{1 + a^2}}. \quad (1)$$

Let the two points on the line be (x_1, y_1) and (x_2, y_2). Then the angle of the line θ is

$$\theta = \tan^{-1} \frac{y_2 - y_1}{x_2 - x_1}. \quad (2)$$

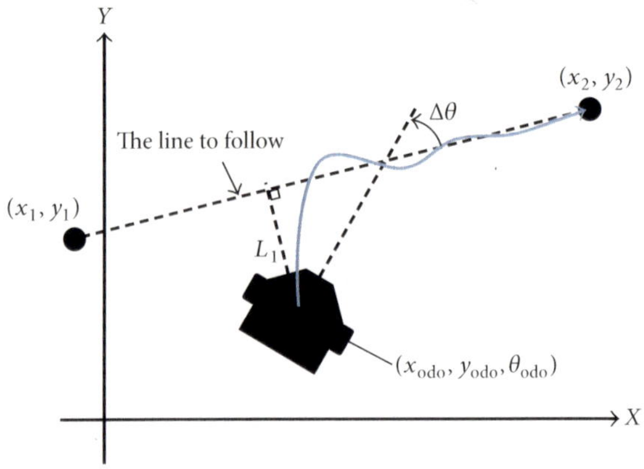

FIGURE 5: The line-following method.

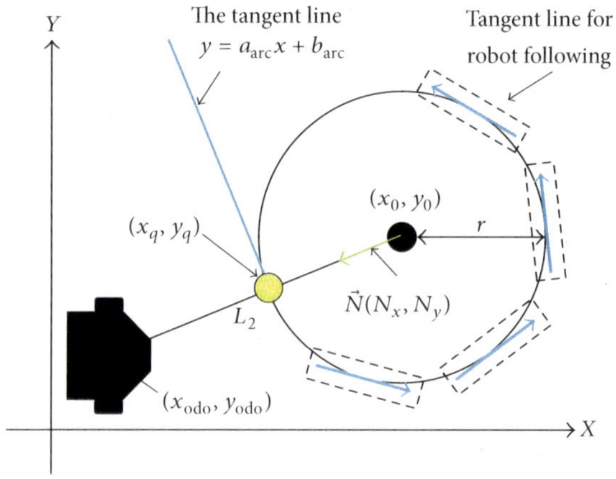

FIGURE 6: The circle-following method.

Then the angular difference $\Delta\theta$ is calculated using the angle of the robot θ_{odo}, as follows.

$$\Delta\theta = \theta - \theta_{odo}. \tag{3}$$

Next, we control V_L and V_R so that L_1 and $\Delta\theta$ become zero. First, let us define

$$\begin{aligned} V_R &= V + \Delta V, \\ V_L &= V - \Delta V, \\ \Delta V &= K_1 L_1 + K_2 \frac{dL_1}{dt} + K_3 \Delta\theta + K_4 \frac{d\Delta\theta}{dt}. \end{aligned} \tag{4}$$

Here, ΔV denotes (half of) the velocity difference between the left and right wheels. Parameters K_1 and K_3 are used for controlling the constraint to decrease L_1 and $\Delta\theta$. K_2 and K_4 are used to control the convergence of L_1 and $\Delta\theta$. These parameters were determined empirically. The parameter values used in the later experiment "line following" are shown as follows:

K_1: 0.26 s^{-1}

K_2: 0.26

K_3: 0.3 mm/s·deg

K_4: 0.3 mm/deg

3.5. Following a Circle. Next, the method for following a circle is described. Figure 6 shows the circle-following method.

Basically, the circle-following method is the same as the line-following method, where the tangent lines of the circle are used as the lines to follow.

First, consider the line segment between the center of the robot (x_{odo}, y_{odo}) and the center of the circle (x_0, y_0), and the intersection point of the line and the circle is (x_q, y_q). Let the tangent at (x_q, y_q) be $y = a_{arc}x + b_{arc}$. Here L_2 [mm] is the distance between the robot and the center of the circle, calculated as follows.

$$L_2 = \sqrt{(x_0 - x_{odo})^2 + (y_0 - y_{odo})^2}. \tag{5}$$

Then let the vector $\vec{N} = (N_x, N_y)$ be a unit vector having the same direction as the line between the robot and the circle, directing from the circle to the robot. The vector is calculated as

$$\begin{bmatrix} N_x \\ N_y \end{bmatrix} = \begin{bmatrix} \dfrac{x_0 - x_{odo}}{L_2} \\ \dfrac{y_0 - y_{odo}}{L_2} \end{bmatrix} \tag{6}$$

and the intersection point (x_q, y_q) is calculated as

$$\begin{bmatrix} x_q \\ y_q \end{bmatrix} = \begin{bmatrix} x_0 + rN_x \\ y_0 + rN_y \end{bmatrix}, \tag{7}$$

where r is the radius of the circle. And then the tangent line at the intersection point, $y = a_{arc}x + b_{arc}$, is calculated as follows.

$$\begin{aligned} a_{arc} &= -\frac{x_0 - x_{odo}}{y_0 - y_{odo}}, \\ b_{arc} &= y_0 - a_{arc}x_0. \end{aligned} \tag{8}$$

Then the robot follows the tangent line using the algorithm explained in the previous section. The center point of the circle is updated by every measurement of the LRF (as the coordinate of the nearest surface point of the obstacle), and the tangent also changes according to the position of the crowd and the robot.

4. Experiments of the Crowd Avoidance

4.1. Overview. We conducted an experiment to confirm if the robot actually could avoid the crowd. In the experiment, three or four persons blocked the robot and the robot moved to avoid the persons. First, we conducted an experiment where the crowd did not move (the static case) and then conducted another experiment where one person of the crowd moved (the dynamic case).

4.2. Hardware. We used ASAHI2015 as the robot, which is the modified version of ASAHI [19]. This robot is equipped

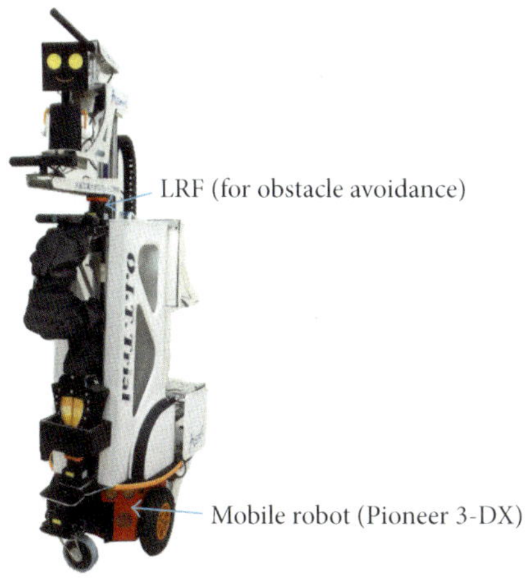

Figure 7: The mobile robot ASAHI 2015.

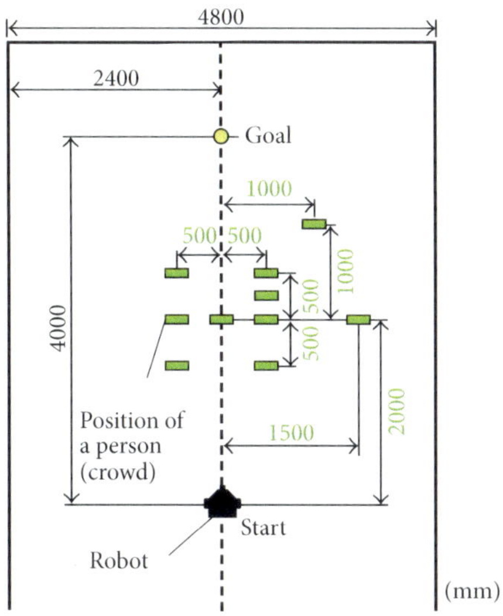

Figure 8: Experimental environment.

with three LRFs at the middle and the bottom of the robot. In the experiment, only the LRF at the middle was used (Hokuyo UTM30-LX). Pioneer 3-DX was used as the moving base. Figure 7 shows the figure of the robot.

4.3. Experimental Conditions. First, when the robot moved to a person in the crowd, we assumed that the person either stayed or moved to avoid the robot, which means that the person did not behave to actively block the robot.

Figure 8 shows the experimental environment. The positions where persons stood were marked using curing tape. First, the participants stood in front of the robot with the seven patterns shown in Figure 9. Second, the participants stood with the two patterns shown in Figure 10, and a person in the crowd was moved during the experiment.

In Figure 9, patterns 1 to 4 were used to check if the proposed method worked for a different arrangement of the crowd. Patterns 5 and 7 were to check if the robot could pass through the crowd when there was a wide gap in the crowd that was enough for the robot to go through. Pattern 6 was used to check if the robot could avoid the crowd if the wide gap was blocked.

Figure 10 shows the dynamic case, where a person in a crowd moved to avoid the robot. In pattern 8, four persons stood in line first, and one person stepped aside to avoid the robot. In pattern 9, a person stepped behind another person to make the gap.

In these experiments, the point of losing the target person was assumed to be known as 700 mm front of the crowd. We also assumed that the robot moves counterclockwise for avoiding the crowd.

4.4. Results. As the results of the experiment, the robot could avoid the crowd for all the static and dynamic patterns.

Figure 11 shows the trajectory of the robot for the static patterns shown in Figure 9. The blue points are the points measured by the LRF. The red triangles show the trajectories of the robot. From Figures 11(a)–11(g), we can confirm that the robot could avoid the crowd properly. Especially, in patterns 5 and 7, the robot could pass through the gap because the gap between persons was 1500 mm, which was enough for a robot to go through. Pattern 4 (Figure 11(d)) was a difficult pattern to avoid because two persons make a gap at the center and the gap is blocked by another person. As shown in Figure 11(d), the robot could avoid the crowd without collision even when the crowd is the "inverted wedge" formation. In this case, the robot starts avoiding the crowd before going between the two persons standing at the front. It was possible because the depth of the person at the center from the two persons at the front was 500 mm, which was comparable to the radius of the avoidance area circle. If the distance were larger, the robot would enter between the two persons and go back to avoid the crowd.

As for the dynamic patterns shown in Figure 10 (patterns 8 and 9), the robot could avoid the moving crowd properly by passing through the gap made by the person's movement. Here, we describe the motion of the robot for pattern 9 in detail. Figure 12 shows the robot's behavior in pattern 9. First, as shown in Figures 12(a) and 12(b), the robot moved toward the crowd along with the hypothetical target line. Figure 13(a) shows the sensors' measurement at this stage, where the red circles show the circular areas for avoidance. The robot focused on the blue circle in front of the robot, which was the nearest point from the robot. The orange point is the center of the robot and the green line is the line segment between the robot and the nearest point of the obstacles. Next, as shown in Figure 12(c), the robot started to avoid the nearest person. At this point, the second person from the right yielded to the robot. Then the robot recognized the gap as Figure 12(d) and passed through the gap as shown in Figure 12(e). The sensor signal is shown in Figure 13(b), where the blue circle is the circle along which the robot was moving. Finally, the robot moved toward the tentative target point, shown in Figure 12(f). Figure 13(c) shows the sensor signal at this phase.

4.5. Discussion. There are several limitations in the proposed method. First, the method assumes that the tentative target point is known. If another crowd were on that point, the

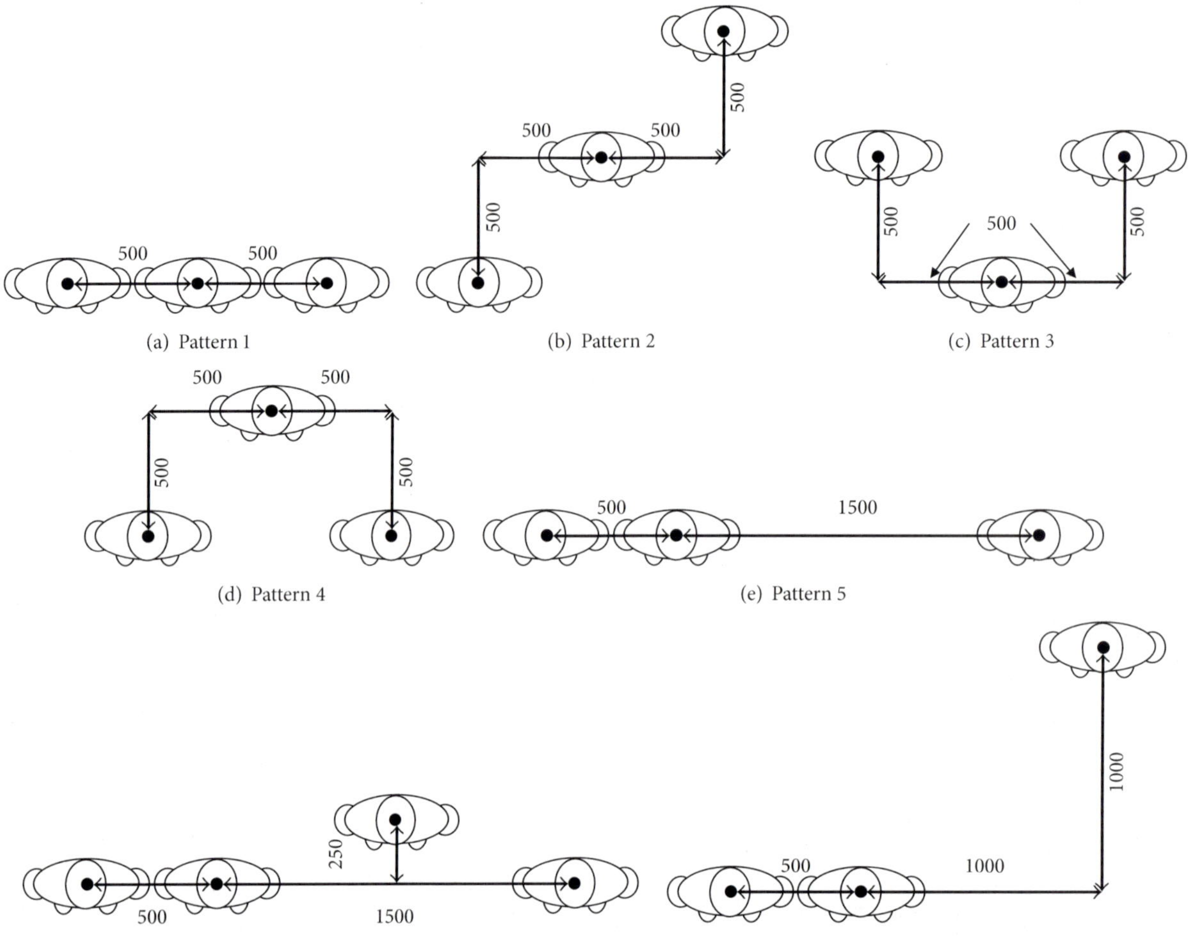

(a) Pattern 1 (b) Pattern 2 (c) Pattern 3 (d) Pattern 4 (e) Pattern 5 (f) Pattern 6 (g) Pattern 7

FIGURE 9: Patterns of the crowd (static case).

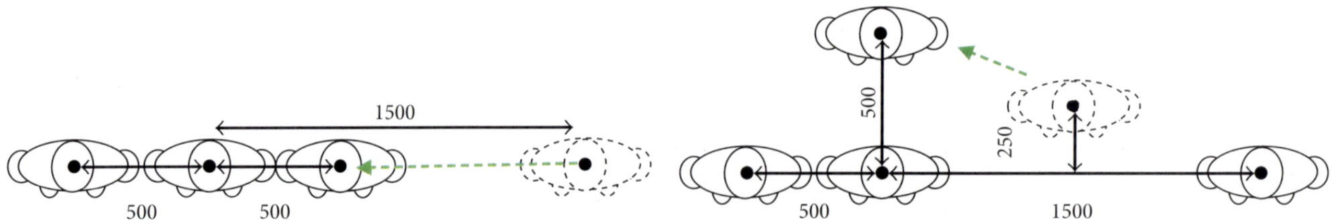

(a) Pattern 8 (b) Pattern 9

FIGURE 10: Pattern of the crowd (dynamic case).

robot could not arrive at the target point unless the crowd yields. Another limitation is that this method does not work when the persons move faster than the scanning speed of the LRF or the moving speed of the robot. The advantage of this method is that this method is based on the measurement obtained one after the other, without using the fixed map. Because of this feature, the robot could pass through the crowd under dynamic environments such as patterns 8 and 9. Another advantage is that the proposed method is simple enough to implement easily because the method is based on the following behavior along simple figures such as a straight line or a circle.

5. Conclusion

In this paper, we proposed a crowd avoidance method that uses the avoidance region. Using this method, the robot could avoid the crowd even when the crowd was in the "inverted wedge" formation and the persons in the crowd were moving. In addition, the robot could pass through the crowd when there was a gap with enough width.

In the experiment, we did not use the person-following method before the avoidance behavior because we wanted to check if the proposed algorithm worked correctly. As a future work, we implement the person-following method to

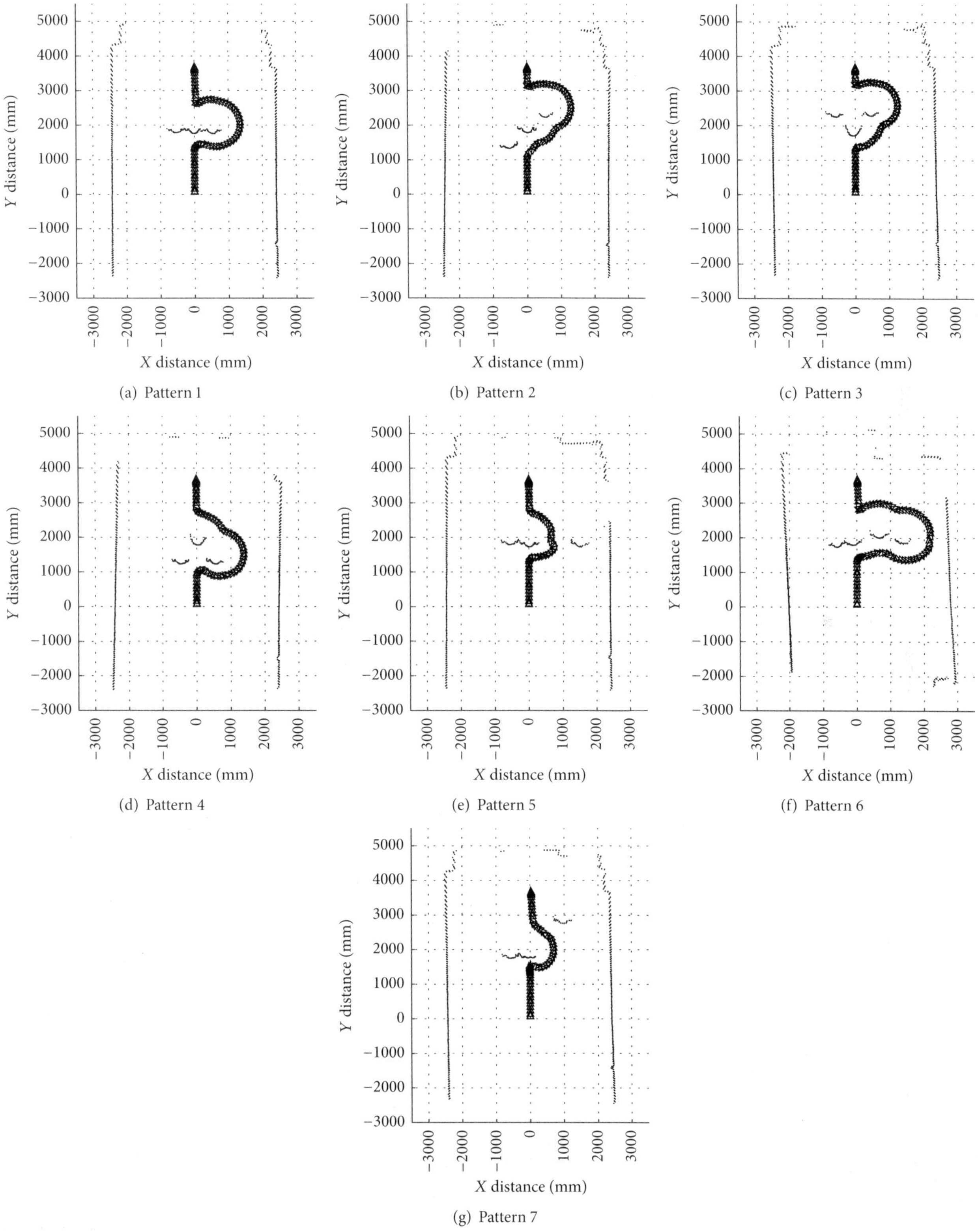

FIGURE 11: Experimental results.

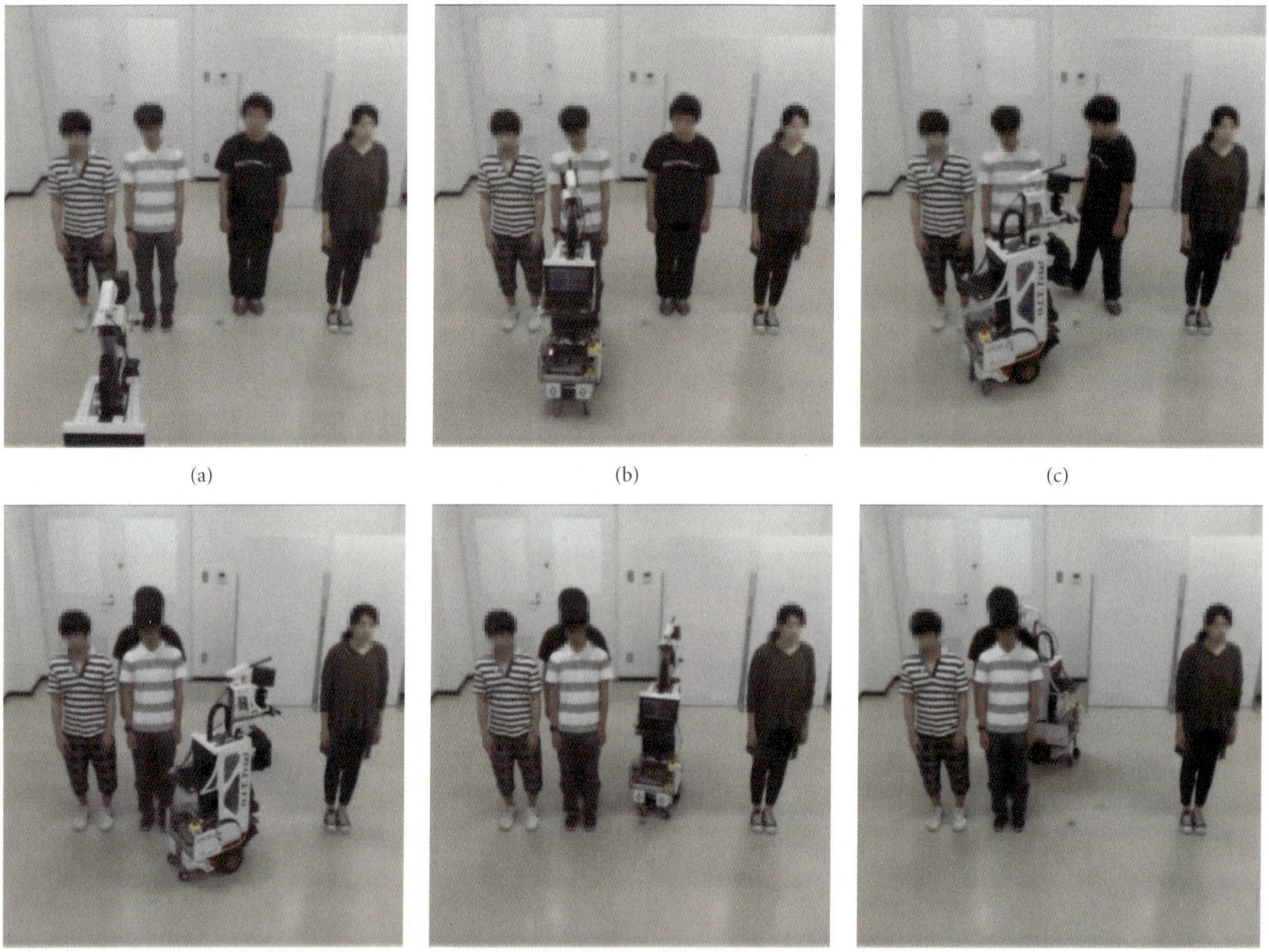

(a) (b) (c)

(d) (e) (f)

FIGURE 12: The avoidance behavior of the robot under pattern 9.

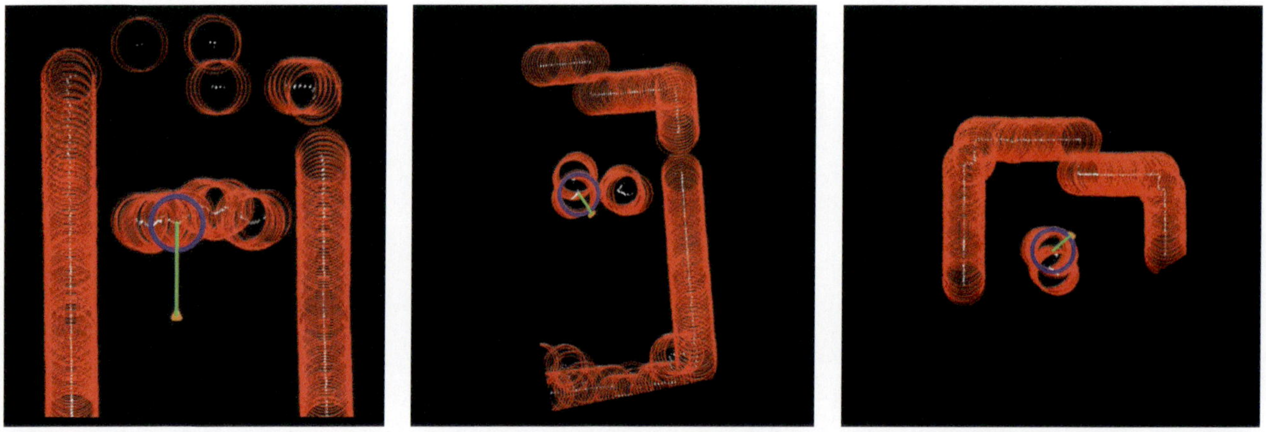

(a) (b) (c)

FIGURE 13: Measurement examples of sensor signals obtained from the robot's avoidance behavior.

the robot and conducted an experiment that combines the person following and the crowd avoidance under the static and dynamic environments.

Another issue is that we need to implement a method to discover the lost target person after avoiding the crowd.

Competing Interests

The authors declare that there are no competing interests regarding the publication of this paper.

Acknowledgments

Part of this work was supported by JSPS KAKENHI JP16K00363.

References

[1] K. Doelling, J. Shin, and D. O. Popa, "Service robotics for the home: a state of the art review," in *Proceedings of the the 7th International Conference on Pervasive Technologies Related to Assistive Environments (PETRA '14)*, pp. 1–8, Rhodes, Greece, May 2014.

[2] D. Holz, L. Iocchi, and T. van der Zant, "Benchmarking intelligent service robots through scientific competitions: The RoboCup@Home approach," in *Proceedings of the AAAI Spring Symposium: Designing Intelligent Robots 2013*, pp. 27–32, Stanford, Calif, USA, March 2013.

[3] K. Hashimoto, F. Saito, T. Yamamoto, and K. Ikeda, "A field study of the human support robot in the home environment," in *Proceedings of the IEEE Workshop on Advanced Robotics and Its Social Impacts (ARSO '13)*, pp. 143–150, Tokyo, Japan, November 2013.

[4] J. Stückler, D. Droeschel, K. Gräve et al., "Increasing flexibility of mobile manipulation and intuitive human-robot interaction in RoboCup@Home," in *Proceedings of the RoboCup 2013: Robot World Cup XVII*, S. Behnke, M. Veloso, A. Visser and, and R. Xiong, Eds., Springer, 2014.

[5] J. Stückler, D. Holz, and S. Behnke, "RoboCup@Home: demonstrating everyday manipulation skills in RoboCup@Home," *IEEE Robotics & Automation Magazine*, vol. 19, no. 2, pp. 34–42, 2012.

[6] G. Medioni, A. R. J. François, M. Siddiqui, K. Kim, and H. Yoon, "Robust real-time vision for a personal service robot," *Computer Vision and Image Understanding*, vol. 108, no. 1-2, pp. 196–203, 2007.

[7] B. Graf, U. Reiser, M. Hägele, K. Mauz, and P. Klein, "Robotic home assistant care-O-bot® 3—product vision and innovation platform," in *Proceedings of the IEEE Workshop on Advanced Robotics and Its Social Impacts (ARSO '09)*, pp. 139–144, Tokyo, Japan, November 2009.

[8] H.-J. Kwak and G.-T. Park, "Study on the mobility of service robots," *International Journal of Engineering and Technology Innovation*, vol. 2, no. 2, pp. 97–112, 2012.

[9] S. Yoon, K. S. Roh, and Y. Shim, "Vision-based obstacle detection and avoidance: application to robust indoor navigation of mobile robots," *Advanced Robotics*, vol. 22, no. 4, pp. 477–492, 2008.

[10] N. Sariff and N. Buniyamin, "An overview of autonomous mobile robot path planning algorithms," in *Proceedings of the 4th Student Conference on Research and Development (SCOReD '06)*, pp. 183–188, Selangor, Malaysia, June 2006.

[11] V. Kunchev, L. Jain, V. Ivancevic, and A. Finn, "Path planning and obstacle avoidance for autonomous mobile robots: a review," in *KES 2006, Part II*, B. Gabrys, R. J. Howlett, and L. C. Jain, Eds., vol. 4252 of *Lecture Notes in Computer Science*, pp. 537–544, Springer, 2006.

[12] I. Al-Taharwa, A. Sheta, and M. Al-Weshah, "A mobile robot path planning using genetic algorithm in static environment," *Journal of Computer Science*, vol. 4, no. 4, pp. 341–344, 2008.

[13] Q. Cao, Y. Huang, and J. Zhou, "An evolutionary artificial potential field algorithm for dynamic path planning of mobile robot," in *Proceedings of the IEEE/RSJ International Conference on Intelligent Robots and Systems (IROS '06)*, pp. 3331–3336, Beijing, China, October 2006.

[14] L. Lapierre, R. Zapata, and P. Lepinay, "Combined path-following and obstacle avoidance control of a wheeled robot," *International Journal of Robotics Research*, vol. 26, no. 4, pp. 361–375, 2007.

[15] D. Soetanto, L. Lapierre, and A. Pascoal, "Adaptive, non-singular path-following control of dynamic wheeled robots," in *Proceedings of the 42nd IEEE International Conference on Decision and Control*, vol. 2, pp. 1765–1770, Maui, Hawaii, USA, December 2003.

[16] G. Antonelli, S. Chiaverini, and G. Fusco, "A fuzzy-logic-based approach for mobile robot path tracking," *IEEE Transactions on Fuzzy Systems*, vol. 15, no. 2, pp. 211–221, 2007.

[17] P. Coelho and U. Nunes, "Path-following control of mobile robots in presence of uncertainties," *IEEE Transactions on Robotics*, vol. 21, no. 2, pp. 252–261, 2005.

[18] T. Yoshimi, M. Nishiyama, T. Sonoura et al., "Development of a person following robot with vision based target detection," in *Proceedings of the IEEE/RSJ International Conference on Intelligent Robots and Systems (IROS '06)*, pp. 5286–5291, Beijing, China, October 2006.

[19] Y. Hiroi and A. Ito, "ASAHI: OK for failure: a robot for supporting daily life, equipped with a robot avatar," in *Proceedings of the International Conference on Human-Robot Interaction*, pp. 141–142, Tokyo, Japan, March 2013.

[20] Y. Hiroi, S. Matsunaka, and A. Ito, "Mobile robot system with semi-autonomous navigation using simple and robust person following behavior," *Journal of Man, Machine and Technology*, vol. 1, no. 1, pp. 44–62, 2012.

[21] L. Huang, "Velocity planning for a mobile robot to track a moving target—a potential field approach," *Robotics and Autonomous Systems*, vol. 57, no. 1, pp. 55–63, 2009.

[22] M. Deng, A. Inoue, K. Sekiguchi, and L. Jiang, "Two-wheeled mobile robot motion control in dynamic environments," *Robotics and Computer-Integrated Manufacturing*, vol. 26, no. 3, pp. 268–272, 2010.

[23] J. Borenstein and Y. Koren, "Real-time obstacle avoidance for fast mobile robots," *IEEE Transactions on Systems, Man and Cybernetics*, vol. 19, no. 5, pp. 1179–1187, 1989.

[24] S. S. Ge and Y. J. Cui, "New potential functions for mobile robot path planning," *IEEE Transactions on Robotics and Automation*, vol. 16, no. 5, pp. 615–620, 2000.

17

Mobile Robot Simultaneous Localization and Mapping based on a Monocular Camera

Songmin Jia,[1,2] Ke Wang,[1,2] and Xiuzhi Li[1,2]

[1]*College of Electronic Information & Control Engineering, Beijing University of Technology, Beijing 100124, China*
[2]*Beijing Key Laboratory of Computational Intelligence and Intelligent System, Beijing 100124, China*

Correspondence should be addressed to Songmin Jia; jsm@bjut.edu.cn

Academic Editor: Shahram Payandeh

This paper proposes a novel monocular vision-based SLAM (Simultaneous Localization and Mapping) algorithm for mobile robot. In this proposed method, the tracking and mapping procedures are split into two separate tasks and performed in parallel threads. In the tracking thread, a ground feature-based pose estimation method is employed to initialize the algorithm for the constraint moving of the mobile robot. And an initial map is built by triangulating the matched features for further tracking procedure. In the mapping thread, an epipolar searching procedure is utilized for finding the matching features. A homography-based outlier rejection method is adopted for rejecting the mismatched features. The indoor experimental results demonstrate that the proposed algorithm has a great performance on map building and verify the feasibility and effectiveness of the proposed algorithm.

1. Introduction

To successfully accomplish autonomous or semiautonomous tasks to assist daily human activities, building the 3D geometry map of the environment is becoming one of the fundamental issues for mobile robotics. In computer vision community, tracking and mapping of the sparse tracked features can be easily conducted in real time by using a hand-held camera. Existing Structure-from-Motion (SFM) [1] and filter-based Simultaneous Localization and Mapping (SLAM) [2] approaches have been investigated in recent years in order to concurrently estimate the pose of a freely moving camera and the sparse spatial structure of the environment [3].

Computer vision-based approaches offer substantial advantages for robot localization and environment recognition. Therefore, a lot of researches were devoted to extend the vision technique for mobile robot platform. Recently, it has been shown that it is possible to estimate the trajectory using a monocular camera in real time. The typical filter-based vision SLAM is proposed by Davison with a single camera. In Davison's work, a probabilistic filtering was used for features of tracking and mapping procedures in a consistent scene map and the camera pose was estimated with a relative coordination in real time [4]. Royer et al. proposed a monocular vision system for mobile robot localization and autonomous navigation with the prevailing SFM SLAM [5]. In Royer et al.'s work, a sparse 3D map was built with a video sequence for the learning step. Then the sparse map was used to localize the robot for its learning path or a slightly different path. In particular, Klein and Murray proposed a famous SFM-based vision SLAM which is named PTAM (parallel tracking and mapping) [6, 7]. In the PTAM, the mapping and tracking were split into two separate tasks and performed in parallel threads. In the tracking thread, the robust pose estimation of camera was calculated with the feature detection and matching. In the mapping thread, the features which are achieved by epipolar searching procedure were triangulated for map building. Recently, the PTAM algorithm has been widely applied for initial guess for some dense mapping systems [8, 9] or the localization method for robotic systems [10–12]. Forster et al. proposed a semidirect monocular visual odometry algorithm which is somehow different from PTAM [13]. In their work, the feature extraction was only required when a key frame is selected to initialize new 3D points. In the tracking step, the correspondence of feature was an implicit result of direct motion estimation as in [14]. For the outlier measurement of triangulated features' depth, a Bayesian filter was used to estimate the depth at feature locations.

Spirited by above-mentioned works, this paper aims to propose a robust algorithm for a mobile robot based on an improved PTAM. The main contributions of this paper are that an initialization method is proposed by using ground features and a homography-based constraint is used for mismatched features rejection. Assume that a camera is fixed on a mobile robot which is moving on a plane at the beginning. In the initialization procedure, the mobile robot pose is roughly localized by using a homography-based pose estimation method with features tracked by frames. When the distance of mobile robot is about up to 5 cm, the ground features-based pose estimation is performed to achieve the accurate localization of robot. And the pose of camera is calculated by using the translation between camera and robot. Then, the initial map is established by triangulating those matched features. After this initialization procedure, the algorithm begins to track the robot by using a weighted projection error function. When a key frame is achieved, the mapping procedure is performed to achieve the matched features by using epipolar searching and all the matched features will be tested by a homography-based constraint for outlier rejection. To optimize the generated map, local and global bundle adjustments are also adapted as in PTAM algorithm. The indoor experimental results demonstrate the feasibility and effectiveness of the proposed algorithm.

The rest of the paper is organized as follows: in Section 2, each part of the proposed algorithm is explained in detail. Section 3 presents the experimental results of our system in real environments. Finally, conclusions are drawn and future work is outlined in Section 4.

2. Algorithm Architecture

The overall structure of the proposed algorithm is illustrated in Figure 1. Following the structure of the original PTAM algorithm, the proposed algorithm is divided into two independent threads which are the tracking thread and the mapping thread. In the tracking thread, a pre-calibrated camera which is fixed on the mobile robot is capturing the images from the experiment scene. Then, a ground feature-based pose estimation algorithm is employed to achieve a high quality initial map for further camera tracking procedure. After the initialization, the robot pose is estimated by minimizing a weighted projection error-based energy function. When a key frame is selected, the mapping procedure will start in the other thread. The new map points are detected by using the epipolar searching between the new coming key frame and its nearest key frame. To get rid of the mismatched features, a homography-based constraint is utilized to reject the outliers and achieve a clear and high quality 3D sparse map building. Finally, the local and global bundle adjustments are used to achieve an optimized map.

2.1. Parameter Definition. Let $p = (x, y, z)^T$ represent the coordinate of a 3D point defined in the camera reference frame and $u_n = (x_n, y_n)^T$ denotes the normalized image projection coordinate. For a pinhole camera, projection of a 3D point p onto the image plane appears as

$$u_n = K\pi(p), \tag{1}$$

where $\pi(p) = (x/z, y/z, 1)$ is the projection function. The term K specifies the intrinsic parameter matrix; that is,

$$K = \begin{bmatrix} f_x & 0 & u_x \\ 0 & f_y & u_y \\ 0 & 0 & 1 \end{bmatrix}, \tag{2}$$

where (f_x, f_y) is the focal length expressed in units of horizontal and vertical pixels, whereas (u_x, u_y) denotes the principal point coordinates of the image plane. It is worth noting that the above linear model fails to describe a wide-angle camera. In this work, conforming to the mathematical model proposed by Devernay and Faugeras [15], geometric mapping constraints between the distorted image projection (with subscript d) and the corresponding undistorted location (with subscript u) are formulated as

$$u_d = \begin{bmatrix} u_0 \\ v_0 \end{bmatrix} + \begin{bmatrix} f_u & 0 \\ 0 & f_v \end{bmatrix} \cdot \frac{r_d}{r_u} \cdot u, \qquad r_d = \frac{1}{\omega}\arctan\left(2r_u \tan\frac{\omega}{2}\right). \tag{3}$$

r is the radial distance; that is, $r_u = \|u\|$. We use the pose of camera with respect to the world frame of reference as

$$T = \begin{bmatrix} R & t \\ 0 & 1 \end{bmatrix}, \tag{4}$$

where $T \in SE(3)$ is a 4×4 matrix describing the camera pose for each frame relative to the world, R denotes the rotation matrix, and t stands for the translation vector. T transforms the point form the camera's frame of reference to that of the world, such that $p_c = Tp_w$.

Assume that the camera is fixed on the mobile robot with a constant transformation described by transformation T_{cr} as is shown in Figure 1. Let T_{rs} be the transformation describing the relative motion of the robot within a time step when its camera captures two images. Since T_{cr} is a constant, the relative motion of the camera between the captured two images in its own frame of reference T_{cs} is related to the motion of the robot and can be parameterized by planar motion of the robot as follows:

$$T_{rs} = T_{rc}T_{cs}T_{cr}, \tag{5}$$

where $T_{cr} = T_{rc}^{-1}$.

2.2. Initialization of Algorithm. As the ground features are sometimes much richer than the upper space in the indoor environment, a ground feature-based localization algorithm is adopted for initializing the improved PTAM algorithm. Two captured images (live I_c and reference I_r) of the same ground plane are related by a plane homography matrix H.

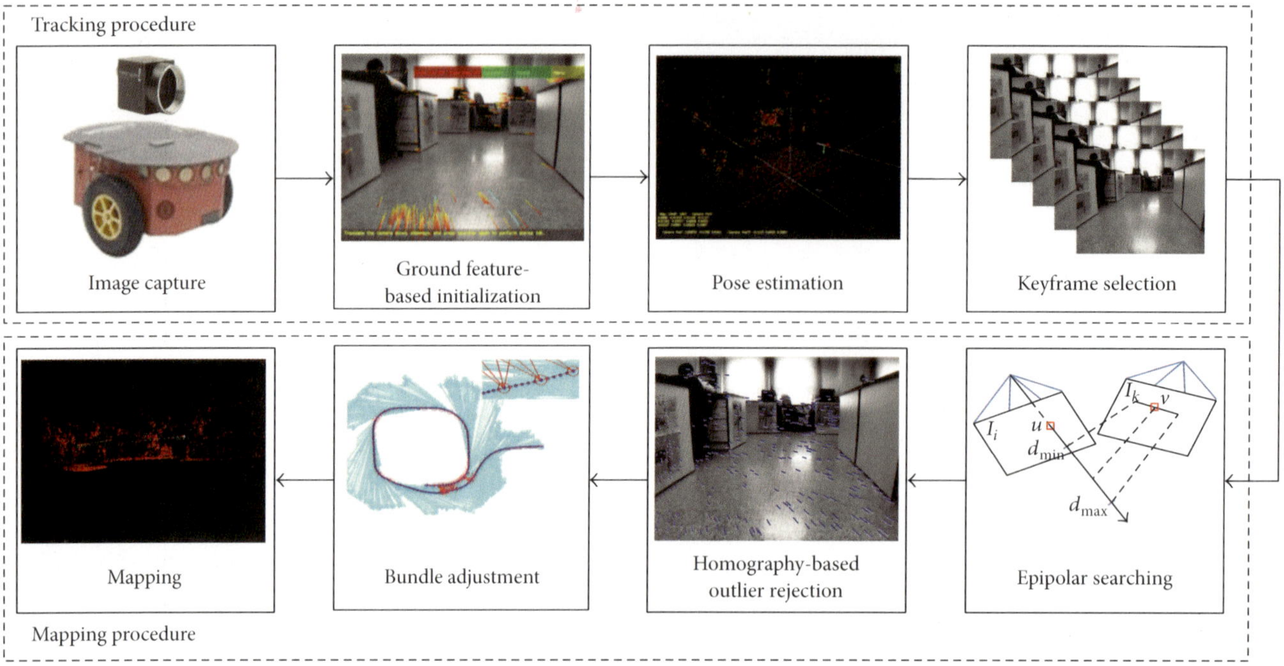

FIGURE 1: The illustration of the proposed algorithm.

For two corresponding features u and v, the homography matrix transforms feature coordinates in the reference image into the pixels in the live image and depends on the camera motion T_{cs} and the parameters n_d of the ground plane in the reference camera frame:

$$u = \alpha H v,$$
$$H = K T_{cr} T_{rs}(\mathbf{x}) T_{rc} (I \mid n_d)^T K^{-1}, \tag{6}$$

where α is the scale factory. n_d and T_{rc} are easily precalculated by using the calibration method for camera.

The pose of camera can be achieved by minimizing the following energy function:

$$\arg\min_{\mathbf{x}} \sum \left\| r_p(\mathbf{x}) \right\|, \tag{7}$$

where $r_p = \pi(H(\mathbf{x})u) - v$ is the homography-based projection error.

As the nonground features have great influence on the pose estimation, a ground feature detection and pose estimation algorithm is expressed as (8) according to the homography-based pose estimation algorithm and a 3D scene flow depth estimation algorithm:

$$\arg\min_{\mathbf{x},d} \sum L\left(d(u) - d_g(u)\right) \left\| r_p(\mathbf{x}) \right\| + \left\| r_d(u,d) \right\|,$$
$$L(x) = \begin{cases} 1 & |x| \le \sigma \\ 0 & \text{others}, \end{cases} \tag{8}$$

where $d_g = 1/n^T u$ is the depth of the ground feature; d is the depth estimated by a 3D scene flow depth estimation algorithm. $L(x)$ indicates whether the feature is placed on the ground. Only when $|d_i - d_g| \le \delta$, $L(x) = 1$. $r_d = K\pi(T_{cs}(\mathbf{x})p(u,d)) - v$ measures the projection error driven by the depth of features. $p(u,d)$ is the coordinate of the 3D point corresponding to the feature u with the depth d.

To optimize this proposed ground feature detection model, an alternating scheme is utilized to update either $\mathbf{x}$ or d in every iteration. The full procedure in iteration is shown as follows:

(a) Fix d; solve

$$\arg\min_{\mathbf{x}} \sum L\left(d - d_g\right) \left\| r_p(\mathbf{x}) \right\|. \tag{9}$$

To optimize this energy function, the first-order Taylor expansion of $r_p(\mathbf{x})$ is performed around $\mathbf{x} = \mathbf{x}_0$. Then, the error function becomes

$$r_p(\xi) = r_p(\mathbf{x}_0) + J\xi, \tag{10}$$

where $J_p = (\partial r_p / \partial \mathbf{x})|_{\mathbf{x}_0}$.

Finally, a Gauss-Newton step is obtained to efficiently optimize that error function iteratively. The solution of energy function becomes

$$\sum J_p{}^T J_p \xi = -\sum J_p{}^T r_p(\mathbf{x}_0),$$
$$\xi = -\left(\sum J_p{}^T J_p\right)^{-1} \sum J_p{}^T r_p(\mathbf{x}_0). \tag{11}$$

The updating procedure of robot pose is applied as follows:

$$T = T_{\text{inc}}(\mathbf{x}) T, \tag{12}$$

where T is the last estimate of the solution and $T_{\text{inc}}(\mathbf{x})$ is a small update to the estimate parameterized by $\mathbf{x} \in \text{se}(2)$.

In this approach, we are required to evaluate the partial derivatives of the cost function at the solution $\mathbf{x} = \mathbf{x}_0$. The derivation of the error function is calculated in each feature point by partial derivatives:

$$\frac{\partial r_p(\mathbf{x})}{\partial x_i} = \frac{\partial \pi(p)}{\partial p}\frac{\partial H(\mathbf{x})}{\partial x_i}p,$$
$$\frac{\partial H(\mathbf{x})}{\partial x_i} = KT_{rc}T_{rs}\frac{\partial T(\mathbf{x})}{\partial x_i}T_r c(I \mid n_d)^T K^{-1}, \tag{13}$$

where x_i is the ith generator of the se(2) group.

(b) Fix $\mathbf{x}$; solve

$$\arg\min_d \sum \left\| r_d(u,d) \right\|. \tag{14}$$

The energy function can be solved by iterating the following equations:

$$\Delta u_i = J_d \nabla d,$$
$$d = d + \nabla d, \tag{15}$$

where J_d is the Jacobian for the projection function evaluated at point u; $J_d = (\partial\pi(p)/\partial p)(\partial p/\partial d)$. Generally, it only needs 3~5 iterations to achieve the convergent solution.

(c) Revalue δ by using $\delta = \max(\gamma, \beta\delta)$.

(d) The iteration will not stop until the ground features are no longer changed or the maximum iteration is arrived at.

(e) When all the ground features are detected, all the ground features are used to achieve more accurate pose estimation with an m-estimation scheme of (9). The weighted energy function is described as follows:

$$\min_{\mathbf{x}} \sum w(r_p) \left\| r_p(\mathbf{x}) \right\|, \tag{16}$$

where w is Tukey's biweight function [16] for r_p and

$$x = \frac{|x_i|}{cx_m},$$
$$w(x) = \begin{cases} \left(1 - \left(\frac{x}{b}\right)^2\right)^2 & |x| \le b \\ 0 & \text{others,} \end{cases} \tag{17}$$

where x_m is the median value of x, $c = 1.4826$ is the robust standard deviation, and $b = 6.6851$ is Tukey specific constant.

Figure 3 illustrates the initialization of the proposed algorithm. Firstly, the algorithm is tracking the features using a patch-based method as is shown in Figure 3(a). All tracked features are regarded as the ground features for raw pose estimation by solving (9). When the translation of mobile robot is about up to 5 cm, the ground feature detection and pose estimation algorithm is performed for the accurate localization of mobile robot. In Figure 3(b), all the ground features are detected by using the proposed algorithm, and reliable and accurate pose estimation is achieved by using the m-estimation scheme. With this accurate localization of mobile robot, all matched features are triangulated for map building and an initial map is established after the epipolar searching procedure, as shown in Figures 3(c) and 3(d). It demonstrates that the proposed algorithm is able to detect all the ground features and achieve high accurate pose estimation during the initialization.

2.3. Tracking Procedure. Accurate camera pose determination is necessitated in a consistent monocular map building, and the goal of this process is to find a rotation matrix R and translation vector t which transform the current camera coordinate frame to that of the global world. In this work, the optimal pose is derived by iterating a robust objective function which minimizes the reprojection error on the image plane used in PTAM algorithm:

$$\xi = \arg\min_{\xi} \sum_i w(e_i) \left\| e_i \right\|, \tag{18}$$

where e_i is the projection error. $w(e_i)$ stands for Tukey's biweight function.

After intense experimental study, we come to realize that the quality of map initialization has a significant impact on the subsequent mapping. Prompted by this requirement, we initiate the 3D point cloud before the dynamic tracking. The method in Section 2.2 is performed for initialization. During this initialization procedure, an image pair is thus gathered at different camera locations which are corresponding to the robot moving and estimated by employing a homography-based method. When the distance of robot moving is about 5 cm and an accurate localization of robot is achieved during the initialization, a good quality initial map guess is provided by triangulating the correspondent points for further tracking and sparse mapping procedure.

Once tracking starts, we attempt to track the FAST-10 corners [17] within the impeding images and they are employed as visual features for tracking in data association procedure. Spatial 6D trajectory of the moving camera could be derived by SFM. As correct interframe correspondence is of vital importance, FAST-10 features are extracted and matched on a four-level coarse-to-fine architecture to boost tracking performance. An 8 × 8-pixel patch template is employed for matching the corner features and the most appealing matched pair is selected by searching within the circular candidate region with the smallest difference ZMSSD (zero-mean SSD) [18] scores. The features on coarse levels of the image pyramid are searched to roughly estimate the relative pose with (18), and then a final accurate pose is optimized by using a thousand of feature correspondences. During the experiment, the tracking is executed with a key frame-based strategy for subsequent incremental mapping. A key frame is selected by taking into account such factors as tracking quality, number of skip frames, proportion of new features, and distance away from the nearest key point.

TABLE 1: The deviation evaluation of original PTAM and proposed algorithm to ground only PTAM.

		x	y	z	Pitch	Yaw	Roll
Proposed algorithm	Mean	−0.0055	0.0052	0.0015	−0.0003	−0.0002	−0.0003
	Variance	0.0025	0.0048	0.0015	0.0006	0.0009	0.0004
Original PTAM	Mean	−0.0231	0.1835	0.0104	−0.0035	0.0057	0.0044
	Variance	0.1166	0.2207	0.1540	0.0272	0.0159	0.0115

2.4. Mapping Procedure. Based on the pose estimation by image features, the matched feature points are triangulated into 3D point cloud. After the sparse map initialization stated previously, the sparse mapping thread begins to query for incorporation of new key frames. Once a new key frame is successfully selected, the extracted FAST features are evaluated by using Shi-Tomas score to find out the most salient points in the new key frame. As Euclidean transformation between current key frame and its closest neighbor, which is reported in tracking routine, is available, new correspondent features can be found in adjacent key frame by using epipolar constraint and ZMSSD criterion is applied for those remaining unmatched feature points in current key frame. Then those new map points are incorporated into the map after triangulation of new feature pairs.

It is worth noting that projection of a 3D point in the sparse map is generally inconsistent with the corresponding feature point due to inaccuracy involved in feature finding, pose estimation, and so forth. A small region around the corresponding features can usually be found in epipolar searching procedure between key frames. With this association among the key frames, a refine map is optimized by using a bundle adjustment routine [19] to simultaneously optimize the pose for all key frames and all map points. In practice, this procedure becomes increasingly expensive in computation as map size and the number of key frames increase. For this reason, a local bundle adjustment is also adopted to perform in this thread. The most recent key frame and its closest neighbors and all of the map points in their view are adjusted by using the local bundle adjustment. According to this local and global bundle adjustment mechanism, the key frame can be integrated in a very short possible time, and it also allows a reasonable rate of camera tracking.

2.5. Homography-Based Outlier Rejection. In the mapping thread, an epipolar searching procedure is employed to achieve some new matched features. The matched features are searched by finding a minimum difference of ZMSSD among features which are around the epipolar. Ideally, the ZMSSD value between two matched patches should be very small. In particular, the ZMSSD which is equal to 0 means that two matched patches are perfectly fitted. As ZMSSD is not an invariant with the rotations, some mismatched features are triangulated and added into map in PTAM. Although a few mismatched features will be rejected during the bundler adjustment optimization, most of them are hardly to be removed. As a projection error-based pose estimation algorithm is utilized in PTAM, the noise will have some effect on the accuracy of localization.

To analyze the performance of features matching in PTAM, the features are detected in the three typical images which are shown in Figure 4(a) by using the FAST-10 algorithm. The ZMSSDs between the original image patch and warped patch which, respectively, rotates and translates along x-axis, y-axis, and z-axis are calculated and drawn in Figures 4(b)–4(d) for all features. The ZMSSD values of some features with different rotation or translation are easily confused for feature matching during the patch searching. Moreover, the epipolar searching procedure is generally hard to establish one-one mapping for feature matching with such confused ZMSSD evaluation. Therefore, a noised map and the low accurate pose estimation are sometimes achieved by the original PTAM algorithm.

To efficiently establish a correct feature matching relationship, a homography-based constraint is adopted for those new detected features. After epipolar searching procedure in ith image level, the homography H_i is estimated by using a RANSAC approach with all the detected features. Only when the projection error r_j is smaller than the threshold δ_i is the detected feature u_j accepted as good matched feature. Consider

$$s\left(u_j^i\right) = \begin{cases} r_j \le \sigma_i & \text{accept} \\ r_j > \sigma_i & \text{reject} \end{cases} \qquad (19)$$

$$r_j = \pi\left(H_i v_j^i\right) - u_j^i,$$

where v_j^i is the reference feature of u_j^i. Generally, the homography in the higher level of the image pyramid can be approximated by the lower one with the following equation:

$$v_j^{i-1} = \begin{bmatrix} \frac{1}{s} & & \\ & \frac{1}{s} & \\ & & 1 \end{bmatrix} v_j^i,$$

$$H_i \begin{bmatrix} \frac{1}{s} & & \\ & \frac{1}{s} & \\ & & 1 \end{bmatrix} v_j^i = s \begin{bmatrix} \frac{1}{s} & & \\ & \frac{1}{s} & \\ & & 1 \end{bmatrix} u_j^i, \qquad (20)$$

$$H_i = \begin{bmatrix} \frac{1}{s} & & \\ & \frac{1}{s} & \\ & & 1 \end{bmatrix} H_{i-1} \begin{bmatrix} s & & \\ & s & \\ & & 1 \end{bmatrix},$$

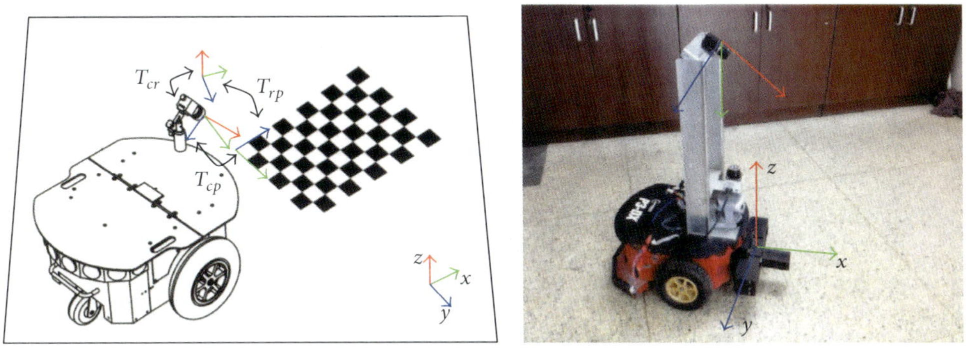

(a) (b)

FIGURE 2: The parameter calibration for mobile robot: (a) is the procedure of calibration of plane parameter, and (b) is the robot platform.

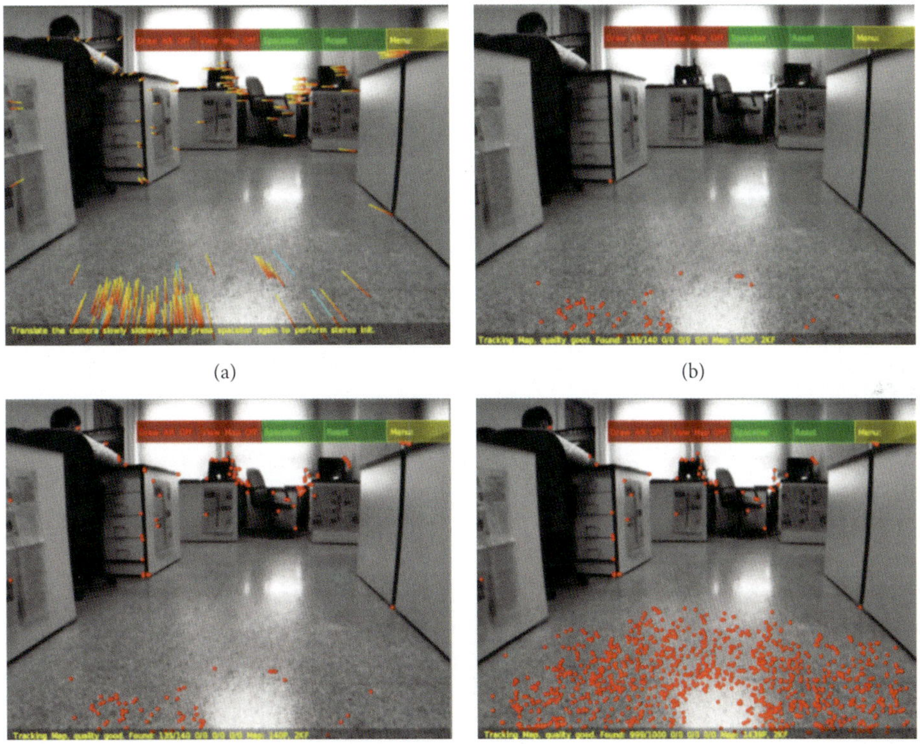

(a) (b) (c) (d)

FIGURE 3: The procedure of initialization: (a) illustrates the tracking of features, the detected ground features are shown in (b), (c) shows all the matched features, and (d) demonstrates the initial map after epipolar searching.

where s is the down-sample rate of image pyramid. During the experiment, the threshold δ_i in the high level of image pyramid is also decreased by the scale of image down-sample rate.

Figure 5 is a compare result between the original PTAM algorithm and the improved PTAM algorithm. The experimental environment is shown in Figure 5(a). Figure 5(b) shows the map generated by the original PTAM algorithm. 7,680 feature points are achieved in the established map and 6 key frames are generated during the camera moving. There are some mismatched features which have been triangulated into the map. Though a few of the mismatched features are rejected after bundle adjustment optimization, some mismatched features still exist which would have an effect on the localization of PTAM algorithm. Figure 5(c) demonstrates the map established by using PTAM with the homography-based constraint. More than 1500 feature points are accepted and added to the map. And the mismatched features are rejected by using the homography-based constraint.

3. Experimental Result

Indoor experiments are conducted to verify the validity and accuracy of the algorithm. Our hardware system setup

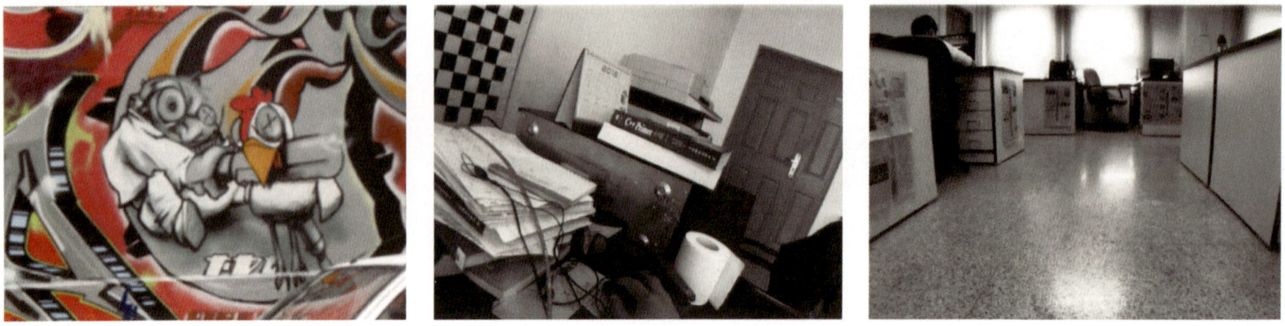

(a)

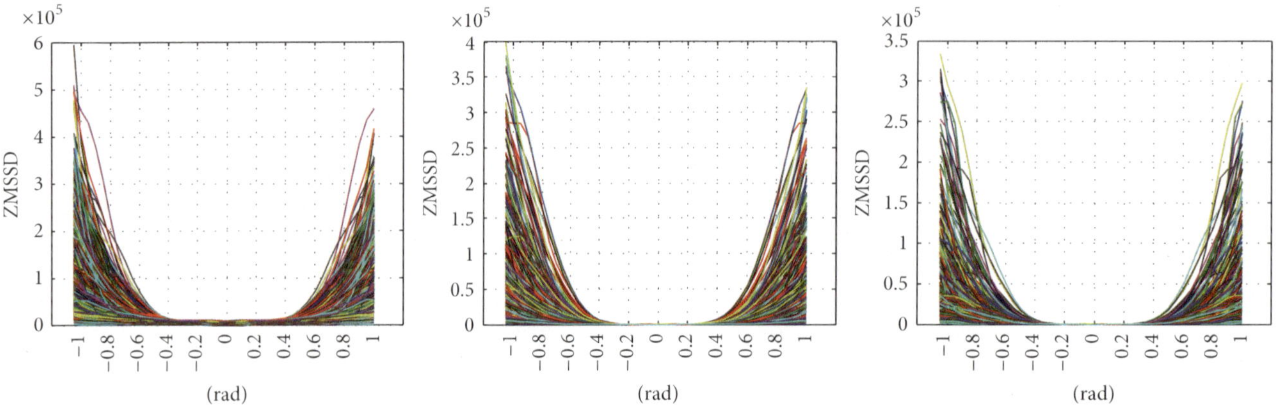

(b)

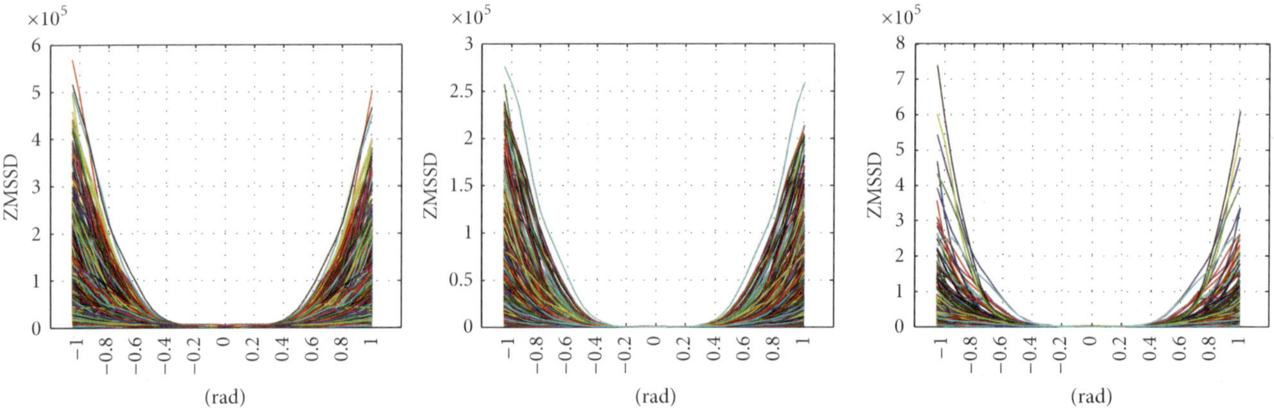

(c)

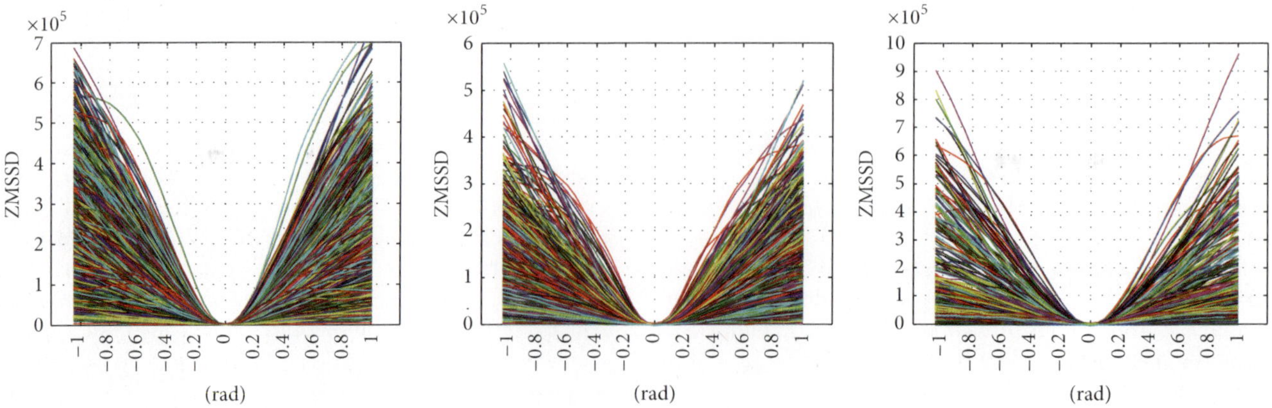

(d)

FIGURE 4: Continued.

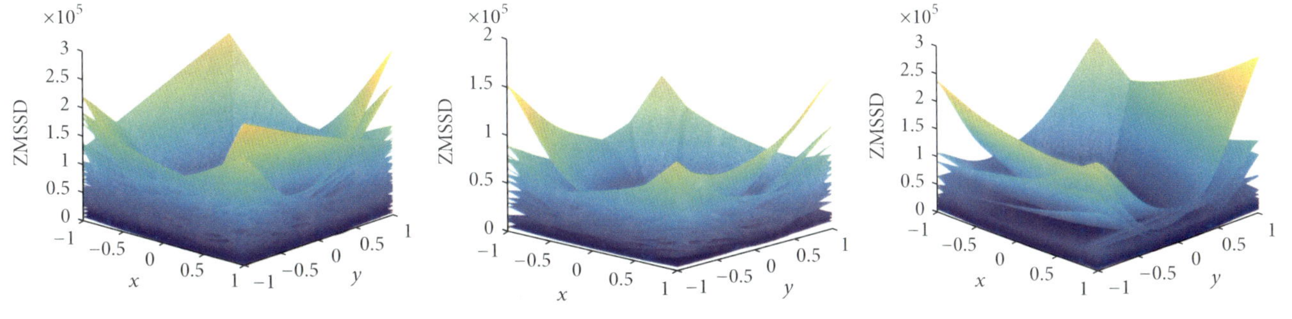

FIGURE 4: The ZMSSDs for rotation and translation of features: (a) shows the images for feature detection; (b), (c), and (d) are the ZMSSDs with rotations around x-axis, y-axis, and z-axis; (d) illustrates the ZMSSDs with the translations along x-axis and y-axis.

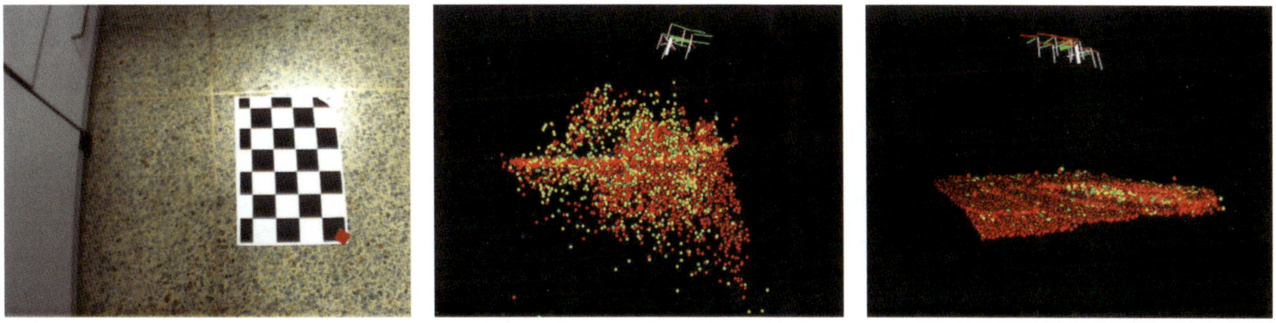

(a) (b) (c)

FIGURE 5: A compare result of the original PTAM and the improved PTAM algorithm: (a) is the experiment environment, (b) shows the map generated by the original PTAM, and (c) shows the map generated by the improved PTAM.

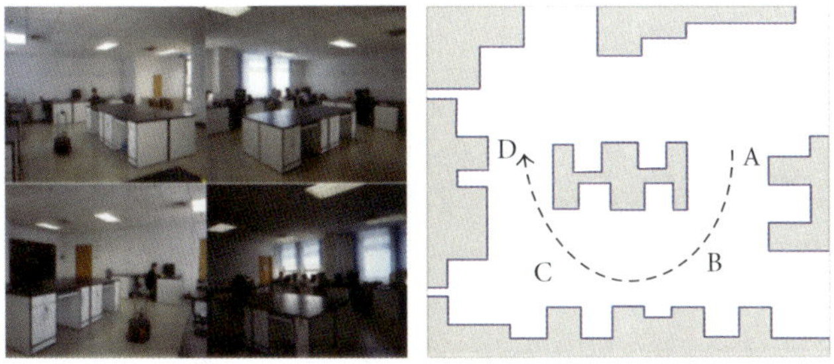

FIGURE 6: The experimental environment.

consists of a Pioneer3-DX platform and a monocular vision system which is depicted in Figure 2. The robot platform is qualified for the indoor research activities like mapping, obstacle avoidance, path planning, and so forth. A forward looking Point Grey Flea2 camera (shown in Figure 1) is fixed on the mobile robot. The focal length of the camera is about 3.5 mm and the field of view is about 65°. The camera is connected via 1394 B to a 2.8 GHz Intel i5 computer with 4 G RAM. The experiments are performed in an indoor environment as shown in Figure 6.

During the experiment, the mobile robot is running following the path from A to D, while the camera is capturing the images. Firstly, the algorithm is initialized by using the proposed ground features-based method. During the initialization procedure, the features are tracked frame by frame. All the features are regarded as the ground features and the coarse localization of mobile robot is estimated by (9). When the distance of robot moving is about up to 5 cm, the ground features detection procedure is performed to achieve accurate localization for the mobile robot. With the translation between the mobile robot and camera, the estimated pose of camera is calculated for initializing the mapping and epipolar searching. Then, all matched features are triangulated and added to the map for further tracking procedure. After the initialization, the algorithm begins to estimate its pose by minimizing a weighted projection error function and the map is updated with the epipolar searching after a key frame is achieved. To evaluate the proposed method, the PTAM algorithm which only uses the ground features in tracking procedure is considered as the ground truth during the experiment. In the ground only PTAM algorithm, the detected features are triangulated to achieve the 3D coordination p_i. And the features are accepted in the algorithm only when $0.95 \le \|n_d^T p_i\| \le 1.05$. As the initialization of the original PTAM algorithm is sometimes unsuitable during the experiment, the ground feature-based pose estimation is employed to initialize the algorithm.

Figure 7 is the mapping result of the proposed algorithm and the original PTAM algorithm. Figures 7(a) and 7(b) are

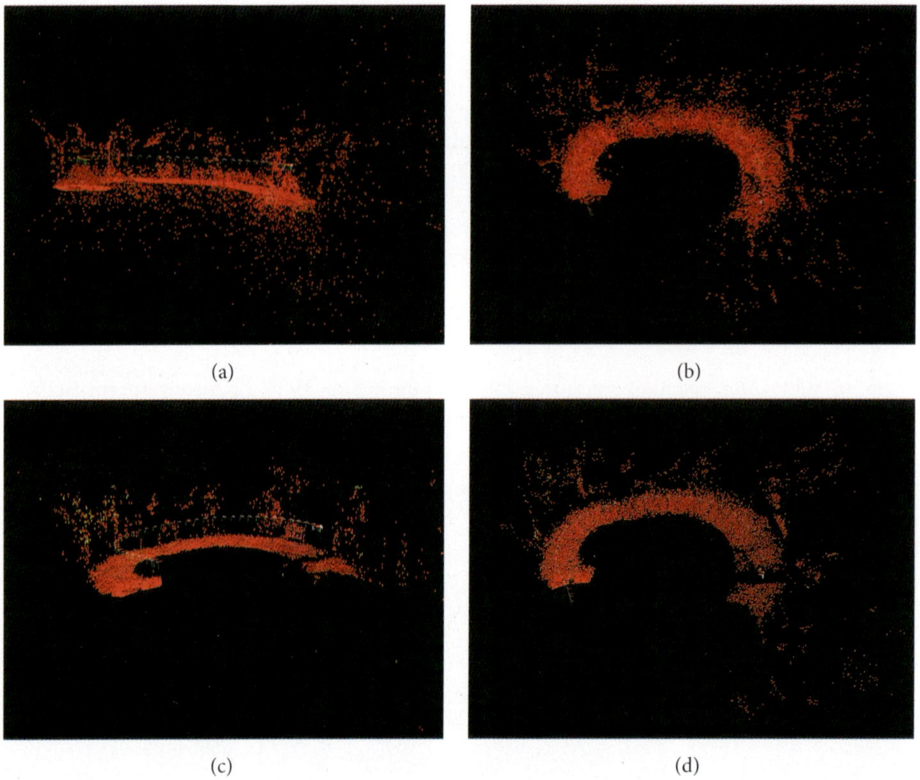

FIGURE 7: The mapping result: (a) and (c) are generated by original PTAM algorithm and proposed algorithm; (b) and (d) are the downlooking of (a) and (c).

the map generated by the original PTAM algorithm in the different viewpoints. Some mismatched features have been added to the map by using the epipolar searching procedure. It would have much influence on the pose estimation with mismatched features. Even worse is the fact that the tracking procedure would make some ridiculous mistakes and fail to track the features, which is hardly to be recovered. To deal with these mismatched features, a homography-based constraint is employed in our improved algorithm. After the epipolar searching procedure, all the matched features which are detected between the new coming key frame and its nearest key frame are used to estimate homography with a RANSAC scheme. The features are filtered with the distance of homography projection errors. In Figures 7(c) and 7(d), the mismatched ground features are rejected and a clearness map is generated by using the proposed method. The experiment demonstrates that the proposed algorithm is able to achieve a clear and high quality 3D feature map.

Figure 8 shows the compare result of the localization among the original PTAM, ground only PTAM, and our proposed algorithm during the mapping experiment. As is shown in Figure 8, the estimated parameters *x*, *y*, *z*, *pitch*, *yaw*, and *roll* of the proposed algorithm and ground only PTAM are highly consistent with each other. But in the original PTAM algorithm, the estimated parameters deviate far from the ground only PTAM algorithm due to the noised map. Table 1 illustrates the mean and the variance of the error which is achieved by the original PTAM and the proposed algorithm. The estimated trajectories of three methods are shown in Figure 9. The proposed algorithm is highly consistent with the ground only PTAM. All above experimental results demonstrate that the proposed algorithm has a high performance on mismatching features rejection and pose estimation. Those results also verify the feasibility and effectiveness of the proposed algorithm.

4. Conclusions

This paper proposes a parallel tracking and mapping algorithm with a monocular camera for mobile robot. The camera is fixed on the mobile robot with a consistent translation, and the robot is moving in an indoor experiment. A ground feature-based pose estimation method is employed to initialize the algorithm. And an initial map is built by triangulating the matched features for further tracking procedure. To establish a sparse 3D map, an epipolar searching procedure is utilized for finding the matching features when it comes to a key frame. Moreover, a mismatched feature rejection method is adopted based on homography for achieving a clearness map. The indoor experimental results demonstrate that the proposed algorithm has a high performance on sparse feature-based mapping and verify the feasibility and effectiveness of the proposed algorithm.

As future work, we will further improve the performance of the proposed algorithm to deal with the mapping in the lighting changed scene. And we also plan to find way to extend our work to a dense visual SLAM system with a monocular camera.

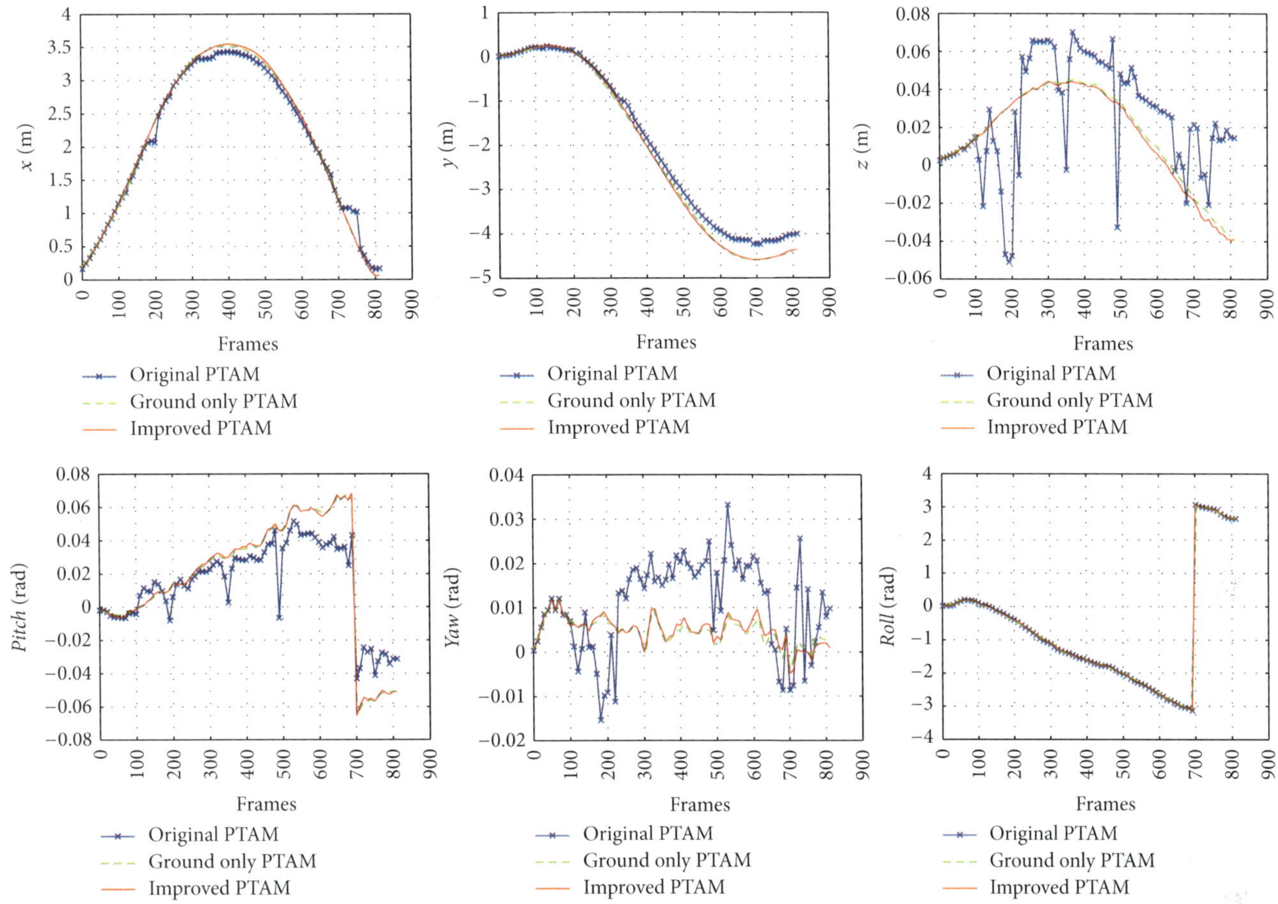

FIGURE 8: The compare result among the original PTAM, ground only PTAM, and our proposed algorithm.

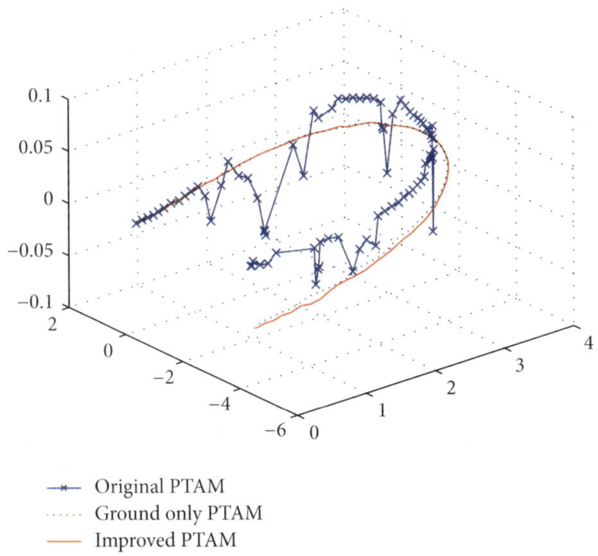

FIGURE 9: The estimated trajectories of mobile robot.

Competing Interests

The authors declare that there are no competing interests regarding the publication of this paper.

Acknowledgments

The research work is financially supported by Promotion Project on Intelligent Robot in Beijing University of Technology "Key Technologies on Modularization and Practicalization of Intelligent Service Robot" and the Natural Science Foundation of China (61175087 and 61105033).

References

[1] M. Tomono, "3-D localization and mapping using a single camera based on structure-from-motion with automatic baseline selection," in *Proceedings of the IEEE International Conference on Robotics and Automation (ICRA '05)*, pp. 3342–3347, April 2005.

[2] A. J. Davison, *Mobile Robot Navigation Using Active Vision*, University of Oxford, Oxford, UK, 1998.

[3] H. Strasdat, J. M. M. Montiel, and A. J. Davison, "Visual SLAM: why filter?" *Image and Vision Computing*, vol. 30, no. 2, pp. 65–77, 2012.

[4] A. J. Davison, "Real-time simultaneous localisation and mapping with a single camera," in *Proceedings of the 9th IEEE International Conference On Computer Vision*, pp. 1403–1410, October 2003.

[5] E. Royer, M. Lhuillier, M. Dhome, and J.-M. Lavest, "Monocular vision for mobile robot localization and autonomous

navigation," *International Journal of Computer Vision*, vol. 74, no. 3, pp. 237–260, 2007.

[6] G. Klein and D. Murray, "Parallel tracking and mapping for small AR workspaces," in *Proceedings of the 6th IEEE and ACM International Symposium on Mixed and Augmented Reality (ISMAR '07)*, pp. 225–234, Nara, Japan, November 2007.

[7] G. Klein and D. Murray, "Improving the agility of keyframe-based SLAM," in *Proceedings of the 10th European Conference on Computer Vision (ECCV '08)*, pp. 802–815, October 2008.

[8] J. Stühmer, S. Gumhold, and D. Cremers, "Parallel generalized thresholding scheme for live dense geometry from a handheld camera," in *Trends and Topics in Computer Vision*, pp. 450–462, Springer, Berlin, Germany, 2012.

[9] R. A. Newcombe and A. J. Davison, "Live dense reconstruction with a single moving camera," in *Proceedings of the IEEE Computer Society Conference on Computer Vision and Pattern Recognition (CVPR '10)*, pp. 1498–1505, June 2010.

[10] M. Blösch, S. Weiss, D. Scaramuzza, and R. Siegwart, "Vision based MAV navigation in unknown and unstructured environments," in *Proceedings of the IEEE International Conference on Robotics and Automation (ICRA '10)*, pp. 21–28, Anchorage, Alaska, USA, May 2010.

[11] S. Weiss, M. W. Achtelik, S. Lynen et al., "Monocular vision for long-term micro aerial vehicle state estimation: a compendium," *Journal of Field Robotics*, vol. 30, no. 5, pp. 803–831, 2013.

[12] J. Sheng, S. Tano, and S. Jia, "Mobile robot localization and map building based on laser ranging and PTAM," in *Proceedings of the IEEE International Conference on Mechatronics and Automation (ICMA '11)*, pp. 1015–1020, Beijing, China, August 2011.

[13] C. Forster, M. Pizzoli, and D. Scaramuzza, "SVO: fast semi-direct monocular visual odometry," in *Proceedings of the IEEE International Conference on Robotics and Automation (ICRA '14)*, pp. 15–22, Hong Kong, June 2014.

[14] J. Engel, J. Sturm, and D. Cremers, "Semi-dense visual odometry for a monocular camera," in *Proceedings of the 14th IEEE International Conference on Computer Vision (ICCV '13)*, pp. 1449–1456, Sydney, Australia, December 2013.

[15] F. Devernay and O. Faugeras, "Straight lines have to be straight," *Machine Vision and Applications*, vol. 13, no. 1, pp. 14–24, 2001.

[16] P. J. Huber, *Robust Statistics*, Springer, Berlin, Germany, 2011.

[17] E. Rosten and T. Drummond, "Machine learning for high-speed corner detection," in *Computer Vision—ECCV 2006: 9th European Conference on Computer Vision, Graz, Austria, May 7-13, 2006. Proceedings, Part I*, vol. 3951 of *Lecture Notes in Computer Science*, pp. 430–443, Springer, Berlin, Germany, 2006.

[18] J. Martin and J. L. Crowley, "Comparison of correlation techniques," in *Proceedings of the International Conference on Intelligent Autonmous Systems*, pp. 86–93, Karlsruhe, Germany, 1995.

[19] R. Hartley and A. Zisserman, *Multiple View Geometry in Computer Vision*, Cambridge University Press, 2003.

18

A Novel Method of Dexterity Analysis for a 5-DOF Manipulator

Wanjin Guo,[1] Ruifeng Li,[1] Chuqing Cao,[2,3] and Yunfeng Gao[1]

[1] *State Key Laboratory of Robotics and System, Harbin Institute of Technology, 92 West Dazhi Street, Nangang, Harbin 150001, China*
[2] *School of Mechanical Engineering, Nanjing University of Science and Technology, Xiaolingwei 200, Nanjing 210094, China*
[3] *Wuhu HIT Robot Technology Research Institute Co., Ltd., Electronic Industrial Park, Block E, JiuJiang District, Wuhu 241007, China*

Correspondence should be addressed to Wanjin Guo; guowanjin0102@gmail.com

Academic Editor: Yuan F. Zheng

The dexterity characteristics and dexterous workspace features are applied as the considerable and key kinematics properties to evaluate the manipulator performance. In this paper, a novel method of the dexterity analysis with the character of graphic dexterity and a new concept of the spatial mapping envelope surface are proposed for a 5-degree of freedom (DOF) manipulator. Additionally, for an arbitrary selected point and a dexterous workspace, the dexterity characteristics and the dexterity conditions are drawn, respectively. Finally, the dexterous workspace of this 5-DOF manipulator is obtained and it can be utilized as the basis research of this manipulator in future such as the structure optimization and motion control.

1. Introduction

The most obvious consideration in designing a manipulator is to feature a set of its workspace characteristics [1, 2]. The kinematic dexterity characteristics and reachable workspace features are applied to demonstrate the fundamental properties of a manipulator. Several studies investigating kinematic dexterity indices and measures have been carried out for manipulators. The representative issues of these performance criteria for manipulators are researched.

To examine the kinematic manipulability of manipulators, several attempts have been made. The Yoshikawa manipulability ellipsoid measure was proposed to discuss the manipulating ability of robotic mechanisms in positioning and orienting end-effectors [3]. A homogeneous manipulability measure by means of the power concept was proposed for the dexterity analysis of a manipulator [4–6]. An extension to the Yoshikawa manipulability ellipsoid measure was proposed to quantify a manipulator performance in terms of dexterity and maneuverability [7]. In [8], two dexterity indices, angularity and axiality, were used to estimate the motion sensitivity of the mobile platform of a parallel manipulator.

The dexterity measures and evaluations for manipulators have been investigated. Two dexterity measures were presented for planar manipulators by Gosselin [9], which are based on the redundant kinematic equations and the minimum number of kinematic parameters, respectively. In [10], dexterity measures were presented to analyze the dexterity of an underactuated manipulator and the fault tolerance of a redundant manipulator. Two kinematic dexterity measures were proposed by Tchón and Zadarnowska [11] to determine optimal configurations and geometries of a mobile manipulator through the endogenous configuration space approach. An approach of workspace modeling with Euler parameters to explore the dexterity evaluation was proposed by Bai et al. [12] for spherical parallel manipulators with symmetrical architecture. A general and systematic approach was presented by Liu et al. [13] to evaluate the dexterity of parallel manipulators which possess mixed rotational and translational movement capabilities. In [14], weight service sphere and weight service region were defined to evaluate the dexterity of a radial symmetrical hexapod robot. Meanwhile, the dexterity of single point in workspace and the dexterity distribution in vertical and horizontal projection plane were demonstrated. In [15], a directional manipulability transmission ratio was presented to reflect the manipulability capability along the task direction for the dexterity analysis of an omnidirectional wheeled mobile manipulator.

For optimizing the kinematics dexterity of manipulators, several surveys have been published. A global conditioning index (GCI) was proposed for the kinematic optimization of robotic manipulators [16]. A mathematical theory was developed to optimize the kinematic dexterity of robotic mechanisms [17]. In [18], the local/global performance indexes of dexterity and manipulability are mathematically modeled and mapped to investigate the comprehensive features for the multiobjective optimization of a hybrid robotic machine tool.

Although the above studies investigating the dexterity indexes or measures are applied commonly to evaluate quantitatively the kinematic dexterity characteristics, they are not able to obviously demonstrate the topological considerations between the manipulation workspace and the performance evaluation. These studies focus on numerical or algebraic methods to characterize the relationship between the manipulator and the performance evaluation. This relationship, which cannot be revealed intuitively and obviously in the practical application, is viewed as one of the main obstacles that these numerical or algebraic methods face in attempting to feature the dexterity characteristics. More-general design principles or insights are difficult to come by using these approaches.

In this paper, the forward kinematics and the reachable workspace of a 5-DOF manipulator are derived by the Denavit and Hartenberg convention. Based on the mapping and envelope theories, a novel method with the character of graphic dexterity for dexterity analysis is proposed to evaluate this manipulator's kinematic performance. Additionally, the relationship between the dexterity characteristics and influence structure parameters can be demonstrated evidently. The dexterous workspace of this manipulator can be basically derived from the results of the dexterity analysis (these results are obtained by the proposed method) and the reachable workspace.

The proposed dexterity analysis method is developed to address the dexterity analysis problem for this 5-DOF manipulator. This method can also be applied to other manipulator structures for the dexterity analysis. This proposed method is a topological approach with clear and definite graphic dexterity character which can clearly demonstrate the dexterity characteristics and the reachable workspace features through the topological figures of the spatial mapping envelope surface and the relationship of the mapping and the orientation.

The remainder of this paper is organized as follows. The structure of a 5-DOF manipulator and its forward kinematics model are presented in Section 2. The reachable workspace of this manipulator is described in Section 3. A novel dexterity analysis method and a new concept of the spatial mapping envelope surface are proposed for the dexterity analysis of this manipulator in Section 4. The reachable workspace and the dexterous workspace of this manipulator are demonstrated in Section 5. Finally, conclusions are drawn in Section 6.

2. Structure Description and the Forward Kinematics Model

The structure and drive system of this 5-DOF manipulator are described. Then the forward kinematics of this manipulator is modeled based on the Denavit-Hartenberg method.

2.1. Structure Description of the Manipulator. The structure of this 5-DOF manipulator is shown in Figure 1. It is a 3T2R hybrid manipulator, which is composed of a 3-DOF (3T) parallel module and a 2-DOF (2R) serial module. The 3T parallel module is constituted by the axes J_1, J_2, and J_3 (consisting of ball screws), which are driven by motor M_1, M_2, and M_3, respectively. The 2R serial module is formed by the rotational axes J_4 and J_5, which are driven by motors M_4 and M_5, respectively. The axis J_6 is taken as the end-effector which is a high-speed motorized spindle. The lengths of the rod i and the end-effector are denoted by L_i and L_t, respectively. The distance between J_6 and J_4 in the horizontal direction is denoted as e. The nuts on the screws S_1 and S_2 are fixedly connected to the slider assemblies A and C, respectively. The slider assembly B is rotated through a rotation shaft which is connected to A, and the slider assembly D is rotated in the same way. The pairs of slider assemblies B and D are moved along the slide plate which goes across S_1 and S_2. The turntable, which can rotate within a certain range, is fixedly connected to the slide plate. In the meantime, the turntable is moved along the symmetry axis of S_1 and S_2 in the horizontal direction.

It is worth noting that the first two DOFs of this manipulator (i.e., two translational movements in the horizontal plane) are achieved by the parallel mechanism which is a coupled mechanism. It is driven in parallel by M_1 and M_2 as shown in Figure 1. This coupled mechanism achieves simultaneously two translational movements in the horizontal plane by two small power motors. Furthermore, the orientation and the position of the end-effector are regulated by two rotation joints and three ball-screw drives, respectively. More details of the mobility analysis and the structure analysis for this 5-DOF manipulator are described in our paper [19].

2.2. Modeling of the Forward Kinematics. The D-H reference frame [20] is established as shown in Figure 2. The rotation angles of J_1, J_2, J_3, J_4, and J_5 are represented by φ_1, φ_2, φ_3, φ_4, and φ_5, respectively. The strokes of the nut along S_1, S_2, and S_3 are represented by X_1, X_2, and X_3, respectively. The rotation angle of the axis J_t and link 1 is denoted as α. L_{01} represents the distance between the point O_2 and the point O_1 in the vertical direction when the nut is located at the top starting point along S_3. These variables are related by

$$\begin{aligned}
\alpha &= \arctan\left(\frac{(X_2 - X_1)}{L_1}\right), \\
X_1 &= \frac{s_x\varphi_1}{(2\pi)}, \\
X_2 &= \frac{s_x\varphi_2}{(2\pi)}, \\
X_3 &= \frac{s_x\varphi_3}{(2\pi)},
\end{aligned} \tag{1}$$

where the constant s_x represents the pitch of S_1 and S_2 and the constant s_z represents the pitch of S_3.

Link 1 is driven in parallel by M_1 and M_2 as shown in Figures 1 and 2. This link is moved along the axis X_1 which

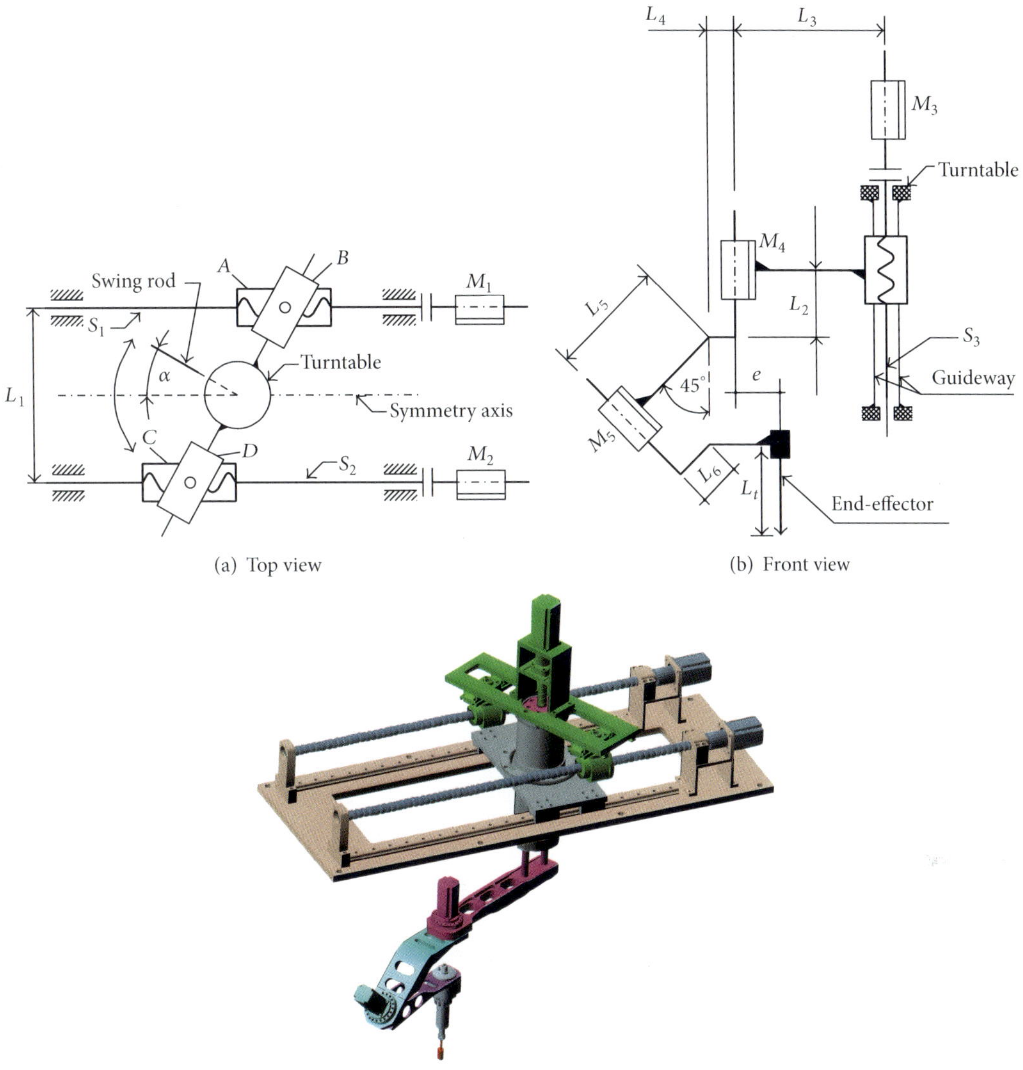

(a) Top view

(b) Front view

(c) 3D model

FIGURE 1: Manipulator structure.

is also the symmetry axis of S_1 and S_2. It is also rotated about J_t with the turntable when the strokes of A and C along the screw are different. Hence, a coupled movement is generated. The movement direction of link 1 along the axis X_1 is always paralleled to X_0 while this link is rotated about Z_1 and J_t. According to this movement relationship, the homogeneous transformation matrix ${}^0T_1$ which relates the spatial displacement of the reference frame 1 to the base reference frame 0 is directly given as

$$ {}^0T_1 = \begin{bmatrix} \cos\alpha & -\sin\alpha & 0 & \dfrac{(X_1 + X_2)}{2} \\ \sin\alpha & \cos\alpha & 0 & 0 \\ 0 & 0 & 1 & 0 \\ 0 & 0 & 0 & 1 \end{bmatrix}. \quad (2) $$

The geometric parameters of this manipulator for links 2, 3, 4, and 5 (tool) shown in Figures 1 and 2 are listed in Table 1.

The homogeneous transformation matrix ${}^{i-1}T_1$ is conducted to provide a representation of reference frame i relative to reference frame $i-1$ according to the D-H reference frames and the geometric parameters. It is given as

$$ {}^{i-1}T_1 = \begin{bmatrix} \cos\theta_i & -\cos\alpha_i \cdot \sin\theta_i & \sin\alpha_i \cdot \sin\theta_i & a_i \cdot \cos\theta_i \\ \sin\theta_i & \cos\alpha_i \cdot \cos\theta_i & -\sin\alpha_i \cdot \cos\theta_i & a_i \cdot \sin\theta_i \\ 0 & \sin\alpha_i & \cos\alpha_i & d_i \\ 0 & 0 & 0 & 1 \end{bmatrix}, \quad (3) $$

where $i = 2, \ldots, 5$, $i \in N$.

By successive multiplications of the transformation matrices, the equivalent homogeneous transformation matrix ${}^0T_5$ which relates the spatial displacement of the tool frame to the base reference frame is obtained as

$$ {}^0T_5 = {}^0T_1 \, {}^1T_2 \, {}^2T_3 \, {}^3T_4 \, {}^4T_5. \quad (4) $$

TABLE 1: Geometric parameters of links.

Link i	θ_i	α_i	d_i	a_i
2	0	0	$X_3 + L_{01}$	L_3
3	$-\pi/2 + \varphi_4$	$\pi/4$	$L_2 + L_4 + \sqrt{2}L_5$	0
4	$\pi + \varphi_5$	$\pi/4$	$\sqrt{2}e$	0
5	0	0	0	0

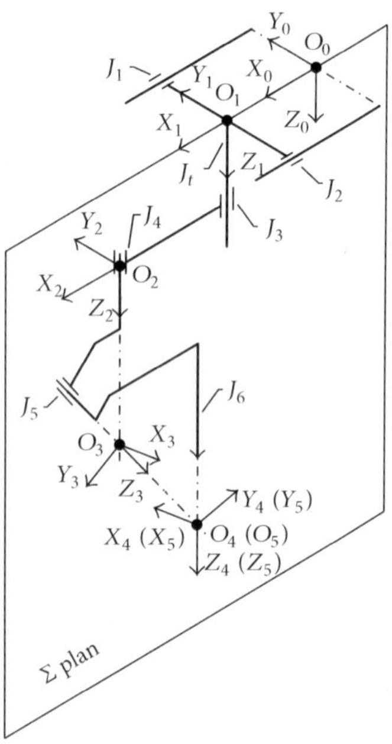

FIGURE 2: Denavit-Hartenberg reference frame.

3. Representation of the Reachable Workspace

More generally, the workspace of a manipulator can be represented as a total volume swept out by the end-effector as the manipulator executes all possible motions [21]. The workspace is determined by the geometry of the manipulator and the limits of the joint motions. In order to analyze the dexterity characteristics and the dexterous workspace in the following sections, the reachable workspace corresponding to the operation point of the end-effector is derived by the following steps.

Suppose that a position vector $\vec{P} = [x_P, y_P, z_P]^T$ represents the operation point P in the tool reference frame. According to the position of P in the tool reference frame, $x_P = 0$, $y_P = 0$, and $z_P = L_P$ are obtained, where L_P defined as the tool compensation which is the value of the tool operating point in Z_5 axis is given as

$$L_P = L_t - \left(L_4 + \sqrt{2}L_6 + e\right). \tag{5}$$

The reachable workspace of the operation point P is obtained based on the forward kinematics model. By successive multiplication of the transformation matrices $[x_0, y_0, z_0, 1]^T = {}^0T_1 \cdot {}^1T_2 \cdot {}^2T_3 \cdot {}^3T_4 \cdot {}^4T_5 \cdot [x_P, y_P, z_P, 1]^T$, the expression of this reachable workspace is derived. In the base reference frame, a space manifold $\Sigma_{O_0\text{-}P}$ is obtained as

$$\Sigma_{O_0\text{-}P}: \begin{cases} x_0 = \dfrac{X_1 + X_2}{2} - \dfrac{\sqrt{2}}{2} L_P \sin(\alpha + \varphi_4)\sin\varphi_5 - \cos(\alpha + \varphi_4)\left(\dfrac{1 - \cos\varphi_5}{2} L_P + e\right) + L_3 \cos\alpha \\ y_0 = \dfrac{\sqrt{2}}{2} L_P \cos(\alpha + \varphi_4)\sin\varphi_5 - \sin(\alpha + \varphi_4)\left(\dfrac{1 - \cos\varphi_5}{2} L_P + e\right) + L_3 \sin\alpha \\ z_0 = X_3 + L_{01} + \left(L_2 + L_4 + \sqrt{2}L_5\right) + e + \dfrac{1 + \cos\varphi_5}{2} L_P. \end{cases} \tag{6}$$

4. The Proposed Method for Dexterity Analysis

In this section, based on the mapping and envelope theory, a novel dexterity analysis method with the character of graphic dexterity and a new concept of the spatial mapping envelope surface are proposed for the dexterity analysis of this manipulator. The proposed dexterity analysis method can be also applied to other manipulator structures to address the dexterity analysis problem. The brief outline of the proposed method is stated as follows.

The reachable workspace of the reference point is partitioned into a regional workspace volume and an orientation workspace volume based on the effect of the orientation; that is, the space manifold of the reachable workspace is decomposed into a regional portion and an orientation portion. Then, an arbitrary point is selected in this reachable workspace and this point is mapped to a spatial mapping envelope surface with a matching relationship. By analyzing the mapping envelope features and the geometry meanings of this spatial mapping envelope surface, the dexterity characteristics of this arbitrary selected point are drawn. Finally, the condition that the arbitrary selected point is a strict dexterity point with the strict dexterity performance is deduced. Meanwhile, the condition that a dexterous workspace is a strict dexterous workspace with the strict dexterity performance is also derived.

4.1. Decomposition of the Reachable Workspace. The workspace can be defined by an envelope which is essentially of one or two types, either a manifold or a surface that is smooth almost everywhere, that is, smooth everywhere except for a set of points of measure zero in the Lebesgue sense [22]. In this paper, based on the influence factors for reachable workspace, the space manifold $\Sigma_{O_0\_P}$ corresponding to the reachable workspace of P is decomposed into a regional portion $\Sigma^{r}_{O_0\_P}$ and an orientation portion $\Sigma^{o}_{O_0\_P}$, and it is given as

$$\Sigma_{O_0\_P} = \Sigma^{r}_{O_0\_P} + \Sigma^{o}_{O_0\_P}, \tag{7}$$

where

$$\Sigma^{r}_{O_0\_P}: \begin{cases} x_{0-T} = \dfrac{X_1 + X_2}{2} + L_3 \cos\alpha \\ y_{0-T} = L_3 \sin\alpha \\ z_{0-T} = X_3 + L_{01} + \left(L_2 + L_4 + \sqrt{2}L_5\right) + e, \end{cases} \tag{8}$$

$$\Sigma^{o}_{O_0\_P}: \begin{cases} x_{0-R} = -\dfrac{\sqrt{2}}{2} L_P \sin(\alpha + \varphi_4)\sin\varphi_5 - \cos(\alpha + \varphi_4)\left(\dfrac{1 - \cos\varphi_5}{2} L_P + e\right) \\ y_{0-R} = \dfrac{\sqrt{2}}{2} L_P \cos(\alpha + \varphi_4)\sin\varphi_5 - \sin(\alpha + \varphi_4)\left(\dfrac{1 - \cos\varphi_5}{2} L_P + e\right) \\ z_{0-R} = \dfrac{1 + \cos\varphi_5}{2} L_P. \end{cases} \tag{9}$$

According to the expressions of the regional portion and the orientation portion, it can be found that the regional workspace volume of the reachable workspace is mostly provided by the former, and a trivial or even negligible share is provided by the latter. Meanwhile, the orientation workspace volume is entirely provided by the latter. Hence, the dexterity of an arbitrary operation point in the reachable workspace is completely determined by the latter.

The envelope surface of $\Sigma^{o}_{O_0\_P}$ is obtained by eliminating the parameter variables α, φ_4, and φ_5 from its expression. r_1 and z_{0-R} are defined as the radius and the axis of the circular cross section for $\Sigma^{o}_{O_0\_P}$, respectively. Q_1 is defined as the geometric center of the envelope surface $\Sigma^{o}_{O_0\_P}$ and the center of the maximum circular cross-section. $r_{1-\max}$ and h_1 are defined as the radius of the maximum circular cross-section and the height of the envelope surface in the z_{0-R} direction, respectively. The expression of the envelope circular cross-section in the z_{0-R} direction is given as

$$r_1^2 = x_{0-R}^2 + y_{0-R}^2 = (L_P + e)^2 - (z_{0-R} + e)^2 + e^2,$$
$$z_{0-R} \in [0, |L_P|] \ (L_P \geq 0) \ \text{or} \ z_{0-R} \in [-|L_P|, 0] \ (L_P < 0), \tag{10}$$

where $|L_P|$ represents the absolute value of the parameter L_P and the value of parameter e is nonnegative.

To illustrate the envelope surface shape of $\Sigma^{o}_{O_0\_P}$, an example of $L_P = -70$ mm and $e = 30$ mm is given. $r_{1-\max} = 50$ mm, $h_1 = 70$ mm, and $Q_1 = [0, 0, -30]^T$ are derived. The shape of this envelope surface is shown in Figure 3.

4.2. Dexterity Analysis of an Operation Point P. Several conventions for the dexterity properties are made. A reference operation point P_1 within the reachable workspace is called a strict dexterity point with the strict dexterity when the end-effector can reach P_1 in any arbitrary orientation. Likewise, P_2 is called a finite dexterity point with the finite dexterity when the end-effector can only reach P_2 in a certain range of orientation. A workspace is called a strictly dexterous workspace if any arbitrary point within this workspace is the strict dexterity point.

In order to analyze the kinematic dexterity characteristics of an operation point and the dexterity features of the reachable workspace, an arbitrary operation point P is selected on the envelope surface of the reachable workspace. According to the envelope surface and the shape of $\Sigma^{o}_{O_0\_P}$, the center of the circular cross section at $z_{0-R} = 0$ coincides with the origin O_5 of the tool reference frame. The distance between an arbitrary operation point P on $\Sigma^{o}_{O_0\_P}$ and O_5 is written as

$$R = \sqrt{x_{0-R}^2 + y_{0-R}^2 + z_{0-R}^2} = \sqrt{(L_P + e)^2 - 2ez_{0-R}},$$
$$z_{0-R} \in [0, |L_P|] \ (L_P \geq 0) \ \text{or} \ z_{0-R} \in [-|L_P|, 0] \ (L_P < 0). \tag{11}$$

Based on the concepts of mapping and envelope, a novel spatial mapping envelope surface is established. An arbitrary operation point P is selected on the envelope surface $\Sigma^{o}_{O_0\_P}$ of the reachable workspace. Then, rotating the end-effector at the selected point P in any arbitrarily reachable orientation, a novel spatial mapping envelope surface $\Sigma^{o}_{O_5}$ which consists of mapping points O_5 is mapped by P with the distance $R = \sqrt{(L_P + e)^2 - 2ez_{0-R}}$.

r_2 and z_{0-R} are defined as the radius and the axis of the circular cross section for the spatial mapping envelope, respectively. The circular cross section expression of the mapping envelope in the z_{0-R} direction is given as

$$r_2^2 = x_{O_5}^2 + y_{O_5}^2 = (L_P + e)^2 - (z_{0-R} + e)^2 + e^2,$$
$$z_{0-R} \in [0, |L_P|] \ (L_P \geq 0) \ \text{or} \ z_{0-R} \in [-|L_P|, 0] \ (L_P < 0). \tag{12}$$

To demonstrate the shape of a mapping envelop surface $\Sigma^{o}_{O_5}$ and the relationship between the mapping and the orientation, two examples are provided.

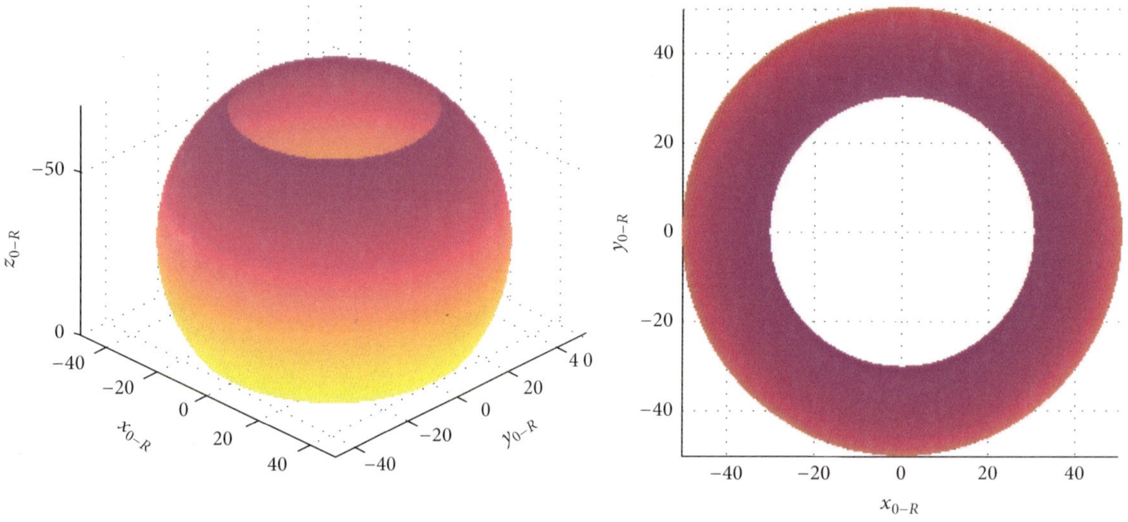

FIGURE 3: Shape of envelope surface.

Example 1 ($L_P = -70$ mm, $e = 30$ mm). According to the mapping relationship, a spatial mapping envelope surface is obtained, as shown in Figure 4(a). There exists a cavity region on $\Sigma^o_{O_5}$. An arbitrary mapping point O_5 which is located on $\Sigma^o_{O_5}$ can be mapped by P in an orientation. However, an arbitrary mapping point O_5 which is situated in the cavity region does not have a counterpart on $\Sigma^o_{O_5}$; that is, the corresponding orientations of the end-effector are unreachable. Hence, the condition of P with dexterity properties in a certain orientation is that the mapping point O_5 has a counterpart on $\Sigma^o_{O_5}$ in this orientation. The relationship between the mapping and the orientation is shown in Figure 4(b).

Example 2 ($L_P = -70$ mm, $e = 0$ mm). Similarly, another spatial mapping envelope surface is given, as shown in Figure 5(a). It is an entire and continuous hemisphere without a cavity region. An arbitrary mapping point O_5 which is located on $\Sigma^o_{O_5}$ can be mapped by P in an orientation; that is, the corresponding orientations of the end-effector are reachable. The relationship between the mapping and the orientation is shown in Figure 5(b).

Through the proposed method, by analyzing the mapping envelope features and mapping figures, the relationship between the manipulator and the performance evaluation and the relationship between the mapping and the orientation, with the character of graphical dexterity, are obviously demonstrated, respectively. Based on the proposed dexterity analysis method, the conditions that the arbitrary selected point is a strict dexterity point with the strict dexterity performance and that a dexterous workspace is a strict dexterous workspace with the strict dexterity performance are also derived, respectively.

(i) When a cavity region exists on $\Sigma^o_{O_5}$, some corresponding orientations of the end-effector are unreachable; P can only be reached by the end-effector within a certain range in an arbitrary orientation; that is, P is a finite dexterity point with finite dexterity. However, when the entire and continuous hemisphere $\Sigma^o_{O_5}$ without a cavity region is mapped, all corresponding orientations of the end-effector are reachable; P can be reached by the end-effector in any arbitrary orientation; that is, this is the necessary and sufficient condition that P is a strict dexterity point with the strict dexterity.

(ii) A strictly dexterous workspace exists if and only if an entire and continuous hemisphere $\Sigma^o_{O_5}$ without a cavity region can be mapped by any arbitrary point in this workspace; that is, any arbitrary point in this workspace is a strict dexterity point.

TABLE 2: The assumed manipulator structure parameters (mm).

L_1	L_2	L_3	L_4	L_5	L_6	e	L_t	L_{01}	s_x/s_z
420	50	450	180	180	80	0	225	400	10

5. Workspace Analysis

The manipulator workspace is classified into two components, that is, the reachable workspace and the dexterous workspace [23]. In this section, the reachable workspace corresponding to the selected point P for this 5-DOF manipulator is obtained; then the dexterous graphic workspace can be derived on the basis of the results of the dexterity analysis (these results are obtained by the proposed method) and the reachable workspace.

5.1. Depiction of the Reachable Workspace. The reachable workspace of this manipulator is obtained through MATLAB simulation based on the forward kinematics solution with the assumed structure parameters (Table 2). The reachable workspace of the selected point P is shown in Figure 6. The top view and the front view of Figure 6 show the range in

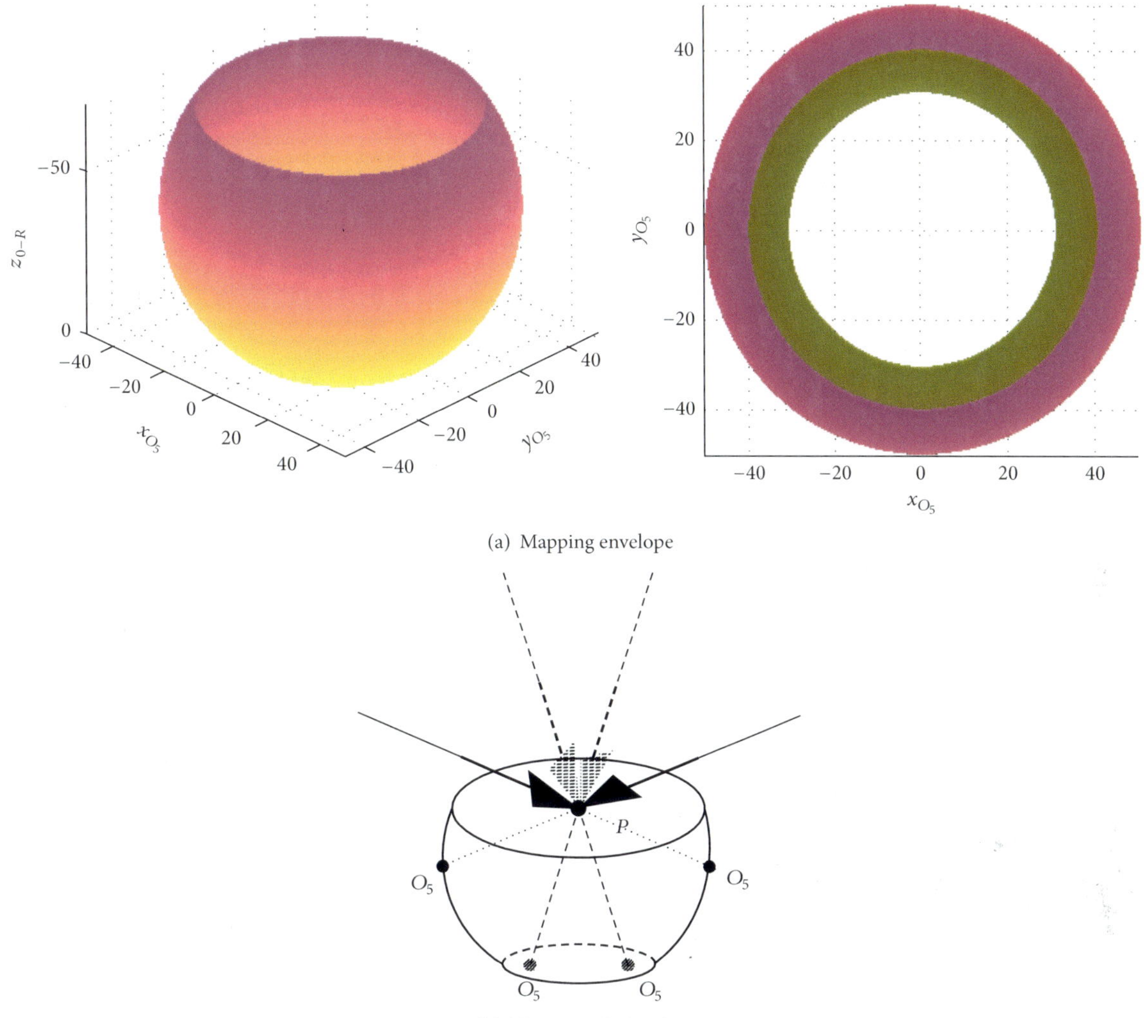

(a) Mapping envelope

(b) Mapping relationship

FIGURE 4: The example of $L_P = -70$ and $e = 30$.

the X, Y, and Z direction of this manipulator reachable workspace.

Note that the shapes of the left end and the right end (in the top view of Figure 6) are different. This is caused by the sway of the swing rod. The sway values are $L_3 \cos\alpha$ and $L_3 \sin\alpha$ in the X and Y directions, respectively. In the start phase, α increases from zero to the maximum attainable value, and the sway values in the X and Y directions gradually decrease and increase, respectively. In the end phase, α decreases from the maximum attainable value to zero, and the sway values in the X and Y directions gradually increase and decrease, respectively. The mutual variation relationship is shown in Figure 6(a).

5.2. Determination of the Dexterous Workspace. When a strictly dexterous workspace exists, the dexterous workspace is derived on the basis of the above-obtained reachable workspace and the results of the dexterity analysis. The process is carried out as the following steps.

The boundary of the reachable workspace of P is contracted towards the inner area with two times of $r_{1-\max}$ in the top view, as shown in Figure 7(a). Then, in the front view, the boundaries of the left and right edges and the boundaries of the upper and lower edges are contracted towards the inner area with two times of $r_{1-\max}$ and two times of h_1, respectively, as shown in Figure 7(b).

6. Conclusions

In this paper, a dexterity analysis method is proposed to address the dexterity analysis and the dexterity characteristics for a 5-DOF manipulator, through a geometric and topological approach with the character of graphic dexterity. Based on the proposed method, the relationship of the manipulator and the performance evaluation and the relationship of the mapping and the orientation are demonstrated obviously with the character of graphical dexterity, respectively. To summarize, the major contributions of this paper are as follows.

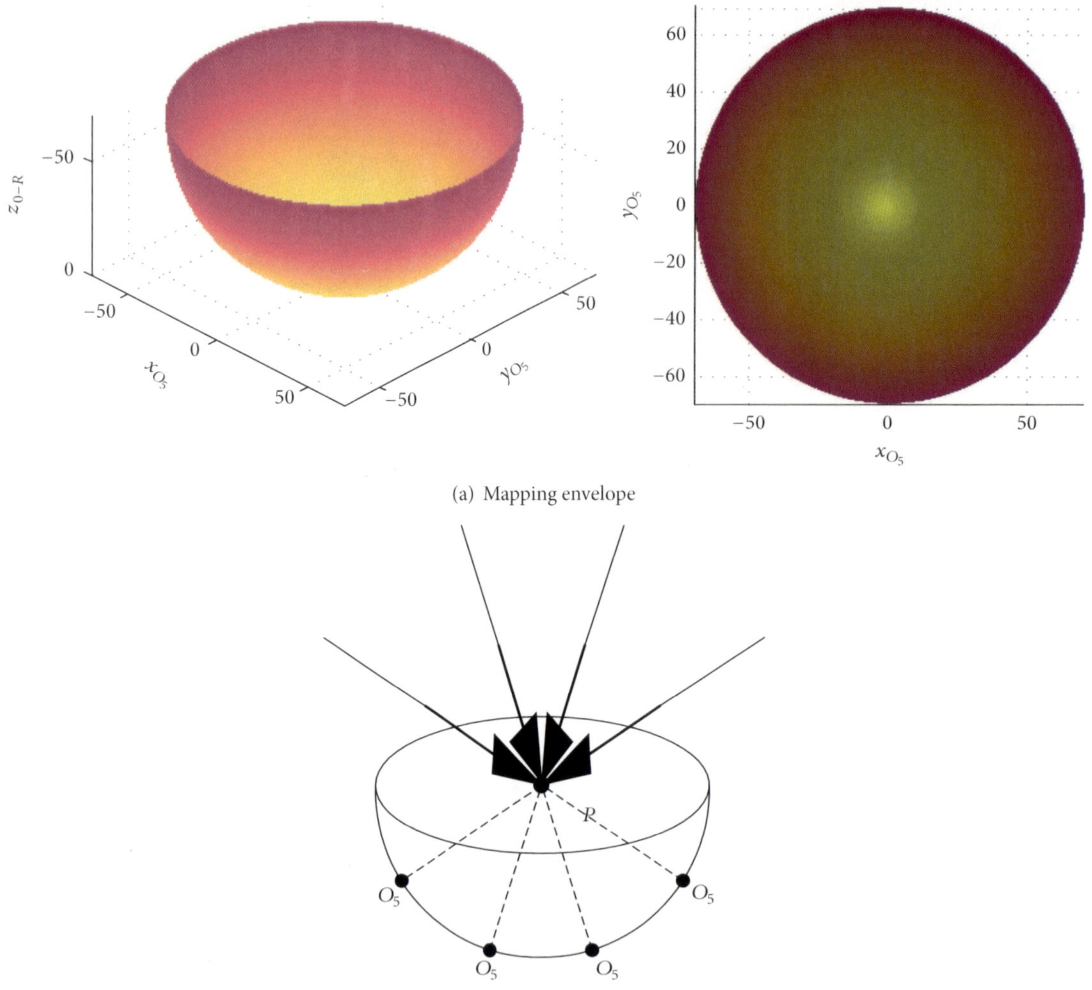

(a) Mapping envelope

(b) Mapping relationship

FIGURE 5: The example of $L_P = -70$ and $e = 0$.

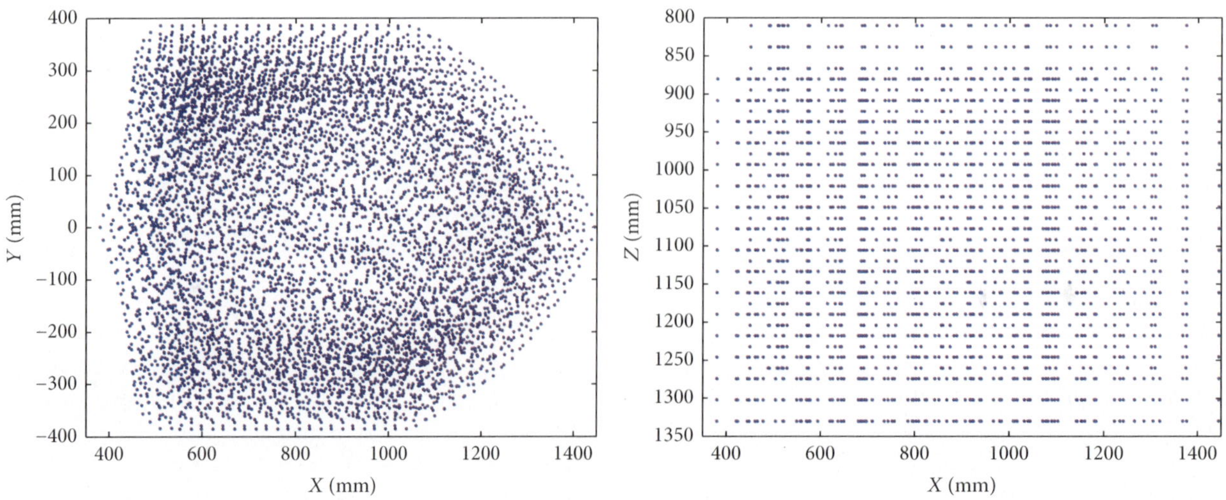

(a) Top view

(b) Front view

FIGURE 6: Reachable workspace.

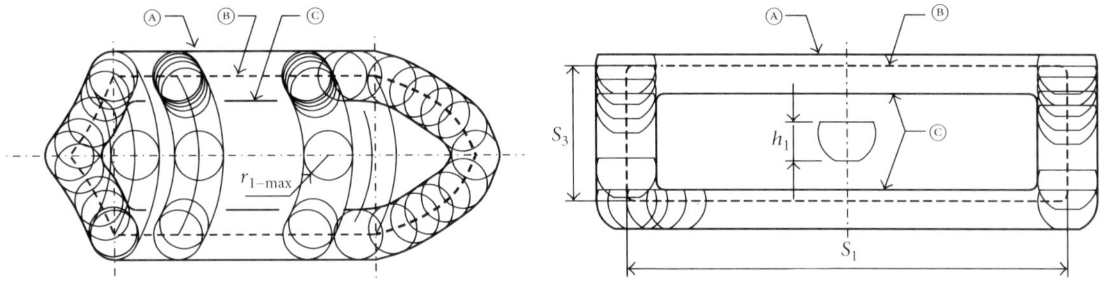

(a) Top view

(b) Front view

FIGURE 7: Projection of dexterous workspace of P. The symbols (A), (B), and (C) represent the boundary of the reachable workspace of P, the locus of the geometric center Q_1, and the boundary of the dexterous workspace of the selected point P, respectively.

(1) The structure of a 5-DOF manipulator and its forward kinematics model are derived. The reachable workspace of this manipulator is presented for the dexterity analysis and dexterous workspace determination.

(2) A novel method of the dexterity analysis with the character of graphic dexterity and a new concept of the spatial mapping envelope surface are proposed for this 5-DOF manipulator. The proposed method can be also applied to other manipulator structures for the dexterity analysis problem. The dexterity characteristics of an arbitrary selected point are given. Meanwhile, the conditions that the arbitrary selected point is a strict dexterity point with the strict dexterity performance and that a dexterous workspace is a strict dexterous workspace with the strict dexterity performance are derived, respectively.

(3) The dexterous graphic workspace of this 5-DOF manipulator is determined. It might serve as the base research of this manipulator for the structure optimization and motion control.

In future work, the structure optimization, dynamics problem, and motion control will be further considered.

Conflict of Interests

The authors declare that there is no conflict of interests regarding the publication of this paper.

Acknowledgment

This work was supported by the National High-Tech R&D Program (863 Program) of China (no. 2013AA040902-2).

References

[1] D. Lau, J. Eden, D. Oetomo, and S. K. Halgamuge, "Musculoskeletal static workspace analysis of the human shoulder as a cable-driven robot," *IEEE/ASME Transactions on Mechatronics*, vol. 20, no. 2, pp. 978–984, 2015.

[2] S. Seriani, M. Seriani, and P. Gallina, "Workspace optimization for a planar cable-suspended direct-driven robot," *Robotics and Computer-Integrated Manufacturing*, vol. 34, pp. 1–7, 2015.

[3] T. Yoshikawa, "Manipulability of robotic mechanisms," *The International Journal of Robotics Research*, vol. 4, no. 2, pp. 3–9, 1985.

[4] I. Mansouri and M. Ouali, "A new homogeneous manipulability measure of robot manipulators, based on power concept," *Mechatronics*, vol. 19, no. 6, pp. 927–944, 2009.

[5] I. Mansouri and M. Ouali, "The power manipulability—a new homogeneous performance index of robot manipulators," *Robotics and Computer-Integrated Manufacturing*, vol. 27, no. 2, pp. 434–449, 2011.

[6] I. Mansouri and M. Ouali, "Quaternion representation of the power manipulability," *Transactions of the Canadian Society for Mechanical Engineering*, vol. 35, no. 2, pp. 309–336, 2011.

[7] N. Vahrenkamp and T. Asfour, "Representing the robot's workspace through constrained manipulability analysis," *Autonomous Robots*, vol. 38, no. 1, pp. 17–30, 2015.

[8] J. J. Cervantes-Sánchez, J. M. Rico-Martínez, and V. H. Pérez-Muñoz, "Two natural dexterity indices for parallel manipulators: angularity and axiality," *Journal of Mechanisms and Robotics*, vol. 6, no. 4, Article ID 041007, 13 pages, 2014.

[9] C. M. Gosselin, "Dexterity indices for planar and spatial robotic manipulators," in *Proceedings of the IEEE International Conference on Robotics and Automation*, vol. 1, pp. 650–655, IEEE, Cincinnati, Ohio, USA, May 1990.

[10] R. G. Roberts, "The dexterity and singularities of an underactuated robot," *Journal of Robotic Systems*, vol. 18, no. 4, pp. 159–169, 2001.

[11] K. Tchón and K. Zadarnowska, "Kinematic dexterity of mobile manipulators: an endogenous configuration space approach," *Robotica*, vol. 21, no. 5, pp. 521–530, 2003.

[12] S. Bai, M. R. Hansen, and T. O. Andersen, "Modelling of a special class of spherical parallel manipulators with Euler parameters," *Robotica*, vol. 27, no. 2, pp. 161–170, 2009.

[13] H. Liu, T. Huang, and D. G. Chetwynd, "A method to formulate a dimensionally homogeneous jacobian of parallel manipulators," *IEEE Transactions on Robotics*, vol. 27, no. 1, pp. 150–156, 2011.

[14] K. Xu and X. Ding, "Gait analysis of a radial symmetrical hexapod robot based on parallel mechanisms," *Chinese Journal of Mechanical Engineering*, vol. 27, no. 5, pp. 867–879, 2014.

[15] B. Du, J. Zhao, and C. Song, "Dexterity analysis for omni-directional wheeled mobile manipulator based on double

quaternion," *Chinese Journal of Mechanical Engineering (English Edition)*, vol. 26, no. 3, pp. 585–593, 2013.

[16] C. Gosselin and J. Angeles, "A global performance index for the kinematic optimization of robotic manipulators," *Journal of Mechanical Design*, vol. 113, no. 3, pp. 220–226, 1991.

[17] F. C. Park and R. W. Brockett, "Kinematic dexterity of robotic mechanisms," *The International Journal of Robotics Research*, vol. 13, no. 1, pp. 1–15, 1994.

[18] Z. Gao and D. Zhang, "Performance analysis, mapping, and multiobjective optimization of a hybrid robotic machine tool," *IEEE Transactions on Industrial Electronics*, vol. 62, no. 1, pp. 423–433, 2015.

[19] W. J. Guo, R. F. Li, C. Q. Cao, and Y. F. Gao, "Kinematics analysis of a novel 5-DOF hybrid manipulator," *International Journal of Automation Technology*, vol. 9, no. 6, pp. 765–774, 2015.

[20] J. Denavit, "A kinematic notation for lower-pair mechanisms based on matrices," *Transactions of the ASME. Journal of Applied Mechanics*, vol. 22, pp. 215–221, 1955.

[21] B. Siciliano and O. Khatib, *Springer Handbook of Robotics*, Springer, Berlin, Germany, 2008.

[22] K. Hoffman, *Analysis in Euclidean Space*, Courier Dover Publications, Mineola, NY, USA, 2013.

[23] A. Kumar and K. J. Waldron, "The workspaces of a mechanical manipulator," *Journal of Mechanical Design*, vol. 103, no. 3, pp. 665–672, 1981.

19

Robot Obstacle Avoidance Learning based on Mixture Models

Huiwen Zhang,[1,2] Xiaoning Han,[1,2] Mingliang Fu,[1,2] and Weijia Zhou[1]

[1]*State Key Laboratory of Robotics, Shenyang Institute of Automation, Shenyang 110016, China*
[2]*University of Chinese Academy of Sciences, Beijing, China*

Correspondence should be addressed to Huiwen Zhang; zhanghuiwen@sia.cn

Academic Editor: Keigo Watanabe

We briefly surveyed the existing obstacle avoidance algorithms; then a new obstacle avoidance learning framework based on learning from demonstration (LfD) is proposed. The main idea is to imitate the obstacle avoidance mechanism of human beings, in which humans learn to make a decision based on the sensor information obtained by interacting with environment. Firstly, we endow robots with obstacle avoidance experience by teaching them to avoid obstacles in different situations. In this process, a lot of data are collected as a training set; then, to encode the training set data, which is equivalent to extracting the constraints of the task, Gaussian mixture model (GMM) is used. Secondly, a smooth obstacle-free path is generated by Gaussian mixture regression (GMR). Thirdly, a metric of imitation performance is constructed to derive a proper control policy. The proposed framework shows excellent generalization performance, which means that the robots can fulfill obstacle avoidance task efficiently in a dynamic environment. More importantly, the framework allows learning a wide variety of skills, such as grasp and manipulation work, which makes it possible to build a robot with versatile functions. Finally, simulation experiments are conducted on a Turtlebot robot to verify the validity of our algorithms.

1. Introduction

With the development of robot technology, especially the development of computer science and cognitive science, some promising technologies come out, such as deep learning and computer vision. These technologies greatly improve the intelligence level of robots and also laid the foundation for the modern intelligent robots. Unlike the traditional mature industrial robots, intelligent robots are expected to coexist with people, such as service robots. To achieve this purpose, robots are required to have strong environmental adaptability and autonomous decision-making ability. Obstacle avoidance and planning are basic skills of robots, which are also research focus in the field of robotics. There have been a lot of mature algorithms for obstacle avoidance, but how to develop obstacle avoidance algorithm with strong robustness, especially algorithms that can make robots fulfill task under unstructured environment, is still worthy of further research.

The obstacle avoidance algorithms can be classified into traditional obstacle avoidance algorithms and intelligent algorithms based on the developments history. In [1] the authors concluded them as classical approaches and heuristic approaches. In [2] the authors called them classical algorithms and evolutionary algorithms. Here the traditional approaches cover the same content, while the intelligent algorithms combine some machine learning methods applied to robot obstacle avoidance in recent years, which is different from the terms coined by other authors and covers a wider range. Classical obstacle avoidance algorithms can be categorized into three classes based on their main idea: (i) methods based on geometric model, (ii) methods based on virtual field, and (iii) methods based on mathematical programming. Details of classic obstacle avoidance algorithms can be found in [3]. Methods based on geometric models can be divided into two steps. First build the geometric model of the environment, and then choose the right kind of search algorithms. The main idea of virtual field methods is to assume that all robots stay in a virtual field, the change of virtual field reflects the feasible solution for the path. This concept is first put forward and used in [4, 5]. Mathematical programming methods represent constraints for obstacle as some parameterized inequities in configuration space, so a geometric problem is turned into a mathematical problem; the best obstacle-free path is a solution for the inequities which can be solved by optimization methods. All these methods have been successfully used, but the traditional method generally is only applicable to the static environment and easy to fall into a local optimal solution. To solve these problems, more and more intelligent approaches are used. By 2003, more than half of articles use heuristic algorithms to solve obstacle avoidance problem [1].

The core idea of heuristic algorithms is to simulate some intelligent behaviors of animals or people. Algorithms are designed by bionic mechanism. There are a large number of representative algorithms, such as Ant Colony Optimization (ACO) algorithm, Particle Swarm Optimization (PSO) Algorithm, and Genetic Algorithm (GA). In [6] ACO is used as a global path planning method to avoid obtaining a local optimal solution when using potential field method. In [7] distance and memory feature are introduced to speed up the computation of the ACO. The emergence of PSO derived from the simulation of birds migrating behavior. Applications of particle swarm optimization to solve the path planning problem include [8–10]. GA, derived from the theory of evolution and genetics, is a kind of global probabilistic optimization algorithm. In [11] a special encoding technology is used to speed up the path generation. The fitness function is designed according to security and distance relative to a target, and then a barrier-free path is solved through a genetic algorithm. In [12], a variable-length chromosome genetic algorithm is designed for path planning. In [13], the authors use artificial network to solve the obstacle avoidance problem of a redundant manipulator. An improved Grossberg neural network method is used in [14] to solve the UAV path planning under a "danger mode." Neural network can also realize the obstacle avoidance in conjunction with other methods, such as fuzzy neural network learning [15]. Among all the intelligent algorithms, there are broad categorical approaches which are based on the learning. A representative work is using reinforcement learning. Reinforcement learning derives from the psychology. The main idea is through constant interaction with the environment and "trial and error" to maximize the reward defined by a reward function [16]. Based on this concept, a lot of obstacle avoidance algorithms are proposed [17–22]. It is worth noting that, with a great improvement of hardware and the development of deep learning theory, some methods can directly learn obstacle avoidance strategy from the sensor data, called end-to-end learning [23]. Intelligent algorithms usually can obtain the global optimal solution and have good ability to adapt to the environment. But often these methods bear high computational complexity and bad real-time performance.

In this paper, we proposed a kind of obstacle avoidance method based on teaching and learning. First, learning sample is obtained by showing robots how to avoid the obstacle and then using probabilistic mixture model for extraction and encoding task constraints autonomously. Finally, obstacle-free path is generated by GMR. Our approach differs from previous work in the following: (i) it is more in line with the habit of human cognition, drawing lessons from imitation learning and (ii) the algorithm has good generalization ability and can adapt to dynamic unstructured environment. Besides, the computational complexity is low, and in particular the path generation process can be realized online efficiently; (iii) the learning framework not only can be used to learn obstacle avoidance but also can be used for learning other skills, which is a key point to achieve generalized artificial intelligence (GAI).

The main contributions of the paper are as follows: (i) the first contribution is using LfD for the obstacle avoidance task for the first time, making up for the disadvantages of other methods; (ii) due to the nature of LfD, nontechnical people can programme the robots to realize obstacle avoidance function; (iii) obstacle avoidance learning using mixture models is able to adapt the environmental changes and dynamic obstacles and at the same time shares low computational complexity.

The remainder of the paper is organized as follows. In Section 2, some background knowledge is introduced, including the concepts of LfD and GMM. Section 3 introduces parameters solving methods for GMM. Section 4 gives a detailed introduction of the obstacle avoidance learning architecture. In Section 5, simulation experiments based on a Turtlebot robot are conducted. In the end, conclusion is made and further improvements are introduced.

2. Learning from Demonstration

LfD is a paradigm which allows robots to automatically acquire new skills. Unlike the traditional robots, to fulfill a task, a previous task decomposition work and programming for each action need to be done by humans. LfD adopts the point that robot's skills can be acquired by observing the human operation. This approach has the advantages of simple programming and even people who do not understand programming technique can programme robots through teaching. Besides, due to the nature of LfD, robots are more advantageous to adapt to new environment. LfD is also named robot Programming by Demonstration (PbD) (Imitation Learning and Apprenticeship Learning), Learning by Demonstration (LbD), Learning by Showing, Learning by Watching, Learning from Observation, and so forth. It is worth noting that some authors think that imitation learning and LfD have different meanings; in [24], the authors distinguish imitation learning and LfD according to how learning data is recorded and how the imitator uses the data in the learning process.

2.1. Formulation of LfD Problem. Solving process for LfD can be divided into two steps. First, acquire training data; second, use the obtained data to derive a control policy. We use S and A to represent state set and action set, respectively. The probability distribution function $T(s' \mid s, a) : S \times A \times S \rightarrow [0, 1]$ represents the transfer probability from state (s, a) to state s'. An experience dataset D should include this information. Actually, in most situations, the true states S are unknown, the observable states are Z, and then a learning problem can be defined as deriving a policy π mapping form Z to A. For an obstacle avoidance task, the problem can

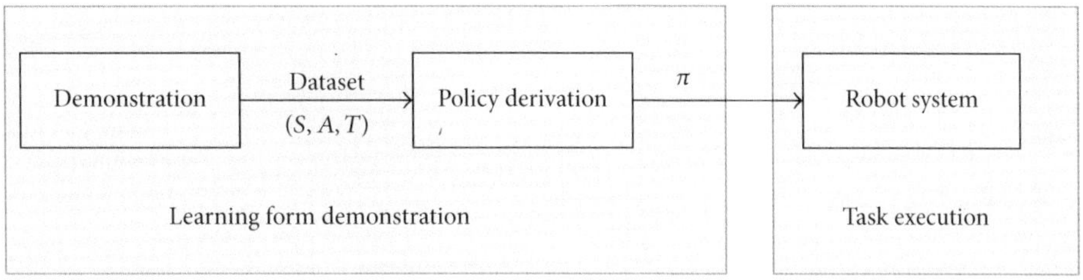

FIGURE 1: Framework of learning form demonstration (LfD).

be described as finding the more general barrier-free path (policy) from the demonstrated barrier-free path (sample D). The main process for LfD framework is illustrated in Figure 1.

2.2. Main Research Topics in LfD. Nehaniv and Dautenhahn [25] summarized the main problems in LfD as four issues: (1) What to imitate? (2) How to imitate? (3) When to imitate? (4) Whom to imitate? So far, only the former two problems are solved. For the problem of robot obstacle avoidance, what to imitate means the robot needs to extract the right constraints which can characterize the task, and problem of how to imitate refers to generating a trajectory that satisfies the constraints obtained in the previous step.

In order to solve the above problems (mainly the former two problems), there are three main ideas:

(i) Directly get a map function $\pi = f(z)$ from training set $D = \{(z^i, a^i)\}$, where z stands for the observed state. This idea treats the problem as supervised learning problem.

(ii) Learn a system model, namely, the return function and state transition probability distribution, form training set, and then derive a policy using MDP methods.

(iii) Learn the pre- or postcondition of each action, and then obtain control sequence using programming methods.

For the obstacle avoidance problem, through learning the barrier-free path from teachers to produce a generalized path can be seen as a supervised regression problem. GMM is a probabilistic model which is suitable to model the natural variations in human and robot motions. Theoretically, GMM can model any probabilistic density functions (PDFs) by increasing the number of Gaussian components, so GMM has strong encoding capacity. Finally, movements recognizing, predicting, and generating can be integrated together within the same GMM and GMR encoding strategy. So it is very favorable for us to use GMM encoding our prior demonstration information and use GMR as the solution to the regression problem mentioned above. Details of GMM theory are described below.

3. Gaussian Mixture Model

For a probability density estimation problem, if, in addition to the known observation variables, there may exist some unobservable latent variables, but we know the distribution of the latent variable and the conditional distribution of the observable variable given the latent variable, then we can use mixture models to estimate the parameters of the distribution, so as to get the approximate probability distribution of the known data. Before introducing GMM, we deviate to talk about Gaussian distribution.

3.1. Marginal and Conditional Distribution of a Multivariate Gaussian Distribution. Assume that we have a random variable

$$x = \begin{bmatrix} x_1 \\ x_2 \end{bmatrix}, \quad (1)$$

in which $x_1 \in \mathbb{R}^r$, $x_2 \in \mathbb{R}^s$, $x \in \mathbb{R}^{r+s}$. $x \sim \mathcal{N}(\mu, \Sigma)$, where

$$\mu = \begin{bmatrix} \mu_1 \\ \mu_2 \end{bmatrix}, \qquad \Sigma = \begin{bmatrix} \Sigma_{11} & \Sigma_{12} \\ \Sigma_{21} & \Sigma_{22} \end{bmatrix}. \quad (2)$$

And $\mu_1 \in \mathbb{R}^r$, $\mu_2 \in \mathbb{R}^s$, $\mu \in \mathbb{R}^{r+s}$, $\Sigma_{11} \in \mathbb{R}^{r\times r}$, and $\Sigma_{12} \in \mathbb{R}^{r\times s}$. Noting that the covariance matrix is symmetric, so we have $\Sigma_{12} = \Sigma_{21}{}^T$.

Because the marginal distribution of a Gaussian is also a Gaussian distribution, it is easy to get that $E[x_1] = \mu_1$, $\mathrm{cov}(x_1) = \Sigma_{11}$, $E[x_2] = \mu_2$, and $\mathrm{cov}(x_2) = \Sigma_{22}$, and we have $x_1 \sim \mathcal{N}(\mu_1, \Sigma_{11})$ and $x_2 \sim \mathcal{N}(\mu_2, \Sigma_{22})$. As a step forward, we can prove $x_1 \mid x_2 \sim \mathcal{N}(\mu_{1|2}, \Sigma_{1|2})$, where

$$\mu_{1|2} = \mu_1 + \Sigma_{12}\Sigma_{22}^{-1}(x_2 - \mu_2), \qquad \Sigma_{1|2} = \Sigma_{11} - \Sigma_{12}\Sigma_{22}^{-1}\Sigma_{21}. \quad (3)$$

3.2. Formulation of GMM. Assuming that we have a training set $D = \{x^{(1)}, x^{(2)}, \ldots, x^{(m)}\}$ containing m samples, every sample $x^{(i)}$ is a vector. We are trying to find a density function to describe the data with the least error, which is equal to maximizing the probability of the data. In order to represent this probability, we need to make some assumptions. First we assume that the latent variable $z^{(i)} \sim \text{Multinomial}(\phi)$, in which $\phi_j \geq 0$, $\sum_{j=1}^{k} \phi_j = 1$; besides we suppose $x^{(i)} \mid z^{(i)} = j \sim \mathcal{N}(\mu_j, \Sigma_j)$. So we can model the data in our hands by

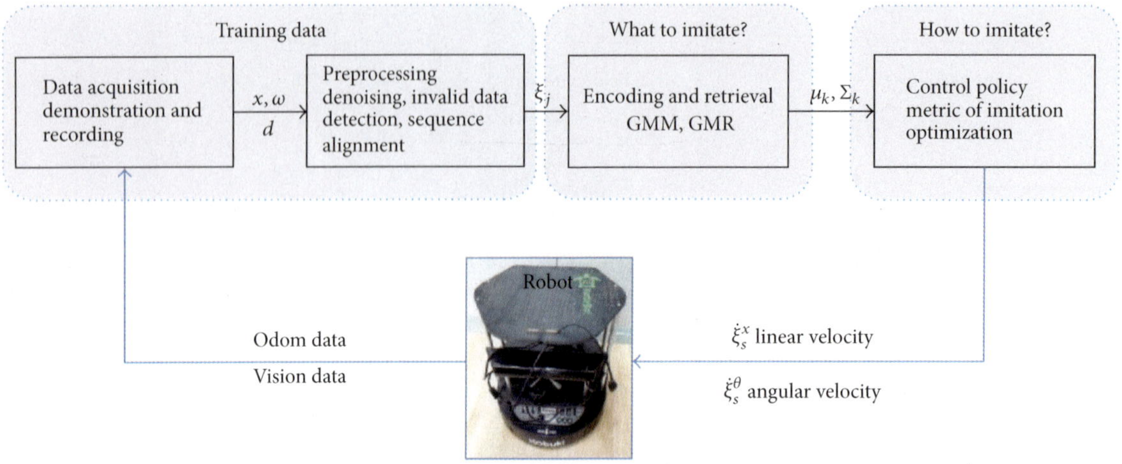

FIGURE 2: System architecture of obstacle avoidance learning.

adjusting parameters (ϕ, μ, Σ). The likelihood function can be written as

$$\begin{aligned} l(\phi, \mu, \Sigma) &= \sum_{i=1}^{m} \log p\left(x^{(i)}; \phi, \mu, \Sigma\right) \\ &= \sum_{i=1}^{m} \log \sum_{z^{(i)}=1}^{k} p\left(x^{(i)} \mid z^{(i)}; \mu, \Sigma\right) p\left(z^{(i)}; \phi\right). \end{aligned} \tag{4}$$

Maximize (4), and then we can find the right model parameters. To realize this purpose, typically we find the partial derivatives for each of the parameters and set them equal to zero, but actually we find that we cannot get a closed-loop solution for this problem using the above method. The main reason is that z is unknown; if z is known, the model degrades to a Gaussian discriminant model; following the previously mentioned steps, we can obtain a closed-loop solution. We will introduce a kind of commonly used iterative algorithm to solve the problem in (4).

3.3. EM for GMM. Expectation Maximization (EM) is an iterative algorithm, mainly including two steps, which are called E-step and M-step. The main idea is to find a tight lower bound function of the objective function and then maximize the lower bound function so that the objective function can be maximized indirectly. The reason for introducing a lower bound function is that finding the extreme value of the lower bound function is much easier compared to the original function. For GMM, the E-step tries to guess the value of $z^{(i)}$, equivalent to finding out which components the current sample belongs to, and then the model parameters are updated based on our conjecture. The algorithm can be summarized as follows.

E-step: for each i and j, compute the poster-probability of $z^{(i)}$:

$$w_j^i := p\left(z^{(i)} = j \mid x^{(i)}; \phi, \mu, \Sigma\right). \tag{5}$$

M-step: modify our model parameters:

$$\begin{aligned} \phi_j &:= \frac{1}{m} \sum_{i=1}^{m} w_j^{(i)}, \\ \mu_j &:= \frac{\sum_{i=1}^{m} w_j^{(i)} x^{(i)}}{\sum_{i=1}^{m} w_j^{(i)}}, \\ \Sigma_j &:= \frac{\sum_{i=1}^{m} w_j^{(i)} \left(x^{(i)} - \mu_j\right)\left(x^{(i)} - \mu_j\right)^T}{\sum_{i=1}^{m} w_j^{(i)}}. \end{aligned} \tag{6}$$

Note that the value of w_j^i can be computed using the following equation:

$$\begin{aligned} w_j^i &:= p\left(z^{(i)} = j \mid x^{(i)}; \phi, \mu, \Sigma\right) \\ &= \frac{p\left(x^{(i)} \mid z^{(i)} = j; \mu, \Sigma\right) p\left(z^{(i)} = j; \phi\right)}{\sum_{l=1}^{k} p\left(x^{(i)} \mid z^{(i)} = l; \mu, \Sigma\right) p\left(z^{(i)} = l; \phi\right)}, \end{aligned} \tag{7}$$

iterating between (6) and (7) until convergence, which means that the changes of model parameters are less than a threshold ϵ or the value of likelihood function changes in a very narrow range.

4. The Architecture of Obstacle Avoidance Learning

Figure 2 shows the architecture of obstacle avoidance learning and the information flow between different modules. The whole system contains three modules: data acquisition and preprocessing, constraints extraction and trajectory retrieval, and optimal control strategy derivation. Mileage information is collected by built-in sensors, such as odometers, and distance information is acquired by Kinect. Through noise filtering, invalid data reduction, and sequence alignment for the original data, we get the input for the learning machine. Then, the input signal is encoded by GMM and a general

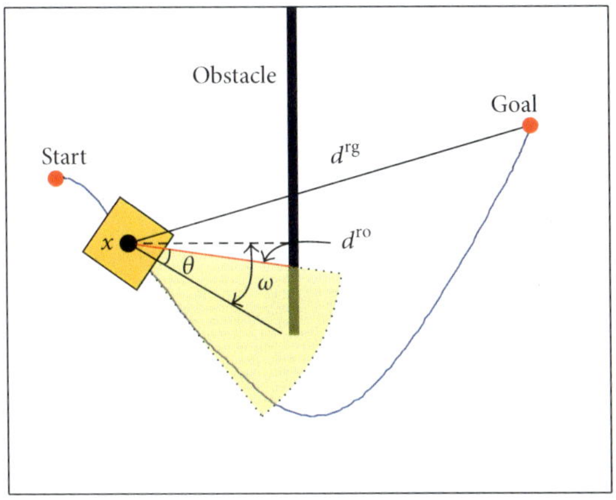

FIGURE 3: Useful data collected by sensors. Line in blue shows the demonstrated trajectory. Sectors with a fan shape in shallow yellow color represent the visual field of the Kinect. $\{x, \omega, d^{ro}, d^{rg}\}$ represent location, orientation, and distance information of the robot, respectively.

trajectory is generated by GMR. The general trajectory satisfies all the constraints for the task. Eventually, a controller is designed to follow the expected input trajectory and output the control commands to the robot. Subsequent sections will describe each part in detail.

4.1. Data Acquisition. The experimental data is acquired form a Turtlebot robot. Turtlebot is a widely used research platform which is compatible with ROS, integrating a variety of sensors, such as odometer and extensible vision sensors. We can develop many kinds of applications by using Turtlebot, such as mapping and navigation. In this paper, we will implement our obstacle avoidance algorithm on the bases of ROS development kit for Turtlebot. The teaching process can be achieved by three ways: (i) using a remote joystick; (ii) with the aid of Turtlebot development kit for the navigation algorithms, simulating some trajectories as training samples; and (iii) by means of the following apps implemented on Turtlebot, the teacher can guide robot to avoid obstacles and record sensor data as samples. In this paper, we adopt the second method, which means that we create simulated environment in software and then make robot execute multiple obstacle avoidance tasks in the simulated environment using built-in navigation algorithms. During this process, linear velocity and angular velocity are obtained by built-in sensors like odometer. Using Kinect we can get the distance between the robot and obstacles, as well as distance from the robot to the target.

We assume that the important sensing information comes from: (i) states of the robot, defined as linear velocity and angular velocity; (ii) the robot's absolute location; (iii) distance between the robot and obstacles, noting that the Kinect returns many pieces of distance information representing the relative position of the obstacle in a specific range, and we choose the shortest distance as our reference value; and (iv) distance between the robot and the target, which can be computed indirectly. As illustrated in Figure 3, we use $(x, \omega, d^{ro}, d^{rg})$ to represent the sensor information mentioned above, respectively. Here the location information and velocity information are redundant. We think it is necessary to take both kinds of information into consideration for the reason that location information is very important and direct for an obstacle avoidance task. In addition, choosing variables relating to the task as many as possible is generally very necessary in the initial stage of a task.

In the experiment, demonstration is conducted n times. In each demonstration, T data points are recorded, so the dataset contains $N = n \times T$ points in total. Each sample point is a collection of variables, represented as $D = \{x, \omega, d\}$, in which x stands for the location of robot, ω stands for the current orientation, and d stands for distance. Each variable contains a one-dimensional temporal value and $(d - 1)$-dimensional spatial values; that is, $x = \{x_s, x_t\}$, $\omega = \{\omega_s, \omega_t\}$, $d^{ro} = \{d_s^{ro}, d_t^{ro}\}$, and $d^{rg} = \{d_s^{rg}, d_t^{rg}\}$. $x_s \in \mathbb{R}^{N\times 2}$ represent location, $\omega \in \mathbb{R}$ represent orientation, $d_s^{ro} \in \mathbb{R}^{N\times 2}$ represent distance between robot and obstacles, and $d_s^{rg} \in \mathbb{R}^{N\times 2}$ represent distance between robot and goal state. Supposing coordinates of the goal are $o_{s,i}$, then we have

$$d_s^{rg} = x_{s,i,j} - o_{s,i}, \quad \forall i \in \{1, \ldots, n\}, \ \forall j \in \{1, \ldots, T\}, \tag{8}$$

where i indicates the trajectory number and j indicates time in a trajectory. We can rewrite the above equation for velocities:

$$\dot{d}_s^{rg} = \dot{x}_{s,i,j} - \dot{o}_{s,i}. \tag{9}$$

4.2. Data Preprocessing. Because of noise and bad demonstration quality, there exist serious mismatch and invalid data in the original data. As shown in Figure 4, there is no change at the beginning and end parts of these curves, so this part of data is useless. To reduce computation cost, it is a wise choice to abandon these data. In addition, each trajectory has different length; we should resample them to a uniform length so that we can handle them easier in the subsequent steps. So before inputting the data to our algorithms, preprocessing steps including noise filter, useless data reduction, and resampling are conducted. Figure 4 displays differences between the original data and the processed data.

4.3. Trajectory Encoding and Retrieval. Considering a datapoint in training set $\xi_j = \{\xi_{t,j}, \xi_{s,j}\}_{j=1}^{N} \in D$, ξ_j is a D-dimensional vector. We encode the data with GMM, supposing we have K Gaussian components. The probability items in (4) can be represented as

$$\begin{aligned} p(k) &= \pi_k, \\ p(\xi_j \mid k) &= \mathcal{N}(\xi_j; \mu_k, \Sigma_k) \\ &= \frac{1}{\sqrt{(2\pi)^D |\Sigma_k|}} e^{-(1/2)((\xi_j-\mu_k)^T \Sigma_k^{-1}(\xi_j-\mu_k))}, \end{aligned} \tag{10}$$

in which $\{\pi_k, \mu_k, \Sigma_k\}$ are model parameters, representing prior-probability, mean value, and variance, respectively.

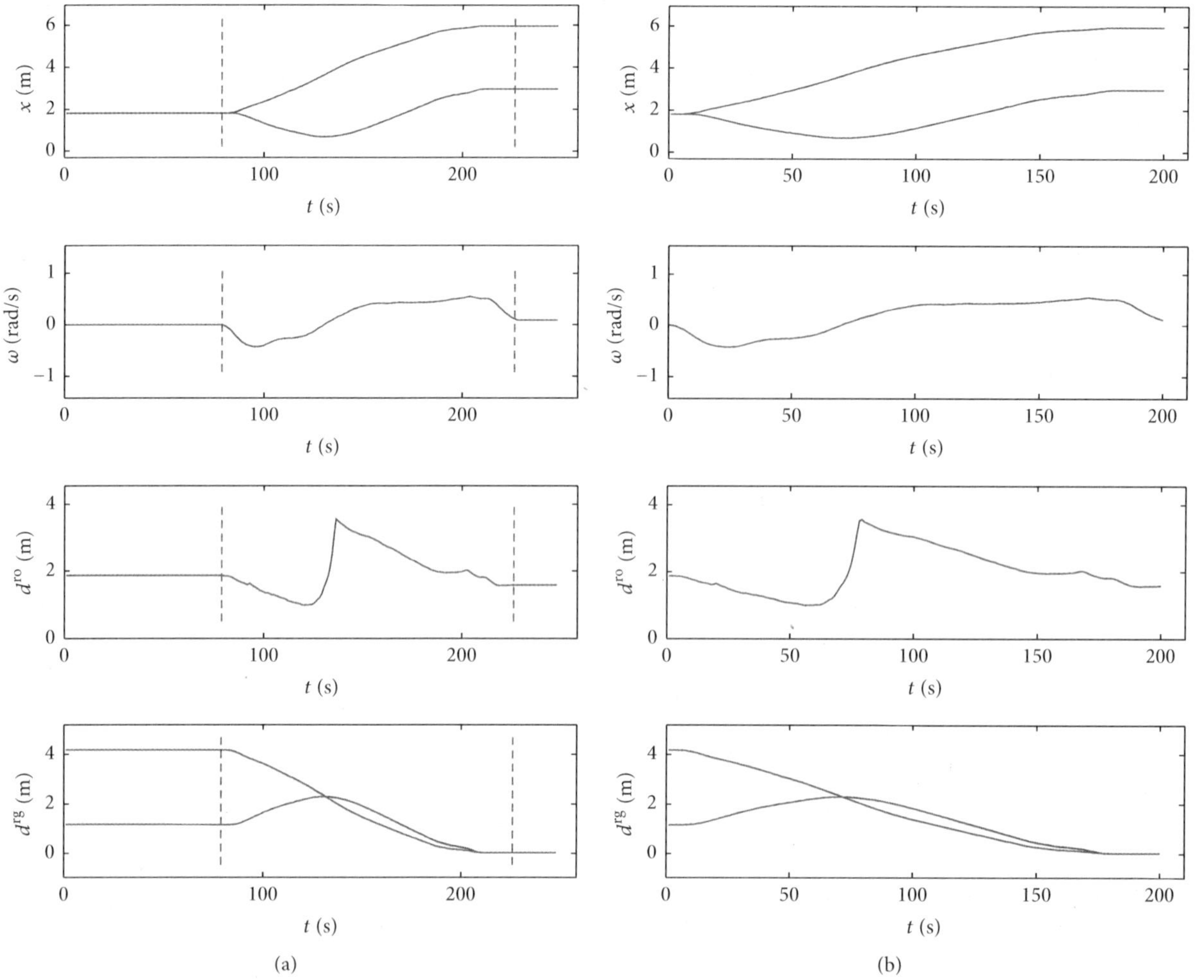

FIGURE 4: Comparison of original data and processed data. (a) shows the original data. Lines in red represent cutting position. (b) displays valid data after processing.

To solve these parameters, the EM algorithm introduced in the previous section can be used. When EM algorithm converges to a given value, we get the right $\{\pi_k, \mu_k, \Sigma_k\}$; that is, we encode the trajectory data successfully.

In order to get a general barrier-free trajectory, which is given an input, which can be temporal signals or other kinds of sensor data, then a barrier-free path is generated according to the given input, a natural idea is to divide the variable ξ into two parts $(\xi_t^{\mathcal{I}}, \xi_t^{\mathcal{O}})$, which represents input and output, respectively, and then trajectory retrieval problem can be formulized as follows: given $\xi \sim \sum_{i=1}^{K} \pi_i \mathcal{N}(\mu_i, \Sigma_i)$ and $\xi_t^{\mathcal{I}}$, solve the density probability distribution for $\xi_t^{\mathcal{O}}$. Solution to this problem is called Gaussian mixture regression. Regarding $\xi_t^{\mathcal{I}}$ as query points, $\xi_t^{\mathcal{O}}$ can be estimated using regression methods. In this paper, we use superscripts $\mathcal{I}$ and $\mathcal{O}$ standing for the input and output part for each variable. With this notation, a block decomposition of the data-point ξ_t, vectors μ_k, and matrices Σ_t can be written as

$$\xi_t = \begin{bmatrix} \xi_k^{\mathcal{I}} \\ \xi_k^{\mathcal{O}} \end{bmatrix},$$

$$\mu_k = \begin{bmatrix} \mu_k^{\mathcal{I}} \\ \mu_k^{\mathcal{O}} \end{bmatrix},$$

$$\Sigma_k = \begin{bmatrix} \Sigma_k^{\mathcal{I}} & \Sigma_k^{\mathcal{I}\mathcal{O}} \\ \Sigma_k^{\mathcal{O}\mathcal{I}} & \Sigma_k^{\mathcal{O}} \end{bmatrix}. \quad (11)$$

Based on (3), for a time step t, the mean vector and covariance matrix for the kth component are

$$\begin{aligned} \hat{\mu}_k^{\mathcal{O}}\left(\xi_t^{\mathcal{I}}\right) &= \mu_k^{\mathcal{O}} + \Sigma_k^{\mathcal{O}\mathcal{I}} \Sigma_k^{\mathcal{I}-1} \left(\xi_t^{\mathcal{I}} - \mu_k^{\mathcal{I}}\right), \\ \hat{\Sigma}_k^{\mathcal{O}} &= \Sigma_k^{\mathcal{O}} - \Sigma_k^{\mathcal{O}\mathcal{I}} \Sigma_k^{\mathcal{I}-1} \Sigma_k^{\mathcal{I}\mathcal{O}}. \end{aligned} \quad (12)$$

Taking K components into consideration, the mean vector and covariance matrix for distribution $\xi_t^{\mathcal{O}}$ given $\xi_t^{\mathcal{I}}$ are

$$P\left(\xi_t^{\mathcal{O}} \mid \xi_t^{\mathcal{I}}\right) \sim \sum_{k=1}^{K} h_k\left(\xi_t^{\mathcal{I}}\right) \mathcal{N}\left(\hat{\mu}_k^{\mathcal{O}}\left(\xi_t^{\mathcal{I}}\right), \hat{\Sigma}_k^{\mathcal{O}}\right), \quad (13)$$

where

$$h_k\left(\xi_t^{\mathcal{I}}\right) = \frac{\pi_i \mathcal{N}\left(\xi_t^{\mathcal{I}} \mid \mu_i^{\mathcal{I}}, \Sigma_i^{\mathcal{I}}\right)}{\sum_{k=1}^{K} \pi_k \mathcal{N}\left(\xi_t^{\mathcal{I}} \mid \mu_k^{\mathcal{I}}, \Sigma_k^{\mathcal{I}}\right)}. \quad (14)$$

Equation (13) represents a multimodal distribution, which can be approximated by the following Gaussian distribution:

$$P\left(\xi_t^{\mathcal{O}} \mid \xi_t^{\mathcal{I}}\right) = \mathcal{N}\left(\widehat{\mu}_t^{\mathcal{O}}, \widehat{\Sigma}_t^{\mathcal{O}}\right), \text{ with}$$

$$\begin{aligned} \mu_t^{\mathcal{O}} &= \sum_{i=1}^{K} h_i\left(\xi_t^{\mathcal{I}}\right) \mu_i^{\mathcal{O}}\left(\xi_t^{\mathcal{I}}\right) \\ \widehat{\Sigma}_t^{\mathcal{O}} &= \sum_{i=1}^{K} h_i\left(\xi_t^{\mathcal{I}}\right)\left(\widehat{\Sigma}_i^{\mathcal{O}} + \widehat{\mu}_i^{\mathcal{O}}\left(\xi_t^{\mathcal{I}}\right) \widehat{\mu}_i^{\mathcal{O}}\left(\xi_t^{\mathcal{I}}\right)^T\right) \\ &\quad - \widehat{\mu}_t^{\mathcal{O}} \widehat{\mu}_t^{\mathcal{O}T}. \end{aligned} \quad (15)$$

Substituting $\xi_t^{\mathcal{I}}$ into $\xi_t^{\mathcal{O}}$ using equations described above, a general trajectory is obtained. The trajectory not only can avoid obstacles but also can be executed by robots due to its good smooth property.

4.4. Finding an Optimum Control Policy. Through the previous section, we can obtain a general trajectory, and this is a planning problem. How to combine planning with control? In other words, how to find a set of control inputs that can make the robot follow the general trajectory accurately? For the time-based GMR, $\xi_t^{\mathcal{O}}$ is equal to ξ_s, and s represents spatial location information. Using $\widehat{\xi}_s$ represents the desired trajectory and ξ_s represents the actual reached location. In order to evaluate the tracking performance, a simple idea is using the weighted Euclidean distance:

$$\sum_{i=1}^{D-1} \omega_i \left(\xi_{s,i} - \widehat{\xi}_{s,i}\right)^2 = \left(\xi_s - \widehat{\xi}_s\right)^T W \left(\xi_s - \widehat{\xi}_s\right). \quad (16)$$

Naturally, the above equation can be applied to multiple variables. For this paper, we have four variables. Let $\{\widehat{\xi}^x, \widehat{\xi}^\omega, \widehat{\xi}^{d^{ro}}, \widehat{\xi}^{d^{rg}}\}$ represent the desired trajectory; the actual tracking performance P can be expressed as

$$\begin{aligned} P &= \left(\dot{\xi}_s^x - C_x\right)^T W^x \left(\dot{\xi}_s^x - C_x\right) \\ &\quad + \left(\dot{\xi}_s^\omega - C_\omega\right)^T W^\omega \left(\dot{\xi}_s^\omega - C_\omega\right) \\ &\quad + \left(\dot{\xi}_s^{d^{ro}} - C_{d^{ro}}\right)^T W^{d^{ro}} \left(\dot{\xi}_s^{d^{ro}} - C_{d^{ro}}\right) \\ &\quad + \left(\dot{\xi}_s^{d^{rg}} - C_{d^{rg}}\right)^T W^{d^{rg}} \left(\dot{\xi}_s^{d^{rg}} - C_{d^{rg}}\right), \end{aligned} \quad (17)$$

where

$$\begin{aligned} &C_{*,i,j} = \widehat{\xi}_{s,i,j}^{*} - \xi_{s,i,j-1}^{*}, \\ &\forall i \in \{1, \ldots, n\},\ \forall j \in \{1, \ldots, T\},\ \forall * \in \{x, \omega, d^{ro}, d^{rg}\}. \end{aligned} \quad (18)$$

According to performance indicator P, finding a desired control policy is equivalent to maximizing P. Since P is a quadratic function, an analytical solution can be easily obtained by gradient method. But noting that there exist constraints between variables, as shown in (9), maximizing (17) turns into a constrained optimization problem. Usually, this kind of problem can be solved by constructing a Lagrange function, defined as follows:

$$L\left(\dot{\xi}_s^x, \dot{\xi}_s^\omega, \dot{\xi}_s^{d^{ro}}, \dot{\xi}_s^{d^{rg}}, \lambda\right) = P + \lambda^T \left(\dot{\xi}_s^x - \dot{\xi}_s^{d^{rg}}\right), \quad (19)$$

where λ is Lagrange multiplier. Letting $\nabla L = 0$, solving partial derivatives for $\dot{\xi}_s^x, \dot{\xi}_s^{d^{rg}}$, we find

$$\begin{aligned} 2W^x \left(\dot{\xi}_s^x - C_x\right) + \lambda &= 0, \\ 2W^{d^{rg}} \left(\dot{\xi}_s^{d^{rg}} - C_{d^{rg}}\right) - \lambda &= 0. \end{aligned} \quad (20)$$

Adding (20), we find

$$2W^{d^{rg}} \left(\dot{\xi}_s^{d^{rg}} - C_{d^{rg}}\right) + 2W^x \left(\dot{\xi}_s^x - C_x\right) = 0. \quad (21)$$

Substituting $\dot{\xi}_s^x = -\dot{\xi}_s^{d^{rg}}$ into the above equation, we find

$$\dot{\xi}_s^x = \left(W^x - W^{d^{rg}}\right)^{-1} \left(W^x C_x + W^{d^{rg}} C_{d^{rg}}\right). \quad (22)$$

Weight matrices in (22) can be computed by solving the inverse matrix of the current covariance matrix; namely,

$$W = \left(\Sigma_t^{\mathcal{O}}\right)^{-1}. \quad (23)$$

5. Simulation Experiments

Experiments are implemented by using a Turtlebot robot. In addition to the robot's own hardware system, a Kinect is integrated to obtain distance information. The size of the map used in the simulation is 7.5 × 5 m. Two different maps are created to verify the learning effects. For the first map, the obstacle locates in the middle of the map, and a small gap lies in the lower side, and the robot can reach the target place through the gap. The second map is much similar to the first one, except that there exist two obstacles in the map. To get the destination, the robot must avoid every obstacle in the map. Figure 5 shows the shape of the map. Figures 5(a) and 5(b) represent the 2D map, and Figures 5(c) and 5(d) show a 3D simulation environment, which corresponds to the 2D maps. The color maps in maps (a) and (b) stand for cost cloud, which provides reference for path planning. The warm color, such as yellow and red, stands for a lower cost. To the contrary, the cool color tune stands for a higher cost. A path planer always tends to choose a trajectory with the least cost. The cost value is calculated by built-in functions in ROS package. For the convenience of description, we use task 1 and task 2 to represent the two different obstacle avoidance tasks in maps (a) and (b). It is worth mentioning that we select relatively simple environment for both tasks for the reason that

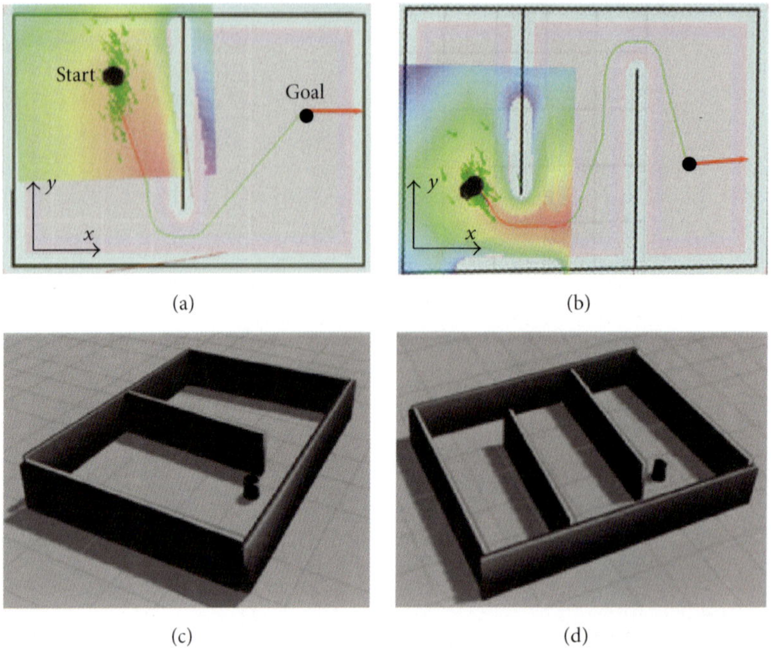

FIGURE 5: Two different maps used in our simulation. (a) and (b) are actual maps generated by software. (c) and (d) are simulation environment corresponding to maps above them.

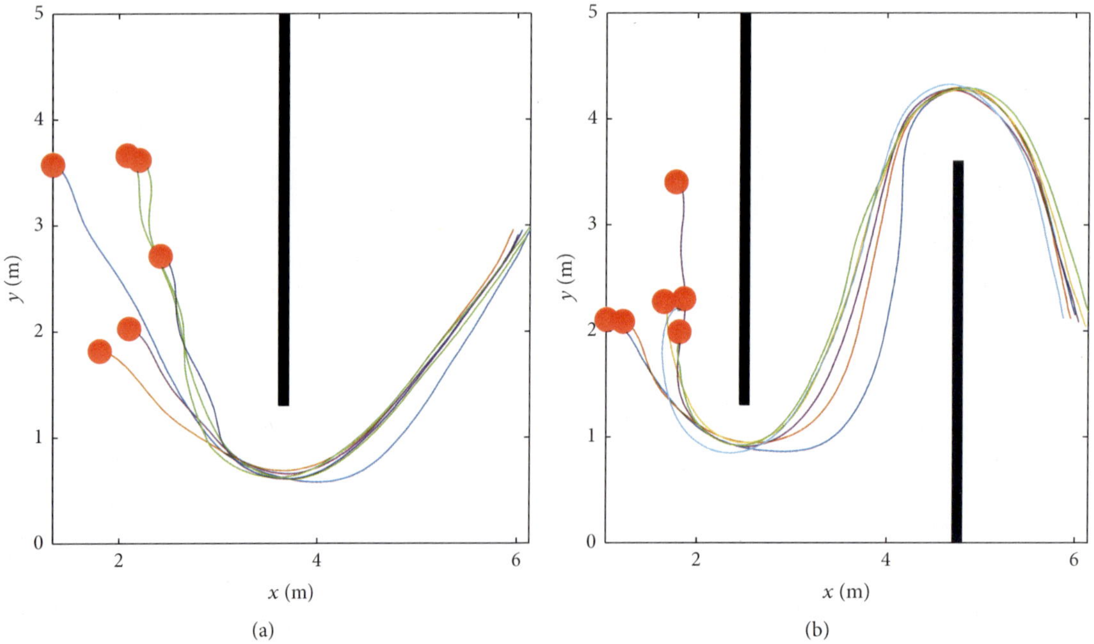

FIGURE 6: Demonstrated trajectories in two situations. (a) represents sampled data in task 1. (b) represents sampled data in task 2.

we put more emphasis on the mechanism behind the obstacle avoidance, which means we can learn skills from simple environment, but these skills can be reused in complex situations.

Figure 6 shows the demonstrated trajectories in two different situations. Coordinates of goal states are set to $(6, 3)$ and $(6, 2)$ for task 1 and task 2, respectively. The start points for task 1 are randomly generated by sampling from uniform distributions between 0.5 and 2.5 for X-coordinates and 2 to 4 for Y-coordinates. The maximum range is 2 m, which can guarantee a good generalization performance for obstacle avoidance. For task 2, the start points are generated by sampling from uniform distributions from 0.5 to 2 and 2 to 3.5.

5.1. Experiments for Trajectory Encoding and Retrieval. Figure 7 shows the encoding and retrieval results for task 1. The encoded variables are location information x and distance

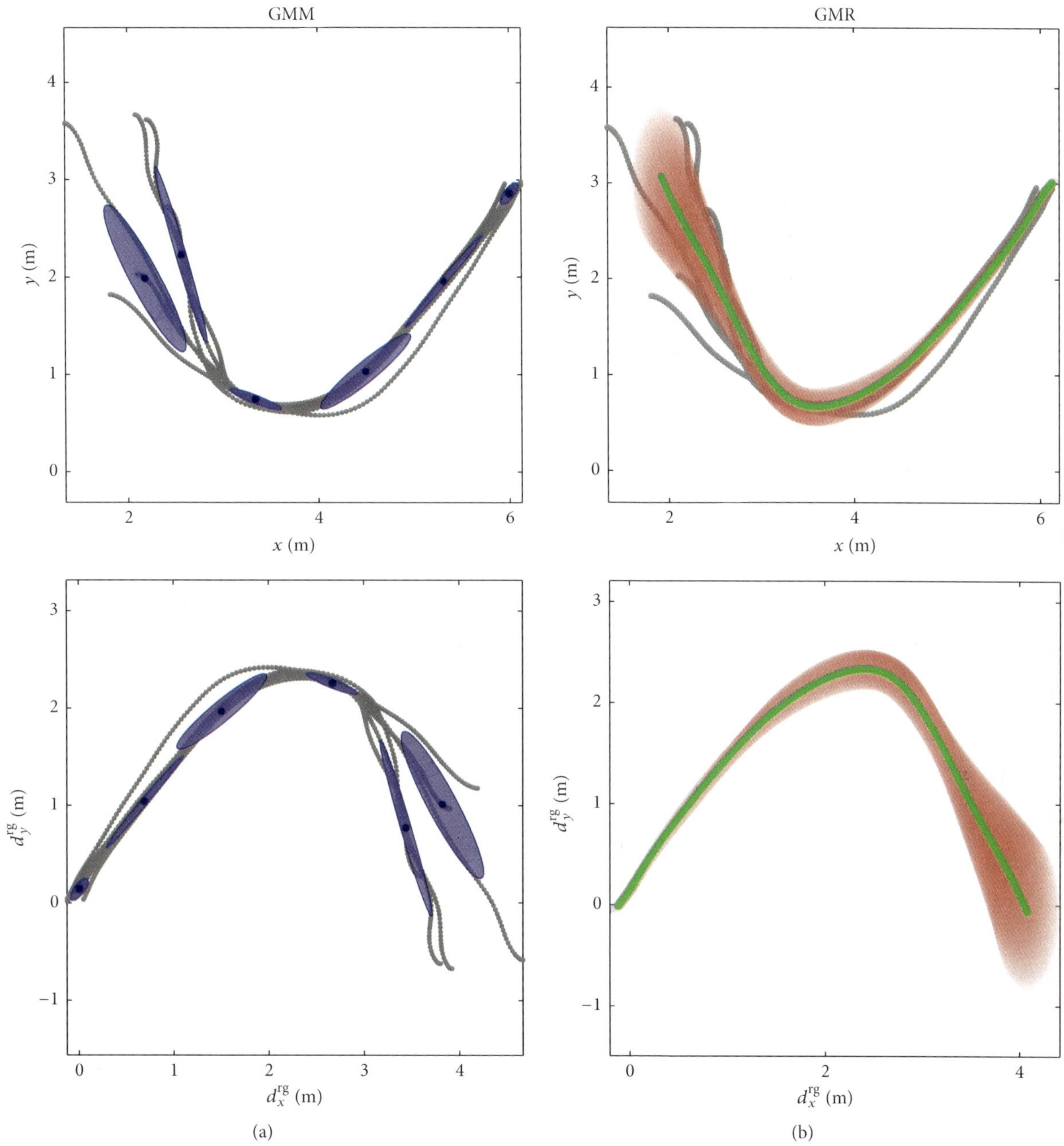

FIGURE 7: Trajectory encoding and retrieval using GMM and GMR for task 1. (a) represents model encoding with GMM. Ellipse in dark blue represents projections of Gaussian components. (b) represents general trajectory generated by GMR. Shades in red color represent generation region.

information d^{rg}, and both variables are 2-dimensional vectors. Number of Gaussian components is set to 6. It is worth noting that the optimal number of components K in a model may not be known beforehand so far. A common method consists of estimating multiple models with increasing number of components and selecting an optimum based on some model selection criterion, such as Cross Validation, Bayesian Information Criterion (BIC), or Minimum Description Length (MDL). In our paper, we decide K by using Cross Validation criterion. As the figures show, variance for starting position is large, corresponding to a big retrieval error. It indicates that there are no strict requirements for starting position. Variance for end point is pretty small, which indicates that the robot can get the goal state accurately. Thus, they show that GMM can successfully extract useful constraints for characterization task. GMR can retrieve barrier-free trajectory by using information encoded by GMM, as shown in the second column of Figure 5. Figures in the second row show encoding and retrieval results for variable d^{rg}. As can be seen from the figures, d^{rg} shares similar distribution properties with x. This is because the two variables have a specific relationship between each other, so

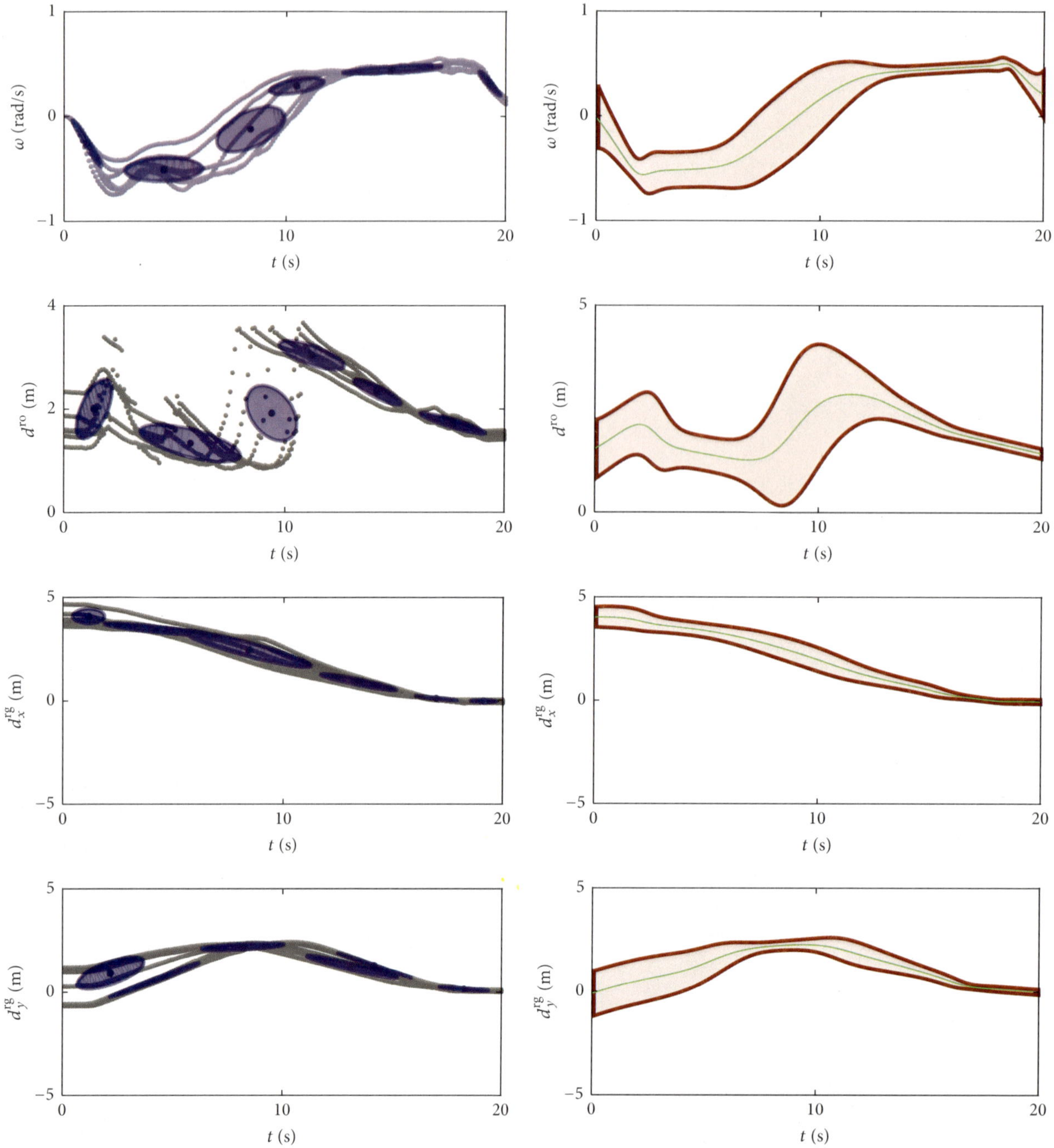

FIGURE 8: Encoding and retrieval for orientation and distance information in task 1, sharing the same meaning with charts in Figure 7.

do mean value and covariance, which verify the validity of the method indirectly.

Figure 8 shows encoding results for orientation and distance information. Figures in the last two rows display encoding for d^{rg} in X and Y direction. As we see, coordinate values in X direction monotone decrease to 0. This is because the robot keeps moving from left to right in the map, while coordinates in Y direction keep going up at the beginning and then drop to zero. This is because the robot encounters obstacles in the middle moment; in order to avoid obstacles, the robot has to go away from the goal state temporarily. Pay attention to the final distance value which is close to $(0, 0)$, which means the robot reaches the target successfully.

Figures 9 and 10 display encoding and retrieval results for task 2. Compared with task 1, task 2 is more complicated considering the law behind the data. In order to encode these laws, more Gaussian components are required. In our experiments K is set to 10 by using the method described in task 1. As illustrated in these figures, no matter how complex the environment is, this method still has very satisfactory effect.

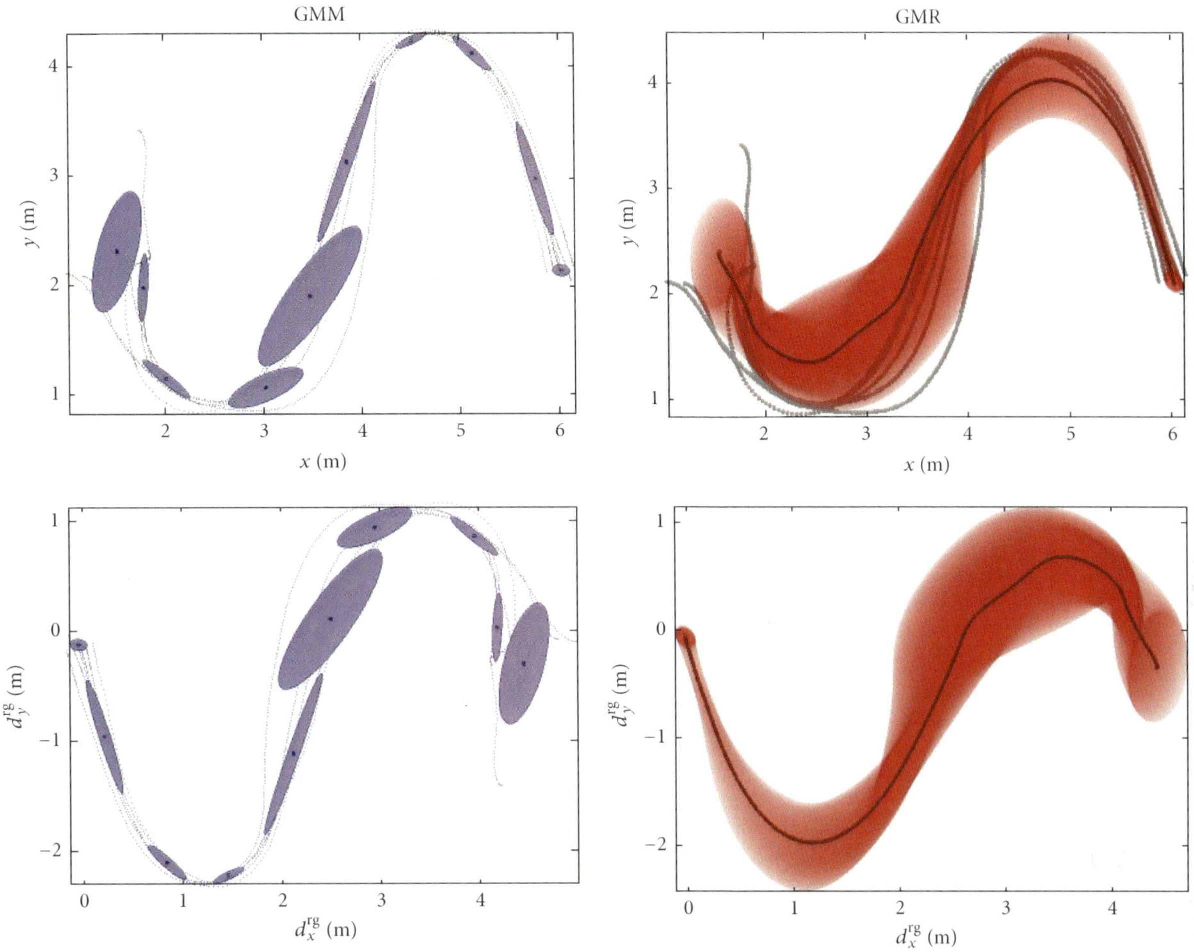

FIGURE 9: Trajectory encoding and retrieval using GMM and GMR for variables x and d in task 2.

5.2. Experiments for Finding Control Strategy. Experiments in this part are mainly conducted to verify the derived control policy. For a Turtlebot robot, the control input is linear velocity $\dot{\xi}_s^x$ and angular velocity $\dot{\xi}_s^\theta$. Assuming we have a desired trajectory (the trajectory can be given arbitrarily; in our experiments, in order to find a reference, we choose the second trajectory in our samples), using theories derived in previous sections, we can compute $\dot{\xi}_s^x$ and $\dot{\xi}_s^\theta$. Iterating on every time, we can get a control sequence. Comparing the calculated value with the sensor recorded value, we can verify the validity of the derived results.

Figure 12 compares the calculated linear velocity and angular velocity with the desired value. Linear velocity is divided into velocity in X direction and Y direction. As the figure in the second row shows, expected velocity in Y direction remains close to 0. This is because the control inputs for Turtlebot are a linear velocity in the forward direction, that is, X direction, and an angular velocity. The line graph for velocity Y reveals that there is a minor fluctuation at the beginning and end stage, while the overall trend is consistent with the fact. Velocity x follows the expected value perfectly. Considering Figure 12(b), it can be seen that the calculated value remains consistent with the actual one. It is worth noting that the length of the calculated sequence differs from the actual one, since the frequency of the readings from robot's sensors is lower than the frequency of commands. In this case, to compare the similarity of the two signals, sequence alignment is an indispensable step. In this paper, we use dynamic time warping (DTW) to achieve this purpose. Signals after alignments show a high correlation, which means the calculated values are correct.

Figure 11 shows the location information obtained by integrating velocity signals. The calculated control inputs allow the robot to follow the given trajectory accurately, except a minor fluctuation at the beginning time.

6. Conclusion

In this paper, based on the idea of the imitation learning, we propose a novel idea that the robot's obstacle avoidance skill can be learned by showing the robot how to avoid an obstacle. Under this consideration, task constraints extraction is a key issue, that is, how to make robots understand the task. In order to achieve this purpose, the Gaussian mixture model is used to encode the data. GMR is used to retrieve satisfactory

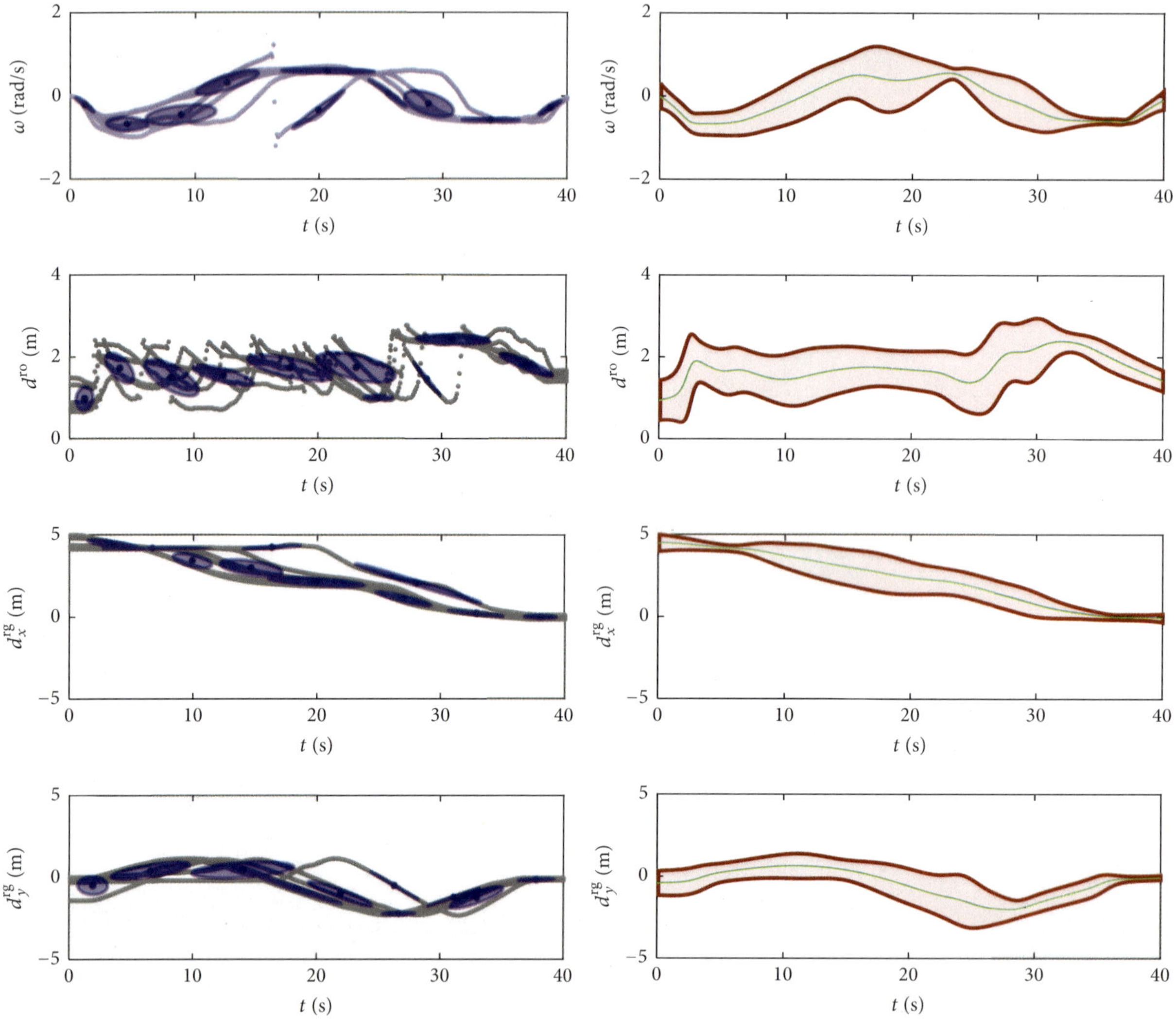

FIGURE 10: Encoding and retrieval for orientation and distance information in task 2.

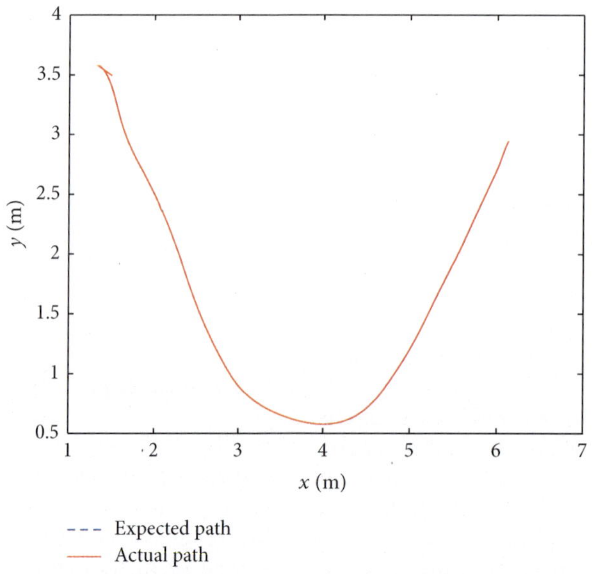

FIGURE 11: Tracking performance using the derived policy.

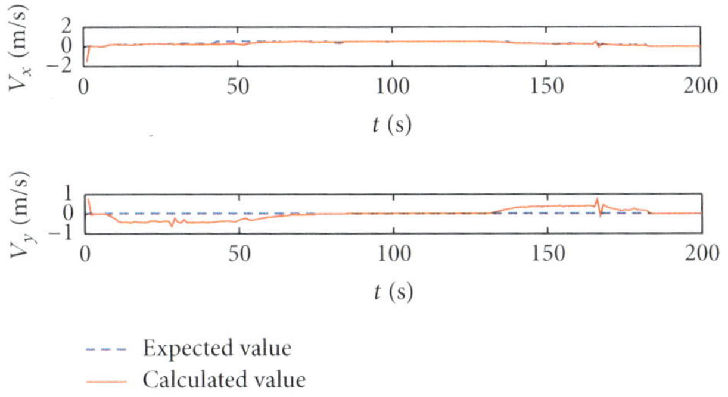

(a) Linear velocity

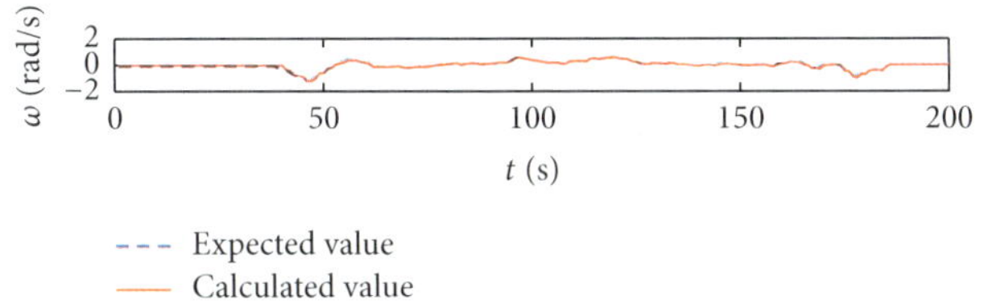

(b) Angular velocity

FIGURE 12: Comparisons between the expected angular velocity and calculated value.

trajectories. Eventually, to obtain the control strategy, a multiobjective optimization problem is constructed. The control strategy is derived by solving this problem. Simulation experiments verified our algorithms.

For the future, improvement work mainly includes the following aspects: (i) in order to avoid obstacles online, online incremental GMM algorithm should be used; (ii) in this paper, a time-based trajectory is generated by GMR, that is, use of time signal as input. A more reasonable idea is using state information as input, so that the robot can change its decision according to the current environment, for example, using distance information as input and velocity as output; thus the control inputs can be directly generated; (iii) reinforcement learning algorithms can be integrated to strengthen robots' exploration ability, so that better generalization performance is achieved.

Competing Interests

The authors declare that the grant, scholarship, and/or funding do not lead to any conflict of interests. Additionally, the authors declare that there is no conflict of interests regarding the publication of this manuscript.

Acknowledgments

This work is supported by National Natural Science Foundation of China (Grant no. 51505470) and the Dr. Startup Fund in Liaoning province (20141152).

References

[1] E. Masehian and D. Sedighizadeh, "Classic and heuristic approaches in robot motion planning—a chronological review," *World Academy of Science, Engineering and Technology*, vol. 23, pp. 101–106, 2007.

[2] P. Raja and S. Pugazhenthi, "Optimal path planning of mobile robots: a review," *International Journal of Physical Sciences*, vol. 7, no. 9, pp. 1314–1320, 2012.

[3] J. C. Latombe, *Robot Motion Planning*, Springer, 2012.

[4] J. Borenstein and Y. Koren, "Real-time obstacle avoidance for fast mobile robots," *IEEE Transactions on Systems, Man and Cybernetics*, vol. 19, no. 5, pp. 1179–1187, 1989.

[5] O. Khatib, "Real-time obstacle avoidance for manipulators and mobile robots," *International Journal of Robotics Research*, vol. 5, no. 1, pp. 90–98, 1986.

[6] H. Mei, Y. Tian, and L. Zu, "A hybrid ant colony optimization algorithm for path planning of robot in dynamic environment," *International Journal of Information Technology*, vol. 12, no. 3, pp. 78–88, 2006.

[7] M. A. P. Garcia, O. Montiel, O. Castillo, R. Sepúlveda, and P. Melin, "Path planning for autonomous mobile robot navigation with ant colony optimization and fuzzy cost function evaluation," *Applied Soft Computing*, vol. 9, no. 3, pp. 1102–1110, 2009.

[8] C. Lopez-Franco, J. Zepeda, N. Arana-Daniel, and L. Lopez-Franco, "Obstacle avoidance using PSO," in *Proceedings of the 9th International Conference on Electrical Engineering, Computing Science and Automatic Control (CCE '12)*, pp. 1–6, Mexico City, Mexico, September 2012.

[9] H.-Q. Min, J.-H. Zhu, and X.-J. Zheng, "Obstacle avoidance with multi-objective optimization by PSO in dynamic environment," in *Proceedings of the International Conference on Machine Learning and Cybernetics (ICMLC '05)*, vol. 5, pp. 2950–2956, IEEE, Guangzhou, China, August 2005.

[10] S. Dutta, "Obstacle avoidance of mobile robot using PSO-based neuro fuzzy technique," *International Journal of Computer Science and Engineering*, vol. 2, no. 2, pp. 301–304, 2010.

[11] W.-G. Han, S.-M. Baek, and T.-Y. Kuc, "Genetic algorithm based path planning and dynamic obstacle avoidance of mobile robots," in *Proceedings of the IEEE International Conference on Systems, Man, and Cybernetics*, vol. 3, pp. 2747–2751, Orlando, Fla, USA, October 1997.

[12] J. Tu and S. X. Yang, "Genetic algorithm based path planning for a mobile robot," in *Proceedings of the IEEE International Conference on Robotics and Automation (ICRA '03)*, vol. 1, pp. 1221–1226, September 2003.

[13] Y. Zhang and J. Wang, "Obstacle avoidance for kinematically redundant manipulators using a dual neural network," *IEEE Transactions on Systems, Man, and Cybernetics, Part B: Cybernetics*, vol. 34, no. 1, pp. 752–759, 2004.

[14] X. Wang, V. Yadav, and S. N. Balakrishnan, "Cooperative UAV formation flying with obstacle/collision avoidance," *IEEE Transactions on Control Systems Technology*, vol. 15, no. 4, pp. 672–679, 2007.

[15] J. Godjevac, "A learning procedure for a fuzzy system: application to obstacle avoidance," in *Proceedings of the International Symposium on Fuzzy Logic*, pp. 142–148, Zurich, Switzerland, 1995.

[16] R. S. Sutton and A. G. Barto, *Reinforcement Learning: An Introduction*, MIT Press, Boston, Mass, USA, 1998.

[17] J. Michels, A. Saxena, and A. Y. Ng, "High speed obstacle avoidance using monocular vision and reinforcement learning," in *Proceedings of the 22nd International Conference on Machine Learning (ICML '05)*, pp. 593–600, ACM, Bonn, Germany, 2005.

[18] C. Ye, N. H. C. Yung, and D. Wang, "A fuzzy controller with supervised learning assisted reinforcement learning algorithm for obstacle avoidance," *IEEE Transactions on Systems, Man, and Cybernetics, Part B: Cybernetics*, vol. 33, no. 1, pp. 17–27, 2003.

[19] B.-Q. Huang, G.-Y. Cao, and M. Guo, "Reinforcement learning neural network to the problem of autonomous mobile robot obstacle avoidance," in *Proceedings of the International Conference on Machine Learning and Cybernetics (ICMLC '05)*, vol. 1, pp. 85–89, August 2005.

[20] C. F. Touzet, "Neural reinforcement learning for behaviour synthesis," *Robotics and Autonomous Systems*, vol. 22, no. 3-4, pp. 251–281, 1997.

[21] W. D. Smart and L. P. Kaelbling, "Effective reinforcement learning for mobile robots," in *Proceedings of the IEEE International Conference on Robotics and Automation (ICRA '02)*, vol. 4, pp. 3404–3410, Washington, DC, USA, May 2002.

[22] K. Maček, I. Petrović, and N. Perić, "A reinforcement learning approach to obstacle avoidance of mobile robots," in *Proceedings of the 7th International Workshop on Advanced Motion Control (AMC '02)*, pp. 462–466, IEEE, Maribor, Slovenia, July 2002.

[23] U. Muller, J. Ben, E. Cosatto et al., "Off-road obstacle avoidance through end-to-end learning," in *Advances in Neural Information Processing Systems*, pp. 739–746, 2005.

[24] B. D. Argall, S. Chernova, M. Veloso, and B. Browning, "A survey of robot learning from demonstration," *Robotics and Autonomous Systems*, vol. 57, no. 5, pp. 469–483, 2009.

[25] C. L. Nehaniv and K. Dautenhahn, "Like me?-measures of correspondence and imitation," *Cybernetics & Systems*, vol. 32, no. 1-2, pp. 11–51, 2001.

20

Hybrid Type-2 Fuzzy-Sliding Mode Controller for Navigation of Mobile Robot in an Environment Containing a Dynamic Target

Dorra Ayedi, [1] **Maïssa Boujelben,** [2] **and Chokri Rekik** [2]

[1] *Control and Energy Management Lab (CEM Lab), University of Sousse, Sousse Engineering School, BP 264 Sousse Erriadh 4023, Tunisia*
[2] *Control and Energy Management Lab (CEM Lab), University of Sfax, Sfax Engineering School, BP W, 3038 Sfax, Tunisia*

Correspondence should be addressed to Dorra Ayedi; dorra.ayedii@yahoo.fr

Academic Editor: Yangmin Li

We are interested in developing a new control architecture to command a mobile robot in a partially known environment. For that purpose, a hybrid control architecture is adopted, combining the reactive and the deliberative method. The reactive method consists of the hierarchical fuzzy controllers based on Type-2 Fuzzy Logic System (T2-FLS), dedicated to commanding the robot towards a mobile target while avoiding unexpected obstacles. A comparative study is made to show the efficiency of T2-FLS against Type-1 Fuzzy Logic System (T1-FLS). Additionally, the used deliberative method is the sliding mode, allowing the robot to track the mobile goal trajectory. Simulation results are given finally to test the proposed architecture.

1. Introduction

Controlling a mobile robot in an unknown environment is an attractive aspect in the field of robotic, as it is used in many applications. Therefore, researchers have developed several methods of controlling non-holonomic robot in a real environment. Those methods can be classified into three groups: reactive approaches, deliberative approaches, and hybrid approach, which is a combination between both of them [1]. The choice between those classes depends on the robot's knowledge of the environment [2].

The reactive approach does not require a prior knowledge about the environment. The robot receives at each instant information about its local environment and then moves autonomously to reach a desired point [2, 3]. This approach is adaptable to the dynamic or unexpected environment, but may have some problems such as nonconvergence or local minima [4]. Neural networks [5], potential field [6], and fuzzy logic [7] are among the methods which illustrate the reactive approach.

Unlike the reactive category, the deliberative method is based on trajectory planning, which needs to consider a complete knowledge about the environment [2]. Although it is considered as a fast and stable navigation method, it can not be adapted in a completely unknown environment [4]. Many control methods were proposed in the domain of deliberative approach. Among those methods, we cite the PID controllers [8] and the sliding mode control [9].

As it was explained, many limitations can be faced while applying only reactive or deliberative methods [10]. They can not solve individually the problems faced in the navigation domain [11]. Therefore, those two methods are combined to form the hybrid approach. The favor of this mix is to benefit from the advantages of the two approaches [2] and provides the possibility of using the reactive behavior as well as global vision of the environment.

In the literature, Arkin and Murphy were the first who have made a combination between the reactive and the deliberative methods, which leads to the hybrid approach [1]. Zhu has used this combination in his work [12], letting the reactive approach playing a dominant role in case of conflict. Tian has designed in [13] a hybrid navigation control strategy for rescue robot. In this paper, the idea of developing a hybrid control architecture is adopted, as it suits with the environment in which the robot moves.

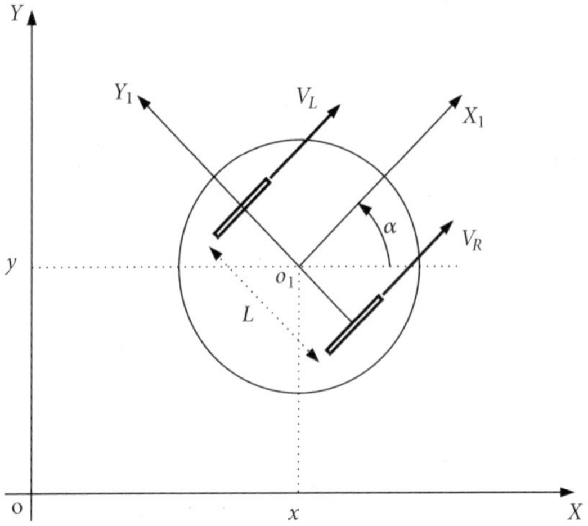

FIGURE 1: The schematic model of the Khepera II robot.

The used robot must have a considerable number of sensors distributed on its circumference. At first, the robot moves from an initial point to reach a dynamic target, which the trajectory is predefined. The use of the reactive method becomes necessary for avoiding unexpected obstacles. Fuzzy logic system was adopted for this task, as it is able to conceive robust controllers [14, 15]. The main problem which faces us is the huge number of obtained rules, due to the big number of information received by each sensor. In [16], the authors have proposed the hierarchical fuzzy controller in order to solve this problem.

On the other hand, the robot can face perturbations due to several sources. Among those sources, we mention the linguistic uncertainties, imprecise sensor information, and uncertainties in inputs or outputs [17, 18]. T1-FLS can not really handle all those uncertainties [15]. An extension of the T1-FLS have been introduced by Zadeh [19]. This new type is characterized by a third dimension which makes it able to handle all the perturbations mentioned [17].

Once the robot becomes close enough from the dynamic goal, it follows the trajectory of the target, which is a predefined safe path. In this case, the deliberative method copes better with this situation. Therefore, the sliding mode was adopted for trajectory tracking, for his fast response, and his robustness against the variation of parameters [3, 20].

The main contribution of this paper is to construct a suitable control architecture based on the association of deliberative and reactive method. Hierarchical T2-FLS and T1-FLS were implemented in the reactive controllers method and tested on the robot. A comparative study is made to show the outperformance of type 2 against type 1. For the deliberative controller, the sliding mode is proposed as a local method, to track the dynamic target until reaching a desired point.

This paper is organized as follows: Section 2 presents the robot control approach. On Section 3, reactive navigation controllers are detailed. Path planning is described on Section 4. At least, conclusion is given.

2. Robot Control Approach

The mobile robot is requested to navigate from an initial point to reach a dynamic target while avoiding obstacles. In this context, the Khepera II robot is used. Our robot is a nonholonomic robot, equipped with two independent driving wheels located on the same axis. Eight sensors are distributed on the circumference of the robot in a way that permits the detection of obstacles, regardless of their positions in the environment. A schematic model of this robot is presented in Figure 1.

2.1. Robot Kinematic Model. Our purpose is to control the wheels speed to command the robot. For that, the kinematic model of the used robot is given by the following system:

$$\begin{aligned} \dot{x} &= \frac{V_R + V_L}{2}\cos\alpha \\ \dot{y} &= \frac{V_R + V_L}{2}\sin\alpha \\ \dot{\alpha} &= \frac{V_R - V_L}{L} \end{aligned} \tag{1}$$

where

(i) V_L and V_R are, respectively, the left and right wheels velocities.

(ii) α is the angle which separates the robot direction from the X-axis.

(iii) L is the distance between the left and the right wheels.

2.2. The Proposed Control Architecture. In this paper, a hybrid control is proposed. This approach combines reactive and deliberative methods. It permits benefiting from the advantages of the two methods [21]. Firstly, the robot moves in an unknown environment towards the mobile target, which needs the application of the reactive method. When the robot is close enough from the mobile target, path planning is sufficient, since the trajectory of the goal is known. Figure 2 presents the proposed strategy of control.

Figure 2 shows the presence of three controllers: the attraction to dynamic target controller allowing the robot to reach the mobile goal, the obstacle avoidance controller which enables it to get away from obstacles, and the path planning controller devoted for trajectory tracking.

According to the information given in each step by the environment (position of the robot, the goal, and the obstacles) and sent to the robot through the sensors, the action selection block chooses the suitable controller to command the robot. If the robot is far away from the target, the reactive method [15] is activated. In this case, the distance which separates the robot from the obstacle defines the choice between the attraction to dynamic target controller and the obstacle avoidance controller. Both of them work in a similar way and give as outputs the right and left wheels velocities V_R and V_L, except that the velocities obtained by the obstacle avoidance controller are inverted.

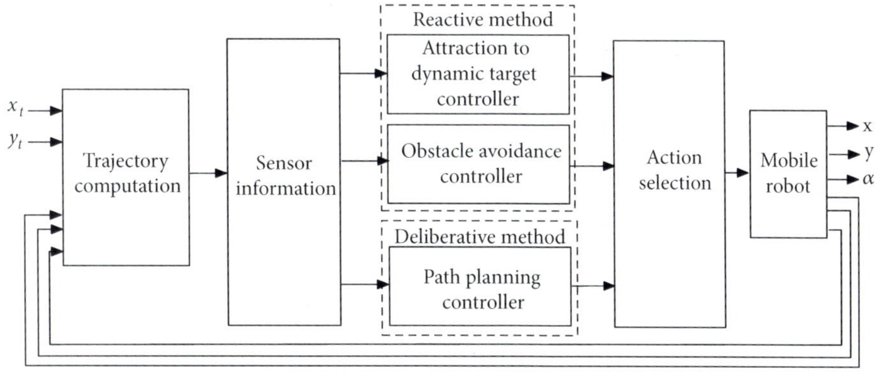

FIGURE 2: The proposed architecture of the mobile robot navigation.

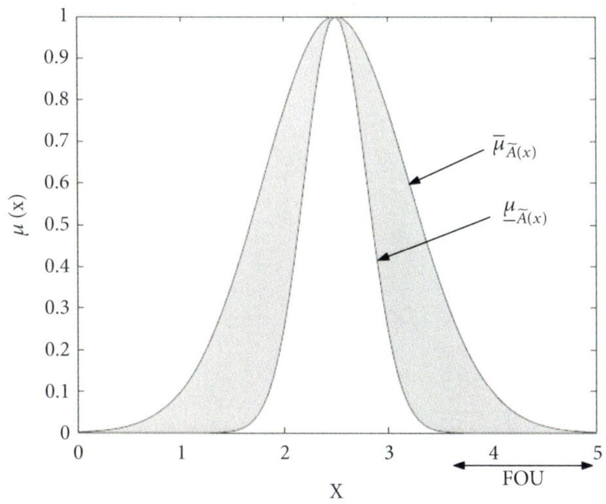

FIGURE 3: Gaussian type-2 fuzzy system.

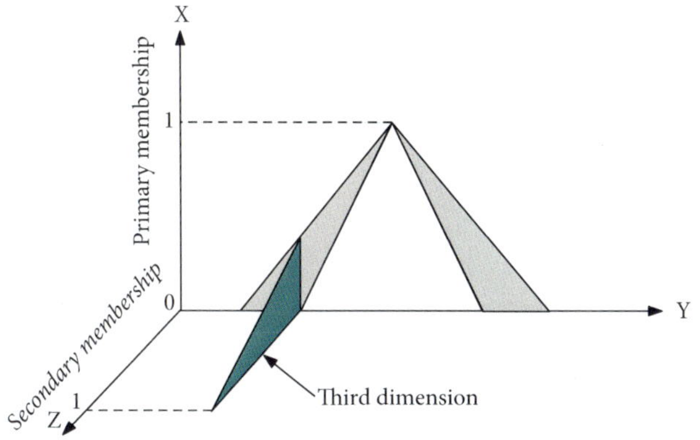

FIGURE 4: The third dimension of a triangular T2-FLS.

At each instant, the goal and robot positions are compared. When the robot becomes close to the goal position, it should follow it until reaching the desired point. In consequence, the deliberative method (path planning controller) is activated.

As a result, we obtain the right and left wheels velocities V_R and V_L, which control the robot.

3. Reactive Navigation

The environment in which the robot moves is an unknown area of navigation. Based on its sensor information, the robot must go towards the target, following a secure path by avoiding the obstacles.

The adopted approach to conceive the controllers which are dedicated for the reactive method is Type-2 Fuzzy Logic System (Type-2 FLS). This approach has become very well known in the field of control and automation, as it permits realizing robust controllers [15].

3.1. Type-2 Fuzzy Logic System. Based on Zadeh's theory, the T2-FLS is an extension of the T1-FLS. The membership function of a T2-FLS is a fuzzy set, unlike the T1-FLS which the membership function is a crisp number [22].

The T2-FLS is characterized by the following:

(i) A footprint of uncertainty as is shown in Figure 3, bounded by a lower and an upper membership function $\underline{\mu}_{\widetilde{A}(x)}$ and $\overline{\mu}_{\widetilde{A}(x)}$ [23].

(ii) A third dimension presented in Figure 4. Indeed, each variable x is associated with a primary membership function and a secondary membership function.

Those particularities guarantee additional degree of freedom [17], which push us to opt for the T2-FLS, to handle the maximum of uncertainties that the robot can face while moving towards the goal.

Mathematically speaking, the definition of a generalized T2-FLS is

$$\widetilde{A} = \{((x,u), \mu_{\widetilde{A}}(x,u)) \mid \forall x \in X,\ \forall u \in J_x \subseteq [0,1]\} \quad (2)$$

where X and J_x are, respectively, the primary and the secondary domain [24, 25].

Due to the complexity of the generalized T2-FLS, the interval T2-FLS is adopted, where all $\mu_{\widetilde{A}}(x,u)$ are equal to 1 [25]. A simple illustration of the third dimension of the T1-FLS, the generalized T2-FLS, and the interval T2-FLS are presented in Figure 5.

3.1.1. Inference System. Figure 6 shows the structure of a T2-FLS. Comparing to the model of a T1-FLS, the difference

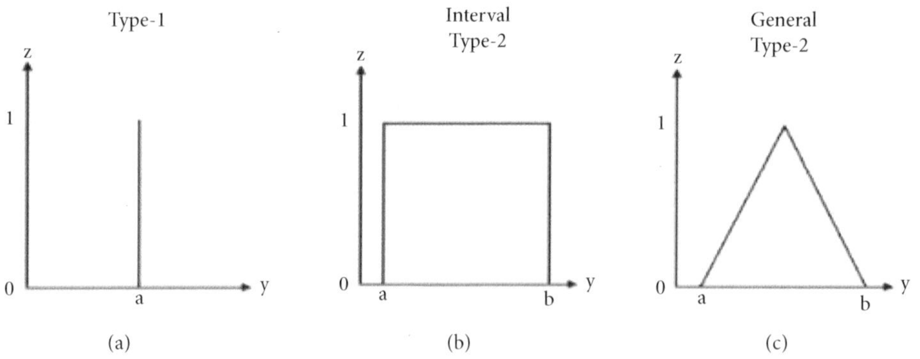

FIGURE 5: A clear view of the third dimension. (a) Type-1 fuzzy set. (b) Interval type-2 fuzzy set. (c) General type-2 fuzzy set.

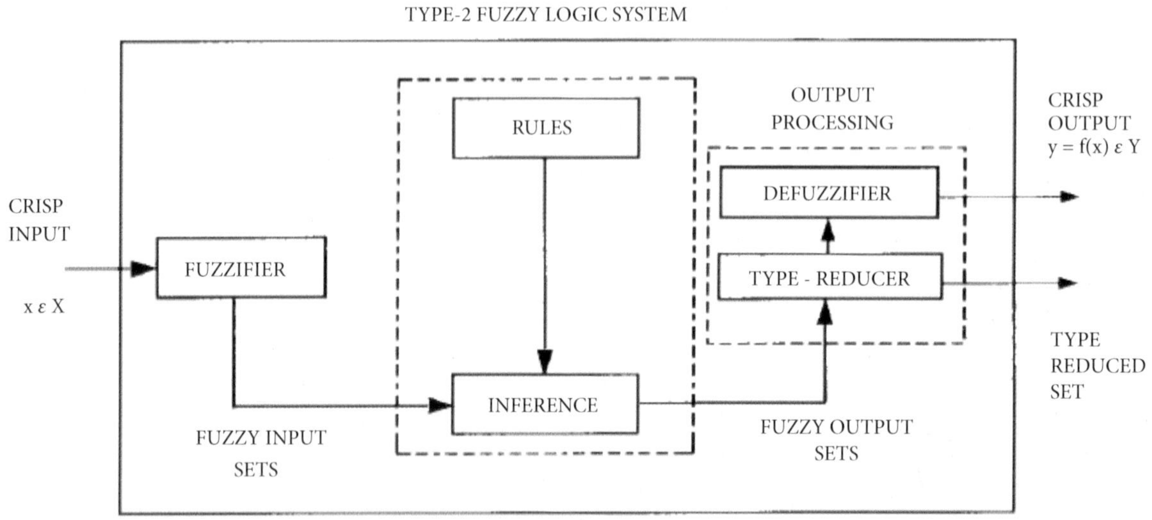

FIGURE 6: The structure of type-2 fuzzy inference system.

lies on the output processor. In addition to the defuzzifer block which exists on the T1-FLS structure, the block of type reducer is joined. Indeed, this block is dedicated to reducing the type of type-2 set to generate type-1 set [25].

Similarly to the T1-FLS, the interval T2-FLS is also based on the IF-THEN rules, except the category of the antecedents and the consequences which differs from type-1. For the case of the proposed architecture to command the Khepera II robot, the ith rule is given as follows:

If (d is $\widetilde{A}_i$) and (φ is $\widetilde{B}_i$) then ($V_R = [\underline{y_i}, \overline{y_i}]$) and ($V_L = [\underline{z_i}, \overline{z_i}]$),

where $i = 1, 2 \ldots, N$, $\widetilde{A}_i$ and $\widetilde{B}_i$ are the interval type-2 fuzzy systems, $\overline{y_i}$, $\underline{y_i}$, $\overline{z_i}$, and $\underline{z_i}$ are, respectively, the upper and lower values of V_R and V_L.

The inference engine has as role to link rules and form firing interval set W_i

$$W_i = [\underline{w}_i, \overline{w}_i] \tag{3}$$

with $\underline{w}_i = [\underline{\mu}_{\widetilde{A}_i}(d) \times \underline{\mu}_{\widetilde{B}_i}(\varphi)]$

and $\overline{w}_i = [\overline{\mu}_{\widetilde{A}_i}(d) \times \overline{\mu}_{\widetilde{B}_i}(\varphi)]$

where × represents the t-norm used, which can be the minimum or the product operation [26].

3.1.2. Output Processing. In this step, a fuzzy type-2 set goes through a type reducer to become a fuzzy type-1 set, then through a defuzzifier to obtain a crisp value.

According to [27, 28], different methods of type reduction have been mentioned and detailed. The center of sets (COS) method was adopted in this paper. The type 1 fuzzy set expression is given in the following equation.

$$Y_{cos} = [y_l, y_r] \tag{4}$$

where Y_{cos} is an interval set bounded by two limits: y_l and y_r.

To compute those two values, Karnik and Mendel have proposed an algorithm [23, 29], explained in the following steps:

KM for Computing y_l

(1) Sort rule consequence $\underline{y}_i$ $(i = 1, 2 \ldots, M)$ in increasing order then match each w_i with his respective $\underline{y}_i$, so their index corresponds to the renumbered $\underline{y}_i$.

(2) Initialize w_i as the following equation:

$$w_i = \frac{\underline{w}_i + \overline{w}_i}{2}, \quad i = 1, 2, \ldots, M \tag{5}$$

then compute

$$y = \frac{\sum_{i=1}^{M} \underline{y}_i w_i}{\sum_{i=1}^{M} w_i} \tag{6}$$

(3) Find the switch point k $(1 \le k \le M - 1)$ such that

$$\underline{y}_i \le y \le \underline{y}_{i+1} \tag{7}$$

(4) Set

$$y_i = \begin{cases} \overline{w}_i, & i \le k \\ \underline{w}_i, & i > k \end{cases} \tag{8}$$

then compute

$$y' = \frac{\sum_{i=1}^{M} \underline{y}_i w_i}{\sum_{i=1}^{M} w_i} \tag{9}$$

(5) Check if $y' = y$. If yes, stop and set $y_l = y$ and $K = L$. If no, set $y = y'$ and go to step (3)

KM for Computing y_r

(1) Sort rule consequence $\overline{y}_i$ $(i = 1, 2 \ldots, M)$ in increasing order then match each w_i with his respective $\overline{y}_i$, so their index corresponds to the renumbered $\overline{y}_i$.

(2) Initialize w_i as the following equation:

$$w_i = \frac{\underline{w}_i + \overline{w}_i}{2}, \quad i = 1, 2, \ldots, M \tag{10}$$

then compute

$$y = \frac{\sum_{i=1}^{M} \overline{y}_i w_i}{\sum_{i=1}^{M} w_i} \tag{11}$$

(3) Find the switch point k $(1 \le k \le M - 1)$ such that

$$\overline{y}_i \le y \le \overline{y}_{i+1} \tag{12}$$

(4) Set

$$y_i = \begin{cases} \underline{w}_i, & i \le k \\ \overline{w}_i, & i > k \end{cases} \tag{13}$$

then compute

$$y' = \frac{\sum_{i=1}^{M} \overline{y}_i w_i}{\sum_{i=1}^{M} w_i} \tag{14}$$

(5) Check if $y' = y$. If yes, stop and set $y_r = y$ and $K = R$. If no, set $y = y'$ and go to step (3)

At last, the values calculated in this algorithm pass through the defuzzification block to obtain the final defuzzified output:

$$Y = \frac{y_l + y_r}{2} \tag{15}$$

Type-2 FLS is used in the development of the attraction to dynamic target controller and the obstacle avoidance controller.

3.2. Design of the Proposed Type-2 Fuzzy Logic Controllers. The robot should find a safe path to reach the mobile goal. For this task, the attraction to dynamic target controller and the obstacle avoidance controller are activated.

The first controller has as inputs

(i) d: the distance between the robot and the mobile target,

(ii) φ: the angle separating the robot orientation and the mobile target.

For the second controller, the inputs are

(i) d_o: the distance between the robot and the obstacle,

(ii) φ_o: the angle separating the robot orientation and the obstacle.

Those parameters are presented in Figure 7.

For the attraction to dynamic target controller, five membership functions are chosen for the distance d, while seven membership functions define the angle, as is detailed in [14].

If the action selection block chooses the second controller dedicated for obstacle avoidance, the use of the eight sensors of the Khepera II mobile robot is requested. It permits detecting the obstacles in different directions. However, a disadvantage occurs if we use one standard fuzzy controller: the number of obtained rules is huge, which increases the simulation time. To solve this problem, the hierarchical fuzzy system is used.

3.3. Hierarchical Type-2 Fuzzy Logic System. The hierarchical fuzzy system consists of attributing each sensor information to a controller, then regrouping them as is shown in Figure 8.

The choice of this structure is based on gathering the robot sensors according to angle measurement (Figure 9).

The first layer of the hierarchical fuzzy system contains eight controllers associated with the sensors measurements. Each controller gives the left and right wheels velocities, and another index denoted I_o^i. This index is dedicated to providing a collision degree between the robot and the obstacle.

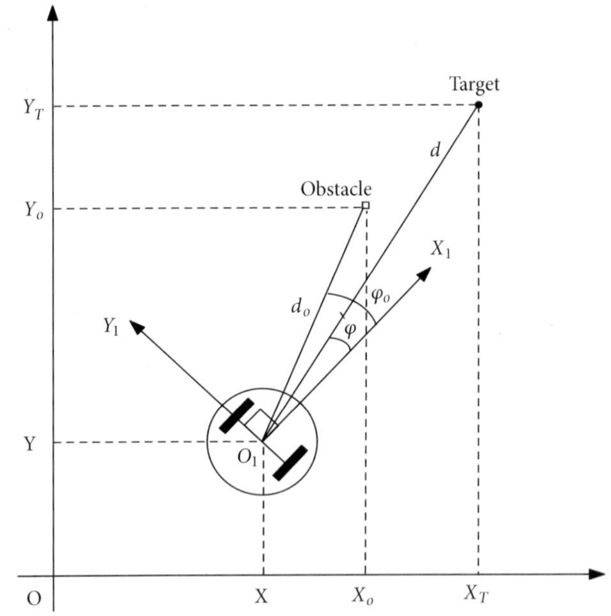

FIGURE 7: Representation of the parameters.

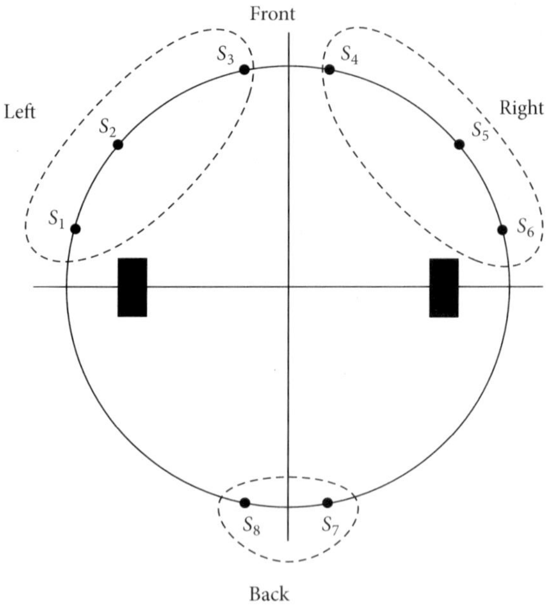

FIGURE 9: Decomposition of the eight sensors of the robot.

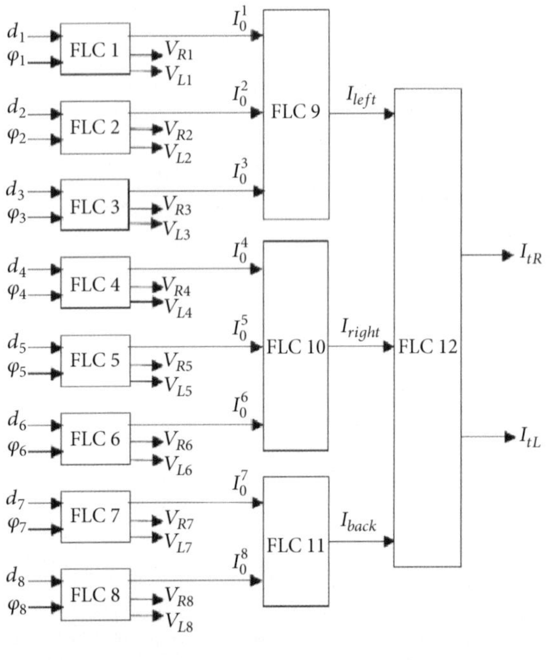

FIGURE 8: Block diagram of the hierarchical system.

According to Figure 9, those indexes are regrouped into three groups (according to sensor's positions in the robot) and are considered as inputs to three other controllers which form the second layer. As a consequence, we obtain three indexes: I_{left}, I_{right}, and I_{back}, which have the same meaning of the other indexes. The third and last layer provides as outputs I_{tR} and I_{tL}. Those indicators help for the robot guidance as they illustrate the absence degree of obstacles. More details and inference tables are given in [16, 30].

4. Path Planning

Once the robot is close enough from the mobile goal, the action selection block passes from the reactive approach to the deliberative approach. The trajectory of the goal is determined, and the robot should just follow the path of the target until reaching the desired point. For that, we have proposed the sliding mode method to conceive the path planning controller [9, 20].

The purpose of the sliding mode controller is to compute the appropriate linear and angular velocities so that the robot can follow exactly the target [3] by converging the tracking error to zero [31].

For that we consider the reference position $p_r = (x_r, y_r, \alpha_r)^T$ and the current position $p = (x, y, \alpha)^T$. The position error between p_r and p is denoted as $p_e = (x_e, y_e, \alpha_e)^T$, and it is expressed as

$$\begin{bmatrix} x_e \\ y_e \\ \alpha_e \end{bmatrix} = \begin{bmatrix} \cos\alpha & \sin\alpha & 0 \\ -\sin\alpha & \cos\alpha & 0 \\ 0 & 0 & 1 \end{bmatrix} \times \begin{bmatrix} x_r - x \\ y_r - y \\ \alpha_r - \alpha \end{bmatrix} \quad (16)$$

This tracking error is illustrated in Figure 10.

The velocity error is then obtained by deriving the error tracking p_e [20], which gives

$$\begin{bmatrix} \dot{x}_e \\ \dot{y}_e \\ \dot{\alpha}_e \end{bmatrix} = \begin{bmatrix} y_e w - v + v_r \cos\alpha_e \\ -x_e w + v_r \sin\alpha_e \\ w_r - w \end{bmatrix} \quad (17)$$

where v_r and w_r are, respectively, the linear and the angular velocities of the mobile target.

Equation (1), which represents the adopted Khepera II mobile robot model, is a multiple-input nonlinear system.

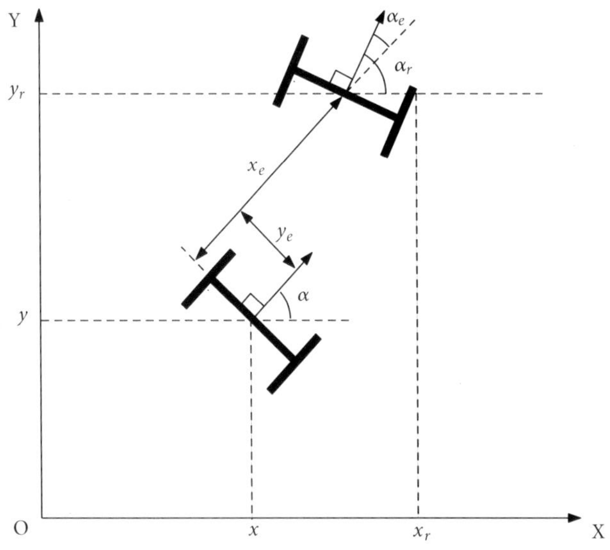

FIGURE 10: Tracking error.

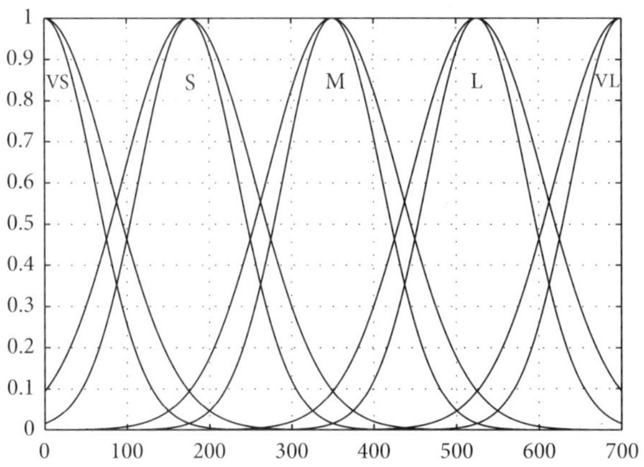

FIGURE 11: Type-2 membership function of the input d (in mm).

According to [3, 20], x_e is chosen equal to zero at the first switching surface to simplify the problem. Therefore, the Lyapunov candidate function proposed in [3, 20] and adopted in this paper is

$$V = \frac{1}{2} y_e^2 \tag{18}$$

where the time derivative is expressed by

$$\begin{aligned} \dot{V} &= y_e \dot{y}_e = y_e \left(-x_e w + v_r \sin(\alpha_e)\right) \\ &= -x_e y_e w - v_r y_e \sin\left(\arctan\left(v_r y_e\right)\right) \end{aligned} \tag{19}$$

The switching function condition is defined as $\alpha_e = -\arctan(v_r y_e)$. We ensure that $v_r y_e \sin(\arctan(v_r y_e)) \geq 0$ which always satisfies the global condition $\dot{V} \leq 0$. The proof is detailed in [20]. Consequently, the vector of the switching surface is expressed in the following equation.

$$s = \begin{bmatrix} s_1 \\ s_2 \end{bmatrix} = \begin{bmatrix} x_e \\ \alpha_e + \arctan\left(v_r y_e\right) \end{bmatrix} \tag{20}$$

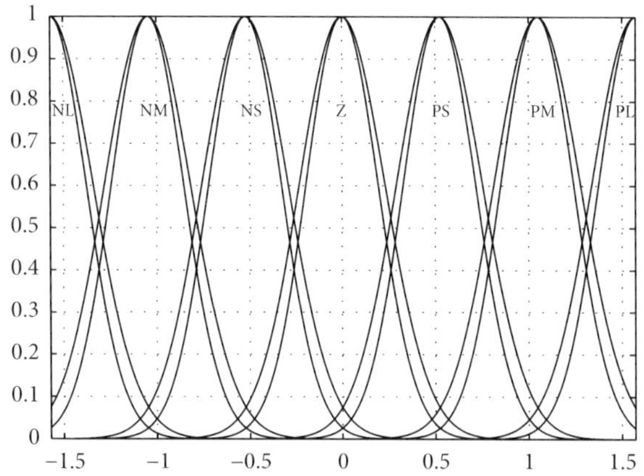

FIGURE 12: Type-2 membership function of the input φ (in rad).

s_1 converges to 0 when x_e converges to 0. In a similar way, s_2 converges to 0 when $\alpha_e + \arctan(v_r y_e)$ converges to 0, which means that α_e converges to $-\arctan(v_r y_e)$. Consequently, we obtain the convergence of α_e and y_e to 0.

To design a control law, the switching surface is expressed as a saturation function [32], supposing that $\gamma = \arctan(v_r y_e)$:

$$\begin{aligned} \dot{s} &= \begin{bmatrix} \dot{s}_1 \\ \dot{s}_2 \end{bmatrix} = \begin{bmatrix} -k_1 \operatorname{sat}(s_1) \\ -k_2 \operatorname{sat}(s_2) \end{bmatrix} = \begin{bmatrix} \dot{x}_e \\ \dot{\alpha}_e + \dfrac{\partial \gamma}{\partial v_r} \dot{v}_r + \dfrac{\partial \gamma}{\partial y_e} \dot{y}_e \end{bmatrix} \\ &= \begin{bmatrix} y_e w - v + v_r \cos \alpha_e \\ w_r - w + \dfrac{\partial \gamma}{\partial v_r} \dot{v}_r + \dfrac{\partial \gamma}{\partial y_e} \left(-x_e w + v_r \sin \alpha_e\right) \end{bmatrix} \end{aligned} \tag{21}$$

The obtained control law is

$$\begin{bmatrix} v \\ w \end{bmatrix} = \begin{bmatrix} y_e w + v_r \cos \alpha_e + k_1 \operatorname{sat}(s_1) \\ w_r + \dfrac{\partial \gamma}{\partial v_r} \dot{v}_r + \dfrac{\partial \gamma}{\partial y_e} \left(v_r \sin \alpha_e\right) + k_2 \operatorname{sat}(s_2) \end{bmatrix} \tag{22}$$

where $\partial\gamma/\partial v_r = y_e/(1+(v_r y_e)^2)$ and $\partial\gamma/\partial y_e = v_r/(1+(v_r y_e)^2)$.

5. Simulation Results

In this part, we present the inputs simulation and some simulations results to test the validity of the control architecture proposed and to compare the performance of T1-FLS and T2-FLS.

5.1. Inputs Simulations. As was explained previously, we have two inputs: the distance d and the angle of direction φ. d varies between 0 and 700 mm and is divided into five fuzzy subsets: VS (very small), S (small), M (medium), L (large), and VL (very large). φ is defined in the interval $[-\pi/2, \pi/2]$, and seven fuzzy subsets are associated with it which are NL (negative large), NM (negative medium), NS (negative small), Z (zero), PS (positive small), PM (positive medium), and PL (positive large) [3]. Figures 11 and 12 illustrate the simulations of those two inputs.

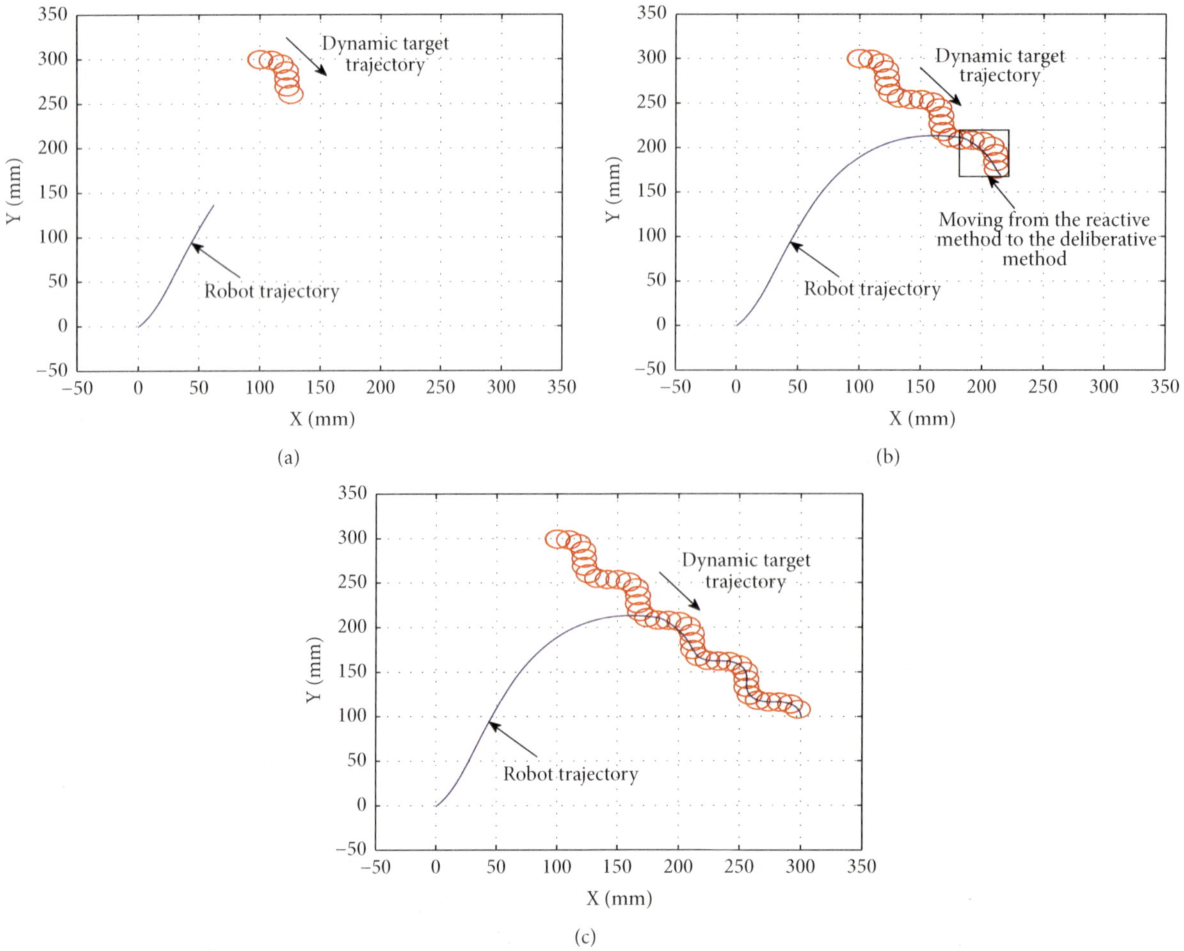

Figure 13: Path planning for the mobile robot using T1-FLS. (a)The robot tries to reach the target using the reactive approach. (b) The transition between the reactive and the deliberative approach. (c) Trajectory tracking applying the deliberative approach.

5.2. Trajectory Simulation. The objective in this part is to test the efficiency of the proposed architecture. For that purpose, we consider a robot which starts from an initial point $(X, Y) = (0, 0)$ and tries to get closer to a dynamic target until reaching the desired final point. Firstly, we have considered a safe environment, using T1-FLS as a reactive approach. Figure 13 illustrates clearly the steps of progress of the robot.

It is obvious that the robot can move towards the target and tracks the goal's trajectory until reaching the final point, which proves the validity of the proposed control architecture.

In the next simulations (Figures 14 and 15), the robot moves in environments containing obstacles and tries to reach a moving target using the T1-FLS and the T2-FLS.

In Figure 14, the robot starts from the initial point $(X, Y) = (0, 0)$. It succeeds to avoid the obstacles and to reach the circular trajectory of the target using the FLS (type 1 and type 2). Then, the sliding mode is applied for path tracking, until attaining the final point. Figure 15 treated the case of sinusoidal trajectory of the dynamic target, with changing the obstacles position and number. The robot adapts with those changes and achieves the navigation and the tracking task properly.

Table 1: Comparison between distances covered with T1-FLS and T2-FLS.

	Figure 14	Figure 15
T1-FLS	459.1522	527.1141
T2-FLS	447.8955	515.9508
Relative errors (%)	2.5	2.2

The comparison between the distances covered by the robot after implanting T1-FLS and T2-FLS is presented as numerical values in Table 1. This table shows that T2-FLS always provides better results comparing to type 1, giving optimized distances.

6. Conclusion

In this paper, we have proposed a hybrid control law to command a mobile robot in a real environment containing

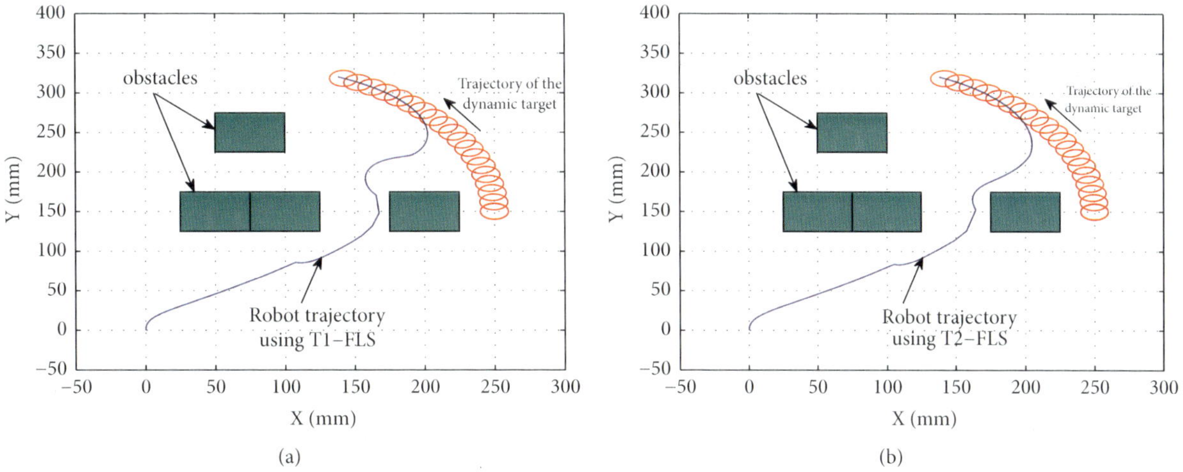

FIGURE 14: Simulation result with circular trajectory of the dynamic target using T1-FLS (a) and T2-FLS (b).

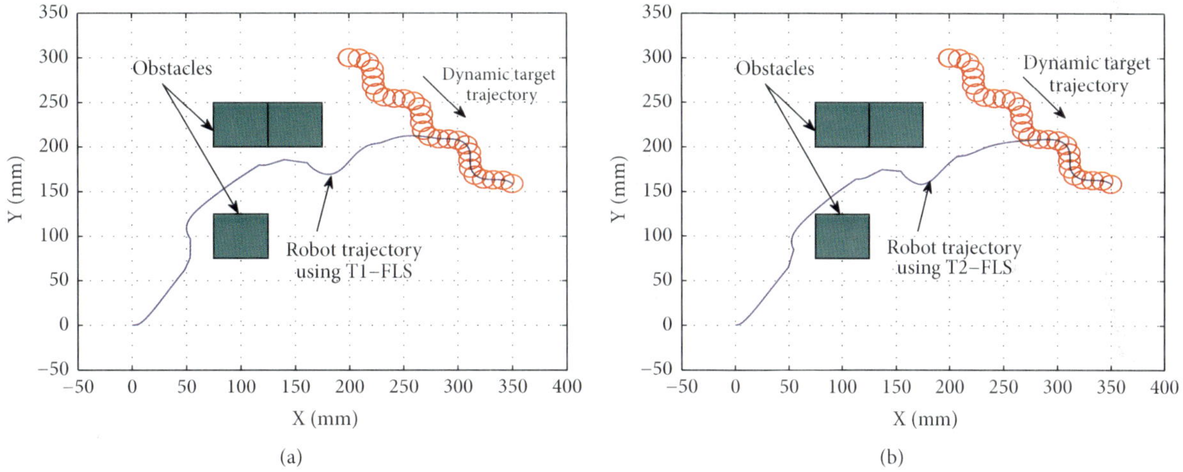

FIGURE 15: Simulation result with sinusoidal trajectory of the dynamic target using T1-FLS (a) and T2-FLS (b).

static obstacles and dynamic target. The control architecture adopted combines reactive approach and deliberative approach. Indeed, the hierarchical fuzzy logic system is applied to bring the robot from an initial portion to become close to the trajectory of the goal. The robot goes after the dynamic target using the sliding mode. Simulations results are presented to test the validity of this algorithm and to prove the outperformance of the T2-FLS comparing to the T1-FLS in the reactive controllers.

Conflicts of Interest

The authors declare that they have no conflicts of interest.

References

[1] D. Nakhaeinia, S. Hong Tang, S. B. Mohd Noor, and O. Motlagh, "A review of control architectures for autonomous navigation of mobile robots," *International Journal of Physical Sciences*, vol. 6, no. 2, pp. 169–174, 2011.

[2] N. Hacene and B. Mendil, "Toward safety navigation in cluttered dynamic environment: A robot neural-based hybrid autonomous navigation and obstacle avoidance with moving target tracking," in *Proceedings of the 3rd International Conference on Control, Engineering and Information Technology, CEIT 2015*, pp. 1–6, May 2015.

[3] M. Boujelben, C. Rekik, and N. Derbel, "A hybrid fuzzy-sliding mode controller for a mobile robot," *International Journal of Modelling, Identification and Control*, vol. 25, no. 3, pp. 155–164, 2016.

[4] M. Hank and M. Haddad, "Hybrid control architecture for mobile robots navigation in partially known environments," in *Proceedings of the 11th International Conference on Informatics in Control, Automation and Robotics, ICINCO 2014*, pp. 513–521, Austria, September 2014.

[5] I. Engedy and G. Horváth, "Artificial neural network based mobile robot navigation," in *Proceedings of the WISP 2009 - 6th IEEE International Symposium on Intelligent Signal Processing*, pp. 241–246, Hungary, August 2009.

[6] O. Khatib, "Real-time obstacle avoida nce for manipulators and mobile robots," *International Journal of Robotics Research*, vol. 5, no. 1, pp. 90–98, 1986.

[7] H. Maaref and C. Barret, "Sensor-based navigation of a mobile robot in an indoor environment," *Robotics and Autonomous Systems*, vol. 38, no. 1, pp. 1–18, 2002.

[8] R. Rojas and A. G. Förster, "Holonomic control of a robot with an omnidirectional drive," *KI-Künstliche Intelligenz*, vol. 20, no. 2, pp. 12–17, 2006.

[9] R. Solea, A. Filipescu, and U. Nunes, "Sliding-mode control for trajectory-tracking of a wheeled mobile robot in presence of uncertainties," in *Proceedings of the 7th Asian Control Conference (ASCC '09)*, pp. 1701–1706, Hong Kong, August 2009.

[10] M. Hank and M. Haddad, "A hybrid approach for autonomous navigation of mobile robots in partially-known environments," *Robotics and Autonomous Systems*, vol. 86, pp. 113–127, 2016.

[11] D. Nakhaeinia, P. Payeur, T. S. Hong, and B. Karasfi, "A hybrid control architecture for autonomous mobile robot navigation in unknown dynamic environment," in *Proceedings of the 11th IEEE International Conference on Automation Science and Engineering, CASE 2015*, pp. 1274–1281, Sweden, August 2015.

[12] Y. Zhu, T. Zhang, J. Song, and X. Li, "A new hybrid navigation algorithm for mobile robots in environments with incomplete knowledge," *Knowledge-Based Systems*, vol. 27, pp. 302–313, 2012.

[13] F. Tian, S. Ge, and H. Zhu, "A navigation control strategy with hybrid architecture for rescue robot," in *Proceedings of the 2010 IEEE International Conference on Intelligent Computing and Intelligent Systems (ICIS 2010)*, pp. 675–680, Xiamen, China, October 2010.

[14] M. Boujelben, C. Rekik, and N. Derbel, "Hierarchical fuzzy controller for a nonholonomic mobile robot," in *Proceedings of the 2012 20th Mediterranean Conference on Control and Automation, MED 2012*, pp. 341–347, Spain, July 2012.

[15] D. Ayedi, M. Boujelben, and C. Rekik, "Interv al type-2 TSK fuzzy approach for autonomous mobile robot control in presence of uncertainties," in *Proceedings of the 2017 14th International Multi-Conference on Systems, Signals & Devices (SSD)*, pp. 280–286, March 2017.

[16] M. Boujelben, C. Rekik, and N. Derbel, "Hierarchical fuzzy controller to avoid mobile obstacle for a mobile robot," in *Proceedings of the 2013 10th International Multi-Conference on Systems, Signals and Devices, SSD 2013*, pp. 1–8, March 2013.

[17] H. A. Hagras, "A hierarchical type-2 fuzzy logic control architecture for autonomous mobile robots," *IEEE Transactions on Fuzzy Systems*, vol. 12, no. 4, pp. 524–539, 2004.

[18] I. Maalej, C. Rekik, D. B. H. Abid, and N. Derbel, "Interval type-2 Takagi-Sugeno-Kang fuzzy logic approach for three-tank system modeling," in *Proceedings of the 2014 IEEE 23rd International Symposium on Industrial Electronics, ISIE 2014*, pp. 144–149, IEEE, June 2014.

[19] L. A. Zadeh, "The concept of a linguistic variable and its application to approximate reasoning I," *Information Sciences*, vol. 8, pp. 199–249, 1975.

[20] J. H. Lee, C. Lin, H. Lim, and J. M. Lee, "Sliding mode control for trajectory tracking of mobile robot in the RFID sensor space," *International Journal of Control, Automation, and Systems*, vol. 7, no. 3, pp. 429–435, 2009.

[21] A. Benzerrouk, L. Adouane, and P. Martinet, "Stable navigation in formation for a multi-robot system based on a constrained virtual structure," *Robotics and Autonomous Systems*, vol. 62, no. 12, pp. 1806–1815, 2014.

[22] K. Tai, A.-R. El-Sayed, M. Biglarbegian, C. I. Gonzalez, O. Castillo, and S. Mahmud, "Review of recent type-2 fuzzy controller applications," *Algorithms*, vol. 9, no. 2, p. 19, 2016.

[23] Q. Liang and J. M. Mendel, "Interval type-2 fuzzy logic systems: theory and design," *IEEE Transactions on Fuzzy Systems*, vol. 8, no. 5, pp. 535–550, 2000.

[24] J. M. Mendel and R. I. B. John, "Type-2 fuzzy sets made simple," *IEEE Transactions on Fuzzy Systems*, vol. 10, no. 2, pp. 117–127, 2002.

[25] L. Amador-Angulo, O. Mendoza, J. R. Castro, A. Rodríguez-Díaz, P. Melin, and O. Castillo, "Fuzzy sets in dynamic adaptation of parameters of a bee colony optimization for controlling the trajectory of an autonomous mobile robot," *Sensors*, vol. 16, no. 9, p. 1458, 2016.

[26] N. Karnik and J. Mendel, "Introduction to type-2 fuzzy logic systems," in *Proceedings of the 1998 IEEE International Conference on Fuzzy Systems IEEE World Congress on Computational Intelligence*, vol. 2, pp. 915–920, IEEE, Anchorage, AK, USA.

[27] N. N. Karnik, J. M. Mendel, and Q. Liang, "Type-2 fuzzy logic systems," *IEEE Transactions on Fuzzy Systems*, vol. 7, no. 6, pp. 643–658, 1999.

[28] N. N. Karnik and J. M. Mendel, "Type-2 fuzzy logic systems: type-reduction," in *Proceedings of the 1998 IEEE International Conference on Systems, Man, and Cybernetics*, vol. 2, pp. 2046–2051, IEEE, 1998.

[29] M. Manceur, *Commande robuste des systèmes non linéaires complexes [Ph.D. thesis]*, Reims, 2012.

[30] C. Rekik, M. Jallouli, and N. Derbel, "Optimal trajectory of a mobile robot using hierarchical fuzzy logic controller," *International Journal of Computer Applications in Technology*, vol. 53, no. 4, pp. 348–357, 2016.

[31] M. N. Dharmaweera, S. Y. Khoo, and Z. Man, "Navigation of four-wheel-steering mobile robots using robust fault-tolerant sliding mode control," in *Proceedings of the 11th International Conference on Control, Automation, Robotics and Vision, ICARCV 2010*, pp. 1375–1380, IEEE, December 2010.

[32] I. González, S. Salazar, and R. Lozano, "Chattering-free sliding mode altitude control for a quad-rotor aircraft: Real-time application," *Journal of Intelligent & Robotic Systems*, vol. 73, no. 1-4, pp. 137–155, 2014.

Permissions

All chapters in this book were first published in JR, by Hindawi Publishing Corporation; hereby published with permission under the Creative Commons Attribution License or equivalent. Every chapter published in this book has been scrutinized by our experts. Their significance has been extensively debated. The topics covered herein carry significant findings which will fuel the growth of the discipline. They may even be implemented as practical applications or may be referred to as a beginning point for another development.

The contributors of this book come from diverse backgrounds, making this book a truly international effort. This book will bring forth new frontiers with its revolutionizing research information and detailed analysis of the nascent developments around the world.

We would like to thank all the contributing authors for lending their expertise to make the book truly unique. They have played a crucial role in the development of this book. Without their invaluable contributions this book wouldn't have been possible. They have made vital efforts to compile up to date information on the varied aspects of this subject to make this book a valuable addition to the collection of many professionals and students.

This book was conceptualized with the vision of imparting up-to-date information and advanced data in this field. To ensure the same, a matchless editorial board was set up. Every individual on the board went through rigorous rounds of assessment to prove their worth. After which they invested a large part of their time researching and compiling the most relevant data for our readers.

The editorial board has been involved in producing this book since its inception. They have spent rigorous hours researching and exploring the diverse topics which have resulted in the successful publishing of this book. They have passed on their knowledge of decades through this book. To expedite this challenging task, the publisher supported the team at every step. A small team of assistant editors was also appointed to further simplify the editing procedure and attain best results for the readers.

Apart from the editorial board, the designing team has also invested a significant amount of their time in understanding the subject and creating the most relevant covers. They scrutinized every image to scout for the most suitable representation of the subject and create an appropriate cover for the book.

The publishing team has been an ardent support to the editorial, designing and production team. Their endless efforts to recruit the best for this project, has resulted in the accomplishment of this book. They are a veteran in the field of academics and their pool of knowledge is as vast as their experience in printing. Their expertise and guidance has proved useful at every step. Their uncompromising quality standards have made this book an exceptional effort. Their encouragement from time to time has been an inspiration for everyone.

The publisher and the editorial board hope that this book will prove to be a valuable piece of knowledge for researchers, students, practitioners and scholars across the globe.

List of Contributors

Minghong She
School of Automation, Chongqing University, Chongqing, China
Computer College, Chongqing College of Electronic Engineering, Chongqing, China

Liyu Tian
Computer College, Chongqing College of Electronic Engineering, Chongqing, China

Karima Benzaid and Noura Mansouri
LARC Laboratory, University Fréres Mentouri Constantine 1, Constantine, Algeria

Ouiddad Labbani-Igbida and Romain Marie
X-Lim Institute, UMR CNRS 7252, University of Limoges, Limoges, France

Zhengcang Chen
State Key Laboratory of Robotics, Shenyang Institute of Automation, The Chinese Academy of Sciences (CAS), Shenyang 110016, China
University of Chinese Academy of Science, Beijing, China

Weijia Zhou
University of Chinese Academy of Science, Beijing, China

Yan Chen, Wenzhuo Chen and Bo Li
Department of Petroleum Supply Engineering, Logistical Engineering University, Chongqing 401331, China

Ken Chen
Department of Mechanical Engineering, Tsinghua University, Beijing 100084, China

Hongling Wang, Chengjin Zhang and Bao Pang
School of Control Science and Engineering, Shandong University, Jinan 250101, China

Yong Song
School of Mechanical, Electrical and Information Engineering, Shandong University at Weihai, Weihai 264209, China

Rami Al-Jarrah
Mechanical Engineering Department, Hashemite University, Zarqa 13133, Jordan

Mohammad Al-Jarrah
Department of Electrical and Computer Engineering, Khalifa University of Science and Technology, Abu Dhabi 127788, UAE

Hubert Roth
Automatic Control Engineering Institute (RST), Siegen University, Siegen, Germany

Xiaoyun Lei, Zhian Zhang and Peifang Dong
Key Laboratory of Intelligent Ammunition Technology, School of Mechanical Engineering, Nanjing University of Science and Technology, Nanjing 210094, China

Ruiqing Mao and Xiliang Ma
Xuzhou Institute of Technology, Xuzhou, Jiangsu 221111, China
School of Mechanical and Electrical Engineering, China University of Mining and Technology, Xuzhou, Jiangsu 221116, China

Songqiao Tao and Juan Tan
School of Mechanic and Electronic Engineering, Wuhan Technical College of Communications, Wuhan 430065, China

Mehwish Tahir, Nadia Kanwal and Fatima Anjum
Department of Computer Science, Lahore College for Women University, Jail Road, Lahore 54000, Pakistan

Haoyuan Cai, Dongming Fang, Hui Huang and Chang Liu
State Key Laboratory of Transducer Technology, Institute of Electronics, Chinese Academy of Sciences, Beijing 100190, China

Tianyang Cao
State Key Laboratory of Transducer Technology, Institute of Electronics, Chinese Academy of Sciences, Beijing 100190, China
The University of Chinese Academy of Sciences, Beijing 100190, China

Jorge Luis Aroca Trujillo and Ruthber Rodriguez Serrezuela
University Corporation of Huila, Corhuila, Neiva, Colombia

Alexander Pérez-Ruiz
Escuela Colombiana de Ingeniería Julio Garavito, Bogotá D.C., Colombia

Daisuke Endo and Atsushi Watanabe
Graduate School of Engineering, Tohoku University, Aramaki-aza Aoba 468-1, Aoba-ku, Sendai 980-0845, Japan

Keiji Nagatani
New Industry Creation Hatchery Center, Tohoku University, Aramaki-aza Aoba 6-6-10, Aoba-ku, Sendai 980-8579, Japan

Yaguang Zhu, Baomin Yi and Tong Guo
Key Laboratory of Road Construction Technology and Equipment of MOE, Chang'an University, Xi'an 710064, China

Weitian Wang, Rui Li, Longxiang Guo, Z. Max Diekel and Yunyi Jia
Department of Automotive Engineering, Clemson University, Greenville, SC 29607, USA

Kohei Morishita and Yutaka Hiroi
Osaka Institute of Technology, 5-16-1 Omiya, Asahi-ku, Osaka 535-8585, Japan

Akinori Ito
Tohoku University, 6-6-5 Aramaki Aza Aoba, Aoba-ku, Sendai 980-8579, Japan

Songmin Jia, Ke Wang and Xiuzhi Li
College of Electronic Information & Control Engineering, Beijing University of Technology, Beijing 100124, China
Beijing Key Laboratory of Computational Intelligence and Intelligent System, Beijing 100124, China

Wanjin Guo, Ruifeng Li and Yunfeng Gao
State Key Laboratory of Robotics and System, Harbin Institute of Technology, 92West Dazhi Street, Nangang, Harbin 150001, China

Chuqing Cao
School of Mechanical Engineering, Nanjing University of Science and Technology, Xiaoling wei 200, Nanjing 210094, China
Wuhu HIT Robot Technology Research Institute Co., Ltd., Electronic Industrial Park, Block E, Jiu Jiang District, Wuhu 241007, China

Weijia Zhou
State Key Laboratory of Robotics, Shenyang Institute of Automation, Shenyang 110016, China

Huiwen Zhang, Xiaoning Han and Mingliang Fu
State Key Laboratory of Robotics, Shenyang Institute of Automation, Shenyang 110016, China
University of Chinese Academy of Sciences, Beijing, China

Dorra Ayedi
Control and Energy Management Lab (CEM Lab), University of Sousse, Sousse Engineering School, BP 264 Sousse Erriadh 4023, Tunisia

Maïssa Boujelben and Chokri Rekik
Control and Energy Management Lab (CEM Lab), University of Sfax, Sfax Engineering School, BPW, 3038 Sfax, Tunisia

Index